Edited by
David T. Pierce and
Julia Xiaojun Zhao

**Trace Analysis
with Nanomaterials**

Related Titles

Blackledge, R. D. (ed.)

Forensic Analysis on the Cutting Edge

New Methods for Trace Evidence Analysis

2007
ISBN: 978-0-471-71644-0

Xu, X.-H. N. (ed.)

New Frontiers in Ultrasensitive Bioanalysis

Advanced Analytical Chemistry Applications in Nanobiotechnology, Single Molecule Detection, and Single Cell Analysis

2007
E-Book
ISBN: 978-0-470-11949-5

Otto, M.

Chemometrics

Statistics and Computer Application in Analytical Chemistry

2007
ISBN: 978-3-527-31418-8

Hillenkamp, F., Peter-Katalinic, J. (eds.)

MALDI MS

A Practical Guide to Instrumentation, Methods and Applications

2007
ISBN: 978-3-527-31440-9

Schalley, C. A. (ed.)

Analytical Methods in Supramolecular Chemistry

2007
ISBN: 978-3-527-31505-5

Edited by David T. Pierce and Julia Xiaojun Zhao

Trace Analysis with Nanomaterials

WILEY-VCH Verlag GmbH & Co. KGaA

The Editors

Dr. David T. Pierce
University of North Dakota
Department of Chemistry
151 Cornell Street, Stop 9024
Grand Forks, ND 58202-9024
USA

Dr. Julia Xiaojun Zhao
University of North Dakota
Department of Chemistry
151 Cornell Street, Stop 9024
Grand Forks, ND 58202-9024
USA

All books published by **Wiley-VCH** are carefully produced. Nevertheless, authors, editors, and publisher do not warrant the information contained in these books, including this book, to be free of errors. Readers are advised to keep in mind that statements, data, illustrations, procedural details or other items may inadvertently be inaccurate.

Library of Congress Card No.: applied for

British Library Cataloguing-in-Publication Data
A catalogue record for this book is available from the British Library.

Bibliographic information published by the Deutsche Nationalbibliothek
The Deutsche Nationalbibliothek lists this publication in the Deutsche Nationalbibliografie; detailed bibliographic data are available on the Internet at <http://dnb.d-nb.de>.

© 2010 WILEY-VCH Verlag GmbH & Co. KGaA, Weinheim

All rights reserved (including those of translation into other languages). No part of this book may be reproduced in any form – by photoprinting, microfilm, or any other means – nor transmitted or translated into a machine language without written permission from the publishers. Registered names, trademarks, etc. used in this book, even when not specifically marked as such, are not to be considered unprotected by law.

Composition Toppan Best-set Premedia Limited, Hong Kong
Printing and Bookbinding betz-druck GmbH, Darmstadt
Cover Design Adam Design, Weinheim

Printed in the Federal Republic of Germany
Printed on acid-free paper

ISBN: 978-3-527-32350-0

Contents

Preface *XIII*
List of Contributors *XVII*

Part 1 Biological and Chemical Analysis *1*

1 Photoswitchable Nanoprobes for Biological Imaging Applications *3*
Zhiyuan Tian, Wuwei Wu, and Alexander D.Q. Li
1.1 Introduction *3*
1.2 Photoswitchable Fluorescent Nanoprobes *4*
1.2.1 Single-Color (On-Off) Fluorescent Nanoprobes *5*
1.2.1.1 Fluorescence Modulation of Semiconductor Nanocrystals *5*
1.2.1.2 Isomerization of Photochromic Spiropyrans *7*
1.2.1.3 Isomerization of Photochromic Diarylethenes *9*
1.2.1.4 Structural Conversion of a Photoswitchable Protein *11*
1.2.2 Dual-Color Fluorescence Nanoprobes *13*
1.2.2.1 Green to Red Fluorescence Conversion with Proteins *13*
1.2.2.2 FRET-Based Fluorescence Photoswitching *15*
1.2.2.3 Photoswitchable Nanoparticles Based on a Single Dye *17*
1.2.3 Photoswitchable Fluorescent Nanoparticles for Bioimaging *19*
1.3 Photoswitchable Magnetic Nanoparticles *22*
1.3.1 Nanoparticles with Photoswitchable Magnetization *22*
1.3.2 Magnetic Nanoparticles with Photoswitchable Fluorescence *24*
1.4 Future Perspectives *25*
Acknowledgments *26*
References *26*

2 Applications of Semiconductor Quantum Dots in Chemical and Biological Analysis *31*
Xingguang Su and Qiang Ma
2.1 Introduction *31*
2.2 History *32*
2.3 Classifications *32*

2.3.1	II–VI Quantum Dots 32
2.3.2	III–V Quantum Dots 33
2.3.3	IV–VI Quantum Dots 34
2.3.4	Core–Shell Quantum Dots 34
2.3.5	Alloyed Quantum Dots 35
2.4	Characteristics 35
2.4.1	Electronic Properties 36
2.4.2	Unique Optical Properties 38
2.5	Synthesis and Surface Chemistry 40
2.5.1	Organometallic Approaches 40
2.5.2	Aqueous Phase Colloidal Synthesis 41
2.5.3	Modification of Surface Chemistry 41
2.6	Trace Analysis Using Quantum Dots 42
2.6.1	Determinations Based on Direct Fluorescence Response 42
2.6.2	Fluorescence Resonance Energy Transfer (FRET) Analysis Using QDs 44
2.6.3	Room-Temperature Phosphorescence Detection 45
2.6.4	Near-Infrared Detection Using QDs 46
2.6.5	Rayleigh Light Scattering (RLS) Analysis 48
2.6.6	Chemiluminescence Analysis 48
2.6.7	Electrochemical Analysis 50
2.6.8	Chemosensors and Biosensors 51
2.6.9	"Nano-On-Micro" Assay 52
2.7	Summary 54
	Acknowledgments 55
	References 55
3	**Nanomaterial-Based Electrochemical Biosensors and Bioassays** 61
	Guodong Liu, Xun Mao, Anant Gurung, Meenu Baloda, Yuehe Lin, and Yuqing He
3.1	Introduction 61
3.2	Nanomaterial Labels Used in Electrochemical Biosensors and Bioassays 63
3.2.1	Metal Nanoparticles 63
3.2.1.1	Metal NP Labels for Electrochemical Detection of DNA 64
3.2.1.2	Metal NP Labels for Electrochemical Immunoassays and Immunosensors 68
3.2.2	Semiconductor Nanoparticles 70
3.2.2.1	QD Labels for Electrochemical Detection of DNA 70
3.2.2.2	QD Labels for Electrochemical Immunoassay 72
3.2.3	Carbon Nanotubes 73
3.2.4	Apoferritin Nanovehicles 75
3.2.5	Liposomes 77
3.2.6	Silica Nanoparticles 79
3.2.7	Nanowires and Nanorods 81
3.3	Nanomaterial-Based Electrochemical Devices for Point-of-Care Diagnosis 82

3.4	Conclusions *84*	
	Acknowledgments *85*	
	References *85*	

4	**Chemical and Biological Sensing by Electron Transport in Nanomaterials** *89*	
	Jai-Pil Choi	
4.1	Introduction *89*	
4.2	Electron Transport through Metal Nanoparticles *90*	
4.2.1	Coulomb Blockade Effect and Single Electron Transfer *91*	
4.2.2	Voltammetry of Metal Nanoparticles in Solutions *92*	
4.2.3	Electron Transport through Metal Nanoparticle Assemblies *95*	
4.3	Sensing Applications Based on Electron Transport in Nanoparticle Assemblies *97*	
4.3.1	Chemical Sensors *97*	
4.3.1.1	Sensors Based on Metal Nanoparticle Films *98*	
4.3.1.2	Sensors Based on Semiconducting Oxide Nanoparticles *101*	
4.3.2	Biosensors *103*	
4.4	Concluding Remarks *106*	
	Acknowledgments *107*	
	References *108*	

5	**Micro- and Nanofluidic Systems for Trace Analysis of Biological Samples** *111*	
	Debashis Dutta	
5.1	Introduction *111*	
5.2	Nucleic Acid Analysis *112*	
5.2.1	Miniaturization of PCR Devices *112*	
5.2.2	Integration of PCR with Separation, DNA Hybridization, and Sample Preparation *115*	
5.2.3	Novel Micro- and Nanofluidic Tools for DNA Analysis *116*	
5.3	Protein Analysis *118*	
5.3.1	Protein Separations *118*	
5.3.2	On-Chip Protein Pre-concentration *121*	
5.3.3	Integrated Microfluidic Devices for Protein Analysis *122*	
5.4	Microfluidic Devices for Single-Cell Analysis *123*	
5.5	Conclusion *127*	
	References *128*	

Part 2 Environmental Analysis *133*

6	**Molecularly Imprinted Polymer Submicron Particles Tailored for Extraction of Trace Estrogens in Water** *135*	
	Edward Lai, Anastasiya Dzhun, and Zack De Maleki	
6.1	Introduction *135*	
6.2	Principle of Molecular Recognition by Imprinting *138*	

6.2.1	Monomers, Crosslinkers, and Porogen Solvents	139
6.2.2	Rebinding of Target Analytes	140
6.2.3	Computational Modeling	141
6.3	Analytical Application of MIPs for Biopharmaceuticals and Toxins	143
6.4	Preparation of MIP Submicron Particles	146
6.5	Binding Properties of MIP Submicron Particles with E2	148
6.5.1	Models of E2 Binding with MIP Submicron Particles	149
6.5.2	Kinetics of MIP Binding with E2	150
6.6	Trace Analysis of E2 in Wastewater Treatment	150
6.7	Current Progress	152
6.8	Recent Advances in MIP Technology for Continuing Development	153
	Acknowledgments	156
	References	156

7 Trace Detection of High Explosives with Nanomaterials 161

Wujian Miao, Cunwang Ge, Suman Parajuli, Jian Shi, and Xiaohui Jing

7.1	Introduction	161
7.2	Techniques for Trace Detection of High Explosives	164
7.2.1	Electrochemical Sensors	164
7.2.1.1	Nanomaterial Modified Electrodes	165
7.2.1.2	"Artificial Peroxidase"-Modified Electrodes Based on Prussian Blue	166
7.2.2	Electrogenerated Chemiluminescence	167
7.2.3	Fluorescence-Based Sensors	169
7.2.3.1	Quenching Sensors Based on Fluorescent Polymer Porous Films	169
7.2.3.2	Quenching Sensors Based on Fluorescent Nanofibril Films	171
7.2.3.3	Quenching Sensors Based on Quantum Dots	172
7.2.3.4	Quenching Sensors Based on Organic Supernanostructures	175
7.2.3.5	Fluoroimmunoassays Using QD-Antibody Conjugates	176
7.2.3.6	Displacement Immunosensors	176
7.2.4	Microcantilever Sensors	178
7.2.5	Metal Oxide Semiconductor (MOS) Nanoparticle Gas Sensors	180
7.2.6	Surface-Enhanced Ramam Scattering Spectroscopy	180
7.3	Conclusions	181
	Acknowledgments	182
	References	182

8 Nanostructured Materials for Selective Collection of Trace-Level Metals from Aqueous Systems 191

Sean A. Fontenot, Timothy G. Carter, Darren W. Johnson, R. Shane Addleman, Marvin G. Warner, Wassana Yantasee, Cynthia L. Warner, Glen E. Fryxell, and John T. Bays

8.1	Introduction	191
8.2	Sorbents for Trace-Metal Collection and Analysis: Relevant Figures of Merit	192

8.3	Thiol-Functionalized Ordered Mesoporous Silica for Heavy Metal Collection *194*	
8.3.1	Performance Comparisons of Sorption Materials for Environmental Samples *194*	
8.3.2	Performance Comparisons of Sorption Materials for Biological Samples *197*	
8.4	Surface-Functionalized Magnetic Nanoparticles for Heavy Metal Capture and Detection *200*	
8.5	Nanoporous Carbon Based Sorbent Materials *206*	
8.5.1	Chemically-Modified Activated Carbons *207*	
8.5.2	Templated Mesoporous Carbons *209*	
8.6	Other Nanostructured Sorbent Materials *212*	
8.6.1	Zeolites *212*	
8.6.2	Ion-Imprinted Polymers *214*	
8.7	Concluding Thoughts *215*	
	Acknowledgments *217*	
	References *217*	

9 Synthesis and Analysis Applications of TiO$_2$-Based Nanomaterials *223*
Aize Li, Benjamen C. Sun, Nenny Fahruddin, Julia X. Zhao, and David T. Pierce

9.1	Introduction *223*	
9.2	Synthesis of TiO$_2$ Nanostructures *225*	
9.2.1	TiO$_2$ Nanoparticles *225*	
9.2.2	Mesoporous TiO$_2$ *225*	
9.2.3	TiO$_2$ Nanotubes *225*	
9.2.4	TiO$_2$-Based Nanohybrids *226*	
9.2.4.1	TiO$_2$-Metal Nanoparticle Hybrids *227*	
9.2.4.2	TiO$_2$–SiO$_2$ Hybrids *228*	
9.2.5	Fabrication of TiO$_2$ Nanofilms *228*	
9.3	Applications of TiO$_2$-Based Nanomaterials for Chemical Analysis *229*	
9.3.1	Analysis of Gas-Phase Samples *229*	
9.3.1.1	Hydrogen *230*	
9.3.1.2	Carbon Monoxide *232*	
9.3.1.3	Oxygen *233*	
9.3.1.4	Water Vapor *233*	
9.3.2	Analysis of Aqueous Samples *235*	
9.3.2.1	Ion Detection and Sensing *235*	
9.3.2.2	Metal Ion Extraction *235*	
9.3.2.3	Organic Compounds *237*	
9.3.3	Biosensors *240*	
9.3.3.1	Voltammetric Biosensors *240*	
9.3.3.2	Optical Biosensors *245*	

9.4	Conclusions	246
	Acknowledgments	247
	References	247

10   Nanomaterials in the Environment: the Good, the Bad, and the Ugly   255
Rhett J. Clark, Jonathan G.C. Veinot, and Charles S. Wong

10.1	Introduction	255
10.2	The Good: Nanomaterials for Environmental Sensing	256
10.2.1	Colorimetric Detection	256
10.2.1.1	Noble Metal Nanoparticles	256
10.2.1.2	DNAzymes	260
10.2.1.3	Monolithic Nanoporous Sensors	263
10.2.2	Fluorescence-Based Detection	264
10.2.3	Fluorescence Quenching	267
10.3	The Bad: Environmental Fate of Nanomaterials	269
10.3.1	Environmental Fate	270
10.3.1.1	Factors Affecting Aggregation	270
10.3.1.2	Nanoparticles in Porous Media	272
10.3.2	Toxicity	273
10.4	The Ugly: Detection of Nanomaterials in the Environment	275
10.5	Conclusions	278
	Acknowledgments	279
	References	279

Part 3   Advanced Methods and Materials   283

11   Electroanalytical Measurements at Electrodes Modified with Metal Nanoparticles   285
James A. Cox and Shouzhong Zou

11.1	Introduction	285
11.2	Modification of Electrodes with Nanoparticles	286
11.2.1	Fabrication of Two-Dimensional Arrays of Nanoparticles	286
11.2.1.1	Seed-Mediated Formation of a Two-Dimensional Array of Nanoparticles	286
11.2.1.2	Direct Deposition of Nanoparticles on Bare Electrodes	289
11.2.2	Deposition of Three-Dimensional Films Containing Metal Nanoparticles	293
11.2.2.1	Layer-by-Layer Electrostatic Assemblies Containing Metal Nanoparticles	294
11.2.2.2	Fabrication of Conducting Polymer Films Doped with Metal Nanoparticles	295
11.3	Geometric Factors in Electrocatalysis by Nanoparticles	296
11.3.1	Particle Size Effects on Electrocatalysis	296
11.3.2	Particle Shape Dependence	299

11.3.3	Particle Composition Dependence	*302*
11.4	Analytical Applications of Electrodes Modified with Metal Nanoparticles	*304*
11.4.1	Determination of Inorganic Analytes	*305*
11.4.2	Determination of Organic and Biologically Important Analytes	*310*
11.5	Conclusions	*313*
	References	*314*
12	**Single Molecule and Single event Nanoelectrochemical Analysis**	*319*
	Shanlin Pan and Gangli Wang	
12.1	Introduction	*319*
12.2	Basic Concepts	*320*
12.2.1	Electrochemistry	*320*
12.2.2	Nanoelectrodes	*320*
12.3	Single-Molecule Electrochemistry	*321*
12.3.1	Single-Molecule Electrochemistry Using Nanoelectrodes	*321*
12.3.2	Single-Molecule Spectroelectrochemistry	*323*
12.4	Single-Nanoparticle Electrochemical Detection	*326*
12.4.1	Single-Nanoparticle Detection Using Nanoparticle Collision at a Microelectrode	*326*
12.4.2	Single-Nanoparticle Electrochemistry Using Single-Molecule Spectroscopy	*328*
12.5	Nanoelectrodes for Ultrasensitive Electrochemical Detection and High-Resolution Imaging	*328*
12.5.1	Nanoelectrode Fabrication	*328*
12.5.2	Mass Transfer near a Nanoelectrode	*329*
12.5.3	Combined Optical and Electrochemical Imaging	*330*
12.6	Electrochemical Detection in Nanodomains of Biological Systems	*333*
12.7	Localized Delivery and Imaging by Using Single Nanopipette-Based Conductance Techniques	*333*
12.8	Final Remarks	*335*
	Acknowledgments	*336*
	References	*337*
13	**Analytical Applications of Block Copolymer-Derived Nanoporous Membranes**	*341*
	Takashi Ito and D.M. Neluni T. Perera	
13.1	Introduction	*341*
13.2	Monolithic Membranes Containing Arrays of Cylindrical Nanoscale Pores	*341*
13.3	BCP-Derived Monoliths Containing Arrays of Cylindrical Nanopores	*344*
13.4	Surface Functionalization of BCP-Derived Cylindrical Nanopores	*346*
13.5	Investigation of the Permeation of Molecules through BCP-Derived Nanoporous Monoliths and their Analytical Applications	*347*

13.5.1	Permeation of Small Molecules through PS-*b*-PMMA-Derived Nanoporous Monoliths	347
13.5.2	Regulation of Molecular Permeability Based on Electrostatic Interactions	348
13.5.3	Influence of Supporting Electrolyte Concentration to Effective Nanopore Diameter	350
13.5.4	Permeation of Nanoparticles, Polymers, and Biomacromolecules through BCP-Derived Nanopores	351
13.6	Conclusions	355
	Acknowledgments	356
	References	356

14 Synthesis and Applications of Gold Nanorods 359
Carrie L. John, Shuping Xu, Yuhui Jin, Shaina L. Strating, and Julia Xiaojun Zhao

14.1	Introduction	359
14.2	Au Nanorod Synthesis	360
14.2.1	Electrochemical Synthesis	360
14.2.1.1	Electrochemical Synthesis Employing a Hard Template	360
14.2.1.2	Electrochemical Synthesis Employing a Soft Template	362
14.2.2	Photochemical Synthesis	364
14.2.3	Seed-Mediated Growth	364
14.3	Signal Enhancement	367
14.3.1	Plasmon Resonance	367
14.3.2	Surface-Enhanced Raman Scattering	368
14.3.3	Luminescence Enhancement of Dye Molecules	369
14.3.4	Enhanced Luminescence of Au Nanorods	370
14.4	Applications of Au Nanorods in Trace Analysis	372
14.4.1	Fabrication of Au Nanorod-Based Sensors	372
14.4.1.1	Fabrication of SERS and LSPR Sensors	372
14.4.1.2	Fabrication of Luminescence Sensors	372
14.4.2	Bioimaging and Bioanalysis Based on Optical Measurements	374
14.4.3	Bioanalysis Based on Electrochemical Measurements	376
14.5	Applications of Au Nanorods in Other Fields	377
14.5.1	Au Nanorods as Supporting Material for Electrocatalyts	377
14.5.2	Au Nanorod-Based Photothermal Therapy	377
14.6	Conclusions	378
	Acknowledgments	379
	References	379

Index 383

Preface

Trace analysis is an important topic that impacts various areas ranging from environmental monitoring to national security, food safety, clinical diagnosis, and forensic investigation. The need for sensitive and robust determinations in these areas has driven a rapid development of novel nanomaterials as well as new methodologies with which to implement them. Unfortunately, essential details of these new nanomaterials and approaches for their use have largely remained in wide-ranging journals and specialized compilations. Our goal for this book has been to provide an introduction to these new methods and materials in one source and thereby encourage the development of new, cross-disciplinary ideas among scientists from many different fields. Accordingly, this book includes a broad cross-section of nanomaterial-based methodologies and applications. Selected topics are reviewed in 14 chapters and organized in three sections.

Section I (Chapters 1–5) is dedicated to "Biological and Chemical Analysis." The performance of biosensors and bioassays has been aided by the rapid development of nanotechnology and the application of various nanomaterials. Chapters 1 and 2 are focused on photoactive nanomaterials. Chapter 1 summarizes recent advances of photoswitchable nanoprobes that feature changes in fluorescence and magnetization. Contributions of these nanoprobes to super-resolution fluorescence imaging and acquisition of quantitative information of biological targets are reviewed. Semiconductor quantum dots (QDs) have attracted considerable attention in the fields of chemistry and biology over the past decade. Based on these recent advances, Chapter 2 reviews the fundamental properties, characteristic advantages, and synthetic methods for various semiconductor QDs. Several successful analytical applications of QDs in the field of chemistry and biology are discussed.

Chapters 3 and 4 emphasize nanomaterials-based electrochemical biosensors and bioassays. The enormous signal enhancement associated with the use of nanomaterial labels and the formation of nanomaterial–biomolecule complexes provides the basis for ultrasensitive electrochemical detection of disease-related gene or protein markers, biothreat agents, or infectious agents. Chapter 3 discusses various nanomaterials for bioanalysis, including nanoparticles, nanowires, nanotubes, and nanocarriers. The surface-dependent electron transport properties of nanomaterials are often used to develop chemical and biological sensors for trace analysis. Specifically, physical and/or chemical sorption of analytes on a nanomaterial surface may affect the rates of electron transport through a nanomaterial assembly, resulting in detectable change in its electronic conductivity. In Chapter 4, the authors review chemical and biological sensing applications based on electron transport through nanoparticle assemblies.

In addition to nanomaterials, nanodevices based on fluidics have reduced significantly the time and costs involved in chemical/biochemical experimentation. They have also permitted the study of several physical/chemical/biological systems at a fundamentally higher level. In Chapter 5, the authors describe applications of micro- and nanofluidic technology to the trace analysis of biological samples with a focus on assays involving nucleic acids, proteins/peptides, and biological cells.

Section II (Chapters 6–10) is dedicated to "Environmental Analysis." Nanomaterials have shown great potential for improving the detection and extraction of trace contaminants in the environment, but they may themselves pose an environmental hazard if released. Chapter 6 focuses on the analysis of water contaminated with endocrine-disrupting chemicals by the use of molecularly imprinted polymers (MIPs). This chapter gives an overview of significant achievements to improve the performance of MIPs in solid-phase extraction using particles during the last two years. Chapter 7 reviews current research on trace detection and quantification of nitrated and peroxide-based high explosives with various techniques involving nanomaterials. In particular, sensors based on electrochemistry, fluorescence, microcantilevers, and metal oxide semiconductive nanoparticles are discussed.

Hazardous metals are another important class of environmental pollutants. The authors of Chapter 8 survey the use of nanostructured materials for the selective collection of trace-level metals from aqueous systems. It has been shown that, when correctly constructed, these nanomaterials are superior sorbents and can be used to enhance trace level analysis. In addition to newly developed nanomaterials, some traditional nanomaterials, such as TiO_2, have demonstrated new applications for trace detection, particularly in environmental analysis. Several TiO_2-based nanostructures important in analysis applications are introduced in Chapter 9, including colloidal and mesoporous TiO_2 nanoparticles, TiO_2 nanotubes, TiO_2-based hybrids, and TiO_2 nanofilms. Also described are sample pretreatment and analyte preconcentration methods based on the strong adsorption of organic and inorganic species onto TiO_2 nanomaterials.

In Chapter 10 the authors bring an interesting perspective to the use of engineered nanoparticles (ENPs). Not only are recent advances in the use of ENPs for environmental sensing described, but the environmental fate and toxicity of ENPs are also discussed.

Section III (Chapters 11–14) is dedicated to "Advanced Methods and Materials." Here the development of new nanomaterials and new methods for trace analysis are discussed. In Chapter 11, a wide variety of analytical methods employing electrodes modified with nanoparticles are summarized. Chapter 12 focuses on the analysis of single molecules or single events using nanoelectrodes and combined optical and electrochemical methods.

Chapters 13 and 14 focus on the development and application of several new nanomaterials. Membranes derived from block copolymers (BCPs) and containing arrays of cylindrical nanoscale pores are described in Chapter 13. Recent achievements indicate that BCP-derived nanoporous monoliths are promising materials to develop highly efficient separation membranes for biomolecules and detection devices with high selectivity and sensitivity. Chapter 14 introduces gold nanorods (AuNRs) and a broad range of applications that transcend the now-familiar gold nanoparticle. The methods of synthesis and unique physicochemical characteristics of AuNRs are described in detail and their most recent uses in trace analysis are discussed.

We hope that you find this book useful during your research and that it proves helpful in developing new avenues for trace analysis. We certainly look forward to receiving your feedback. Finally, and most especially, we wish to thank each of the contributors to this book. Without their dedication and expertise, this work would never have been possible.

Grand Forks, USA, January 2010

David T. Pierce
Julia Xiaojun Zhao

List of Contributors

R. Shane Addleman
Pacific Northwest National Laboratory
P.O. Box 999
Richland, WA 99352
USA

Meenu Baloda
North Dakota State University
Department of Chemistry and
Molecular Biology
Fargo, ND 58102
USA

John T. Bays
University of Oregon
P.O. Box 999
Eugene, OR 97403
USA

Timothy G. Carter
University of Oregon
Department of Chemistry and
Materials Science Institute
Eugene, OR 97403-1253
USA

Jai-Pil Choi
California State University–Fresno
Department of Chemistry
2555 East San Ramon Avenue M/S SB70
Fresno, CA 93740-8034
USA

Rhett J. Clark
University of Alberta
Department of Chemistry
11227 Saskatchewan Dr.
Edmonton, Alberta
Canada T6G 2G2

James A. Cox
Miami University
Department of Chemistry and
Biochemistry
501 East High Street
Oxford, OH 45056
USA

Zack De Maleki
Carleton University
Department of Chemistry
1125 Colonel By Drive
Ottawa, Ontario
Canada K1S 5B6

Debashis Dutta
University of Wyoming
Department of Chemistry
1000 East University Avenue
Laramie, WY 82071
USA

Trace Analysis with Nanomaterials. Edited by David T. Pierce and Julia Xiaojun Zhao
Copyright © 2010 WILEY-VCH Verlag GmbH & Co. KGaA, Weinheim
ISBN: 978-3-527-32350-0

List of Contributors

Anastasiya Dzhun,
Carleton University
Department of Chemistry
1125 Colonel By Drive
Ottawa, Ontario
Canada K1S 5B6

Nenny Fahruddin
University of North Dakota
Department of Chemistry
151 Cornell Street
Grand Forks, ND 58202-9024
USA

Sean A. Fontenot
University of Oregon
Department of Chemistry and
Materials Science Institute
Eugene, OR 97403-1253
USA

Glen E. Fryxell
Pacific Northwest National Laboratory
P.O. Box 999
Richland, WA 99352
USA

Cunwang Ge
Nantong University
School of Chemistry and Chemical
Engineering
Nantong
226007
China

Anant Gurung
North Dakota State University
Department of Chemistry and
Molecular Biology
Fargo, ND 58102
USA

Yuqing He
Guangzhou Institute of Dermatology
Department of Dermatology
Guangzhou
510095
China

Takashi Ito
Kansas State University
Department of Chemistry
111 Willard Hall
Manhattan, KS 66506-0401
USA

Yuhui Jin
University of North Dakota
Department of Chemistry
151 Cornell Street
Grand Forks, ND 58202-9024
USA

Xiaohui Jing
Nantong University
School of Chemistry and Chemical
Engineering
Nantong
226007
China

Carrie L. John
University of North Dakota
Department of Chemistry
151 Cornell Street
Grand Forks, ND 58202-9024
USA

Darren W. Johnson
University of Oregon
Department of Chemistry and
Materials Science Institute
Eugene, OR 97403-1253
USA

Edward Lai
Carleton University
Department of Chemistry
1125 Colonel By Drive
Ottawa, Ontario
Canada K1S 5B6

Aize Li
University of North Dakota
Department of Chemistry
151 Cornell Street
Grand Forks, ND 58202-9024
USA

Alexander D.Q. Li
Washington State University
Department of Chemistry
Pullman, WA 99164
USA

Yuehe Lin
Pacific Northwest National Laboratory
Richland, WA 99351
USA

Guodong Liu
North Dakota State University
Department of Chemistry and
Molecular Biology
Fargo, ND 58102
USA

Qiang Ma
Jilin University
Department of Analytical Chemistry
College of Chemistry
Changchun
130012
China

Xun Mao
North Dakota State University
Department of Chemistry and
Molecular Biology
Fargo, ND 58102
USA

Wujian Miao
The University of Southern
Mississippi
Department of Chemistry and
Biochemistry
118 College Drive
Hattiesburg, MS 39406
USA

Shanlin Pan
The University of Alabama
Department of Chemistry
Tuscaloosa, AL 35487-0336
USA

Suman Parajuli
The University of Southern
Mississippi
Department of Chemistry and
Biochemistry
118 College Drive
Hattiesburg, MS 39406
USA

D.M. Neluni T. Perera
Kansas State University
Department of Chemistry
111 Willard Hall
Manhattan, KS 66506-0401
USA

David T. Pierce
University of North Dakota
Department of Chemistry
151 Cornell Street
Grand Forks, ND 58202-9024
USA

Jian Shi
Nantong University
School of Chemistry and Chemical
Engineering
Nantong 226007
China

Shaina L. Strating
University of North Dakota
Department of Chemistry
151 Cornell Street
Grand Forks, ND 58202-9024
USA

Xingguang Su
Jilin University
Department of Analytical Chemistry
College of Chemistry
Changchun, 130012
China

Benjamen C. Sun
University of North Dakota
Department of Chemistry
151 Cornell Street
Grand Forks, ND 58202-9024
USA

Zhiyuan Tian
Washington State University
Department of Chemistry
Pullman, WA 99164
USA

Jonathan G.C. Veinot
University of Alberta
Department of Chemistry
11227 Saskatchewan Dr.
Edmonton, Alberta
Canada T6G 2G2

Gangli Wang
Georgia State University
Department of Chemistry
Atlanta, GA 30302-4098
USA

Cynthia L. Warner
Pacific Northwest National Laboratory
P.O. Box 999
Richland, WA 99352
USA

Marvin G. Warner
Pacific Northwest National Laboratory
P.O. Box 999
Richland, WA 99352
USA

Charles S. Wong
University of Winnipeg
Richardson College for the
Environment, Environmental Studies
Program and Department of
Chemistry
515 Portage Ave.
Winnipeg Manitoba
Canada R3B 2E9

Wuwei Wu,
Washington State University
Department of Chemistry
Pullman, WA 99164
USA

Shuping Xu
Jilin University
State Key Laboratory of
Supramolecular Structure and
Materials
Changchun 130012
China

Wassana Yantasee
Pacific Northwest National Laboratory
P.O. Box 999
Richland, WA 99352
USA

Julia X. Zhao
University of North Dakota
Department of Chemistry
151 Cornell Street
Grand Forks, ND 58202-9024
USA

Julia Xiaojun Zhao
University of North Dakota
Department of Chemistry
151 Cornell Street
Grand Forks, ND 58202-9024
USA

Shouzhong Zou
Miami University
Department of Chemistry and
Biochemistry
501 East High Street
Oxford, OH 45056
USA

Part 1
Biological and Chemical Analysis

1
Photoswitchable Nanoprobes for Biological Imaging Applications

Zhiyuan Tian, Wuwei Wu, and Alexander D.Q. Li

1.1
Introduction

Fluorescence imaging is widely used to study biological processes because it provides abundant information non-invasively regarding various biological mechanisms. The diffraction limit, however, restricts the best resolution of conventional fluorescence imaging techniques to a level two of orders of magnitude coarser than nano-sized molecules, leaving many intracellular organelles and molecular structures unresolvable. For example, fluorescent probes under conventional fluorescence microscope cannot be localized to accuracy better than 250 nm because of the diffraction limit barrier. Most biological targets such as RNA, DNA, and proteins form nanometer scale structures in cells, thus higher resolution beyond the diffraction limit of ca. 250 nm is essential to detect and monitor biological mechanisms at the single-molecule level.

Far-field fluorescence microscopy techniques with increased spatial resolution have the potential to convert microscopy into nanoscopy and thus enable near-molecular scale spatial resolution [1, 2]. Specifically, several super-resolution far-field fluorescence imaging techniques with 20 to 30 nm lateral and 50 to 60 nm axial imaging resolutions have been developed to circumvent the diffraction limit. These techniques include stimulated emission depletion (STED) microscopy [3–6] and reversible saturable optically linear fluorescent transitions (RESOLFT) microscopy [7–9], stochastic optical reconstruction (STORM) microscopy [10–12], photoactivated localization (PALM) microscopy [13–17], photoactuated unimolecular logical switching attained reconstruction (PULSAR) microscopy [18], and other methods using similar principles [19]. Each of these techniques share a common underlying principle: images with high-resolution are obtained based on the switching of the fluorescent probes between two distinctively different fluorescent states, either fluorescent "ON" and "OFF" states or two fluorescent states with distinct color. In other words, the fluorescent probes employed must be actively modulated (usually via photoswitching or photoactivation) in time to ensure that only an optically resolvable subset of fluorophores is activated at any time in a diffraction-limited region, thereby allowing their localization with high accuracy. From this viewpoint, the

Trace Analysis with Nanomaterials. Edited by David T. Pierce and Julia Xiaojun Zhao
Copyright © 2010 WILEY-VCH Verlag GmbH & Co. KGaA, Weinheim
ISBN: 978-3-527-32350-0

development of fluorescence microscopy with ultrahigh spatial resolution depends, to a great extent, on the construction of photoswitchable fluorescent probes.

In addition to the contribution to super-resolution imaging, a dual-color photoswitchable fluorescent probe can highlight the biological target from its background despite the presence of autofluorescence in the target matrix [20, 21]. Besides photoswitching, high emission intensity of the fluorescent probe is also highly desirable. For example, both PULSAR and STORM rely on the detection of a single fluorescent probe and the location of fluorescent probe with high accuracy. The total number of photons that one fluorophore site can emit before photobleaching is an important photophysical parameter and ultimately determines the spatial resolution of the obtained fluorescence image. With single-molecule fluorophores as labeling reagents, rapid photobleaching and limited emission intensity per labeled site greatly hinder further improvement in imaging resolution. With fluorescent nanoparticles as labeling reagents, the overall fluorescence signal output from individual nanoparticle is proportional to the product of fluorescence efficiency and loading density of the encapsulated dyes. Because a large number of fluorophores is encapsulated inside a single nanoparticle, nanoparticles produce strong emission when properly excited [22, 23]. Newly developed photoswitchable fluorescent nanoparticles are expected to overcome limits of single-molecule fluorophores and therefore contribute to the further development of fluorescence imaging.

Several types of photoswitchable nanoprobes, including inorganic semiconductor nanoparticles, metal/metal oxide nanoparticles, polymeric nanoparticles, vesicle-like probes, and cloned photoactivatable proteins have been developed to date for various applications. Among them, nanoprobes with photoswitchable fluorescence [24–41] and magnetization [42–49] have drawn considerable attention for their important applications in biological fluorescence imaging and magnetic resonance imaging. Hence, in this chapter we outline the major classes of photoswitchable nanoprobes developed recently with an emphasis on fluorescent and magnetic nanoparticles and their roles in biological imaging. Although other photoswitchable features are also interesting, such as photoswitchable conductance [50–53] and surface wettability [54], we limit our scope to photoswitchable nanoprobes for biological applications.

1.2
Photoswitchable Fluorescent Nanoprobes

Among various stimuli-responsive signaling modes, fluorescence emission is promising, not only because fluorescence signals can be readily and sensitively detected but also because they have many identifying characteristics for monitoring, such as signal intensity, wavelength (color), and lifetime. When combined with novel types of fluorescence microscopy, photoswitchable fluorescent nanoprobes have contributed significantly to super-high resolution fluorescence imaging. For the different types of photoswitchable fluorescent nanoprobes developed to date, light-induced isomerization of fluorophores plays a key role. Additionally, fluorescence quenching based on Förster resonance energy transfer (FRET) is also involved in some cases. In this section, we outline different types of nanoprobes with photoswitchable fluorescence and discuss the mechanism involved in the photoswitching.

1.2.1
Single-Color (On-Off) Fluorescent Nanoprobes

A typical single-color type of photoswitchable fluorescent nanoprobe adopts either a fluorescent (on) or a non-fluorescent (off) state. It may be selectively and reversibly transferred between these two states by irradiation with appropriate wavelengths of light. In this case, fluorescence signal intensity, rather than wavelength, is the key parameter being modulated.

1.2.1.1 Fluorescence Modulation of Semiconductor Nanocrystals

Semiconductor nanocrystals (quantum dots, QDs) display unique photophysical features as their size approaches the nanoscale [55, 56], making them attractive candidates in diverse biological imaging and sensing applications [57]. Because of their intrinsic properties, QDs themselves do not possess photoswitching ability, which restricts their application in ultrahigh resolution biological imaging. Thus, photochromism must be integrated into QDs to empower such functionality of fluorescence photoswitching.

By attaching photochromic spiropyran moieties to the surfaces of QDs, Li and coworkers have developed QDs-based hybrid nanoparticles with photoswitchable fluorescence emission [25]. As demonstrated in Figure 1.1a, spiropyran moieties were anchored on the surface of a core–shell CdSe/ZnS QDs via a thiol–metal linkage. As a photochromic material, colorless spiropyran (SP) undergoes photoinduced ring-opening reactions upon ultraviolet irradiation and consequently yields isomeric merocyanine (MC) forms that have a strong visible absorption band in the range 500–600 nm. This band arises from the extended π-conjugation that develops between the indolene and the pyran rings following rupture of the C(spiro)–O bond. Back-conversion from the MC form of the dye into the SP forms occurs thermally and is accelerated by visible light illumination. As shown in Figure 1.1b, these hybrid nanoparticles exhibited reversible photochromism that was very similar to that of unbound spiropyran dyes.

In these nanoparticles, the QDs acted as the dominant emitters as well as the FRET donor while neither the SP nor MC forms emitted light strongly. Exposure to UV light catalyzed the reversible photoconversion of the photochromic moiety from the colorless SP form into the colored MC form, which functioned as the FRET acceptor. When the photochromic moiety was in its SP form, green fluorescence (546 nm) from the nanocrystals was observed, showing the fluorescence "ON" state. When the photochromic moiety was transferred to its isomeric merocyanine using ultraviolet irradiation, fluorescence from the nanocrystals was seriously quenched by the MC form of the dye owing to a good overlap of the

(a)

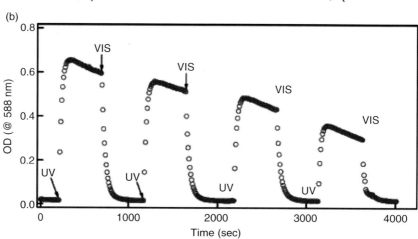

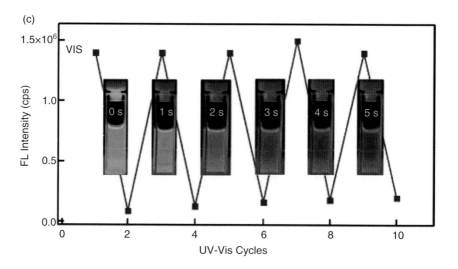

absorption band of the MC form and the fluorescence emission band of the nanocrystals. This quenching produced the fluorescence "OFF" state. A FRET process was involved in the light-induce fluorescence switching, with QDs acting as the donor and the MC form of the dye as the acceptor (fluorescence quencher). Because the light-induced conversion between the spiro form and the merocyanine form is reversible, the conversion between the "ON" and "OFF" states was therefore reversible (Figure 1.1c).

Similar fluorescence photoswitching systems have also been developed in other groups. Medintz and coworkers have constructed such composite photoswitchable fluorescent nanoparticles by attaching spiropyran-labeled maltose binding proteins (MBP) to the surface of individual QDs [26, 27]. In Jares-Erijman's group, a biotinylated diheteroarylethene derivative was bound to the surfaces of QDs bearing conjugated streptavidin and, thus, composite nanoparticles with photoswitchable fluorescence features were constructed [28, 29]. In addition, a dithiolane anchoring group was used in Raymo's group [30]; and thermal and Cu(I)-mediated "click" reactions were employed in Binder's group [31] to attach photochromic spiropyran to the surfaces of QDs with the purpose of constructing photoswitchable fluorescent composite nanoparticles. A FRET process was involved in the fluorescence photoswitching performance of these systems.

1.2.1.2 Isomerization of Photochromic Spiropyrans

FRET-based fluorescence photoswitchable systems typically require two chromophores: a fluorescence donor and a fluorescence acceptor. The emission band of the former must overlap with the absorption band of the latter [58–61]. Additionally, the donor and acceptor must approach to each other within the Förster distance. However, these stringent requirements can be circumvented if a single dye can carry out fluorescence switching.

Recently, Li and coworkers have constructed a fluorescence photoswitchable particulate system using a single chromophore, spiropyran, as the photoactive unit [32]. Hydrophobic–hydrophilic core–shell nanoparticles were first prepared by a modified microemulsion copolymerization. This method used N-isopropylacrylamide and styrene as the primary monomeric units, minor amounts of divinylbenzene as the crosslinker, and 5-(1,3-dihydro-3,3-dimethyl-6-nitrospiro

Figure 1.1 (a) Quantum dot with anchored spiropyrans on its surface showing a fluorescence photoswitching function; (b) photochromism of 6 nm CdSe/ZnS core–shell nanocrystals at 2.6×10^{-7} M concentration with a SP/NC ratio of 105 under alternating cycles of filtered UV and visible illumination; (c) photoswitchable fluorescence emission at 546 nm with alternating cycles of UV and visible light (squares and lines); the size of core–shell CdSe/ZnS QDs is 5 nm, the QDs concentration is 3.5×10^{-7} M, and the SP/NC ratio is 150. Overlay: six cuvettes showing the response time at 1 s intervals for conversion from the spiro into the mero form under 365-nm UV illumination. Reproduced with permission from Reference [25]. Copyright 2005 American Chemical Society.

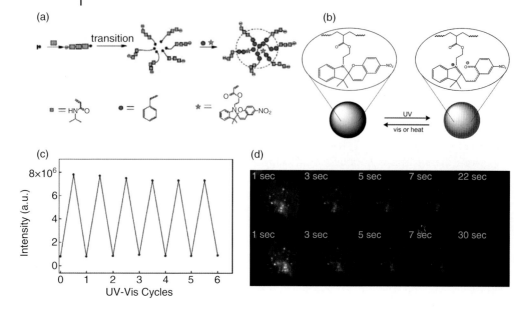

Figure 1.2 (a) Approach to constructing hydrophilic–hydrophobic core–shell nanoparticles with spiropyran moieties in the hydrophobic cavity of the nanoparticle; (b) nanoparticles with embedded spiropyrans show fluorescence photoswitching; (c) fluorescence photoswitching of 68-nm photochromic nanoparticle under 420-nm excitation, using alternating UV and visible irradiation; (d) fluorescence imaging of optically switchable nanoparticles in a single HEK-293 cell. Red fluorescence (bright spots) of mero-particles is elicited with a 10-s UV light pulse, followed by fluorescence imaging using 488-nm excitation and a liquid N_2-cooled CCD detector. The fluorescence is slowly switched off by visible 488-nm light, and the bright spots fade with increasing exposure time (top panel with time labels). Application of a second 10-s UV pulse regenerates the red fluorescence (lower panel with time labels), which again fades under 488-nm illumination with identical kinetics. Reproduced with permission from Reference [32]. Copyright 2006 American Chemical Society.

[2H-1-benzopyran-2,2′-2H]-indole)ethyl acrylate (SP) as the photoactive unit (Figure 1.2a).

Notably, neither the SP nor MC form of the photochromic dye fluoresces appreciably in water, although the MC form possesses weak fluorescence in polar organic solvents and within self-assembled films. However, in the hydrophobic cavities of water-soluble polymeric nanoparticles, the MC form fluoresced strongly in the red region 600–750 nm. The quantum yield of the MC form was found to be 0.24 by using a Rhodamine B standard. Merocyanine residing in a hydrophobic nanoparticle was found to fluoresce at least 200 times more strongly than in aqueous solution. The nanoparticle microphase environment was responsible for the enhancement of fluorescence intensity, with several factors involved. The fluorophores confined within the hydrophobic cavities were isolated from nonradiative decay pathways or electron-transfer pathways generated by collisions with solution components. Additionally, molecules trapped in the polymer environment had restricted conformational flexibility, which minimized nonradiative relaxation through internal motions.

The fluorescence photoswitching features were generated by light-induced formation and rupture of the spiro bond (Figure 1.2b). Upon irradiation with UV-blue light, SP molecules within nanoparticles were converted into the MC forms while the back conversion from the MC forms into SP forms was accomplished through lower-energy irradiation ($\lambda > 450$ nm). Specifically, the SP forms were not fluorescent while the MC forms emitted strong red fluorescence. Thus, reversible ON/OFF fluorescence photoswitching of the polymeric nanoparticles was achieved by alternating illumination with UV and visible light. Additionally, the photoswitching displayed better reversibility than other systems, including the one in which spiropyran molecules were covalently attached on the surfaces of QDs (Figure 1.2c). Covalent attachment by polymerization more effectively isolated individual spiropyran molecules within the nanoparticles and thereby minimized bimolecular degradative reactions. The photoswitchable fluorescent nanoparticles were also tested in living cell for *in vivo* optical imaging (Figure 1.2d). The results demonstrated that fluorophores residing in the hydrophobic core of the nanoparticle were largely unquenched by components of the biological milieu and displayed reversible, less-fatigued fluorescence photoswitching features. Since Li and coworkers' studies on photoswitchable fluorescent nanoparticles prepared via microemulsion copolymerization strategy, similar strategies have been employed in the groups of Zeng [33], Tong [34], Kawaguchi [35], Zhang [36] and Liu [37], and similar particulate systems with photoswitching fluorescence features were developed. For instance, spiropyran-linked methacrylate monomers were copolymerized with other monomers into polymeric nanoparticles with photoswitchable fluorescence in Zeng's group via one-step mini-emulsion polymerization approach; and enhanced photoreversibility, photostability, and relatively fast photoresponsive characteristics of the as-prepared photoswitchable nanoparticles were confirmed [33].

1.2.1.3 Isomerization of Photochromic Diarylethenes

As classical photochromic compounds, diarylethene derivatives have drawn considerable attention due to their thermal stability and excellent fatigue resistance [62–66]. Recently, Yi and coworkers have developed vesicle-like nanostructures with photoswitchable fluorescence features between two distinct states [38]. The photoactive unit developed was an amphiphilic photochromic diarylethene molecule with hydrophobic and hydrophilic chains at two ends of a rigid diarylethene core (Figure 1.3a).

Unlike spiropyran, UV light irradiation induced diarylethene ring-closing rather than ring opening. Moreover, UV light converted the diarylethene derivatives from their fluorescence ON-state into an OFF-state, whereas the spiropyran switched in opposite fashion. These vesicle-like nanostructures dispersed in aqueous solution

(a)

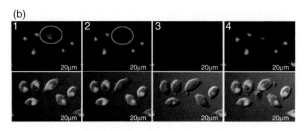

(b)

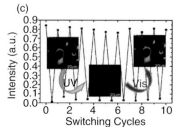

(c)

Figure 1.3 (a) Photochromism of an amphiphilic diarylethene; (b) confocal laser scanning microscopy (CLSM) image (above) and the overlay image (bottom) of KB living cells incubated with the open isomer of the diarylethene for 20 min at 25 °C: (1) in original state, (2) irradiated by 405-nm light (2 mW) for a single cell, (3) all cells, and (4) recovered by 633-nm light (0.7 mW) (λ_{ex} = 633 nm, 0.15 mW); (c) fluorescence switching of fixed KB cells by alternating UV (405 nm, 2 mW, 10 s/time) and visible (633 nm, 0.7 mW, 20 min/time) light illumination. Inset: images of one cycle. Reproduced with permission from Reference [38]. Copyright 2008 American Chemical Society.

and remained unchanged for more than a month in a certain concentration range. The vesicle-like nanostructures displayed a classical photochromic behavior that resulted in fluorescence ON/OFF switching. Upon irradiation with 365 nm light, the colorless diarylethene moiety was converted from an open structure into closed structure, resulting in fluorescence quenching and an accompanying fluorescence OFF state. Under visible light irradiation, the colored diarylethene moiety was completely converted into the open isomer with recovered fluorescence. The fluorescence photoswitching feature of this vesicle-like nanostructure was also confirmed in a biological system using confocal laser scanning microscopy. As shown in Figure 1.3b, all cells after incubated with the aqueous solution of vesicles were switched between fluorescence ON and OFF states upon alternating irradiation of 405 and 633 nm light. Additionally, photoswitching of the vesicles in cells displayed excellent reversibility and stability. As shown in Figure 1.3c, no apparent "fatigue" effect or photobleaching were detected after many photoswitching cycles.

For the aforementioned photoswitchable particulate system, the photochromic diarylethene component is the only photoactive unit and functions as both the photochromophore and the fluorophore simultaneously. Specifically, the open structure of the diarylethene component represents the fluorescence ON state while the closed structure corresponds to the fluorescence OFF state. Unlike such systems, another type of diarylethene-based photoswitchable particulate system has also been developed with a FRET mechanism involved. For instance, Bossi and coworkers have developed photoswitchable fluorescent silica nanoparticles with Rhodamine dye as the fluorophore and diarylethene as the photochromic quencher [39]. Upon light-driven isomerization of diarylethene and the controlled FRET between Rhodamine dye and the closed status of the diarylethene moiety, the fluorescence of Rhodamine dye was modulated and fluorescence ON/OFF photoswitching was achieved. Similar composite photoswitchable fluorescent nanoparticles with perylene derivative as fluorophore and diarylethene derivative as photochromophore have also been developed in Liu's group [37].

1.2.1.4 Structural Conversion of a Photoswitchable Protein

The discovery of green fluorescent protein (GFP) and subsequent developments have provided revolutionary and indispensable research tools for molecular biology and cell biology [67–81]. Fluorescent proteins can be easily fused to any protein of interest, enabling real-time monitoring of biochemical processes in living cells, including protein expression, localization, and movement. As an emerging new class of fluorescent proteins, reversible switchable fluorescent proteins (RSFPs) have attracted widespread interest owing to their potential usage in diverse fields, such as data storage, *in vivo* study of protein behavior, and high-resolution optical imaging [82–88].

Dronpa is a GFP-like, monomeric RSFP with favorable switching properties and a remarkable fluorescence quantum yield of 0.85. At equilibrium, Dronpa is in its fluorescence ON-state, fluorescing bright green light with an emission maximum at 518 nm. Illumination with intense light of 488 nm converts Dronpa into its fluorescence OFF-state while subsequent minimal irradiation with 400-nm light drives Dronpa back into its on-state and restores fluorescence. Miyawaki and coworkers have successfully used Dronpa as a powerful tool to analyze the regulation of fast protein dynamics [73]. Excellent photochromic behavior of Dronpa in fixed cells was confirmed by using 490-nm light to inactivate the protein fluorescence and 400-nm light to recover it. Additionally, high reversibility and stability of Dronpa were observed, with 75% of the fluorescence intensity retained even after 100 photocycles. To obtain insight into the underlying molecular mechanism of Dronpa's photochromic behavior, Jakobs and coworkers generated Dronpa crystals and analyzed the structure by X-ray crystallography [89–91]. As shown in Figure 1.4a, a truncated diamond-shaped crystal of Dronpa reversibly switched between fluorescent and non-fluorescent states while alternating between blue (488 nm) and UV (405 nm) irradiation. This light-induced fluorescence switching was repeated over 20 cycles and ~2% photobleaching per cycle was observed (Figure 1.4c). From the crystal structure, a folded structure of Dronpa was deduced, which closely resembled to that of GFP and related fluorescent proteins.

RSFP studies so far indicate that the primary event responsible for photoswitching is a light-induced cis–trans isomerization of the chromophore. The well ordered and relatively planar *cis*-conformation of the chromophore represents the

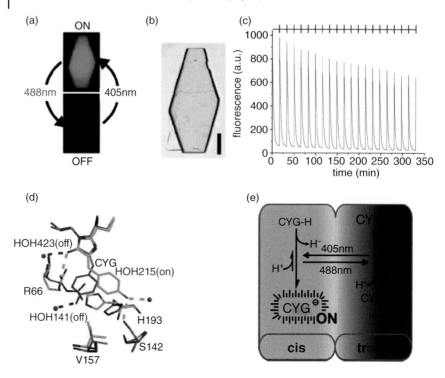

Figure 1.4 Reversible photoswitching of Dronpa protein crystals: (a) fluorescence images of a Dronpa crystal in the fluorescent (upper) and non-fluorescent (lower) form and (b) bright-field image of the same crystal (scale bar: 50 μm). (c) Fluorescence of the Dronpa protein crystal recorded over 20 switching cycles using the same intensities as in (a). Each repeated excitation sequence consists of 16.5-min blue light illumination followed by 4 s of blue and UV light. (d) Detailed comparison of the chromophores and important environment in their vicinities. The chromophores and residues are shown as sticks, color-coded by atom type (on-state carbon, green; off-state carbon, light blue; oxygen, red; nitrogen, blue; sulfur, yellow). Water molecules are displayed as blue spheres. Hydrogen bonds that disappear in the opposite state are indicated by green and blue dotted lines, respectively. (e) Model for reversible photoswitching in Dronpa, connecting light-induced intramolecular changes with different protonation states of the chromophore. Reproduced with permission from Reference [91]. Copyright 2007 National Academy Sciences USA.

fluorescence ON-state. Conversely, the chromophore in the fluorescence off-state adopts a trans-conformation – a more twisted and partially disordered conformation likely to favor non-irradiative decay of the exited chromophore [89–91]. When Dronpa is in its fluorescence ON-state, the key chromophore is formed from the Cys-62-Tyr-63-Gly-64 (CYG) tripeptide that resides in the "β-can" cavity with a cis-conformation. In sharp contrast, the OFF-state adopts a trans-conformation in which four more residues (Arg-66, Ser-142, Val-157, and His-193) close to the chromophore have changed their position relative to the ON-state structure (Figure

1.4d). Protonation–deprotonation may also play a key role in the fluorescence switching of Dronpa (Figure 1.4e). At equilibrium, the CYG chromophore is predominantly deprotonated due to the cis cavity environment and is therefore fluorescent. This deprotonated form exhibits a strong absorption peak at 503 nm with negligible absorption at 390 nm, enabling blue light-induced cis-to-trans isomerization of the chromophore. However, in its trans-conformation, the chromophore is predominantly protonated due to the influence of the cavity and therefore nonfluorescent. Such loss of fluorescence upon protonation originates from a more twisted and partially disordered conformation, which likely opens further channels for nonradiative decay. As a concurrent result, protonation of the *trans*-CYG also changes the absorption spectrum of the protein and a strong absorption peak in UV region is observed. Apparently, such a change in absorption provides an opportunity to convert the fluorescence OFF-state back into the initial ON-state using UV light irradiation.

1.2.2
Dual-Color Fluorescence Nanoprobes

In contrast to single-color photoswitchable fluorescent nanoprobes, dual-color photoswitchable fluorescent nanoprobes alternately emit two distinct fluorescence colors. Irradiation with a specific wavelength can switch one fluorescence color to the other and vice versa. For biological fluorescence imaging, cell autofluorescence can mask signals from the labeled targets, making it difficult to distinguish the labeled targets from the background. Because cell autofluorescence has a broad spectrum from blue to red, one may prefer the obvious advantage of dual-color photoswitchable fluorescent nanoprobes over single-color photoswitchable fluorescent nanoprobes. The former can provide two-channel imaging that helps to unambiguously confirm the site of interest from a false-positive signal originating from an interfering fluorophore within the biological sample.

1.2.2.1 Green to Red Fluorescence Conversion with Proteins

Several fluorescent proteins based on GFP have been developed, providing biologists with fluorescent markers having many different emission hues. As GFP homologues, a family of proteins isolated from nonbioluminescent *Anthozoa* has attracted considerable attention because its emissions in the red spectral range [92–96]. In particular, a fluorescent protein isolated from the scleractinian coral *Trachyphyllia geoffroyi*, named Kaede, in Miyawaki'group [68, 76, 97], and another fluorescent protein cloned from the scleractinian coral *Lobophyllia hemprichii*, named EosFP, in Nienhaus' group [75], were found to be distinctly different from most mutants of GFP. Instead of enabling fluorescence emission by photoactivation, Kaede and EosFP can be irreversibly converted from green- into red-emitting forms by near-UV irradiation, making them particularly useful for tracking and localization experiments in living cells.

EosFp and Kaede were cloned from different coral species; the sequence of the former is closest to that of the latter, with 84% identical residues. EosFP was found

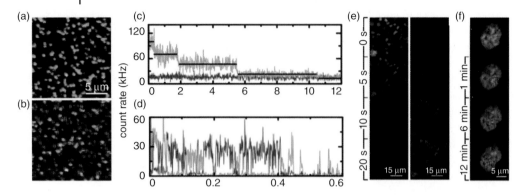

Figure 1.5 (a) Confocal scan image of EosFP immobilized on a BSA surface (488-nm excitation); (b) confocal scan image of EosFP immobilized on a BSA surface (488-nm excitation) after an additional 400-nm irradiation; (c) fluorescence emission traces from an individual EosFP tetramer (488-nm excitation) showing sequential bleaching of the four subunits; (d) an additional EosFP tetramer (488-nm excitation plus 400-nm irradiation) showing green-to-red switching and sequential bleaching; (e) the fusion protein of EsoFP with the intracellular domain of mouse Notch-1 protein, localized in the nuclei of stably transfected HEK 293 cells, is converted from green into red under microscope; (f) tracking of movement of the fusion protein of EosFP with the recombination signal-binding protein in the nucleus by localized photoconversion, observed after 0.5-s irradiation with 400-nm light (1 µW) in the single-molecule setup. Reproduced with permission from Reference [75]. Copyright 2004 National Academy of Sciences, USA.

to emit bright green fluorescence with maximum wavelength at 516 nm and a quantum yield of 0.7, but changed efficiently to a bright and stable red fluorescence with peak at 581 nm and a quantum yield of 0.55 upon irradiation with near-UV light around 390 nm. Spectral analysis indicated that this green-to-red photoconversion was caused by structural change. For both EosFP and Kaede, a light-induced dissociation of the peptide backbone accompanied by an excited-state proton transfer is believed to be involved in the green-to-red photoconversion. Figure 1.5a displays a confocal image of wild-type EosFP molecules immobilized on an albumin surface with excitation of 4-µW 488-nm light. While this imaged showed mainly green spots, irradiation with 400-nm light and excitation with 488-nm light turned most spots to red (Figure 1.5b). Timed analysis by single-molecule spectroscopy indicated that a single EosFP contains four subunits, showing four-step decay because of sequential bleaching of the fluorophore (Figure 1.5c). Additionally, a time trace taken with 400-nm irradiation revealed extensive switching between green and red emission (Figure 1.5d). Whenever the red emission turned on, the green emission essentially switched off, indicating efficient interconversion.

These photoswitchable fluorescent proteins have been actively tested in live cells as biological markers when fused to targeted proteins. For example, two fusion proteins, mNotch-1-IC-EosFP and RBP-2N-d2EosFP, have been prepared using the intracellular domain of mouse Notch-1 protein and the recombination signal-binding protein, respectively. Figure 1.5e shows the photoconversion of a stable HEK 293 cell line expressing mNotch1-IC-EosFP fusion protein with 400-nm light as irradiation source. The red fluorescence intensity increased with time at the expense of the green fluorescence; the green fluorescence was completely converted into the red one within 20 s. Figure 1.5f displays the fluorescence of RBP-2N-d2EosFP in the nucleus of a HEK 293 cell. Upon irradiation with tightly focused 400-nm light for 0.5 s, local green-to-red photoconversion was clearly observed with micrometer-sized scale. Additionally, after irradiation for 12 min the red fluorescence was found evenly distributed throughout the nucleus, which confirmed the presence of EosFP labeled proteins.

1.2.2.2 FRET-Based Fluorescence Photoswitching

Reversibility is particularly desirable for dual-color fluorescent nanoprobes in biological labeling applications. However, the dual-color fluorescent proteins developed so far, such as Kaede and EosFP, can only be photoconverted from one color into another in an irreversible manner, which is an impediment to accurate detection. Recently, Li and coworkers have developed new nanoprobes that can be repeatedly and stably photoswitched between green and red fluorescence [40]. These novel nanoparticles were constructed by covalently encapsulating perylene diimide (PDI) and photochromic spiropyrans (SP) in the hydrophobic core of core–shell polymeric nanoparticles using a microemulsion copolymerization (Figure 1.6a). In a hydrophobic environment, PDI has a high fluorescence quantum yield of 0.9 associated with a strong green emission with maximum wavelength at 535 nm and a strong overtone at 575 nm. As a photochromic species, SP undergoes photo-induced ring-opening reactions upon ultraviolet irradiation and consequently yields its isomeric merocyanine (MC) form. The MC form has a strong visible absorption with a maximum around 588 nm. This absorption perfectly matches the emission band of the PDI dye. Again, neither SP nor its MC form fluoresces appreciably in water, although the mero form possesses weak fluorescence in polar organic solvents and within self-assembled films. However, in the hydrophobic cavities of water-soluble nanoparticles, the MC form acquires a strong red fluorescence with a quantum yield of 0.18, estimated by comparison to a Rhodamine B standard. When the photochromic moiety is in its SP form, bright green fluorescence from the PDI is observed. When the photochromic moiety is transferred to its isomeric MC form using ultraviolet irradiation, a typical FRET process occurs between the matched PDI excited-state and MC ground-state. Accordingly, green fluorescence of PDI is strongly quenched while MC receives energy transferred from excited-state PDI species and fluoresces a vivid red. Upon illumination with visible light, the MC form of the photochromic dye is converted back into the SP form, which is no longer able to quench the PDI fluorescence. In this situation, the PDI dye can only function as the fluorescent acceptor, not the donor. Therefore, the green fluorescence of PDI is recovered.

Because the photochromic reaction of spiropyran is a reversible process, photoswitching between green and red fluorescence in these nanoparticles was

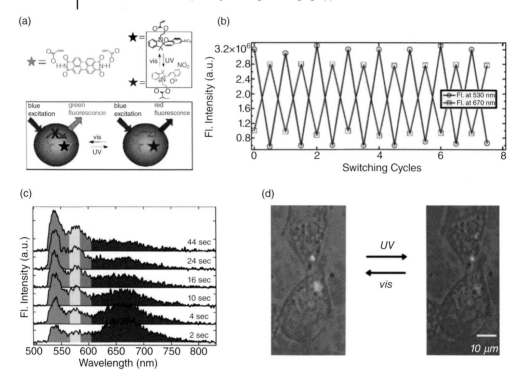

Figure 1.6 (a) Reversible photoswitching of dual-color fluorescent nanoparticles based on FRET, with embedded perylene diimide (PDI) as donor and spiropyran (SP) as acceptor; (b) fluorescence switching between red and green was achieved upon alternating UV and visible illumination followed by fluorescence measurement with 420-nm excitation (UV light illumination period is 2 min, and visible light illumination period is 20 min); (c) fluorescence spectra obtained upon 488-nm excitation of a single polymer nanoparticle after pre-illumination of the same sample with a UV lamp for 2 min; (d) dual-color photoswitchable fluorescent nanoparticles were transported into HEK-293 cells using liposomes as delivery vehicles; these internalized polymer nanoparticles can be selectively highlighted with either green or red fluorescence. A short UV pulse switches the highlighted green spots (left) to vivid red fluorescence (right), while visible light reverses the process. Reproduced with permission from Reference [40]. Copyright 2007 American Chemical Society.

reversible (Figure 1.6b). In addition to desired dual-color photoswitching properties, these nanoprobes displayed perfect photochemical stability with no noticeable fatigue or photobleaching after several photoswitching cycles. The hydrophobic core of core–shell structured nanoparticle provided a protective hydrophobic nanoenvironment for the chromophores and significantly depressed bimolecular degradative reactions. Time-resolved fluorescence photoswitching was also observed at the single nanoparticle level (Figure 1.6c).

Being water-soluble, these nanoparticles can be delivered into living cells and they show no apparent cytotoxicity during the cell-imaging experiment. The fluorescence photoswitching and the photoswitching reversibility of the nanoparticles remained intact in living cells. Figure 1.6d displays a typical fluorescence image of the nanoparticles residing in HEK 293 cells. Because these fluorescent nanoparticles can be selectively highlighted with either green- or red-fluorescence, they should prove to be useful nanoprobes in biological systems that exhibit high background autofluorescence or possess fluorophores that interfere with one-color detection.

Most of the FRET-based photoswitchable fluorescent nanoparticles developed so far are fluorescence ON/OFF photoswitching type with the photochromic moieties merely acting as the controlled energy quencher to quench the fluorescence from the fluorophores within the nanoparticles [25–31, 34–37, 39]. In contrast, the photochromic spiropyran moiety within the nanoparticles developed in Li's group [40] acted as energy acceptor to quench the fluorescence of the energy donor and subsequently emitted distinct red-fluorescence. As a result, dual-color reversible photoswitching fluorescence features were achieved, making such a system unique.

1.2.2.3 Photoswitchable Nanoparticles Based on a Single Dye

As introduced in last section, FRET provides one avenue for dual-color fluorescence photoswitching. However, FRET itself has a limited range of distances over which it is sensitive and it depends on several variables, including molecular orientation, dielectric constant, spectral overlap, and excited-state lifetime [98, 99]. From this viewpoint, new strategies for photoswitchable nanoprobes are needed to effectively circumvent the stringent requirements of FRET.

As an extension to the FRET-based dual-color photoswitchable fluorescent nanoprobes, a series of single-chromophore fluorescent nanoparticles have been developed recently in Li's group [41]. These novel nanoparticles were constructed by a microemulsion copolymerization with various photochromic spiropyran derivatives as photoactive units (Figure 1.7a). In a typical protocol, major monomers acrylamide (A), styrene (ST), and butyl acrylate (BA) were polymerized with minor functional monomers, including optically switchable spiropyran derivatives, crosslinker divinylbenzene (DVB), and surface-decorating water-soluble acrylic acid (AA). These nanoprobes contained only one type of dye molecule. Yet, a single photoswitchable fluorophore alternately emitted two distinct colors and thus no FRET was needed. Substituted groups played a key role in determining the dual fluorescence emissions of the polymeric nanoparticles. Light-induced ismerization of MSP (5'-methoxy-3',3'-dimethyl-6-nitro-1'-(4-vinylbenzyl) spiro[chromene-2,2'-indoline]) nanoparticles, for example, resulted in a dual-color fluorescence photoswitching between green (530 nm) and red (665 nm) fluorescence; while MCSP (5'-methoxy-3',3'-dimethyl-6-cyano-1'-(4-vinylbenzyl) spiro[chromene-2,2'-indoline]) nanoparticles showed a dual-color fluorescence photoswitching between blue (470 nm) and red (665 nm) color. The fluorescence spectra in Figure 1.7b demonstrate that the fluorescence switching of MSP nanoparticles was from nearly pure red color to pure green with a conspicuous fluorescence isosbestic point, suggesting the fluorescence switching was based on a clean one-to-one photochemical conversion. In MSP nanoparticles, the red

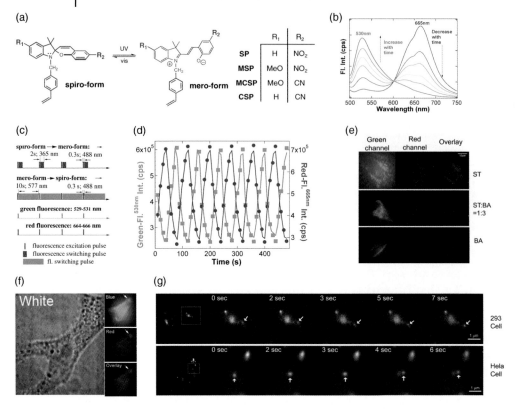

Figure 1.7 (a) Chemical structures illustrate the spiro- and mero-forms of the fluorescent dyes. (b) Fluorescence spectra evolution of MSP nanoparticles that were pre-photoswitched to the mero-form using 365-nm illumination ($\lambda_{ex} = 488$ nm). (c) Pulse sequences that elicited the dual fluorescence color switching used three colors. (d) The pulse sequences in (c) generate red (665 nm)- and green (530 nm)-fluorescence oscillation. Symbols represent data and modulation fitting yields oscillation frequency and phase difference up to the fourth harmonic. (e) Live-cell imaging uses MSP nanoparticles: nanoparticles only adhere to the cells when the monomer A and BA are out of balance, but undergo no endocytosis. The overlay images confirm the signals are from nanoparticles, not interference. (f) Functionalized by HMGA1a protein, MCSP-nanoparticles also bind to live cells and undergo endocytosis. The most intensive red-fluorescence spot (arrow) is unveiled to be a false positive because its signal cannot be corroborated in the blue channel. (g) When A monomer was removed from the nanoparticle formula, live cells readily endocytose the resulting MSP-nanoparticles. Reproduced with permission from Reference [41]. Copyright 2009 American Chemical Society.

fluorescence exponentially decreased while the green fluorescence exponentially increased and both processes followed first-order kinetics. A stimulated sequence of green- and red-fluorescence from MSP nanoparticles is plotted against time in Figure 1.7c. Fitting of these modulated fluorescence intensity, using a series of periodic sine functions up to the fourth harmonic, revealed that the red and green fluorescence were synchronized at the same frequency with an almost 180° phase shift (Figure 1.7d). This anti-phase synchronization indicated that the spiro-mero inter-conversion occurred near simultaneously with either no intermediate or extremely short-lived intermediates.

In biomedical imaging, the nanoprobes must interact specifically with the target (e.g., disease tissue) but should have minimal non-specific interactions with their surrounding environment (e.g., normal tissue). Li and coworkers found that the copolymer sequence and composition effectively controlled non-specific interactions of single-chromophore fluorescent nanoparticles with live cells [41]. For example, the ST-to-BA monomer ratio exerted an obvious influence on how MSP nanoparticles adhered to HeLa cells (Figure 1.7e). In particular, MSP nanoparticles without BA or A monomer bound significantly to HeLa cell membranes and their fluorescence brightened the HeLa cell contour. In sharp contrast, MSP nanoparticles containing balanced A and BA co-monomers hardly bound to HeLa cells. Interestingly, MSP nanoparticles with specific composition could "distinguish" normal HEK 293 cells from cancer (HeLa) cells.

An additional benefit of these dual-color nanoparticles was their strong fluorescence emission for bioimaging. Typically, the MSP nanoparticles fluoresced nearly 38-times brighter than the organic dye fluorescein. Dual-color photoswitchable fluorescent nanoparticles may simultaneously enable two-channel fluorescence imaging and therefore unambiguously distinguish nanoparticles from interfering emitters that fluoresce the same color. In the case of MCSP nanoparticles, for example, only those image pixels that registered both blue and red fluorescence represented probe signals from MCSP-nanoparticles. Other emitters were either from interference or cell auto fluorescence. As shown in Figure 1.7f, for example, the bright red-fluorescence region indicated by the arrow had no blue counterpart and therefore must have been from interfering cell autofluorescence even though its intensity dominated the image. It was also found that MSP nanoparticles prepared without the A monomer not only induce non-specific binding to cell membranes but also triggered endocytosis. Although the mechanism for this interesting action is still unknown, it was found that those endocytic cargoes containing MSP nanoparticles could not mature into late endosomes or lysosomes, even after 16 h from their entry (Figure 1.7g).

1.2.3
Photoswitchable Fluorescent Nanoparticles for Bioimaging

An important application of photoswitchable fluorescent nanoparticles is to explore biological events and mechanisms at the nanoscale level. Taking advantage of the photoswitching features of newly developed fluorescent probes, far-field fluorescence microscopy techniques have been able to achieve images with resolution down to tens of nanometers – far beyond the diffraction limit of conventional microscopy techniques.

Recently, Li and coworkers have demonstrated that photoswitchable fluorescent nanoparticles containing spiropyran can be used as powerful biological markers for photo-actuated unimolecular logical switching attained reconstruction

(PULSAR) microscopy, through which cellular organelles in fixed cells can be imaged down to 10–40 nm [18]. The fluorescent nanoprobes employed for PULSAR imaging were core–shell polymeric nanoparticles with photochromic spiropyran covalently embedded in the hydrophobic core. Such nanoparticles can be reversibly switched between fluorescence ON- and OFF-states by alternating irradiation of UV (365 nm) and visible light (577 nm) (Figure 1.8a and b). PULSAR imaging consisted of thousands of single molecule images, with each image divulging a limited number of photoswitchable molecules that were under fluorescence control. Fitting the location of every molecule in each image frame and subsequently summing all the fitted single molecule locations finally yielded the reconstructed high-resolution microscopic image. Figure 1.8c compares traditional fluorescence microscopy images to corresponding high-resolution PULSAR images. While traditional fluorescence microscopy could not resolve four 70-nm nanoparticles arranged in a row, PULSAR microscopy clearly resolved the four juxtaposed nanoparticles. Figure 1.8d displays a sub-cellular area of interest in a fixed HeLa cell in which most nanoparticles resided in lysosomes and the fluorescence of the lysosomes apparently intensified. After the merocyanine emission of the nanoparticles was switched off and most interfering fluorophores were photobleached through illumination of 532-nm light, PULSAR microscopy was used to sequentially deliver a short low-power UV pulse to activate some merocyanine red-fluorescence and a 532-nm beam to image the newly switched-on nanoparticles. By repeating this pulse and probe pattern over many cycles, hundreds of activation events were accumulated and a final high-resolution image was reconstructed (Figure 1.8e). Conventional wide field optical microscopy cannot resolve individual nanoparticles within a lysosome because the organelle size is smaller than the diffraction limit. The PULSAR microscope, however, can resolve the fine structures of nanoparticles inside a single organelle. As shown in Figure 1.8e, the bright spots were not spherical, suggesting multiple nanoparticles residing in a single lysosome. Closer examination of a selected bright spot reveals that there were at least two particles residing in the single lysosome (Figure 1.8f). Furthermore, three parallel lines drawn along the elongated direction of the spot revealed the profiles of fluorescence intensity of these two bright features. Figure 1.8g definitely resolved these two bright features and yielded a measured peak-to-peak distance of 50, 55, and 70 nm, respectively. Fitting the intensity profile (blue line) yielded the projected center-to-center distance of 69 nm between these two bright features. These data unambiguously confirmed the ability of PULSAR microscopy to resolve cellular organelles down to the nanoscale. As another example, van Bossi and coworkers have demonstrated that photoswitchable fluorescent silica nanoparticles played the key role in the super-resolution fluorescence imaging by combining with the reversible saturable optical fluorescence transitions (RESOLFT) [39]. Significantly improved fluorescence imaging resolution when compared to conventional confocal microscopy was obtained. Admittedly, efficient photoswitchable nanoprobes played crucial roles in such super-resolution fluorescence imaging methods. To summarize the crucial roles that photoswitchable probes played, Heilemann and Sauer et al. have proposed to unify all subdiffraction-resolution

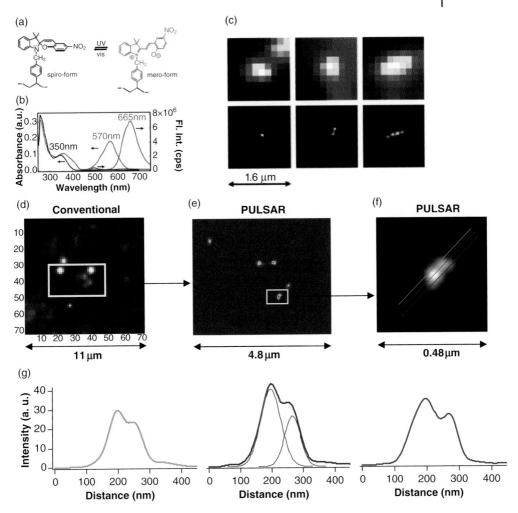

Figure 1.8 (a) Photochemical conversion between spiropyran (spiro-form) and merocyanine (mero-form) occurs within a single particle (upper); (b) optical absorption (left and bottom axes) of spiro- (black) and mero-form (blue) and fluorescence spectra (bottom and right axes) of the spiro- (green) and mero-form (red) plotted against wavelength; (c) using 70-nm nanoparticle to compare conventional fluorescent microscopy (upper) and PULSAR microscopy (bottom) reveals the resolution difference in the nanoparticle patterns; (d) conventional wide-field fluorescence imaging reveals that most of the nanostructures of interest consist of photoswitchable nanoparticles; (e) the PULSAR image of these SP nanoparticles produces much higher resolution, separating fuzzy clusters shown in (d) into distinct bright spots; (f) zooming-in on one bright spot in (e) divulges resolved two overlapping spots, corresponding to at least two nanoparticles; (g) lines drawn along the two overlapping spots reveal the fluorescence intensity profile and yield a center-to-center distance of 69 nm. Reproduced with permission from Reference [18]. Copyright 2008 American Chemical Society.

fluorescence microscopy imaging techniques with photoswitching microscopy [100].

So far, most investigations focus on using photoswitchable fluorescent probes and various microscopy techniques to carry out revolutionary roles in super-resolution fluorescence imaging [100–103]. Using a specific photoswitchable fluorescent probe, that is, Alexa 746 labeled antibodies, van de Linde et al. have demonstrated the possibility of unraveling quantitative information of biological targets in small cellular organelles. Specifically, quantitative information about the number and distribution of proteins in the inner membrane of mitochondria were obtained [104]. Furthermore, photoswitching-induced double fluorescence modulation, as demonstrated in the case of particulate system developed by Li et al. [40], revealed the interactive nature between the fluorescent donor and acceptor. Such photoswitching-based secondary fluorescence modulation intrinsically enables a new detection technique. Specifically, controlled light-pulse sequences can be employed to induce the correlated synchronized oscillation between the donor and acceptor, thus separating the information-rich signals from non-oscillating (zero-frequency) noises. It is expected that such a mechanism will work as the rationale for photoswitching-based quantitative and even trace analysis methods in the future.

1.3
Photoswitchable Magnetic Nanoparticles

1.3.1
Nanoparticles with Photoswitchable Magnetization

One of important applications of magnetic nanoparticles is highly sensitive and target-specific magnetic resonance imaging (MRI) of various biological systems, including tumor detection, cell trafficking, and angiogenesis [105–111]. MRI is essentially proton nuclear magnetic resonance spectroscopy performed spatially on tissue. Because proton densities can be very similar, a major limitation of MRI has been its inability to distinguish soft tissue types, such as healthy parts of the liver from diseased lesions. Inspired by the revolutionary role of photoswitchable fluorescent nanoparticles in super-high resolution fluorescence imaging, the development of magnetic nanoparticles with photoswitchable magnetism might circumvent these MRI imaging problems by strengthening the MRI signal at desired locations.

Einaga and coworkers have developed recently several types of nanoparticles with photoswitchable magnetization properties [43–47]. Photoswitchable magnetic composite nanoparticles were constructed with azobenzene-derivatized ligands coated on the surface of magnetic nanoparticles. In all cases of magnetization photoswitching, reversible photoisomerization of azobenzene chromophores was involved. The photoisomerization of the azo moieties led to a geometrically confined structural change, as well as changes in the dipole moment and the

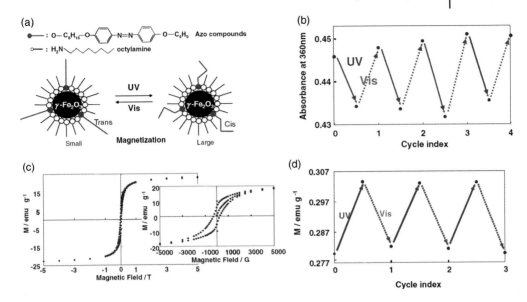

Figure 1.9 (a) Scheme showing the composition of γ-Fe$_2$O$_3$/azo composite nanoparticles and the proposed mechanism for magnetization photoswitching; (b) changes in absorbance at 360 nm by alternating illumination with UV and visible light; (c) magnetization versus applied magnetic field [inset: magnified region (1 G = 10^{-4} T)]; (d) changes in the magnetization of nanoparticles induced by alternating illumination with UV and visible light at 5 K with an external magnetic field of 10 G. Reproduced with permission from Reference [46]. Copyright 2006 The Chemical Society of Japan.

electrostatic field around the core magnetic nanoparticles. As a result, the magnetization response of the composite nanoparticles could be reversibly switched by alternating UV and visible light illumination. Figure 1.9a displays the structure of the composite nanoparticles and the mechanism for the magnetization photoswitching. Among three components, a γ-Fe$_2$O$_3$ particle core provided magnetization, the azo compound provided photochromic activity, and n-octylamine acted as a spacer. Photoisomerization of azobenzene derivatives does not occur in the solid state, because photoisomerization (especially for the trans-to-cis isomerization) is usually accompanied by an increase in molecular volume. For the composite γ-Fe$_2$O$_3$/azo nanoparticles, sufficient free volume was guaranteed by spacing the azo moieties with n-octylamine and the reversible photoisomerization was therefore allowed to take place even in the solid state. Figure 1.9b shows periodic switching of absorbance by alternating UV and visible light illumination, which indicated light-induced isomerization of azo derivatives. This result also confirmed the key role of the n-octylamine spacer in allowing for the photoisomerization of azo moieties on the nanoparticle surface. The magnetization curve of the composite nanoparticles measured in the field-cooled state displayed a small hysteresis loop remanence (5.67 emu g^{-1}) and coercivity (300 G) (Figure 1.9c). This means that the superparamagnetic transition was hindered below the blocking temperature (T_B). Figure 1.9d demonstrates the influence of photo-illumination on the magnetic properties of the composite nanoparticles, revealing reversibly photoswitchable magnetization. UV light increased magnetization and visible light decreased magnetization. These cycles could be repeated a few times with no apparent fatigue effect. Moreover, switching of the magnetic properties at room temperature was confirmed by ESP spectra and ^{57}Fe Mössbauer spectra. It was proposed that photoisomerization of the surface azo moieties caused the electrostatic field around the γ-Fe_2O_3 nanoparticles to change and consequently altered the local magnetic fields and moments.

1.3.2
Magnetic Nanoparticles with Photoswitchable Fluorescence

Multifunctionalized nanoparticles have great potential as probes for detecting specific targets in complicated biological systems, including living cells. For example, combining magnetism and fluorescence in a single nanoparticle provides capabilities that cannot be achieved by using conventional nanoparticles with a single function [112–116]. For this reason, fluorescent magnetic nanoparticles have the potential to significantly improve diagnosis.

Recently, a type of superparamagnetic iron oxide nanoparticle with photoswitchable fluorescence was developed by Lee and coworkers [49]. As shown in Figure 1.10a, iron oxide nanoparticles stabilized by oleic acid were used as core of the composite nanoparticles. Sulfur-oxidized diarylethene derivatives were then attached to their surfaces in the presence of Pluronic copolymer F127, PEO_{99}-PPO_{67}-PEO_{99}. Because the hydrophobic PPO segment was anchored around Fe_3O_4 while the hydrophilic PEO segment extended into the aqueous phase, composite nanostructures with good colloidal stability were obtained. The diarylethene derivatives were typical photochromic compounds that undergo photocyclization reaction upon UV irradiation (resulting in a highly fluorescent closed-ring isomer) and revert back into the non-fluorescent open-ring isomer upon visible light irradiation. A dramatic enhancement in the fluorescence of the composite nanoparticle suspension was observed by 312-nm light illumination, resulting in a fluorescent isomer with fluorescence quantum yield of 0.12. Additionally, light-induced isomerization reaction in the composite nanoparticles could be repeated for at least six cycles without obvious fatigue effects or photobleaching (Figure 1.10b).

Another important feature of these composite nanoparticles was their magnetic properties. Specifically, reversible magnetization and demagnetization of these nanoparticles were confirmed during the measurement of field-dependent magnetization at 300 K, indicating their superparamagnetic behavior. The scheme in Figure 1.10c depicts the reversible fluorescence on-off switching induced by alternating illumination of UV and visible light, as well as the reversible flocculation and dispersion process of the composite nanoparticles controlled by the application of an external magnetic field in aqueous media. As shown in Figure 1.10d, a small magnet was able to drive the nanoparticles along the magnetic gradient and

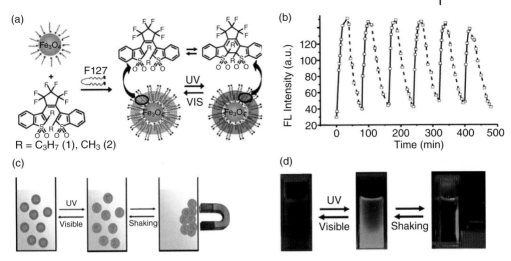

Figure 1.10 (a) Preparation and photoisomerization of the composite nanoparticles; (b) modulation of the fluorescence signal of composite nanoparticles in aqueous suspension, monitored at 505 nm with excitation at 410 nm, upon alternative illumination with UV at 312 nm (solid line) and visible light at 420 nm (dotted line); (c) scheme to show the on-off fluorescence switching and reversible flocculation and dispersion of composite nanoparticles; (d) photographs of the aqueous suspension of composite nanoparticles display that dark suspension under 365-nm excitation is converted into bright fluorescent state at 312-nm illumination. After 312-nm UV illumination for 20 min (middle), a small magnet pulled the particles for 12 h to the side of the vessel (right). Shaking and subsequent illumination with visible light at 420 nm for 40 min returned the appearance sample to that near the initial stage. Reproduced with permission from Reference [49]. Copyright 2008 The Royal Society of Chemistry.

nearly all the nanoparticles were concentrated close to the magnet. Conversely, shaking effectively dispersed the concentrated nanoparticles back to suspension. Undoubtedly, such a unique combination of superparamagnetic feature and photoswitchable fluorescence will make these multifunctional nanoparticles a useful tool in detection, separation, and imaging.

1.4
Future Perspectives

Photoswitchable nanoprobes, as well as most of the applications using these materials, are still in an early stage of technical development and many issues need to be addressed before such nanoprobes can reach full potential in practical biomedical and clinical diagnosis applications. In our opinion, future efforts should address several key issues. First, nanoprobes with smaller size are more suitable for labeling and detection of subcellular components, making intracellular measurements, or tagging a particular protein or a group of proteins in an assembly.

Thus, new synthetic methods that can generate small probes are critically needed. Second, UV illumination is usually applied to photoswitch current nanoprobes. However, this approach can be detrimental for biological applications since long-term exposure to UV radiation induces phototoxicity in living systems. If low-energy sources such as near-IR light can be used for photoswitching irradiation, the phototoxicity to biological systems will be minimized. The introduction of energy-donating fluorophores with two-photon absorption might provide such an opportunity while avoiding UV irradiation [117]. Third, biological probes need to be developed that enhance detection accuracy, minimize nonspecific binding, and improve the binding kinetics and affinities of nanoprobes toward target molecules. Total elimination of nonspecific binding is probably unrealistic, especially when the nanoparticles are used in biological milieu. However, further efforts are needed to develop nanoprobes with "smart" surfaces that are highly specific to target tissue and interact minimally with the surrounding normal tissue. Specifically, optimization of the composition and sequence of the surface structure of the nanoparticle can enable such a "smart" nanoparticle with minimal nonspecific binding [41]. Finally, photoswitchable nanoprobes are expected to play important roles in high-resolution fluorescence imaging and quantitative bioanalysis.

Acknowledgments

The authors acknowledge the support of the National Institute of General Medicine Sciences (GM065306) and the National Science Foundation (CHE-0805547).

References

1 Novotny, L. and Hecht, B. (2006) *Principles of Nano-Optics*, Cambridge University Press, Cambridge.
2 Hell, S.W. (2007) *Science*, **316**, 1153–1158.
3 Hell, S.W. and Wichmann, J. (1994) *Opt. Lett.*, **19**, 780–782.
4 Westphal, V. and Hell, S.W. (2005) *Phys. Rev. Lett.*, **94**, 143903(1)–143903(4).
5 Willig, K.I., Rizzoli, S.O., Westphal, V., Jahn, R., and Hell, S.W. (2006) *Nature*, **440**, 935–939.
6 Westphal, V., Rizzoli, S.O., Lauterbach, M.A., Kamin, D., Jahn, R., and Hell, S.W. (2008) *Science*, **320**, 246–249.
7 Hell, S.W. (2003) *Nat. Biotechnol.*, **21**, 1347–1355.
8 Hell, S.W., Dyba, M., and Jakobs, S. (2004) *Curr. Opin. Neurobiol.*, **14**, 599–609.
9 Hofmann, M., Eggeling, C., Jakobs, S., and Hell, S.W. (2005) *Proc. Natl. Acad. Sci. U. S. A.*, **102**, 17565–17569.
10 Huang, B., Wang, W., Bates, M., and Zhuang, X. (2008) *Science*, **319**, 810–813.
11 Rust, M.J., Bates, M., and Zhuang, X. (2006) *Nat. Methods*, **3**, 793–795.
12 Bates, M., Huang, B., Dempsey, G.T., and Zhuang, X. (2007) *Science*, **317**, 1749–1752.
13 Betzig, E., Patterson, G.H., Sougrat, R., Lindwasser, O.W., Olenych, S., Bonifacino, J.S., Davidson, M.W., Lippincott-Schwartz, J., and Hess, H.F. (2006) *Science*, **313**, 1642–1645.
14 Hess, S.T., Girirajan, T.P.K., and Mason, M.D. (2006) *Biophys. J.*, **91**, 4258–4272.

15 Egner, A., Geisler, C., von Middendorff, C., Bock, H., Wenzel, D., Medda, R., Andresen, M., Stiel, A.C., Jakobs, S., Eggeling, C., SchNnle, A., and Hell, S.W. (2007) *Biophys. J.*, **93**, 3285–3290.
16 Flors, C., Hotta, J., Uji-i, H., Dedecker, P., Ando, R., Mizuno, H., Miyawaki, A., and Hofkens, J. (2007) *J. Am. Chem. Soc.*, **129**, 13970–13977.
17 Bock, H., Geisler, C., Wurm, C.A., von Middendorff, C., Jakobs, S., Schönle, A., Egner, A., Hell, S.W., and Eggeling, C. (2007) *Appl. Phys. B.*, **88**, 161–165.
18 Hu, D.H., Tian, Z.Y., Wu, W.W., Wan, W., and Li, A.D.Q. (2008) *J. Am. Chem. Soc.*, **130**, 15279–15281.
19 Sharonov, A. and Hochstrasser, R.M. (2006) *Proc. Natl. Acad. Sci. U. S. A.*, **103**, 18911–18916.
20 Nechyporuk-Zloy, V., Staock, C., Schilers, H., Oberleithner, H., and Schwab, A. (2006) *Am. J. Physiol. Cell Physiol.*, **291**, C266–C269.
21 Wu, W.W. and Li, A.D.Q. (2007) *Nanomedicine*, **2**, 523–531.
22 Yao, G., Wang, L., Wu, Y.R., Smith, J., Xu, J.S., Zhao, W.J., Lee, E., and Tan, W. (2006) *Anal. Bioanal. Chem.*, **385**, 518–524.
23 Tian, Z.Y., Shaller, A.D., and Li, A.D.Q. (2009) *Chem. Commun.*, **2**, 180–182.
24 Lim, S.J., An, B.K., Jung, S.D., Chung, M.A., and Park, S.Y. (2004) *Angew. Chem. Int. Ed.*, **43**, 6346–6350.
25 Zhu, L.Y., Zhu, M.Q., Hurst, J.K., and Li, A.D.Q. (2005) *J. Am. Chem. Soc.*, **127**, 8968–8970.
26 Medintz, I.L., Trammell, S.A., Mattoussi, H., and Mauro, J.M. (2004) *J. Am. Chem. Soc.*, **126**, 30–31.
27 Medintz, I.L., Clapp, A.R., Trammell, S.A., and Mattoussi, H.M. (2004) *Proc. SPIE-Int. Soc. Opt. Eng.*, **5593**, 300–307.
28 Jares-Erijman, E.A., Giordano, L., Spagnuolo, C., Lidke, K.A., and Jovin, T.M. (2005) *Mol. Cryst. Liq. Cryst.*, **430**, 257–265.
29 Mikoski, S., Giordano, L., Etchelon, M.H., Menendez, G., Lidke, K.A., Hagen, G.M., Jovin, T.M., and Jares-Erijman, E.A. (2006) *Proc. SPIE-Int. Soc. Opt. Eng.*, **6096**, 60960X-1–60960X-8.
30 Tomasulo, M., Yildiz, I., and Raymo, F.M. (2006) *Aust. J. Chem.*, **59**, 175–178.
31 Binder, W.H., Sachsenhofer, R., Straif, C.J., and Zirbs, R. (2007) *J. Mater. Chem.*, **17**, 2125–2132.
32 Zhu, M.Q., Zhu, L.Y., Han, J.J., Wu, W.W., Hurst, J.K., and Li, A.D.Q. (2006) *J. Am. Chem. Soc.*, **128**, 4303–4309.
33 Su, J., Chen, J., Zeng, F., Chen, Q., Wu, S., and Tong, Z. (2008) *Polym. Bull.*, **61**, 425–434.
34 Chen, J., Zeng, F., Wu, S., Chen, Q., and Tong, Z. (2008) *Chem. Eur. J.*, **14**, 4851–4860.
35 Furukawa, H., Misu, M., Ando, K., and Kawaguchi, H. (2008) *Macromol. Rapid Commun.*, **29**, 547–551.
36 Hu, Z., Zhang, Q., Xue, M., Sheng, Q., and Liu, Y. (2008) *Opt. Mater.*, **30**, 851–856.
37 Hu, Z.K., Zhang, Q., Xue, M.Z., Sheng, Q.R., and Liu, Y.G. (2008) *J. Phys. Chem. Solids*, **69**, 206–210.
38 Zou, Y., Yi, T., Xiao, S.Z., Li, F.Y., Li, C.Y., Gao, X., Wu, J.C., Yu, M.X., and Huang, C.H. (2008) *J. Am. Chem. Soc.*, **130**, 15750–15751.
39 Fölling, J., Polyakova, S., Belov, V., Blaaderen, A., van, Bossi, M.L., and Hell, S.W. (2008) *Small*, **4**, 134–142.
40 Zhu, L.Y., Wu, W.W., Zhu, M.Q., Han, J.J., Hurst, J.K., and Li, A.D.Q. (2007) *J. Am. Chem. Soc.*, **129**, 3524–3526.
41 Tian, Z.Y., Wu, W.W., Wan, W., and Li, A.D.Q. (2009) *J. Am. Chem. Soc.*, **131**, 4245–4252.
42 Einaga, Y., Sato, O., Iyoda, T., Fujishima, A., and Hashimoto, K. (1999) *J. Am. Chem. Soc.*, **121**, 3745–3750.
43 Mikami, R., Taguchi, M., Yamada, K., Suzuki, K., Sato, O., and Einaga, Y. (2004) *Angew. Chem. Int. Ed.*, **43**, 6135–6139.
44 Taguchi, M., Yamada, K., Suzuki, K., Sato, O., and Einaga, Y. (2005) *Chem. Mater.*, **17**, 4554–4559.
45 Taguchi, M., Yagi, I., Nakagawa, M., Iyoda, T., and Einaga, Y. (2006) *J. Am. Chem. Soc.*, **128**, 10978–10982.
46 Einaga, Y. (2006) *Bull. Chem. Soc. Jpn.*, **79**, 361–372.

47 Suda, M., Nakagawa, M., Iyoda, T., and Einaga, Y. (2007) *J. Am. Chem. Soc.*, **129**, 5538–5543.

48 Aldoshin, S.M. (2008) *J. Photochem. Photobiol. A Chem.*, **200**, 19–33.

49 Yeo, K.M., Gao, C.J., Ahn, K.H., and Lee, I.S. (2008) *Chem. Commun.*, **38**, 4622–4624.

50 Ikeda, M., Tanifuji, N., Yamaguchi, H., Irie, M., and Matsuda, K. (2007) *Chem. Commun.*, **13**, 1355–1357.

51 Sakano, T., Yamaguchi, H., Tanifuji, N., Irie, M., and Matsuda, K. (2008) *Chem. Lett.*, **37**, 634–635.

52 Matsuda, K., Yamaguchi, H., Sakado, T., Ikeda, M., Tanifuji, N., and Irie, M. (2008) *J. Phys. Chem. C*, **112**, 17005–17010.

53 van der Molen, S.J., Liao, J., Kudernac, T., Agustsson, J.S., Bernard, L., Calame, M., Wees, B.J., Feringa, B.L., and Schönenberger, C. (2009) *Nano Lett.*, **9**, 76–80.

54 Lim, H.S., Han, J.T., Kwak, D., Jin, M., and Cho, K. (2006) *J. Am. Chem. Soc.*, **128**, 14458–14459.

55 Alivisatos, A.P. (1996) *Science*, **217**, 933–937.

56 Michalet, X., Pinaud, F., Lacoste, T.D., Dahan, M., Bruchez, M.P., Alivisatos, A.P., and Weiss, S. (2001) *Single Mol.*, **2**, 261–276.

57 Pinaud, F., Michalet, X., Bentolila, L.A., Tsay, J.M., Doose, S., Li, J.J., Iyer, G., and Weiss, S. (2006) *Biomaterials*, **27**, 1679–1687.

58 Stryer, L. and Haugland, R.P. (1967) *Proc. Natl. Acad. Sci. U. S. A.*, **58**, 719–726.

59 Ha, T., Enderle, Th., Ogletree, D.F., Chemla, D.S., Selvin, P.R., and Weiss, S. (1996) *Proc. Natl. Acad. Sci. U. S. A.*, **93**, 6264–6268.

60 Ha, T., Zhuang, X.W., Kim, H.D., Orr, J.W., Williamson, J.R., and Chu, S. (1999) *Proc. Natl. Acad. Sci. U. S. A.*, **96**, 9077–9082.

61 Giepmans, B.N.G., Adams, S.R., Ellisman, M.H., and Tsien, R.Y. (2006) *Science.*, **312**, 217–224.

62 Irie, M. (2000) *Chem. Rev.*, **100**, 1685–1716.

63 Kawata, S. and Kawatam, Y. (2000) *Chem. Rev.*, **100**, 1777–1788.

64 Tian, H. and Yang, S.J. (2004) *Chem. Soc. Rev.*, **33**, 85–97.

65 Irie, M., Kobatake, S., and Horichi, M. (2001) *Science*, **291**, 1769–1772.

66 Irie, M., Fukaminato, T., Sasaki, T., Tamai, N., and Kawai, T. (2002) *Nature*, **420**, 759–760.

67 Tsien, R.Y. (1998) *Annu. Rev. Biochem.*, **67**, 509–544.

68 Ando, R., Hama, H., Yamamoto-Hino, M., Mizuno, H., and Miyawaki, A. (2002) *Proc. Natl. Acad. Sci. U. S. A.*, **99**, 12651–12656.

69 Patterson, G.H. and Lippincott-Schwartz, J. (2002) *Science*, **297**, 1873–1877.

70 Lippincott-Schwartz, J. and Patterson, G.H. (2003) *Science*, **300**, 87–91.

71 Miyawaki, A., Sawano, A., and Kogure, T. (2003) *Nat. Cell Biol.*, **5** (Suppl.), S1–S7.

72 Lippincott-Schwartz, J., Altan-Bonnet, N., and Patterson, G.H. (2003) *Nat. Cell Biol.*, **5** (Suppl.), S7–S14.

73 Ando, R., Mizuno, H., and Miyawaki, A. (2004) *Science*, **306**, 1370–1373.

74 Shaner, N.C., Campbell, R.E., Steinbach, P.A., Giepmans, B.N., Palmer, A.E., and Tsien, R.Y. (2004) *Nat. Biotechnol.*, **22**, 1567–1572.

75 Wiedenmann, J., Lvanchenko, S., Oswald, F., Schmitt, F., Röcker, C., Salih, A., Spindler, K.D., and Nienhaus, U. (2004) *Proc. Natl. Acad. Sci. U. S. A.*, **101**, 15905–15910.

76 Tsutsui, H., Karasawa, S., Shimizu, H., Nukina, N., and Miyawaki, A. (2005) *EMBO Rep.*, **6**, 233–238.

77 Chudakov, D.M., Lukyanov, S., and Lukyanov, K.A. (2005) *Trends Biotechnol.*, **23**, 605–613.

78 Nam, K.H., Kwon, O.Y., Sugiyama, K., Lee, W.H., Kim, Y.K., Song, H.K., Kim, E.E., Park, S.Y., Jeon, H., and Hwang, K.Y. (2007) *Biochem. Biophys. Res. Commun.*, **354**, 962–967.

79 Kwon, O.Y., Kwon, I.C., Song, H.K., and Jeon, H. (2008) *Biochim. Biophys. Acta*, **1780**, 1403–1407.

80 Biteen, J.S., Thompson, M.A., Tselentis, N.K., Bowman, G.R., Shapiro, L., and Moerner, W.E. (2008) *Nat. Methods*, **5**, 947–949.

81 Andresen, M., Stiel, A.C., Fölling, J., Wenzel, D., Schönle, A., Egner, A., Eggeling, C., Hell, S.W., and Jakobs, S. (2008) *Nat. Biotechnol.*, **26**, 1035–1040.
82 Lukyanov, K.A., Fradkov, A.F., Gurskaya, N.G., Matz, M.V., Labas, Y.A., Savitsky, A.P., Markelov, M.L., Zaraisky, A.G., Zhao, X.N., Fang, Y., Tan, W.Y., and Lukyanov, S.A. (2000) *J. Biol. Chem.*, **275**, 25879–25882.
83 Hell, S.W., Jakobs, S., and Kastrup, L. (2003) *Appl. Phys. A: Mater. Sci. Process.*, **77**, 859–860.
84 Sauer, M. (2005) *Proc. Natl. Acad. Sci. U. S. A.*, **102**, 9433–9434.
85 Chudakov, D.M., Chepurnykh, T.V., Belousov, V.V., Lukyanov, S., and Lukyanov, K.A. (2006) *Traffic*, **7**, 1304–1310.
86 Eggeling, C., Hilbert, M., Bock, H., Ringemann, C., Hofmann, M., Stiel, A.C., Andresen, M., Jakobs, S., Egner, A., Schonle, A., and Hell, S.W. (2007) *Microsc. Res. Tech.*, **70**, 1003–1009.
87 Dedecker, P., Hotta, J., Flors, C., Sliwa, M., Uji-i, H., Roeffaers, M.B., Ando, R., Mizuno, H., Miyawaki, A., and Hofkens, J. (2007) *J. Am. Chem. Soc.*, **129**, 16132–16141.
88 Stiel, A.C., Andersen, M., Bock, H., Hilbert, M., Schilde, J., Schönle, A., Eggeling, C., Egner, A., Hell, S.W., and Jakobs, S. (2008) *Biophys. J.*, **95**, 2989–2997.
89 Andersen, M., Wahl, M.C., Stiel, A.C., Gräter, F., Schäfer, L.V., Trowitzsch, S., Weber, G., Eggeling, C., Grubmüller, H., Hell, S.W., and Jakobs, S. (2005) *Proc. Natl. Acad. Sci. U. S. A.*, **102**, 13070–13074.
90 Stiel, A.C., Trowitzsch, S., Weber, G., Andersen, M., Eggeling, C., Hell, S.W., Jakobs, S., and Wahl, M.C. (2007) *Biochem. J.*, **402**, 35–42.
91 Adersen, M., Stiel, A.C., Trowitzsch, S., Weber, G., Eggeling, C., Wahl, M.C., Hell, S.W., and Jakobs, S. (2007) *Proc. Natl. Acad. Sci. U. S. A.*, **104**, 13005–13009.
92 Matz, M.V., Fradkov, A.F., Labas, Y.A., Savitsky, A.P., Zaraisky, A.G., Markelov, M.L., and Lukyanov, S.A. (1999) *Nat. Biotechnol.*, **17**, 969–973.
93 Wiedenmann, J., Elke, C., Spindler, K.D., and Funke, W. (2000) *Proc. Natl. Acad. Sci. U. S. A.*, **97**, 14091–14096.
94 Wiedenmann, J., Schenk, A., Röcker, C., Girod, A., Spindler, K.D., and Nienhaus, G.U. (2002) *Proc. Natl. Acad. Sci. U. S. A.*, **99**, 11646–11651.
95 Wiedenmann, J., Ivanchenko, S., Oswald, F., and Nienhaus, G.U. (2004) *Mar. Biotechnol.*, **6**, 270–277.
96 Shagin, D.A., Barsova, E.V., Yanushevich, Y.G., Fradkov, A.F., Lukyanov, K.A., Labas, Y.A., Semenova, T.N., Ugalde, J.A., Meyers, A., Nunez, J.M., Widder, E.D., Lukyanov, S.A., and Matz, M.V. (2004) *Mol. Biol. Evol.*, **21**, 841–850.
97 Mizuno, H., Mal, T.K., Tong, K.I., Ando, R., Furuta, T., Ikura, M., and Miyawaki, A. (2003) *Mol. Cell.*, **12**, 1051–1058.
98 Lakowicz, J.R. (1983) *Principles of Fluorescence Spectroscopy*, Plenum Press, New York.
99 Fung, B.K. and Stryer, L. (1978) *Biochemistry*, **17**, 5241–5248.
100 Heilemann, M., Dedecker, P., Hofkens, J., and Sauer, M. (2009) *Laser Photon. Rev.*, **3**, 180–202.
101 Yidiz, I., Deniz, E., and Raymo, F. (2009) *Chem. Soc. Rev.*, **38**, 1859–1867.
102 Tian, Z., Wu, W., and Li, A.-D.-Q. (2009) *ChemPhysChem*, **10**, 1577–1591.
103 Fernández, M. and Ting, A. (2008) *Nat. Rev. Mol. Cell Biol.*, **9**, 929–943.
104 van de Linde, S., Sauer, M., and Heilemann, M. (2008) *J. Struct. Biol.*, **164**, 250–254.
105 Artemov, D., Mori, N., Okollie, B., and Bhujwalla, A.M. (2003) *Magn. Reson. Med.*, **49**, 403–408.
106 Lee, J.H., Huh, Y.M., Jun, Y., Seo, J., Jang, J., Song, H.T., Kim, S., Cho, E.J., Yoon, H.G., Suh, J.S., and Cheon, J. (2007) *Nat. Med.*, **13**, 95–99.
107 Bulte, J.W.M., Douglas, T., Witwer, B., Zhang, S.C., Strable, E., Lewis, B.K., Zywicke, H., Miller, B., van Gelderen, P., Moskowitz, B.M., Duncan, I.D., and Frank, J.A. (2001) *Nat. Biotechnol.*, **19**, 1141–1147.
108 Arbab, A.S., Liu, W., and Frank, J.A. (2006) *Exp. Rev. Med. Dev.*, **3**, 427–439.

109 Seo, W.S., Lee, J.H., Sun, X., Suzuki, Y., Mann, D., Liu, Z., Terashima, M., Yang, P.C., McConnell, M.V., Nishimura, D.G., and Dai, H. (2006) *Nat. Mater.*, **5**, 971–976.

110 Nasongkla, N., Bey, E., Ren, J., Ai, H., Khemtong, C., Guthi, J.S., Chin, S.F., Sherry, A.D., Boothman, D.A., and Gao, J. (2006) *Nano Lett.*, **6**, 2427–2430.

111 Huh, Y.M., Lee, E.S., Lee, J.H., Jun, Y.W., Kim, P.H., Yun, C.O., Kim, J.H., Suh, J.S., and Cheon, J. (2007) *Adv. Mater.*, **19**, 3109–3112.

112 Choi, J., Kim, J.C., Lee, Y.B., Kim, I.S., Park, Y.K., and Hur, N.H. (2007) *Chem. Commun.*, **16**, 1644–1646.

113 Lee, J.H., Jun, Y.W., Yeon, S.I., Shin, J.S., and Cheon, J. (2006) *Angew. Chem., Int. Ed.*, **45**, 8160–8162.

114 Gao, J., Zhang, W., Huang, P., Zhang, B., and Xu, B. (2008) *J. Am. Chem. Soc.*, **130**, 3710–3711.

115 Insin, N., Tracy, J.B., Lee, H., Zimmer, J.P., Westervelt, R.M., and Bawendi, M.G. (2008) *ACS Nano*, **2**, 197–202.

116 Selvan, S.T., Patra, P.K., Ang, C.Y., and Ying, J.Y. (2007) *Angew. Chem., Int. Ed.*, **46**, 2448–2452.

117 Kim, S., Pudavar, H.E., Bonoiu, A., and Prasad, P.N. (2007) *Adv. Mater.*, **19**, 3791–3795.

2
Applications of Semiconductor Quantum Dots in Chemical and Biological Analysis
Xingguang Su and Qiang Ma

2.1
Introduction

Colloidal quantum dots (QDs) have become the most explored nanomaterials [1] since the 1990s. QDs, sometimes referred to as three-dimensional "artificial atoms," are nano-structured materials with typical dimensions ranging from 1 to 10 nm. QDs are composed of a hundreds or thousands of atoms consisting of elements from group II–VI (e.g., CdSe, CdTe), III–V (e.g., GaN, InAs) or IV–VI (e.g., PbSe, PbS) [2, 3]. Usually, an inorganic shell (e.g., ZnS) is used to make core–shell QDs. In addition, surfactant molecules (e.g., trioctylphosphine oxide, mercaptopropionic acid) commonly surround QDs to improve their solubility in applications [4]. Because of their unique nanostructure, QDs are also termed colloidal nanoparticles (NPs), fluorescent semiconductor nanocrystals (NCs), and zero-dimensional materials.

QDs have several unique optical, chemical, and electronic features. Their most fascinating characteristic is the ability to change optical properties based on QD size. When the size of a QD is close to the exciton Bohr radius, quantum confinement effects become evident [5, 6]. In contrast to natural atoms and bulk materials, the size-dependent electronic states of QDs provide modified photophysical properties, including high quantum yields, size tunable emission profiles, and narrow spectral bands [7, 8].

The unique electronic and optical features of QDs have been utilized in nanoelectronics, solar cells, photocatalysis, sensing, and biological applications [9–22]. This chapter reviews methods for synthesis of QDs, the photophysical properties of semiconductor QDs, and the latest applications of QDs in chemical and biochemical analysis.

Trace Analysis with Nanomaterials. Edited by David T. Pierce and Julia Xiaojun Zhao
Copyright © 2010 WILEY-VCH Verlag GmbH & Co. KGaA, Weinheim
ISBN: 978-3-527-32350-0

2.2
History

As far back as ancient Roman times, people found uses for the brilliant color of colloidal Au nanoparticles. Inspired by the preparation of potable Au by Paracelsus, Michael Faraday first introduced the classical synthesis of pure colloidal nanoparticles (called "activated gold") in 1857 [23]. In 1982, Ekimov and Onuschenko [24] reported the first controlled synthesis of semiconductor crystals (CuCl) at the nanometer size by heating supersaturated solutions of copper and chlorine compounds. The nano-crystalline copper chlorides were initially called quantum droplets and later were given other names, including nanoparticles, nanocrystals, nanocrystallites, and QDs. A synthesis for high-quality, monodisperse QDs was developed by Bawendi's group in 1993 [25]. Three key publications appeared in 1996 and 1997, by Bawendi and coworkers [26, 27], Hines and Guyot-Sionnest [28], for synthesis of core–shell QDs.

After pioneering work in the 1980s and the development of effective syntheses in the 1990s, intense research activities were focused on fundamental study of the material and photophysical properties of QDs [4, 7, 26, 29, 30]. However, there were no reports on the application of QDs until the end of the 1990s. In 1998, two breakthrough papers were simultaneously published in *Science* by the Alivisatos and Nie groups [31, 32]. Both groups demonstrated that highly luminescent QDs can be made water soluble and biocompatible by surface modification and bioconjugation. Recently, many practical applications of QDs have emerged in chemistry and biology fields [33–37].

2.3
Classifications

The most common colloidal QDs are formed from groups II (Zn and Cd) and VI (Se, S, and Te). Fewer QDs consist of elements from groups III–V or IV–VI. Considerable efforts have been devoted to the development of core–shell composite QDs and ternary QDs to improve their properties. Table 2.1 lists typical semiconductor QDs with their crystal structures and band gap energies (i.e., energy differences between their respective valence and conduction bands).

2.3.1
II–VI Quantum Dots

The II–VI type QDs are the largest group of quantum dots. Monodisperse CdX QDs (X = sulfur, selenium, tellurium) have become the prototypical semiconductor QDs. The lattice parameter of CdS QDs is about 4.136 Å, whereas those of CdSe and CdTe QDs are about 4.3 and 6.5 Å, respectively. The corresponding densities of CdS, CdSe, CdTe QDs are 4820, 5810, and 5870 kg m^{-3}, respectively. Compared to other types of QDs, CdX nanoparticles have relatively uniform size

Table 2.1 Chemical formula, crystal structure, band gap, and type of quantum dots [38–42].

Quantum dots	Crystal structure	Band gap, E_{gap} (eV)	Type
ZnO	Hexagonal	3.35	II–VI
ZnS	Wurtzite	3.61	II–VI
ZnSe	Zinc blende	2.69	II–VI
ZnTe	Zinc blende	2.28	II–VI
CdS	Wurtzite	2.53	II–VI
CdSe	Wurtzite	1.74	II–VI
CdTe	Zinc blende	1.50	II–VI
HgS	Zinc blende	0.50	II–VI
HgSe	Zinc blende	0.30	II–VI
HgTe	Zinc blende	0.14	II–VI
GaN	Wurtzite	3.44	III–V
GaP	Zinc blende	2.27	III–V
GaAs	Zinc blende	1.43	III–V
GaSb	Zinc blende	0.75	III–V
InN	Wurtzite	0.8	III–V
InP	Zinc blende	1.35	III–V
InAs	Zinc blende	0.35	III–V
InSb	Zinc blende	0.23	III–V
PbS	Sodium chloride	0.37	IV–VI
PbSe	Sodium chloride	0.28	IV–VI
PbTe	Sodium chloride	0.29	IV–VI

and shape, sharp absorption profiles, and perfect emission features at room temperature [43]. More detailed optical characteristics of these and other QDs are presented in Section 2.4.2.

ZnX QDs (X = oxygen, sulfur, selenium, tellurium) have provided particular advantages for medical applications because zinc has lower intrinsic toxicity and better biocompatibility than other heavy metals (e.g., Cd, Hg). The lattice parameters of ZnS, ZnSe, and ZnTe QDs are about 5.41, 5.67, and 6.10 Å, respectively, while the corresponding densities are 4090, 5266, and 5636 kg·m^{-3}, respectively.

Because of environmental concerns and synthetic difficulty, limited work has been performed with mercury chalcogenide QDs (HgX, X = sulfur, selenium, tellurium). However, notably, Hg is a good candidate for making colloidal QDs. Bulk HgTe has an inverted band structure and thus an effectively negative band gap of around −0.15 eV at 295 K [25]. Quantum confinement in HgTe quantum dots should increase their effective bandgap and give rise to near-infrared (NIR) emission.

2.3.2
III–V Quantum Dots

III–V QDs have played a crucial role in the development of optoelectronic devices [44]. In this group, InAs and GaAs QDs have been the most intensively studied due to their potential applications. The photoluminescence or electroluminescence from highly emissive and stable InAs/GaAs (InSb, GaSb) QDs can be observed in the near-infrared (NIR) or even middle-infrared wavelengths [45–52]. NIR emitters are important for *in vivo* biomedical imaging. A NIR signal can penetrate deeply into tissue and background autofluorescence is generally low in this spectral range. For instance, high-quality InAs QDs are among the best candidates for quantum-dot emitters in the NIR window (700–1400 nm) when compared to II–VI (e.g., ZnSe, covering near-UV, purple, and blue), III–V (e.g., InP QDs, covering blue to deep red), and IV–VI QDs [52].

Nitride-based semiconductor QDs (YN, Y = Ga, In, and InGa) have emerged as highly promising QDs for applications involving blue and ultraviolet wavelengths. Heterostructural hexagonal nitride QDs lack inversion symmetry and exhibit piezoelectric effects [53] due to their spontaneous polarization under strain. Conversely, nitride QDs with wurtzite structure have a unique axis, which allows spontaneous polarization even in the absence of strain. Hexagonal III–V nitride QDs also have other important polarization-related properties that make them stand apart from standard zinc blende III–V semiconductors [53].

2.3.3
IV–VI Quantum Dots

Compared to II–VI and III–V QDs, IV–VI semiconductor QDs (PbX, X = S, Se, Te) have attracted less attentions. However, this group of QDs provides unique access to the limit of strong quantum confinement. Both electrons and holes of lead-salt QDs are individually and strongly confined. The excitation Bohr radius of PbS QDs is 18 nm [54], while that PbSe QDs is an outstanding 46 nm – a value eight times larger than that of CdSe QDs [55]. As a consequence, there is a small fraction of atoms at the surface at the same carrier confinement regime of lead-salt QDs. The absorption edges of lead chalcogenide QDs can be size-tuned in the NIR region (e.g., PbS QDs cover the 0.5–2.5 μm wavelength region), which offers important advantages for telecommunication and solar cells [3, 56].

2.3.4
Core–Shell Quantum Dots

Modification of QDs with a shell of a second semiconductor has resulted in improved performance. Such colloidal core–shell QDs contain more than two semiconductor materials, having a well-defined, onion-like structure. Their basic optical properties are controlled by the core nanocrystals and the shell provides a physical barrier between the optically active core and the surrounding medium. Core–shell or core–multiple shell QDs consisting of II–VI, IV–VI, or III–V semiconductor materials have been reported [20, 57, 58]. Core–shell QDs can be categorized as type-I, reverse type-I, and type-II based on the different band gaps of their constituent materials.

In type-I core–shell QDs, the band gap of the shell material is larger than that of the core, making it is possible to confine electrons and holes generated in the core. CdSe/ZnS is a type-I core–shell QD combination. The primary purpose of the ZnS shell is passivation of the core surface. Accompanied with ZnS shell growth on CdSe QDs, peaks in the UV/Vis absorption spectrum and luminescence spectrum of the CdSe core are redshifted (5–10 nm) due to a partial leakage of the exciton into the shell. In reverse type-I core–shell QDs, the band gap of the shell material is smaller than that of the core. With this configuration, generated holes and electrons are at least partially delocalized or completely confined in the shell and the emission wavelength can be tuned by changing the thickness of the shell [59–61]. For example, when a CdSe shell (narrow band gap) is grown on a ZnSe QD core (wider band gap) a significant redshift of the fluorescence peak can be observed to increase with CdSe shell thickness.

In type-II core–shell QDs, a staggered band alignment leads to a smaller effective band gap than any individual core or shell material. The staggered band alignment can induce a spatial separation of holes and the electrons generated in different regions of the core–shell structure [62–65]. Type-II core–shell QDs (e.g., CdTe/CdSe) have been developed as NIR-active materials. Owing to their staggered band alignments, the photoluminescence decay times of type-II core–shell QDs are longer than those of type-I.

2.3.5
Alloyed Quantum Dots

When the structures of the core–shell QDs are significantly mismatched, the formation of structural dislocations can degrade optical properties. To avoid this, a new strategy of alloying is being used to produce highly luminescent and stable QDs. The interface strain of alloyed QDs accumulates dramatically with increasing shell thickness. A rapid alloying process occurs at temperatures above the "alloying point." These highly luminescent alloyed QDs (e.g., ternary QDs such as $Zn_xCd_{1-x}Se$, or doped QDs such ZnSe:Mn) have several advantages, such as larger particle size, higher crystallinity, hardened lattice structure, lower interdiffusion, and spatial compositional fluctuation. The alloying process has been proven to be a practical method for making various high-quality QDs [66–69].

2.4
Characteristics

During the formation QDs, inorganic atoms first nucleate as highly symmetrical cluster compounds that eventually expand into a nanocrystalline core. Sometimes a shell of another combination of inorganic atoms is then grown on the initial core. After growth, QDs are passivated and stabilized by an organic ligand layer that confers solubility and prevents agglomeration (Figure 2.1). As a consequence of these structures, QDs demonstrate several altered characteristics compared to

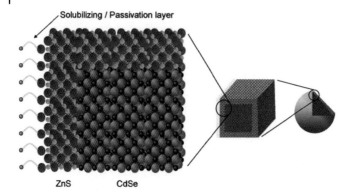

Figure 2.1 Schematic diagram of the structure of core–shell QDs. Each QD consists of an inorganic cadmium core with a thick shell and an organic ligand layer to improve the potential applicability to different situations. Reprinted with permission from Reference [59]. Copyright 2007 Royal Society of Chemistry.

their bulk constituents. The altered electronic and optical properties of QDs are discussed in this section.

2.4.1
Electronic Properties

If QDs absorb energy greater than the band gap of their semiconductor constituent(s), electrons can be promoted to a conduction band (CB), leaving a hole in a valence band (VB) of the material. Depending on the energy gaps and relative positions of these electronic levels, different types of QDs show different properties and can be chosen for particular applications (including photodetectors, solar cells, and photon sources). Figure 2.2 illustrates the band gap energies and relative electronic energy levels of the twelve mostly used QDs [3].

The band gap energy of QDs is closely associated with quantum confinement [2, 24, 70, 71]. Quantum confinement plays a key role in the size-dependent absorption and emission of light by semiconductor QDs. To characterize the effects of this confinement, Brus has applied a particle in a sphere model (approximation to the bulk Wannier Hamiltonian) to develop a relation between size and the electronic band gap energy in QDs [29, 70]. According to the approximation, the lowest eigenvalue in a quantum confined system is given by Equation (2.1):

$$E_g^{eff} = E_g + \hbar^2\pi^2/2R^2 \times (1/m_e + 1/m_h) - 1.8e^2/\varepsilon R + \text{smaller terms} \qquad (2.1)$$

where

R is the QD radius,
E_g is the bulk band gap energy,
E_g^{eff} is the effective bandgap energy,
e is the elementary charge,

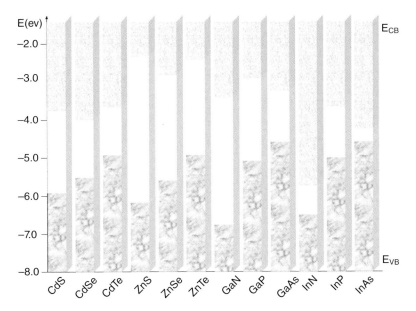

Figure 2.2 Electronic energy levels of selected semiconductor QDs. E_{CB} is the energy of the conduction band; E_{VB} is the energy of the valence band.

ε is the bulk dielectric constant,
m_e and m_h are the effective masses of the electrons and holes, respectively,
$\hbar = h/2\pi$ (h is Planck's constant).

The most evident effect of quantum confinement from this relationship is an increase in the effective band gap energy with a decrease in QD size. This means that as the size (R^2) decreases, shorter emission wavelengths are expected. This quantum confinement phenomenon is borne out in experiments, such as the blue-shift observed for fluorescence wavelength when the dimensions of CdS QDs approach the exciton Bohr radius. To estimate the band gap energy, an effective mass model is commonly used [29]. Band gap energies can be estimated adequately from the first two terms in Equation (2.1):

$$E_g^{eff} = E_g + \hbar^2\pi^2/2\mu R^2 \qquad (2.2)$$

where μ is the effective reduced mass.

The mean size of QDs can be calculated by the Debye–Scherrer formula [72]:

$$D = 0.9\lambda_x/\beta\cos\theta \qquad (2.3)$$

where

D is the diameter of the nanocluster,
λ_x is the wavelength of the incident X-rays,
β is the full-width at half-maximum,
θ is the diffraction angle.

The empirical fitting functions for cadmium chalcogenides QDs have been reported [41] as follows:

$$\text{CdS}: D = (9.8127 \times 10^7)\lambda_{ex}^3 - (1.7147 \times 10^{-3})\lambda_{ex}^2 + 1.0064\lambda_{ex} - 194.84 \quad (2.4)$$

$$\text{CdSe}: D = (1.6122 \times 10^{-9})\lambda_{ex}^4 - (2.6575 \times 10^{-6})\lambda_{ex}^3 + (1.6242 \times 10^{-3})\lambda_{ex}^2 - 0.4277\lambda_{ex} + 41.57 \quad (2.5)$$

$$\text{CdTe}: D = (-6.6521 \times 10^{-8})\lambda_{ex}^3 + (1.9557 \times 10^{-4})\lambda_{ex}^2 - 9.2352 \times 10^{-2}\lambda_{ex} + 13.29 \quad (2.6)$$

In Equations (2.4–2.6), D (nm) is the size of QDs, and λ_{ex} is the wavelength of the first excitonic absorption peak of the corresponding QDs.

2.4.2
Unique Optical Properties

Quantum confinement effects give QDs unique optical properties that include a wide excitation spectrum, a narrow and tunable emission spectrum, high photostability, and a long excited-state lifetime. These properties offer distinct advantages over the corresponding properties of molecular fluorophores (organic dyes, fluorescent proteins, and lanthanide chelates) [73, 74]. First, the emission spectra of QDs can be tuned across a wide range of wavelengths [26, 75, 76, 77]. This allows for the systematic design of QDs that can emit different colors. Their broad excitation spectrum also permits simultaneous excitation of these different QDs by using the same source wavelength. In contrast, excitation spectra of conventional molecular fluorophores are narrow and several organic dyes can only be excited by distinctly different wavelengths [73, 78, 79]. Second, the full-width at half-maximum of most QD emission bands is typically 20 nm, although bandwidths as narrow as 12.7–16.9 nm have been reported [80]. Such narrow emission profiles reduce spectral overlap and enable simultaneous discrimination of QDs designed to emit different colors [81]. Third, the large difference between the excitation and emission wavelengths of QDs (Stokes shift >100 nm) enables the collection of a more complete emission spectra, which can be used to optimize detection sensitivity (Figure 2.3).

Figure 2.4 demonstrates that QDs can be designed to fluoresce at various precise wavelengths – from the ultraviolet (UV) to the infrared (IR) – by changing the size and composition of the semiconductors in a relatively simple manner [82]. With their characteristics of good photostability and long excited-state lifetime (30–100 ns compared to <5 ns for organic dyes at room temperature), QDs typically exhibit higher fluorescence quantum yields. For instance, the fluorescence intensity of a single CdSe QD is roughly 20 times that of a Rhodamine molecule and 100–10 000 times that of standard organic dyes [4, 32, 83]. Unlike organic fluorophores, which bleach after only a few minutes of exposure to irradiation, QDs can withstand hours of exposure while maintaining a high level of brightness and photobleaching threshold [100, 101]. Based on these characteristics, QDs offer new opportunities in analysis through improved detection selectivity and sensitivity.

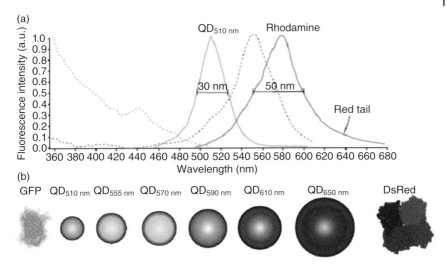

Figure 2.3 Optical properties of quantum dots. (a) The excitation spectrum of QDs is very broad, whereas that of Rhodamine is narrow; the emission spectrum is narrower for QDs than for Rhodamine; (b) as the QD core increases in size the emission shifts to the red end of the spectrum. The combined sizes of the core–shell QDs are in the range of commonly used fluorescent proteins such as green fluorescent protein (GFP) and DsRed. Reprinted with permission from Reference [81]. Copyright 2004 Elsevier.

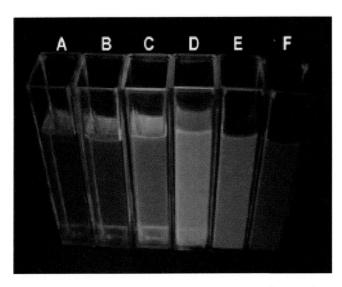

Figure 2.4 Photographs of size-dependent CdTe QDs under UV light. Reprinted with permission from Reference [82]. Copyright 2009 Elsevier.

2.5
Synthesis and Surface Chemistry

Synthetic approaches play a critical role in the applications of QDs since different synthetic methods can alter fundamentally the properties of QDs. Top-down techniques, such as lithography-based synthesis, have been widely used since the 1980s for the preparation of QDs from appropriate bulk substrates [84]. However, poor optical properties resulting from crystal defects and poor reproducibility have rendered such QDs unsuitable for advanced applications. The introduction of a bottom-up colloidal synthesis by Bawendi [25] was a real breakthrough for the development of QDs with consistent properties and toward specific applications. In such wet chemical methods, Cd-chalcogenide QDs can be formed more simply by crystallization and from more available precursors. This is why CdSe and CdTe QDs are most familiar to chemists and nanomaterial researchers. Another landmark in the development of wet chemical routes was the use CdO as precursor instead of $Cd(CH_3)_2$; a method introduced by Peng's group [36]. Compared to CdO, dimethyl cadmium is toxic, explosive, and expensive.

More recent syntheses have been focused on organometallic [87] and aqueous [88] approaches. The synthetic conditions for these routes, such as precursor reagents, temperature, and chelating ligands, have been well studied [25, 85, 86].

2.5.1
Organometallic Approaches

Syntheses of high quality colloidal QDs have been mostly carried out by organometallic approaches. A synthesis starts from organometallic precursors in an organic solvent and proceeds by simple nanocrystal-growth processes. Pyrolysis of organometallic precursors provides hydrophobically capped QDs with a single crystalline structure and well-controlled size and size distribution (ca. 1.2–11.5 nm) [89, 90]. For example, organometallic precursors of cadmium can be formed from dimethylcadmium, cadmium oxide, or cadmium acetate. Under vigorous stirring under vacuum and 300 °C, organometallic precursors of cadmium are rapidly injected into selenium in the organic solvent (trioctylphosphine oxide or trioctylphosphine). Then the mixture is heated to ca. 250–360 °C under an Ar atmosphere. After the nucleation of QDs, the reaction temperature is decreased to ca. 200–300 °C. The injection temperature and growth temperature can decide the size of QDs.

Different sized QDs can be separated by size-selective purification with a mixture 1-butanol and methanol. Specifically, purified QDs are dispersed in anhydrous 1-butanol to form an optically clear solution. When methanol is added dropwise and with stirring, opalescence appears in the clear solution and separation of the flocculate produces a precipitate enriched with the largest crystallites. Dispersion of the precipitate in 1-butanol and size-selective precipitation with methanol can be repeated. Size-selective precipitations have been carried out with various solvent–non-solvent pairs, including pyridine–hexane and chloroform–methanol. Better reproducibility, improved quantum confinement, and high photoluminescence quantum efficiencies (>50%) for the isolated QDs are the main advantages of this method.

2.5.2
Aqueous Phase Colloidal Synthesis

In contrast to organometallic approaches, the direct synthesis of size-controlled and highly luminescent QDs in aqueous solution is less harmful from the standpoint of reagent toxicity and environmental concern. It is also suitable for large-scale production. The preparation of aqueous-compatible CdTe QDs is presented briefly. For precursor preparation, $CdCl_2$ is dissolved in water and then stabilizing agents such as mercapto-alcohols (e.g., 2-mercaptoethanol or 1-thioglicerol) and mercapto-acids (e.g., thioglycolic acid or thiolactic acid) are added while stirring [91]. The nitrogen-saturated precursor solution is then adjusted to pH 9.0 by aqueous NaOH. Freshly prepared and oxygen-free sodium hydrogen telluride (NaHTe), produced by the reaction of sodium borohydride ($NaBH_4$) with tellurium powder, is added to the precursor solution with vigorous stirring and the resultant solution is refluxed at 100 °C to control the dimension of CdTe nanocrystals. QDs prepared in this manner have excellent water-solubility, stability, and biological compatibility [92, 93].

2.5.3
Modification of Surface Chemistry

Surface modification is often an indispensable step in QD synthesis. Surface alteration of QDs can change their physicochemical properties and yield higher luminescence quantum yields [94], improved physical stability [95], stronger interactions with target analytes [96], and more efficient chemical and bioconjugate reactions. There are two major modes of surface modification: direct and post-synthesis [97]. Direct synthesis modification involves capping the QD core with a shell of an inorganic semiconductor during the synthesis process in order to remove surface defects (i.e., surface passivation), as described in Section 2.3.4. The post-synthesis approach involves the exchange of various hydrophobic or hydrophilic ligands (e.g., silica [98–101] or polymer coatings [102, 103]) on the QD for specific surface groups. The ligand-exchange process serves to passivate the QD surface and add functionality. For most hydrophobic QDs, it is necessary to complete a hydrophobic/hydrophilic phase transfer using surface modification techniques in order to prepare them for use in water. For instance, trioctylphosphine oxide capped QDs can be transferred to water by exchanging the TOPO molecules for other ligands that have groups capable of binding to the QDs surface (e.g., thiols, phosphines, carboxylic acids) as well as a hydrophilic head-group (e.g., hydroxyl) [40]. Choice of the right capping layer can lead to better QD morphology in the desired application.

2.6
Trace Analysis Using Quantum Dots

The optical properties of QDs depend strongly on the nature of surface states as well as chemical/physical environment [4]. Because trap states can be filled or energetically moved closer to the band edges by some chemical reactions, the emission spectrum of certain QDs is easily changed by activation. In recent years, great attention has been paid to applications of QDs as probes for inorganic ions, organic compounds, and biological molecules. These applications offer advantages in sensitivity, multiplexing, speed, and cost. In this section, we discuss a series of analytical methods based on QDs in chemistry and biology.

2.6.1
Determinations Based on Direct Fluorescence Response

Several reports have focused on the development of QDs-based probes for metal ions and organic compounds. The direct fluorescence response of QDs to trace analytes can be divided into systems that demonstrate either fluorescence enhancement or quenching. The fluorescence enhancement is generally attributed to passivation of surface trap sites. For example, when Zn and Mn ions are added into solutions of CdS or ZnS QDs [104, 105], a photoluminescence-activation effect is induced that enhances the QD luminescence. Lai and coworkers have modified CdS QDs with bismuthiol II potassium salt as a fluorescence probe for silver ion [106]. The bismuthiol II has a great affinity for "soft" metal ions such as silver ions, due to a complex structure and reactive functional groups containing N and S donor atoms. Under optimum conditions, the fluorescence intensity of CdS QDs was proportional to silver ion concentration from 0.01 to 5.0 $\mu mol \cdot l^{-1}$ with a detection limit of 1.6 $nmol \cdot l^{-1}$.

Compared to fluorescence enhancement, the quenching mechanisms of QDs are more varied and include inner filter effects, non-radiative recombination pathways, electron-transfer processes, and ion-binding interactions. The most studied metal cations in QDs quenching measurements are Hg(II) [108–110], Cu(II) [111–115], and Ag(I) [116, 117]. For example, the luminescence-deactivation of peptide-coated CdS QDs has been reported for optical sensing of Cu(II) and Ag(I) [107]. Based on fluorescence-quenching measurements, many divalent metal cations (e.g., copper, calcium, magnesium, manganese, nickel, and cadmium) [113, 118] and some anions (e.g., iodide [119] or cyanide [120]) have been investigated. However, major problems with the fluorescence-quenching of QDs are the lack of selectivity and divergence of mechanism.

Many research groups have proposed different mechanisms to explain the fluorescence quenching of QDs. A mechanism by which Cu(II) quenches the fluorescence of CdS QDs has been proposed by Isarov and Chrysochoos [121]. They demonstrated that a Cu_xS (x = 1, 2) precipitate or isolated Cu(I) on the surface of CdS QDs quenched the fluorescence of CdS QDs. A mechanism focused on binding interactions between QDs and Cu(II) has been proposed as well [112].

Chen et al. [108] have reported that an electron-transfer process between capping ligands and Hg(II) quenched the fluorescence of QDs. Cai et al. [109] have demonstrated a quenching effect caused by size-quantized HgS particles at the QD surface.

Analytical applications based on QDs have also been used for analysis of organic compounds. Li et al. have developed a new strategy for the efficient recognition and determination of fenamithion and acetamiprid in aqueous solution using the cooperation of p-sulfonato-calixarene via the fluorescence response of CdTe QDs. In this approach, the free CdTe QDs showed remarkable selectivity and sensitivity to acetamiprid and fenamithion [122]. Table 2.2 summarizes some of the trace detection methods that have been developed by using the direct fluorescence response of QDs [16, 113–115, 117–120, 123–127].

Table 2.2 Direct luminescence determinations of some analytes using quantum dots [16, 113–115, 117–120, 123–127].

Analyte	QDs used	Measuring fluorescence signal	Detection limit
Ag(I)	Mercaptopropionic acid-capped CdSe QDs	Quenching	70 nM
Ag(I)	L-cysteine-capped CdS QDs	Enhancement	5.0 nM
Zn(II)	Thioglycerol-capped CdS QDs	Quenching	0.8 mM
Cu(II)	3-Mercaptopropionic acid-capped CdTe QDs	Quenching	$0.19\,\mathrm{ng\,ml^{-1}}$
Cu(II)	Aminophenol-capped CdS QDs	Quenching	0.5 μM
Cu(II)	Polyphosphate-capped CdS QDs	Quenching	0.1 mM
Cu(II)	2-Mercaptoethane sulfonic acid-capped CdSe QDs	Quenching	0.5 μM
Cu(II)	BSA-capped CdSe/ZnS QDs	Quenching	10 nM
CN⁻	2-Mercaptoethane sulfonic acid-capped CdSe QDs	Quenching	1.1 μM
Spirolactone	TOPO-capped CdSe QDs	Quenching	$0.2\,\mathrm{\mu g\,ml^{-1}}$
Tiopronin	Mercaptoacetic acid-capped CdTe QDs	Quenching	$0.15\,\mathrm{\mu g\,ml^{-1}}$
Lysozyme	Mercaptodecanoic acid-capped CdSe QDs	Quenching	$0.115\,\mathrm{\mu g\,ml^{-1}}$
Anthracene	Cyclodextrin-capped QDs	Enhancement	16 nM
p-Nitrophenol	Cyclodextrin-capped QDs	Quenching	7.9 nM
1-Naphthol	Cyclodextrin-capped QDs	Quenching	4.8 nM

2.6.2
Fluorescence Resonance Energy Transfer (FRET) Analysis Using QDs

Fluorescence energy transfer will occur from a donor QD to an acceptor QD if the distance between the particles is smaller than the Förster radius. This process is known as fluorescence resonance energy transfer (FRET) [128]. It is manifested by a reduction in the donor's emission and excited state lifetime, and an increase in the acceptor's emission intensity. Several studies have confirmed that FRET strategies based on QDs are effective for determining the concentration of biological molecules [129], measuring protein conformational changes [130], monitoring protein interactions [131], assaying enzyme activity [132], and performing immunoassays [133].

QDs provide narrow emission profiles, resistance to photobleaching, and an ability to conjugate several acceptor dyes. As such, they are better donors than organic fluorophores for FRET-based applications. One example is the biosensing of maltose [134]. Using a FRET scheme, several active dyes were able to reach the modified QD surface (Figure 2.5). Maltose binding protein (MBP) appended with an oligohistidine tail and labeled with an acceptor dye (Cy3) was immobilized on the nanocrystals via a noncovalent self-assembly scheme. Results showed that substantial acceptor signals were measured upon conjugate formation, indicating efficient nonradiative exciton transfer between QD donors and dye-labeled protein acceptors. It was also observed that conjugation to the QDs did not affect the properties of the biomolecules. With increasing number of MBP-Cy3 attached to a given population of QDs, or with increasing spectral overlap for a given acceptor

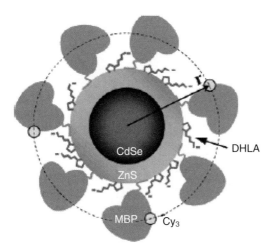

Figure 2.5 Schematic representation of the QD-MBP-dye nano-assembly. The FRET signal is measured for these complexes as a function of both degree of spectral overlap and fraction of dye-labeled proteins in the QD conjugate. The total number of proteins immobilized on each QD surface is ca. 15. Reprinted with permission from Reference [136]. Copyright 2005 American Chemical Society.

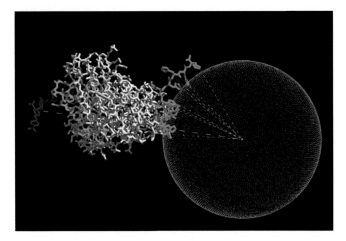

Figure 2.6 Structural model showing TNB-BHQ-10 bound by TNB2-45 conjugated to a QD. Reprinted with permission from Reference [145]. Copyright 2004 American Chemical Society.

dye, a progressive and substantial enhancement in energy transfer efficiency was observed.

QDs have also been used as donors in FRET-based systems for the detection of proteases, adenosine triphosphate, glucose, and metal ions [135–142]. Goldman et al. have used FRET-based QDs for specific detection of the explosive 2,4,6-trinitrotoluene (TNT) in aqueous solution [143]. The FRET system consisted of anti-TNT specific antibody fragments attached to a hydrophilic QD and dye-labeled TNT, which can quenched the QD photoluminescence via FRET. Within the range of examined TNT concentrations (9.8 ng · ml^{-1} to 10 μg · ml^{-1}), the lowest value that produced a measurable signal was 20 ng · ml^{-1}. As reported, the distance between the QD center and the TNB-BHQ-10 (Black Hole Quencher-10) was 72 Å and five of its residues were in close contact with the QD (Figure 2.6).

A few studies have demonstrated the use of QDs as acceptors [144, 145], such as the formation of a bovine serum albumin (BSA)-IgG immunocomplex that induced FRET between the two different QDs [133]. In this work, the BSA antigen was conjugated to red-emitting CdTe QDs and green-emitting QDs were attached to the corresponding anti-BSA antibody (IgG). When FRET occurred, luminescence of the green-emitting QDs (donors) was quenched while emission of the red-emitting QDs (acceptors) was enhanced (Figure 2.7). Luminescence of the donor recovered when the immunocomplex was exposed to an unlabeled antigen, thus demonstrating that competitive FRET inhibition can be used for immunoassay.

2.6.3
Room-Temperature Phosphorescence Detection

Room-temperature phosphorescence (RTP) can be a very useful signal for sensing applications. The triplet excited state that leads to phosphorescence provides better

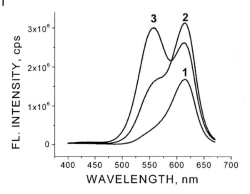

Figure 2.7 FRET between the two different QDs. Reprinted with permission from Reference [133]. Copyright 2002 American Chemical Society.

discrimination between excitation and emission wavelengths and longer emission lifetimes that helps to avoid interference from the short lived autofluorescence and scattering light [146, 147]. However, similar luminescence responses between analyte and coexisting substances have limited applications of RTP.

Recently, quantum dots have attracted considerable attention in the development of novel RTP sensors. The RTP of QDs has allowed detection in biological fluids without interference from autofluorescence and light scattering of matrixes. For example, Yan et al. have reported a new type of RTP sensor based on Mn-doped ZnS QDs that improved selectivity for detecting enoxacin in urine after oral ingestion [148]. The sensor response was due to quenching of the QD phosphorescence emission by interaction with enoxacin (Figure 2.8). In another example, trace pentachlorophenol (PCP) in water was chosen as a target analyte to illustrate the efficiency of QD-based RTP for environmental sensing [149]. The analytical results for river water samples showed good sensitivity (1.1–3.9 µM PCP) and acceptable quantitative recovery ranging from 93% to 106%.

2.6.4
Near-Infrared Detection Using QDs

With the great progress made in developing QD labels, interest in near-infrared (NIR) detection for bioanalytical applications has been growing. Detection in the NIR region (wavelengths >650 nm) can avoid interference from biological media such as tissue autofluorescence and light scattering. However, fewer NIR applications have been developed because of the limited availability of suitable NIR-active materials. Since QDs can be prepared with photoluminescence ranging from visible to near-infrared wavelengths (Figure 2.9), these materials are finding wider applications.

Although most NIR applications of QDs have focused on imaging, QDs have also been explored for NIR detection. Addition of thiol amino acids to CdTe/CdSe QDs was found to enhance NIR fluorescence by reducing most of the Se(IV) and

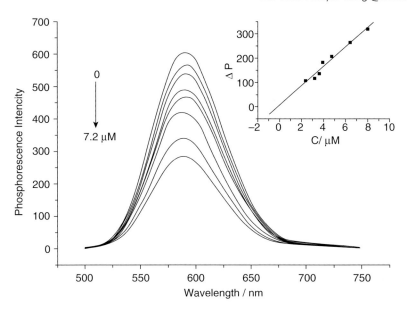

Figure 2.8 Enoxacin concentration-dependent room-temperature phosphorescence (RTP) emission of the synthesized water-soluble Mn-doped ZnS QDs. Reprinted with permission from Reference [150]. Copyright 2008 American Chemical Society.

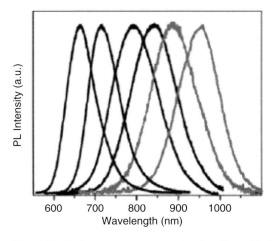

Figure 2.9 Photoluminescence spanning from the red to the NIR (650–975 nm) of QDs. Reprinted with permission from Reference [68]. Copyright 2008 American Chemical Society.

Te(IV) sites on the QD surface and passivating illumination-induced traps [150]. As a result, the photoactivated NIR CdTe/CdSe QDs offered good sensitivity and selectivity for detecting cysteine (Cys), homocysteine (Hcy), and glutathione (GSH) as targets in the presence of interferences. These NIR QDs demonstrated good recovery for the analytes (90–109%) spiked into human urine, plasma, and cell extracts as well as low detection limits: 131, 26, and 20 nM for Cys, Hcy, and GSH, respectively.

In another example, Xia et al. have reported a simple and rapid method for Cu(II) determination using NIR emitting type-II CdTe/CdSe QDs [92]. These QDs were successfully synthesized with high stability and moderate fluorescence quantum yields (10–20%) in aqueous media. There are few reports for NIR fluorescence sensing of Cu(II) ions using organic fluorescent probes. However, NIR fluorescence of the core–shell QDs was markedly quenched by Cu(II), thus enabling quantitative determination to a detection limit for Cu(II) of 2.0×10^{-8} mol·l^{-1}. Physiologically important cations (e.g., Zn, Ca^{2+}, Na$^+$, and K$^+$) had no effect on the fluorescence and therefore posed no interference.

2.6.5
Rayleigh Light Scattering (RLS) Analysis

Rayleigh light scattering (RLS) is a new and attractive method for the analysis of micro-amounts of biomacromolecules. When the wavelength of incident light is close to the molecular absorption band of a chromophore, a special elastic scattering can occur as resonance RLS. This technique can be used to accurately detect aggregates at the molecular level, since larger aggregates enhance the scattering signal. This approach has improved the simplicity, speed, and sensitivity in analysis [151].

QDs are used in RLS analysis as fresh functional particles. For example, the nonspecific interaction between mercaptoacetic acid modified CdSe/ZnS QDs and human immunoglobulin G (IgG) has been investigated by monitoring RLS intensity during conjugation between the QDs and IgG [152]. Figure 2.10 shows typical RLS spectra of the QDs, IgG, and the bioconjugated QDs. The magnitude of the RLS signal confirmed an increased number of QDs conjugated to IgG. The average aggregate size in solution increased quickly and finally surpassed the wavelength of the maximum absorption of QDs (the upper limit for RLS). It was noted that the total RLS intensity probably includes not only Rayleigh scattering of free QDs, free IgG, and small size QDs–IgG conjugates but also non-RLS of large-size QDs–IgG conjugates.

2.6.6
Chemiluminescence Analysis

Chemiluminescence (CL) analysis has the advantages of wide linear range, simple instrumentation, and no background interference. Li et al. have observed, by flow-injection analysis (FIA), the effect of CdTe QDs on CL emitted from the reaction of cerium(IV) with sulfite [153]. Great enhancement of CL was found with addition of CdTe QDs to the cerium(IV)–sulfite system. The possible enhancement mechanism was an efficient CL resonance energy transfer between chemiluminescence donors (SO_2^*) and luminescent quantum-dots as acceptors. The proposed CL

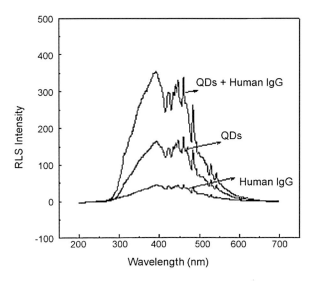

Figure 2.10 Rayleigh light scattering (RLS) spectra of QDs, human IgG, and the bioconjugation. Reprinted with permission from Reference [152]. Copyright 2009 Elsevier.

system may be used to achieve the simultaneous detection of numerous compounds or proteins.

Electrochemiluminescence (ECL) is a special form of chemiluminescence. It has become an important and valuable detection method in analytical chemistry because of its low cost, wide range of analytes, excellent selectivity, and high sensitivity [154]. Many chemiluminescent reagents can be applied in ECL reactions. The first ECL of Si QDs was observed by Bard and coworkers in 2002 [155]. Coupled with the electrochemiluminescence emission of QDs, ECL analytical techniques for several analytes have been rapidly developed, including oxidase substrates, hydrogen peroxide, thiol compounds, proteins, dopamine, and amines [156–163]. These studies have resulted in the development of new QDs-ECL emitters and sensors.

ECL has been employed for the analytical detection of catechol derivatives using mercaptopropionic acid-capped CdTe QDs, which gave anodic ECL behavior at an indium tin oxide (ITO) electrode. Dissolved oxygen and the surface of the semiconductors play important roles in the ECL process, which form QDs with electron-injected radical anions (QD^-). The electron–hole recombination of radical-anion and radical-cation QDs gives rise to excited QDs, leading to ECL emission that can be quenched by adding electrooxidation products of catechol derivatives. ECL-energy transfer from the excited QDs to the analyte was proposed as the quenching mechanism. The detection limits obtained for dopamine and ladrenalin were 50 and 20 nM, respectively [162].

A novel method for the determination of H_2O_2 has been developed, based on the enhancement of light emission from thiol-capped CdTe QDs by H_2O_2 at a

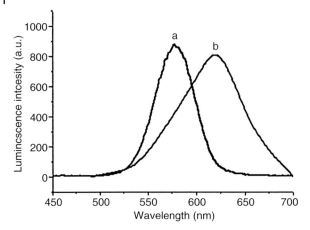

Figure 2.11 Electrochemiluminescence (ECL) and photoluminescence (PL) spectra of CdTe QDs. (a) PL spectrum (emission peak position: around 577 nm); (b) ECL spectrum (peak position: around 620 nm). Reprinted with permission from Reference [164]. Copyright 2007 Elsevier.

negative electrode potential [164]. A proposed mechanism for the ECL was the reaction between electrogenerated reduced species and co-reactants to produce excited states. The QDs produced higher sensitivity for the determination of H_2O_2 (between 2.0×10^{-7} and 1.0×10^{-5} mol·l^{-1}) with a detection limit of 6.0×10^{-8} mol·l^{-1}. Notably, the ECL spectrum of the CdTe QDs exhibited a peak around 620 nm, which was substantially redshifted from the photoluminescence spectrum (Figure 2.11). This suggested that surface states play an important role in this ECL process. In another example, a novel method was described for ECL detection of nitrite. The results indicated that the ECL emission arose from a redox process with QDs core and that sulfite acted as a co-reactant. The quenched ECL emission was used to analyze trace nitrite with a wide linear range (1–500 μM) and low detection limit of 0.1 μM [165].

2.6.7
Electrochemical Analysis

Electrochemical applications offer a great advantage in trace analysis because analytes can be measured easily without addition of mediators. Direct electrochemical reactions based on QDs have been reported in many applications. The amperometric response of QDs with good biocompatibility and conductivity has been observed [166]. The combination of QDs with a cancer marker or DNA as electrode modifiers and signal amplifiers has presented a novel platform for the determination of trace amounts of biomolecules [167–169]. Ho's group have employed CdS QDs as electroactive tags for electrochemical assays of the cancer marker carcinoembryonic antigen (CEA) in standards and human urine samples

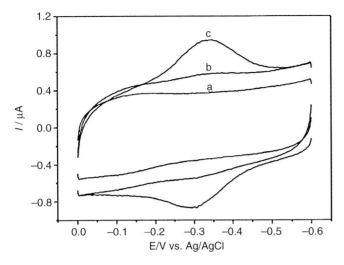

Figure 2.12 Cyclic voltammograms of MCFs (mesopores cellular foam silicate)/GC (a), myoglobin-MCFs/GC (b), and myoglobin-QDs-MCFs/GC (c) electrodes. Reprinted with permission from Reference [171]. Copyright 2007 Elsevier.

[170]. A screen-printed graphite electrode modified with carbon nanoparticles and poly(ethylene imine) was covered with anti-CEA antibodies. CdS QDs were used as biotracers and the carbon nanoparticles amplified the electrochemical signals, which improved the sensitivity and detection limit for CEA. The results indicated that calibration was linear in the range 0.032–10 ng·ml^{-1} CEA and the detection limit was 32 pg·ml^{-1}. This strategy can be further developed for in-home care and self-diagnostics. For example, Li et al. have fabricated hybrid materials with thioglycolic acid-stabilized CdTe QDs assembled in the mesopores of amino-functionalized cellular foam silicate [171]. This material exhibited excellent biocompatibility and a large surface area. When myoglobin was immobilized on the QDs, the hybrid materials not only enhanced direct electron-transfer but also exhibited better electrocatalytic performance to H_2O_2 with better sensitivity and wider linear range (Figure 2.12).

2.6.8
Chemosensors and Biosensors

QDs have frequently been used for making chemical and biological sensors with improved sensitivity, selectivity, and reliability. For example, Wang et al. have developed a sensitive and selective chemical sensor to detect biothiols based on the recovered fluorescence intensity of a QDs–Hg(II) system [172]. As described above, Hg(II) can significantly quench the fluorescence of QDs. Owing to its strong affinity for biothiols such as glutathione(GSH) and cysteine(Cys), Hg(II) forms a Hg(II)–S bond rather than quenching the fluorescence of the QDs. Therefore, recovered fluorescence intensity was observed with increasing amounts of biothiols. A good linear relationship was obtained from 0.6 to 20.0 µM for GSH and from 2.0 to 20.0 µM for Cys, with detection limits of 0.1 and 0.6 µM, respectively.

In another example of a chemosensor, a sensing film has been designed for selective detection of paraoxon by the layer-by-layer technique [126]. Water-soluble CdSe QDs were incorporated together with organophosphorus hydrolase (OPH) in a thin film. An interaction of paraoxon with the OPH in the sensing film induced changes in the QDs photoluminescence emission. This film was proposed for developing a biosensor to detect paraoxon, with detection limits as low as 10 nM [126].

In some biosensors, the unique optical properties of QDs have been combined with the selective catalytic function of an enzyme for detection of organic compounds. For example, phenolic compounds and hydrogen peroxide quenched QD luminescence under horseradish peroxidase (HRP)-catalyzed oxidation by enzymatic catalysis [173]. Relatively low detection limits at $10^{-7}\,mol \cdot l^{-1}$ levels were reported for phenolic compounds. Furthermore, it was found that quinone intermediates quenched the QD luminescence more efficiently, with quenching elevated to more than 100-fold.

In other biosensors, the electrochemical or ECL behavior of QDs has been used. Liu et al. have constructed a QD-bioconjugate sensing platform based on electrochemical measurements [174]. A glassy carbon electrode was modified with CdTe QDs, glucose oxidase, and carbon nanotubes for detection of glucose. A pair of well-defined, quasi-reversible redox peaks was obtained with the modified electrode due to direct electron transfer with glucose oxidase. With a high sensitivity of $1.018\,\mu A \cdot mM^{-1}$, the biosensor displayed a linear response up to 0.7 mM of glucose. Zhu et al. have described an ECL immunosensor based on CdSe quantum dots for the detection of human prealbumin (PAB) [161]. ECL was used to detect formation of the immunocomplex by its blocking of $K_2S_2O_8$ to the surface of the CdSe QD-electrode, which resulted in a decrease of ECL intensity. PAB concentration could be reliably determined in the range 5.0×10^{-10} to $1.0 \times 10^{-6}\,g \cdot ml^{-1}$ and a detection limit of $1.0 \times 10^{-11}\,g \cdot ml^{-1}$ was established.

Taking advantage of single-molecule detection and FRET, Johnson has developed a single-QD-based biosensor [175]. With aptamers on the surface, QDs were able recognize cocaine through both signal-off and signal-on modes (Figure 2.13).

2.6.9
"Nano-On-Micro" Assay

To enable multiplex assays and reduce detection limits, Yoon's group have conjugated quantum dots onto microspheres [176]. Yoon named this configuration "nano-on-micro" or "NOM" in 2007. Two types of QD-microsphere combinations were used. One batch of microspheres was coated with 655 nm emission QDs and mouse IgG. Another batch was coated with 605 nm emission bovine serum albumin (Figure 2.14). Beside the "micro" spheres, a "micro" chip reading device was developed using optical fibers for signal collection. Two different antibodies

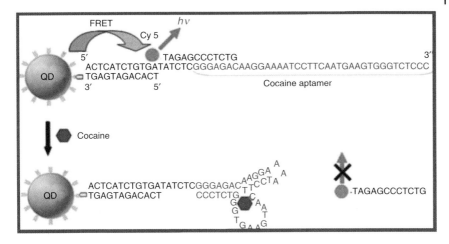

Figure 2.13 Design of the signal-off single-QD-based aptameric sensor. Reprinted with permission from Reference [175]. Copyright 2009 American Chemical Society.

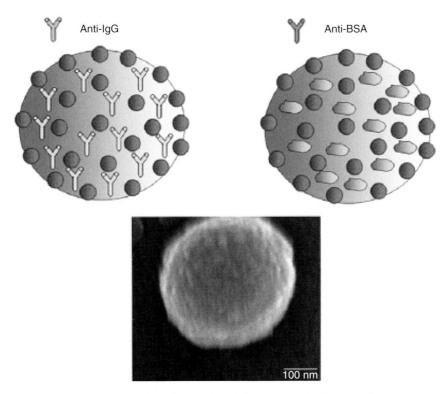

Figure 2.14 Nano-on-micro (NOM) with IgG and QDs. Reprinted with permission from Reference [176]. Copyright 2007 Elsevier.

were detected simultaneously with the NOM detection protocol, with detection limits as low as 50 ng·ml^{-1} (330 pM).

In a similar example, Nie's group has used a swelling procedure to tag polystyrene microbeads with QDs [79]. Incorporating multicolor QDs in microspheres with precisely controlled ratios resulted in QD-encoded microbeads with a certain spectroscopic signature, named a "bar code." In theory, n intensity levels with m colors of QDs can theoretically generate [n^m −1] unique codes beads. The highly emissive and photostable QD-tagged microbeads were proven to be useful in DNA hybridization experiments.

In the two previous examples, more than two types of QDs and microspheres were used for multiplex assay of a complex mixture. Although QDs have several advantages at the nanometer dimension, the NOM strategy can improve the performance of QDs (e.g., toxicity) and enhance other characteristics. For example, if a micro-material (e.g., microspheres) is integrated with molecular recognition and detection elements (e.g., QDs) it could be considered a micro-"chemical lab". Furthermore, when the micro-lab is used with a microfluidic device or microchip, it may be considered a "lab on chip."

Recently, many NOM assays have been reported, such as the development of an acetylcholinesterase biosensor based on CdTe QDs/gold nanoparticles modified with chitosan microspheres at the interface [177]. Xu et al. have described a rapid and ultrasensitive detection method using a microfluidic chip for analyzing 7-aminoclonazepam (7-ACZP) residues in human urine. The whole procedure included a microfluidic chip-based immunoassay with laser-induced fluorescence (LIF) detection, control software, and water-soluble denatured bovine serum albumin-coated CdTe QDs. Under optimal conditions, the assay could be completed within 5 min with a detection limit 0.021 ng·ml^{-1} for 7-ACZP. Compared with ELISA, this detection technique was ultrasensitive and linear regression analysis showed a good correlation. Furthermore, the results were confirmed by high-performance liquid chromatography and tandem mass spectrometry (LC/MS/MS) [178]. Conceivably, the NOM strategy has the capacity for simultaneous detection of a multitude of trace analytes with minimal sample processing and handling. This performance has been demonstrated by using flow cytometric assay technology based on QDs-encoded beads [179] and multicolor quantum dot-encoded microspheres for the detection of biomolecules [180].

2.7
Summary

Advancements in understanding the electronic and optical properties of QDs as well as new strategies for synthesizing QDs have accelerated their use in trace analysis. Precise control over the surface structure and composition of QDs has allowed the successful detection of many different types of analytes, ranging from metal ions and organic molecules to biological systems. QDs are also particularly flexible platforms for analysis because of their many avenues for signaling.

Fluorescence, phosphorescence, near-infrared, Rayleigh light scattering, and chemiluminescence are now common optical methods used with QDs. In addition, the flexible compositional characteristics of QDs have also allowed their use for electrochemical and ECL detection. It is remarkable that most of the development in QD-based analysis has only occurred within past 10 years. Yet with continued development of new synthetic strategies it can be expected that quantum dots will find a wider range of applications in the near future.

Acknowledgments

This work was financially supported by the National Natural Science Foundation of China (No. 20475020, No. 20075009, No. 20875036) and the Development Program of the Ministry of Science and Technology of Jilin Province, China (No. 20080544).

References

1. Peng, X.G. (2009) Nano Res., 2, 425–447.
2. Alivisatos, A.P. (1996) Science, 271, 933.
3. Reiss, P., Protie're, M., and Li, L. (2009) Small, 5, 154–168.
4. Costa-Fernàndez, J.M., Pereiro, R., and Sanz-Medel, A. (2006) Trends Anal. Chem., 25, 207–218.
5. Murphy, C.J. and Coffer, J.L. (2002) Appl. Spectrosc., 56, 16A–27A.
6. Weller, H. (1993) Angew. Chem. Int. Ed. Engl., 32, 41.
7. Brus, L.E. (1983) J. Chem. Phys., 79, 5566–5571.
8. El-Sayed, M.A. (2004) Acc. Chem. Res., 37, 326–333.
9. Davies, J.H. and Long, A.R. (eds (1992) Physics of Nanostructures, Institute of Physics, Philadelphia, PA, pp. 337.
10. Kastner, M.A. (1993) Phys. Today, 46, 24–31.
11. Leute, A. (2004) Phys Unserer Zeit, 35, 206.
12. Galoppini, E. (2004) Coord. Chem. Rev., 248, 1283–1297.
13. Kalyanasundaram, K. (1987) Photochemistry in Microheterogeneous Systems, Academic Press, New York, p. 388.
14. Gratzel, M. (1989) Heterogeneous Photochemical Electron Transfer, CRC Press, Boca Raton, FL, p. 357.
15. Lewis, L.N. (1993) Chem. Rev., 93, 2693–2730.
16. Nazzal, A.Y., Qu, L., Peng, X., and Xiao, M. (2003) Nano Lett., 3, 819–822.
17. Nazzal, A.Y., Wang, X., Qu, L., Yu, W., Wang, Y., Peng, X., and Xiao, M. (2004) J. Phys. Chem. B, 108, 5507–5515.
18. Parak, W.J., Gerion, D., Pellegrino, T., Zanchet, D., Micheel, C., Williams, S.C., Boudreau, R., LeGros, M.A., Larabell, C.A., and Alivisatos, A.P. (2003) Nanotechnology, 14, R15–R27.
19. Hong, R., Fischer, N.O., Verma, A.C., Goodman, M., Emrick, T., and Rotello, V.M. (2004) J. Am. Chem. Soc., 126, 739–743.
20. Pellegrino, T., Kudera, S., Liedl, T., Javier, A.M., Manna, L., and Parak, W.J. (2005) Small, 1, 48–63.
21. Medintz, I.L., Clapp, A.R., Melinger, J.S., Deschamps, J.R., and Mattoussi, H. (2005) Adv. Mater., 17, 2450–2455.
22. Shi, J.J., Zhu, Y.F., Zhang, X.R., Baeyens, W.R.G., and García-Campaña, A.M. (2004) Trends Anal. Chem., 23, 351–360.
23. Biju, V., Itoh, T., Anas, A., Sujith, A., and Ishikawa, M. (2008) Anal. Bioanal. Chem., 391, 2469–2495.

24 Ekimov, A.I., and Onuschenko, A.A. (1982) *Sov. Phys. Semicond.*, **16**, 775–778.
25 Murray, C.B., Norris, D.J., and Bawendi, M.G. (1993) *J. Am. Chem. Soc.*, **115**, 8706–8715.
26 Dabbousi, B.O., Rodriguez-Viejo, J., Mikulec, F.V., Heine, J.R., Mattoussi, H., Ober, R., Jensen, K.F., and Bawendi, M.G. (1997) *J. Phys. Chem. B*, **101**, 9463–9475.
27 Danek, M.K., Jensen, F., Murray, C.B., and Bawendi, M.G. (1996) *Chem. Mater.*, **8**, 173–180.
28 Hines, M.A. and Guyot-Sionnest, P. (1996) *J. Phys. Chem.*, **100**, 468–471.
29 Brus, L.E. (1984) *J. Chem. Phys.*, **80**, 4403–4409.
30 Spanhel, L., Haase, M., Weller, H., and Henglein, A. (1987) *J. Am. Chem. Soc.*, **109**, 5649–5655.
31 Bruchez, M., Moronne, M., Gin, P., Weiss, S., and Alivisatos, A.P. (1998) *Science*, **281**, 2013–2016.
32 Chan, W.C.W. and Nie, S.M. (1998) *Science*, **281**, 2016–2018.
33 Wu, X., Liu, H., Liu, J., Haley, K.N., Treadway, J.A., Larson, J.P., Ge, N., Peale, F., and Bruchez, M.P. (2003) *Nat. Biotechnol.*, **21**, 41–46.
34 Dubertret, B., Skourides, P., Norris, D.J., Noireaux, V., Brivanlou, A.H., and Libchaber, A. (2002) *Science*, **298**, 1759–1762.
35 Gerion, D., Pinaud, F., Williams, S.C., Parak, W.J., Zanchet, D., Weiss, S., and Alivisatos, A.P. (2001) *J. Phys. Chem. B*, **105**, 8861–8871.
36 Peng, Z.A. and Peng, X. (2001) *J. Am. Chem. Soc.*, **123**, 183–184.
37 Parak, W.J., Gerion, D., Zanchet, D., Woerz, A.S., Pellegrino, T., Micheel, C., Williams, S.C., Seitz, M., Bruehl, R.E., Bryant, Z., Bustamante, C., Bertozzi, C.R., and Alivisatos, A.P. (2002) *Chem. Mater.*, **14**, 2113–2119.
38 Rajeshwar, K., Tacconi, N.R., and Chenthamarakshan, C.R. (2001) *Chem. Mater.*, **13**, 2765–2782.
39 Trindade, T., O'Brien, P., and Pickett, N.L. (2001) *Chem. Mater.*, **13**, 3843–3858.
40 Tomczak, N., Jańczewski, D., Han, M., and Vancso, G.J. (2009) *Prog. Polym. Sci.*, **34**, 393–430.
41 Singh, J. (1993) *Physics of Semiconductors and Their Heterostructures*, McGraw-Hill, New York.
42 Yu, W.W., Qu, L.H., Guo, W.Z.X., and Peng, G. (2003) *Chem. Mater.*, **15**, 2854–2860.
43 Miller, A. (1985) *J. Mod. Opt.*, **32**, 507–508.
44 Chen, Z.H., Baklenov, O., Kim, E.T., Mukhametzhanov, I., Tie, J., Madhukar, A., Ye, Z., and Campbell, J.C. (2001) *Infrared Phys. Technol.*, **42**, 479–484.
45 Park, G., Shchekin, O.B., HuIaker, D.L., and Deppe, D.G. (1998) *Appl. Phys. Lett.*, **73**, 3351–3353.
46 Alphandery, E., Nicholas, R.J., Mason, N.J., Zhang, B., Mock, P., and Booker, G.R. (1999) *Appl. Phys. Lett.*, **74**, 2041–2043.
47 Krier, A., Labadi, Z., and Hammiche, A. (1999) *J. Phys. D*, **32**, 2587–2589.
48 Raymond, S., Guo, X., Merz, J.K., and Fafard, S. (1999) *Phys. Rev. B*, **59**, 7624–7631.
49 Fafard, S. (2000) *Appl. Phys. Lett.*, **76**, 2707–2709.
50 Oswald, J., Kuldová, K., Zeman, J., Hulicius, E., Jullian, S., and Potemski, M. (2000) *Mater. Sci. Eng. B*, **69**, 318–323.
51 Xie, R.G. and Peng, X.G. (2008) *Angew. Chem. Int. Ed.*, **47**, 7677–7680.
52 Mokkapati, S. and Jagadish, C. (2009) *Mater. Today*, **12** (4), 22–32.
53 Arakawa, Y. and Kako, S. (2006) *Phys. Stat. Sol. A*, **203**, 3512–3522.
54 Martínez-Castañón, G., Martínez, J.R., Zarzosa, G.O., Ruiz, F., and Sánchez-Loredo, M.G. (2005) *J. Sol-Gel Sci. Technol.*, **34**, 137–145.
55 Dantasa, N.O., Qua, F., Montea, A.F.G., Silvab, R.S., and Moraisb, P.C. (2006) *J. Non-Crys. Solids*, **352**, 3525–3529.
56 Fritz, K.P., Guenes, S., Luther, J., Kumar, S., Sariciftci, N.S., and Scholes, G.D. (2008) *J. Photochem. Photobiol. A*, **195**, 39–46.
57 Medintz, I.L., Uyeda, H.T., Goldman, E.R., and Mattoussi, H. (2005) *Nat. Mater.*, **4**, 435–446.
58 Klostranec, J.M. and Chan, W.C.W. (2006) *Adv. Mater.*, **18**, 1953–1964.
59 Huff, R.F., Swift, J.L., and Crumb, D.T. (2007) *Phys. Chem. Chem. Phys.*, **9**, 1870–1880.

60 Battaglia, D., Li, J.J., Wang, Y.J., and Peng, X.G. (2003) *Angew. Chem. Int. Ed.*, **42**, 5035–5039.
61 Zhong, X.H., Xie, R.G., Zhang, Y., Basché, T., and Knoll, W. (2005) *Chem. Mater.*, **17**, 4038–4042.
62 Sarkar, P., Springborg, M., and Seifert, G. (2005) *Chem. Phys. Lett.*, **405**, 103–107.
63 Li, J.B. and Wang, L.W. (2004) *Appl. Phys. Lett.*, **84**, 3648–3650.
64 Piryatinski, A., Ivanov, S.A., Tretiak, S., and Klimov, V.I. (2007) *Nano Lett.*, **7**, 108–115.
65 Zheng, Y., Yang, Z., and Ying, J.Y. (2007) *Adv. Mater.*, **19**, 1475–1479.
66 Zhong, X., Zhang, Z., Liu, S., Han, M., and Knoll, W. (2004) *J. Phys. Chem. B*, **108**, 15552–15559.
67 Protière, M. and Reiss, P. (2007) *Small*, **3**, 399–403.
68 Peter, M.A. and Moungi, G.B. (2008) *J. Am. Chem. Soc.*, **130**, 9240–9241.
69 Qian, H., Dong, C., Peng, J., Qiu, X., Xu, Y., and Ren, J. (2007) *J. Phys. Chem. B*, **111**, 16852–16857.
70 Brus, L. (1986) *J. Phys. Chem.*, **90**, 2555–2560.
71 Wang, Y. and Herron, N. (1991) *J. Phys. Chem.*, **95**, 525–532.
72 Narayanan, S.S. and Pal, S.K. (2006) *J. Phys. Chem. B*, **110**, 24403–24409.
73 Wang, F., Tan, W.B., Zhang, Y., Fan, X.P., and Wang, M.Q. (2006) *Nanotechnology*, **17**, R1–R13.
74 Jamieson, T., Bakhshi, R., Petrova, D., Pocock, R., Imani, M., and Seifalian, A.M. (2007) *Biomaterials*, **28**, 4717–4732.
75 Mattoussi, H., Mauro, J.M., Goldman, E.R., Anderson, G.P., Sundar, V.C., Mikulec, F.V., and Bawendi, M.G. (2000) *J. Am. Chem. Soc.*, **122**, 12142–12150.
76 Lim, Y.T., Kim, S., Nakayama, A., Stott, N.E., Bawendi, M.G., and Frangioni, J.V. (2003) *Mol. Imaging*, **2**, 50–64.
77 Bailey, R.E. and Nie, S.M. (2003) *J. Am. Chem. Soc.*, **125**, 7100–7106.
78 Chan, W.C.W., Maxwell, D.J., Gao, X.H., Bailey, R.E., Han, M.Y., and Nie, S.M. (2002) *Curr. Opin. Biotechnol.*, **13**, 40–46.
79 Han, M.Y., Gao, X.H., Su, J.Z., and Nie, S. (2001) *Nat. Biotechnol.*, **19**, 631–635.
80 Reiss, P., Quemard, G., Carayon, S., Bleuse, J., Chandezon, F., and Pron, A. (2004) *Mater. Chem. Phys.*, **84**, 10–13.
81 Jaiswal, J.K. and Simon, S.M. (2004) *Trends Cell Biol.*, **14**, 497–504.
82 Liu, F.C., Chen, Y.M., Lin, J.H., and Tseng, W.L. (2009) *J. Colloid Interface Sci.*, **337**, 414–419.
83 Ma, Q., Wang, C., and Su, X.G. (2008) *J. Nanosci. Nanotechnol.*, **8**, 1138–1149.
84 Henini, M., Sanguinetti, S., Brusaferri, L., Grilli, E., Guzzi, M., Upward, M.D., Moriarty, P., and Beton, P.H. (1997) *Microelectron. J.*, **28**, 933–938.
85 Rogach, A.L., Kornowski, A., Gao, M., Eychmüller, A., and Weller, H. (1999) *J. Phys. Chem. B*, **103**, 3065–3069.
86 Talapin, D.V., Rogach, A.L., Kornowski, A., Haase, M., and Weller, H. (2001) *Nano Lett.*, **1**, 207–211.
87 Rogach, A.L., Nagesha, D., Ostrander, J.W., Giersig, M., and Kotov, N.A. (2000) *Chem. Matter.*, **12**, 2676–2685.
88 Zhu, J.J., Palchik, O., Chen, S.G., and Gedanken, A. (2000) *J. Phys. Chem. B*, **104**, 7344–7347.
89 Biju, V., Makita, Y., Nagase, T., Yamaoka, Y., Yokoyama, H., Baba, Y., and Ishikawa, M. (2005) *J. Phys. Chem. B*, **109**, 14350–14355.
90 Biju, V., Makita, Y., Sonoda, A., Yokoyama, H., Baba, Y., and Ishikawa, M. (2005) *J. Phys. Chem. B*, **109**, 13899–13905.
91 Pinaud, F., Michalet, X., Bentolila, L.A., Tsay, J.M., Doose, S., Li, J.J., Iyer, G., and Weiss, S. (2006) *Biomaterials*, **27**, 1679–1687.
92 Xia, Y.S. and Zhu, C.Q. (2008) *Analyst*, **133**, 928–932.
93 Gaponik, N., Talapin, D.V., Rogach, A.L., Hoppe, K., Shevchenko, E.V., Kornowski, A., Eychmuller, A., and Weller, H. (2002) *J. Phys. Chem. B*, **106**, 7177–7185.
94 Peng, X., Schlamp, M.C., Kadavanich, A.V., and Alivisatos, A.P. (1997) *J. Am. Chem. Soc.*, **119**, 7019–7029.
95 Kortan, A.R., Hull, R., Opila, R.L., Bawendi, M.G., Steigerwald, M.L.,

Carroll, P.J., and Brus, L.E. (1990) *J. Am. Chem. Soc.*, **112**, 1327–1332.

96 Murphy, C.J. (2002) *Anal. Chem.*, **74**, 520A–526A.

97 Aldana, J., Wang, Y., and Peng, X. (2001) *J. Am. Chem. Soc.*, **123**, 8844–8850.

98 Bakalova, R., Zhelev, Z., Aoki, I., Ohba, H., Imai, Y., and Kanno, I. (2006) Silica-shelled single quantum dot micelles as imaging probes with dual or multimodality. *Anal. Chem.*, **78**, 5925–5932.

99 Sathe, T.R., Agrawal, A., and Nie, S. (2006) Mesoporous silica beads embedded with semiconductor quantum dots and iron oxide nanocrystals: dual-function microcarriers for optical encoding and magnetic separation. *Anal. Chem.*, **78**, 5627–5632.

100 Zhelev, Z., Ohba, H., and Bakalova, R. (2006) Single quantum dot-micelles coated with silica shell as potentially non-cytotoxic fluorescent cell tracers. *J. Am. Chem. Soc.*, **128**, 6324–6325.

101 Selvan, S.T., Patra, P.K., Ang, C.Y., and Ying, J.Y. (2007) Synthesis of silica-coated semiconductor and magnetic quantum dots and their use in the imaging of live cells. *Angew. Chem. Int. Ed.*, **46**, 2448–2452.

102 Wang, D., Rogach, A.L., and Caruso, F. (2002) Semiconductor quantum dot-labeled microsphere bioconjugates prepared by stepwise self-assembly. *Nano Lett.*, **2**, 857–561.

103 Kuang, M., Wang, D., Bao, H., Gao, M., Möhwald, H., and Jiang, M. (2005) Fabrication of multicolor-encoded microspheres by tagging semiconductor nanocrystals to hydrogel spheres. *Adv. Mater.*, **17**, 267–270.

104 Moore, D.E. and Patel, K. (2001) Q-CdS photoluminescence activation on Zn2+ and Cd2+ salt introduction. *Langmuir*, **17**, 2541–2544.

105 Sooklal, K., Cullum, B.S., Angel, S.M., and Murphy, C.J. (1996) Photophysical properties of ZnS nanoclusters with spatially localized Mn^{2+}. *J. Phys. Chem.*, **100**, 4551–4555.

106 Lai, S.J., Chang, X.J., and Mao, J. (2007) *Ann. Chim.*, **97**, 109–121.

107 Gatas-Asfura, K.M. and Leblanc, R.M. (2003) *Chem. Commun.*, 2684–2685.

108 Chen, B., Yu, Y., Zhou, Z., and Zhong, P. (2004) *Chem. Lett.*, **33**, 1608–1609.

109 Cai, Z.X., Yang, H., Zhang, Y., and Yan, X.P. (2006) *Anal. Chim. Acta*, **559**, 234–239.

110 Chen, J.L., Gao, Y.C., Xu, Z.B., Wu, G.H., Chen, Y.C., and Zhu, C.Q. (2006) *Anal. Chim. Acta*, **57**, 77–84.

111 Chen, Y.F. and Rosenzweig, Z. (2002) *Anal. Chem.*, **74**, 5132–5138.

112 Chen, B. and Zhong, P. (2005) *Anal. Bioanal. Chem.*, **381**, 986–992.

113 Fernández-Argüelles, M.T., Jin, W.J., Costa-Fernández, J.M., Pereiro, R., and Sanz-Medel, A. (2005) *Anal. Chim. Acta*, **549**, 20–25.

114 Xie, H.Y., Liang, J.G., Zhang, Z.L., Liu, Y., He, Z.K., and Pang, D.W. (2004) *Spectrochim. Acta A*, **60**, 2527–2530.

115 Liang, J.G., Ai, X.P., He, Z.K., and Pang, D.W. (2004) *Analyst*, **129**, 619–622.

116 Wang, J.H., Wang, H.Q., Zhang, H.L., Li, X.Q., Hua, X.F., Cao, Y.C., Huang, Z.L., and Zhao, Y.D. (2007) *Anal. Bioanal. Chem.*, **388**, 969–974.

117 Xia, Y.S., Cao, C., and Zhu, C.Q. (2008) *J. Lumin.*, **128**, 166–172.

118 Li, J., Bao, D., Hong, X., Li, D., Li, J., Bai, Y., and Li, T. (2005) *Colloids Surf. A*, 267–271.

119 Lakowicz, J.R., Gryczynski, I., Gryczynski, Z., and Murphy, C.J. (1999) *J. Phys. Chem. B*, **103**, 7613–7620.

120 Sarkar, S.K., Chandrasekharan, N., Gorer, S., and Hodes, G. (2002) *Appl. Phys. Lett.*, **81**, 5045–5047.

121 Isarov, A.V. and Chrysochoos, J. (1997) *Langmuir*, **13**, 3142–3149.

122 Qu, F., Zhou, X., Xu, J., Li, H., and Xie, G. (2009) *Talanta*, **78**, 1359–1363.

123 Jin, W.J., Costa-Fernández, J.M., Pereiro, R., and Sanz-Medel, A. (2004) *Anal. Chim. Acta*, **522**, 1–8.

124 Chen, J.L. and Zhu, C.Q. (2005) *Anal. Chim. Acta*, **546**, 147–153.

125 Constantine, C.A., Gattás-Asfura, K.M., Mello, S.V., Crespo, G., Rastogi, V., Cheng, T.C., DeFrank, J.J., and

Leblanc, R.M. (2003) *J. Phys. Chem. B*, **107**, 13762–13764.

126 Ji, X., Zheng, J., Xu, J., Rastogi, V.K., Cheng, T.C., DeFrank, J.J., and Leblanc, R.M. (2005) *J. Phys. Chem. B*, **109**, 3793–3799.

127 Galian, R.E. and Guardia, M. (2009) *Trends Anal. Chem.*, **28**, 279–291.

128 Riegler, J. and Nann, T. (2004) *Anal. Bioanal. Chem.*, **379**, 913–919.

129 Willard, D.M., Carillo, L.L., Jung, J., and Orden, V.A. (2001) *Nano Lett.*, **1**, 469–474.

130 Heyduk, T. (2002) *Curr. Opin. Biotechnol.*, **13**, 292–296.

131 Day, R.N., Periasamy, A., and Schaufele, F. (2001) *Methods*, **25**, 4–18.

132 Li, J.J. and Bugg, T.D.H. (2004) *Chem. Commun.*, 182–183.

133 Wang, S.P., Mamedova, N., Kotov, N.A., Chen, W., and Studer, J. (2002) *Nano Lett.*, **2**, 817–822.

134 Clapp, A.R., Medintz, I.L., Mauro, M., Fisher, B.R., Bawendi, M.G., and Mattoussi, H. (2004) *J. Am. Chem. Soc.*, **126**, 301–310.

135 Algar, W.R. and Krull, U.J. (2008) *Anal. Bioanal. Chem.*, **391**, 1609–1618.

136 Medintz, I.L. and Mattoussi, H. (2009) *Phys. Chem. Chem. Phys.*, **11**, 17–45.

137 Kim, Y.P., Oh, Y.H., Oh, E., Ko, S., Han, M.K., and Kim, H.S. (2008) *Anal. Chem.*, **80**, 4634–4641.

138 Huang, S., Xiao, Q., He, Z.K., Liu, Y., Tinnefeld, P., Su, X.R., and Peng, X.N. (2008) *Chem. Commun.*, 5990–5992.

139 Suzuki, M., Husimi, Y., Komatsu, H., Suzuki, K., and Douglas, K.T. (2008) *J. Am. Chem. Soc.*, **130**, 5720–5725.

140 Chen, Z., Li, G., Zhang, L., Jiang, J., Li, Z., Peng, Z., and Deng, L. (2008) *Anal. Bioanal. Chem.*, **392**, 1185–1188.

141 Tang, B., Cao, L., Xu, K., Zhuo, L., Ge, J., Li, Q., and Yu, L. (2008) *Chem. Eur. J.*, **14**, 3637–3644.

142 Ruedas-Rama, M.J. and Hall, E.A.H. (2009) *Analyst*, **134**, 159–169.

143 Goldman, E.R., Medintz, I.L., Whitley, J.L., Hayhurst, A., Clapp, A.R., Uyeda, H.T., Deschamps, J.R., Lassman, M.E., and Mattoussi, H. (2005) *J. Am. Chem. Soc.*, **127**, 6744–6751.

144 Härmä, H., Soukka, T., Shavel, A., Gaponik, N., and Weller, H. (2007) *Anal. Chim. Acta*, **604**, 177–183.

145 Cissell, K.A., Campbell, S., and Deo, S.K. (2008) *Anal. Bioanal. Chem.*, **391**, 2577–2581.

146 Kuijt, J., Ariese, F., Brinkman, U.A.T., and Gooijer, C. (2003) *Anal. Chim. Acta*, **488**, 135–171.

147 Sánchez-Barragán, I., Costa-Fernández, J.M., Valledor, M., and Campo, J.C.A. (2006) *Trends Anal. Chem.*, **25**, 958–967.

148 He, Y., Wang, H.F., and Yan, X.P. (2008) *Anal. Chem.*, **80**, 3832–3837.

149 Wang, H.F., He, Y., Ji, T.R., and Yan, X.P. (2009) *Anal. Chem.*, **81**, 1615–1621.

150 Zhang, Y., Li, Y., and Yan, X.P. (2009) *Anal. Chem.*, **81**, 5001–5007.

151 Li, K.A., Ma, C.Q., Liu, Y., Zhao, F.L., and Tong, S.Y. (2000) *Chin. Sci. Bull.*, **45**, 386–394.

152 Liu, J., Zhao, W., Fan, R.L., Wang, W.H., Tian, Z.Q., Peng, J., Pang, D.W., and Zhang, Z.L. (2009) *Talanta*, **78**, 700–704.

153 Sun, C., Liu, B., and Li, J. (2008) *Talanta*, **75**, 447–454.

154 Hua, L.J., Han, H.Y., and Zhang, X.J. (2009) *Talanta*, **77**, 1654–1659.

155 Ding, Z., Quinn, B.M., Haram, S.K., Pell, L.E., Korgel, B.A., and Bard, A.J. (2002) *Science*, **296**, 1293–1297.

156 Jiang, H. and Ju, H.X. (2007) *Chem. Commun.*, 404–406.

157 Zou, G.Z. and Ju, H.X. (2004) *Anal. Chem.*, **76**, 6871–6876.

158 Ding, S.N., Xu, J.J., and Chen, H.Y. (2006) *Chem. Commun.*, 3631–3633.

159 Jiang, H. and Ju, H.X. (2007) *Anal. Chem.*, **79**, 6690–6696.

160 Jie, G.F., Liu, B., Pan, H.C., Zhu, J.J., and Chen, H.Y. (2007) *Anal. Chem.*, **79**, 5574–5581.

161 Jie, G.F., Huang, H.P., Sun, X.L., and Zhu, J.J. (2008) *Biosens. Bioelectron.*, **23**, 1896–1899.

162 Liu, X., Jiang, H., Lei, J.P., and Ju, H.X. (2007) *Anal. Chem.*, **79**, 8055–8060.

163 Zhang, L.H., Zou, X.Q., Ying, E., and Dong, S.J. (2008) *J. Phys. Chem. C*, **112**, 4451–4454.

164 Han, H.Y., Sheng, Z.H., and Liang, J.G. (2007) *Anal. Chim. Acta*, **596**, 73–78.
165 Liu, X., Guo, L., Cheng, L.L., and Ju, H.X. (2009) *Talanta*, **78**, 691–694.
166 Zhang, F., Li, C., Li, X., Wang, X., Wan, Q., Xian, Y., Jin, L., and Yamamoto, K. (2006) *Talanta*, **68**, 1353–1358.
167 Liu, G., Wang, J., Kim, J., Jan, M.R., and Collins, G.E. (2004) *Anal. Chem.*, **76**, 7126–7130.
168 Wang, J., Liu, G., and Merkoci, A.J. (2003) *Am. Chem. Soc.*, **125**, 3214–3215.
169 Wang, J., Liu, G., Wu, H., and Lin, Y. (2008) *Small*, **4**, 82–86.
170 Ho, J.A., Lin, Y.C., Wang, L.S., Hwang, K.C., and Chou, P.T. (2009) *Anal. Chem.*, **81**, 1340–1346.
171 Zhang, Q., Zhang, L., Liu, B., Lu, X.B., and Li, J.H. (2007) *Biosens. Bioelectron.*, **23**, 695–700.
172 Han, B.Y., Yuan, J.P., and Wang, E.K. (2009) *Anal. Chem.*, **81**, 5569–5573.
173 Yuan, J.P., Guo, W.W., and Wang, E.K. (2008) *Anal. Chem.*, **80**, 1141–1145.
174 Liu, Q., Lu, X.B., Li, J., Yao, X., and Li, J.H. (2007) *Biosens. Bioelectron.*, **22**, 3203–3209.
175 Zhang, C.Y. and Johnson, L.W. (2009) *Anal. Chem.*, **81**, 3051–3055.
176 Lucas, L.J., Chesler, J.N., and Yoon, J. (2007) *Biosens. Bioelectron.*, **23**, 675–681.
177 Du, D., Chen, S.Z., Song, D.D., Li, H.B., and Chen, X. (2008) *Biosens. Bioelectron.*, **24**, 475–479.
178 Chen, W., Peng, C.F., Jin, Z.Y., Qiao, R.R., Wang, W.Y., Zhu, S.F., Wang, L.B., Jin, Q.H., and Xu, C.L. (2009) *Biosens. Bioelectron.*, **24**, 2051–2056.
179 Wang, H.Q., Liu, T.C., Cao, Y.C., Huang, Z.L., Wang, J.H., Li, X.Q., and Zhao, Y.D. (2006) *Anal. Chim. Acta*, **580**, 18–23.
180 Ma, Q., Wang, X.Y., Li, Y.B., Shi, Y.H., and Su, X.G. (2007) *Talanta*, **72**, 1446–1452.

3
Nanomaterial-Based Electrochemical Biosensors and Bioassays
Guodong Liu, Xun Mao, Anant Gurung, Meenu Baloda, Yuehe Lin, and Yuqing He

3.1
Introduction

A biosensor is a device that combines a biological component (layer of DNA, antibody, receptor, or enzyme) with a detector component (transducer). The transduction unit is usually electrochemical, optical, magnetic, piezoelectric, or calorimetric in nature [1]. Electrochemical devices have shown great promise for genetic and protein testing and are ideally suited for shrinking the physical size and cost of DNA and protein diagnostics [2, 3]. Typically, electrochemical biosensors are classified as either affinity- or catalytic-based, depending on their different modes of molecular recognition. Most catalytic-based biosensors are enzyme biosensors, in which an enzyme is immobilized on the electrochemical transducer surface and the enzymatic products are monitored by measuring the oxidation or reduction current. Affinity-based biosensors use a specific biological recognition event to trigger an electrochemical response; the recognition elements are usually antibodies, nucleic acids (DNA or RNA), or hormone receptors. In this chapter we focus on the affinity-based electrochemical biosensors and bioassays.

Figure 3.1 presents the basic principles for electrochemical sensing of DNA hybridization and immunoreaction events. To construct such an electrochemical biosensor, a single-stranded DNA probe or antibody is immobilized on the electrode surface to form a recognition layer. DNA hybridization events (or antibody–antigen immunoreactions) allow attachment of signal-generating labels (enzyme, metal ions, electroactive molecules, nanomaterials) to the electrode surface, which can be measured directly or indirectly. The change in electrical or electrochemical signal (current, potential, capacitance, or resistance) resulting from this attachment is then related to the concentration of target analyte(s) in the tested sample.

Electrochemical detection of DNA hybridization is mainly based on differences in the electrochemical behavior of the labels in the absence and presence of target DNA. Generally, there are four different analytical techniques used for electrochemical detection of DNA hybridization: (i) oxidation of electroactive DNA bases such as guanine and/or adenine, (ii) enzymatically amplified oxidation or reduction of an enzyme substrate, (iii) oxidation or reduction of an indicator, which

Trace Analysis with Nanomaterials. Edited by David T. Pierce and Julia Xiaojun Zhao
Copyright © 2010 WILEY-VCH Verlag GmbH & Co. KGaA, Weinheim
ISBN: 978-3-527-32350-0

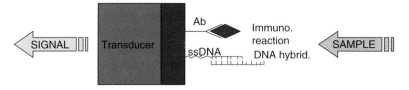

Figure 3.1 Schematic representation of affinity biosensors.

selectively intercalates with double-stranded DNA (dsDNA), and (iv) oxidation or reduction of the captured labels. Electrochemical detection of immunoreactions events can be divided into labeled and label-free methods and can include homogenous or heterogeneous immunoassays. Heterogeneous immunoassays can be competitive or non-competitive. In a competitive immunoassay, the antigen in the unknown sample competes with labeled antigen to bind with antibodies. The amount of labeled antigen bound to the antibody site is then measured. In this method, the response will be inversely proportional to the concentration of antigen in the unknown sample. The greater the response, the less antigen in the unknown was available to compete with the labeled antigen. In non-competitive immunoassays, also referred to as the "sandwich assay," antigen in the unknown is bound to the antibody site, and then labeled antibody is bound to the antigen through the second immunoreaction. The amount of labeled antibody on the site is then measured. Unlike the competitive method, the results of the non-competitive method will be directly proportional to the concentration of the antigen. The labeled-antibody will not bind if the antigen is not present in the unknown sample. Because homogeneous assays do not require this step, these methods are typically faster and easier to perform.

Generally, the sensitivity of label-based electrochemical biosensors and bioassays is higher than that of label-free counterparts. In the traditional electrochemical biosensor and bioassays, electroactive molecules, metal ions, and enzymes (horseradish peroxidase and alkaline phosphatase) have been used as labels to enhance sensitivity. Enzyme-linked electrochemical bioassays and biosensors have attracted considerable interest because of catalytic amplification; the resulting enzymatic products are easily detected by various electrochemical techniques.

With the rapid development of nanoscience and nanotechnology, nanomaterials, including nanoparticles (NPs), nanotubes, and nanowires, have been introduced into the fields of electrochemical biosensors and bioassays. The unique properties of nano-scale materials offer excellent prospects for designing highly-sensitive and selective biosensing of DNA and protein molecules. The creation of such desired nanomaterials for specific sensing applications greatly benefits from the ability to vary the size, composition, and shape of the materials, and hence tailoring their physical properties. These nanomaterials offer elegant approaches for interfacing DNA hybridization events or immunoreactions with electrochemical signal transduction, for dramatically amplifying the resulting electrical response, and for designing novel coding strategies [4–8].

Several excellent reviews have been published on nanomaterial based fluorescence, electrochemical, and piezoelectric bioassays in the past few years [4–14]. Focusing on the most recent developments, the present chapter highlights important contributions toward highly sensitive electrochemical biosensors and bioassays for detection of DNA and proteins. Versatile nanomaterials, such as metal NPs, semiconductor NPs, silica NPs, apoferritin nanovehicles, nanowires, nanorods, carbon nanotubes (CNTs) and liposomes, are described in detail.

3.2 Nanomaterial Labels Used in Electrochemical Biosensors and Bioassays

Many types of nanomaterials, including metal NPs (gold, silver, platinum), semiconductor NPs (CdS, ZnS, PbS), enzyme-loaded carbon nanotubes (CNTs), nanowires, apoferritin nanovehicles, liposomes, and silica NPs, have been used in electrochemical biosensors and bioassays. The applications of nanomaterials can be classified into two categories according to their functions: (i) nanomaterial-based electrochemical transducers and (ii) biomolecule–nanomaterial conjugates (based on DNA probes or antibodies) as labels. Nanomaterials are typically used to modify or construct electrochemical transducers to improve their electron-transfer characteristics or enhance the immobilization and capture efficiency of DNA or antibody probes on the transducer surface. Biomolecule–nanomaterial labels are usually designed in a manner that provides significant signal amplification compared to traditional enzyme and redox probe labels.

This section focuses on both classes of nanomaterial applications – transducers and labels. Its organization is principally based on the types of nanomaterials currently being employed in biosensors and bioassays. For each particular type of nanomaterial, examples are presented that show the latest contexts in which the material is being used for both DNA and immunochemical determinations.

3.2.1
Metal Nanoparticles

Metal nanoparticles (NPs) have attracted extensive interest in versatile fields, especially in electrochemical biosensors and bioassays. The unique stripping voltammetric characteristics of metal ions (high sensitivity and simultaneous detection of multiple metal components) offer an elegant approach to use these metal NPs as labels. The analytical performances of electrochemical bioassay have been enhanced dramatically by using metal NP labels, such as gold, silver, and platinum NPs. The detection limits of DNA and proteins have been lowered by 1 to 3 orders. Applications of metal NP labels include (i) generating electrochemical signals by stripping analysis of the dissolved metals, (ii) causing conductivity changes of electrochemical system, (iii) electrocatalysis and catalysis deposition, and (iv) modifying the electrode surface to construct more sensitive electrochemical transducers. This flexibility in signal generation is a distinct advantage of metal NPs over other nanomaterials. Such metal NPs-based electrochemical devices are expected to have a major impact upon clinical diagnostics, environmental monitoring, security surveillance, and food safety.

3.2.1.1 Metal NP Labels for Electrochemical Detection of DNA

Gold NPs The use of Au NPs (also called colloidal gold) as an electrochemical label for voltammetric monitoring of DNA hybridization was pioneered by several groups [15–17]. Authier et al. have reported an electrochemical DNA detection approach based on Au-NP tags for the sensitive quantification of an amplified 406-base pair human cytomegalovirus DNA sequence (HCMV DNA) [15]. Figure 3.2 presents the typical procedure. The assay relies on (i) the hybridization of the single-stranded target HCMV DNA with an oligonucleotide-modified Au-NP probes, (ii) followed by the release of the gold anchored on the hybrids by oxidative metal dissolution, and (iii) the indirect determination of the solubilized Au^{III} ions by anodic stripping voltammetry at a sandwich-type screen-printed microband electrode (SPMBE). Owing to the enhancement of the Au^{III} mass transfer by nonlinear diffusion during the electrodeposition step, the SPMBE allows sensitive determination of Au^{III} in a small volume of quiescent solution. The combination of sensitive Au^{III} determination at a SPMBE with the large number of Au^{III} released from each Au-NP probe allows detection of as low as 5 pM of the amplified HCMV DNA fragment.

Wang's group have demonstrated electrical detection of DNA hybridization, which used magnetic microbeads and was based on electrochemical stripping detection of the colloidal gold tags [16]. In this work, hybridization of a biotin-functionalized target DNA to magnetic bead-linked oligonucleotide probes was followed by binding of the streptavidin-coated Au NPs to the captured DNA, dissolution of the nanometer-sized Au-NP tag, and potentiometric stripping measurements of the dissolved metal tag at single-use thick-film carbon electrodes. The sensitivity was enhanced by precipitating silver on the captured Au-NP tags and electrochemical stripping analysis of the dissolved silver ions [18]. Picomolar and sub-nanomolar levels of target DNA were thus detected. The sensitivity of

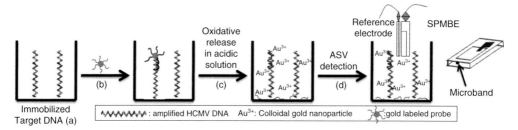

Figure 3.2 Schematic representation of the Au nanoparticle-based electrochemical detection of HCMV-amplified DNA. Reprinted with permission from Reference [15]. Copyright 2001 American Chemical Society.

Au-NP-based electrochemical detection of DNA was further improved by using polymeric beads carrying numerous Au-NP tags [19]. The Au NP-tagged polystyrene beads were prepared by binding biotinylated Au NPs to streptavidin-coated polystyrene spheres. This use of carrier-sphere amplification platforms was combined with catalytic enlargement of the multiple Au-NP tags and an ultrasensitive electrochemical stripping detection of the dissolved Au-NP tags. The result was detection of DNA targets down to 300 atmol – an outstanding capability that holds great promise for ultrasensitive detection of other biorecognition events. A similar protocol has been employed by Cai et al. to develop an electrochemical DNA biosensor using a chitosan-modified glassy carbon electrode and Au-NP tags [17]. The biosensor was able to detect synthesized 32-base complementary oligonucleotides at concentrations down to 1 nM. After silver enhancement, the sensitivity of the biosensor was improved by approximately two orders of magnitude and a detection limit of 50 pM was obtained for complementary oligonucleotides [20].

To avoid dissolution of the captured Au-NP tags, Pumera et al. have reported a magnetically-triggered method for detection of DNA hybridization based on direct electrochemical detection of Au quantum dot tracers [21]. The process was based on binding target DNA (called DNA 1) with Au67 quantum dot in a 1:1 ratio, followed by a genomagnetic hybridization assay between Au67-DNA1 and complementary probe DNA (called DNA 2) marked paramagnetic beads. Differential pulse voltammetry (DPV) was used for direct voltammetric detection of the resulting Au67 quantum dot-DNA1/DNA2-paramagnetic bead conjugate on a magnetic graphite-epoxy composite electrode. The characterization, optimization, and advantages of the direct electrochemical detection assay for target DNA were studied in detail.

Au NPs have also been used as a tag for electrochemical detection of single-nucleotide polymorphisms (SNPs) by Kerman et al.[22]. They described a method to discriminate and code all possible SNP combinations. SNPs were coded by monitoring changes in the electrochemical signal of the monobase-modified Au NPs. In this report, monobase-modified Au NPs showed not only the presence of a SNP, but also identified which bases were involved within the pair. In particular, identification of a transversion SNP, which contains a couple of the same pyrimidine or purine bases, was greatly simplified. To further demonstrate efficacy of the method, a model study was performed using a synthetic 21-base DNA probe related to tumor necrosis factor (TNF-α) along with all of its possible mutant combinations.

Silver NPs Compared with Au-NP tags, Ag NPs require relatively mild conditions for dissolution and the resulting silver(I) ions show excellent electrochemical stripping characteristics. For these reasons, Ag NPs have also been used as tags for sensitive electrochemical DNA detection schemes. For example, Cai et al. [23] have reported an electrochemical DNA biosensor based on Ag-NP tags. After DNA hybridization reactions, the Ag NPs were captured on the electrode surface. Concentrated HNO_3 was chosen to dissolve the Ag-NP tags with an efficiency of 96%, and the aqueous Ag^+ ions were indirectly determined by anodic stripping voltammetry (ASV) at a microelectrode. The detection limit of the biosensor was 0.5 pM. Fu *et al.* have reported a simple and sensitive electrochemical DNA biosensor based on *in situ* DNA amplification with nanosilver and horseradish peroxide (HRP) [24]. A thiolated single-stranded DNA (ssDNA) oligomer was initially immobilized on a gold electrode. With a competitive format, hybridization was carried out by immersing the DNA biosensor into a stirred solution containing different concentrations of the complementary ssDNA and constant concentration of nanosilver-labeled ssDNA, and then further binding with HRP. The amount of HRP adsorbed on the probe surface decreased with the amount of target ssDNA in the sample. The hybridization was monitored by differential pulse voltammetry (DPV) of the H_2O_2 produced by the adsorbed HRP. The reduction current from the enzyme-generated product was related to the number of target ssDNA molecules in the sample. A detection limit of 15 pmol l^{-1} for target ssDNA was obtained with the electrochemical DNA biosensor. The approach discriminated effectively complementary from non-complementary DNA sequence, suggesting that the similar enzyme-labeled DNA assay method holds great promises for sensitive electrochemical biosensors.

More recently, another amplified detection of target DNA has been reported by Hwang *et al.* (Figure 3.3) [25]. In this method, the target DNA and a biotinylated DNA probe were allowed to hybridize with a capture DNA probe tethered to a gold electrode. Neutravidin-conjugated alkaline phosphatase (ALP) was then allowed to conjugate with the biotin of the detection probe and convert the redox-inactive substrate of ALP, *p*-aminophenyl phosphate, into *p*-aminophenol, a reducing agent. The latter, in turn, reduced silver ions in solution, leading to deposition of silver onto the electrode surface and DNA backbone. Anodic stripping of the enzymatically derived silver provided a measure of target oligomer hybridization. This process, named biometallization, leads to a great signal enhancement due to the accumulation of metallic silver by an enzymatically generated product and, thus, to electrochemical amplification of the bioaffinity reaction. This process is highly sensitive, detecting as little as 100 aM of DNA, or 10 zmol of DNA in 100 µL sample solution.

Somasundrum *et al.* [26] have reported a method to prepare polyelectrolyte shells containing Ag NPs and to use these capsules as labels for highly sensitive electrochemical detection. The shells were prepared using layer-by-layer self-assembly on ca 500 nm diameter templates, after which the templates were dissolved. The shells could be opened and closed by adjusting the solution pH. This process was utilized to encapsulate Ag NPs, chiefly by adsorption to the inner walls of the capsules. Based on spectrophotometry, transmission electron microscopy (TEM), and voltammetric measurements, the highest loading achieved was approximately 78 Ag NPs per capsule. The Ag-loaded capsules were used via biotin–avidin binding as labels for the detection of DNA hybridization. Following acid dissolution of the captured labels, the released Ag^+ was determined by ASV. In one demonstration, a 30-mer sequence specific to *Escherichia coli* was measured with a detection limit of ~25 fM using DNA-modified screen-printed electrodes. This level corresponded to the detection of 4.6 fg DNA (~3×10^5 molecules) in the 20 µL analyte sample.

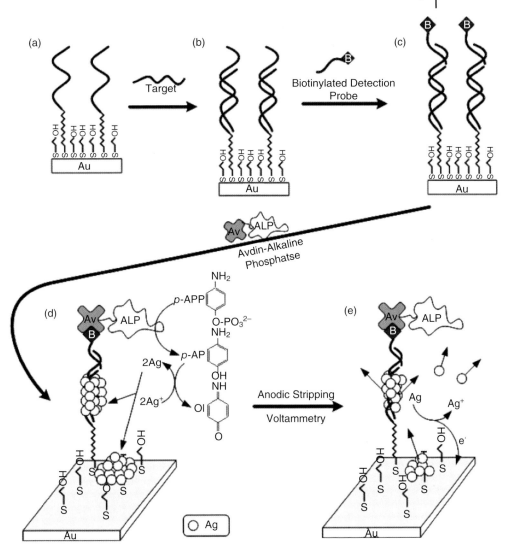

Figure 3.3 Stepwise assembly of a sandwich-type DNA sensing electrode based on enzymatic silver deposition. (a) Formation of mixed SAM on the Au electrode; (b) hybridization with target; (c) hybridization with biotinylated detection probe; (d) association with avidin-alkaline phosphatase and reduction of silver ion by p-AP; (e) dissolution of silver during anodic stripping voltammetry. Reprinted with permission from Reference [25]. Copyright 2005 American Chemical Society.

In a demonstration of the method's selectivity, a 200 fM target containing a single base mismatch gave a significantly lower response (<74%) than 200 fM of a complementary target DNA; while 60 pM of a non-complementary target gave a negligible response.

Platinum NPs Pt NPs are widely used as electrocatalysts for the oxygen reduction reaction in fuel cells due to their high electrocatalytic activity. However, applications of Pt NPs in electrochemical biosensors and bioassays have received little attention because of difficulty in dissolving these NPs. Nevertheless, the excellent electrocatalytic activity offers another avenue to develop highly sensitive electrochemical bioassay approaches. As reported recently [27], nucleic acid-functionalized Pt NPs can act as catalytic labels for the amplified electrochemical detection of DNA hybridization and aptamer/protein recognition. Hybridization of the nucleic acid-modified Pt-NPs with a sensing nucleic acid/analyte DNA complex associated with an electrode enabled the detection of the DNA by the Pt NP electrocatalyzed reduction of H_2O_2 (sensitivity limit, 1×10^{-11} M). Similarly, the association of aptamer-functionalized Pt-NPs to a thrombin aptamer/thrombin complex associated with an electrode allowed detection of thrombin with a sensitivity limit corresponding to 1×10^{-9} M. Liang *et al.* have used hollow Pt nanospheres as electrocatalyst for conductometric immunoassay of interleukin-6 (IL-6) in human serum [28]. In this case, the hollow Pt nanospheres with an average diameter of 20 nm were prepared. The authors demonstrated that the catalytic efficiency of the hollow Pt nanospheres was 100-fold greater than that of solid Pt nanospheres. Using hollow Pt-nanospheres modified anti-IL-6 as *in situ* amplified probes, a sandwich-type conductivity conductometric immunoassay was performed on two parallel electrodes. The detection principle was based on the changes between the two parallel electrodes caused by biochemical reactions in solution before and after antigen–antibody interaction. The results indicated that the method could distinguish IL-6 ranging from 15 to 350 pg ml^{-1} with a relatively low detection limit of 8 pg ml^{-1} at a signal-to-noise ratio of 3.

Modification of Electrochemical Transducers with Metal NPs Metallic NPs have not only been used as labels for sensitive electrochemical detection of DNA, but have also been used to construct electrochemical transducers with enhanced performance. For example, the ability of gold nanoparticles to provide a stable platform for immobilization of biomolecules without degrading their bioactivity is a major advantage in the design of effective biosensors. This is illustrated in the development of DNA biosensors, where immobilization of stable and highly dense single-stranded DNA (ssDNA) monolayers on electrodes is a key aspect. Gold NP films provide a suitable means for ssDNA immobilization. In a typical example, self-assembly of Au NPs onto a gold electrode resulted in easier attachment of an oligonucleotide with a mercaptohexyl group at the 5'-phosphate end and, therefore, in an increased capacity for nucleic acid quantification. The ssDNA surface density on the colloidal Au-modified gold electrode was 1.0×10^{14} molecules cm^{-2}, ca ten-times higher than on a bare gold electrode [29].

3.2.1.2 Metal NP Labels for Electrochemical Immunoassays and Immunosensors

Applications of metallic NPs in immunoassays date back to the early 1970s when 5–50 nm colloidal gold particles were first used as electron-dense probes in electron microscopy and thus enabled sensitive, high-resolution immunocytochemistry

[30]. As recently as a decade ago, it was demonstrated that colloidal gold labels, when involved in an immunoassay, can be sensitively detected by electrochemical methods after oxidative release of Au^{III} ions. Costa-Garcia's group demonstrated the monitoring of streptavidin–biotin interactions down to the 2.5 nM streptavidin level with adsorptive voltammetric measurements of the colloidal gold tags at a carbon-paste electrode [31]. In a related case of the same period, Dequaire et al. reported a sensitive electrochemical immunoassay based on a colloidal gold label that was indirectly determined by anodic stripping voltammetry (ASV) at a single-use screen-printed electrode (SPE) after oxidative gold metal dissolution in an acidic solution [32]. The method was evaluated for a non-competitive heterogeneous immunoassay of an immunoglobulin G (IgG), and a concentration as low as 3×10^{-12} M was determined. This performance was competitive with a colorimetric enzyme-linked immunosorbent assay (ELISA) or with immunoassays based on fluorescent europium-chelated labels. More recently, Liu et al. have developed an electrochemical magnetic immunosensor that was able to avoid the acid-dissolution step and was based on magnetic beads and gold nanoparticle labels [33]. The captured Au-NP labels on the immunosensor surface were directly quantified by square wave voltammetric analysis. A detection limit of $0.02\,\mu g\,ml^{-1}$ of IgG was obtained under optimum experimental conditions.

As described with DNA detection schemes, the nanoparticle-promoted precipitation of sliver on Au-NP labels has also been used to amplify the transduction of antibody–antigen biorecognition events [34–37]. Chu et al. have reported an electrochemical immunoassay based on the deposition of silver on colloidal gold labels. In this case, the deposited silver was indirectly determined by ASV at a glassy-carbon electrode after its dissolution in an acidic solution [34]. A similar potentiometric immunosensor based on gold nanoparticle labels and an ion-selective microelectrode has been reported by Chumbimuni-Torres et al. [35].

Silver-enhanced Au-NP labels have also been used to develop conductive immunosensors. For example, Velev and Kaler have reported a conductive immunosensor using antibody-functionalized latex spheres and a microelectrode gap [36]. A sandwich immunoassay was used to bind a secondary Au-NP-labeled antibody on the latex spheres that were located in the gap. Catalytic deposition of a silver layer bridged the two electrodes and formed conductive paths across interdigitated electrodes. The new paths led to a measurable conductive signal and enabled the ultrasensitive detection of human IgG down to the 0.2 pM level. The method holds promise for creating miniaturized on-chip protein arrays.

Recently, Mao et al. have presented a new method based on cyclic accumulation of Au NPs for detecting human immunoglobulin G (IgG) by ASV [37]. The dissociation reaction between dethiobiotin and avidin in the presence of biotin provided an efficient approach for the cyclic accumulation of Au NPs. These NPs were then used for the final analytical quantification after an acid dissolution. The anodic peak current increased gradually with the increasing accumulation cycles. Five cycles of accumulation were sufficient for the assay. The low background of the method was a distinct advantage, providing determination of at least $0.1\,g\,ml^{-1}$ human IgG. Liao et al. have reported an amplified electrochemical immunoassay by autocatalytic deposition of Au from Au^{3+} onto the Au-NP labels [38]. The enlarged Au-NP labels were monitored by square wave voltammetry (SWV) after an acid dissolution step. The concentration of rabbit immunoglobulin G (RIgG) analyte was proportional to the SWV signal. The detection limit of this method was $0.25\,pg\,ml^{-1}$ (1.6 fM), which was three orders of magnitude lower than that obtained by a conventional immunoassay using the same Au-NP labels. Recently, Das *et al.* reported an ultrasensitive electrochemical immunosensor using the Au-NP labels as electrocatalyts [39]. In this case, the Au-NP labels were attached to the immunosensor surface (indium tin oxide as substrate electrode) by sandwich immunoreaction. Signal amplification was achieved by catalytic reduction of *p*-nitrophenol (p-NP) to *p*-aminophenol (p-AP) and chemical reduction of *p*-quinone-imine (p-QI) to AP by $NaBH_4$. This dual amplification gave a $1\,fg\,ml^{-1}$ detection limit and an outstanding ten-order linear range, from $1\,fg\,ml^{-1}$ to $10\,\mu g\,ml^{-1}$.

Yuan *et al.* have reported recently a new signal amplification strategy based on thionine (TH)-doped magnetic gold nanospheres as labels. The method showed promise to improve the determination of carcinoembryonic antigen (CEA) [40]. This electrochemical immunoassay system was designed around a carbon fiber microelectrode (CFME) covered with a well-ordered anti-CEA/protein A/nanogold architecture. A reverse micelle method was used to prepare the magnetic gold nanospheres, which were subsequently tethered to HRP-bound anti-CEA as a secondary antibody. These labeled bionanospheres were then used in a sandwich-type protocol in which H_2O_2 generated by HRP was reduced at the electrode. Under optimized conditions, the linear range of the immunoassay without HRP as enhancer was $1.2-125\,ng\,ml^{-1}$ CEA. With HRP, the assay sensitivity was enhanced significantly and the linear range was extended from 0.01 to $160\,ng\,ml^{-1}$ CEA.

3.2.2
Semiconductor Nanoparticles

Owing to their unique, size-dependent fluorescence properties, semiconductor (quantum dot, QD) nanoparticles have generated considerable interest in the area of optical bioanalysis [41–45]. However, the intrinsic redox properties of the metals in these nanoparticles (CdS, PbS, and ZnS) and prospects for their sensitive electrochemical stripping analysis have led to their use as labels for electrochemical biosensing and bioassay of DNA and proteins.

3.2.2.1 QD Labels for Electrochemical Detection of DNA
Wang *et al.* have reported electrochemical detection of DNA hybridization by using cadmium sulfide (CdS) NP tracers and electrochemical stripping measurements of cadmium [46]. A NP-promoted cadmium precipitation was used to enlarge the NP tag and thus amplify the stripping DNA hybridization signal. In addition to measurements of the dissolved cadmium(II) ion, the authors demonstrated solid-state measurements following a "magnetic" collection of the magnetic bead/DNA

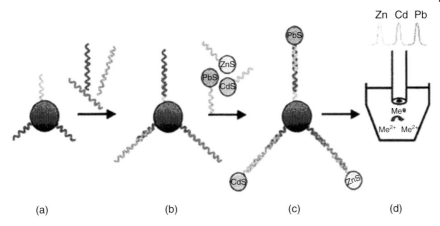

Figure 3.4 Multi-target electrical DNA detection protocol based on different inorganic colloid nanocrystal tracers: (a) introduction of DNA probe-modified magnetic beads; (b) hybridization with the DNA targets; (c) second hybridization with the QD-labeled probes; (d) dissolution of QDs and electrochemical detection. Reprinted with permission from Reference [49]. Copyright 2003 American Chemical Society.

hybrid/CdS tracer onto a thick-film electrode transducer. This protocol combined the amplification features of NP/polynucleotides assemblies and highly sensitive potentiometric stripping detection of cadmium with an effective magnetic isolation of the duplex. The low detection limit (100 fmol) was achieved with good reproducibility (RSD = 6%).

Analogous electrochemical DNA biosensors using CdS [47] and lead sulfide (PbS) [48] semiconductor NPs have been constructed and the detection protocol has been multiplexed. This application relies on the use of different inorganic-colloid nanocrystal tracers (Figure 3.4), whose metal components yield well-resolved stripping voltammetric signals for corresponding DNA targets [49]. Three encoding NPs (ZnS, CdS, and PbS) have thus been used to differentiate the signals of three distinct DNA targets by combining sandwich hybridization with stripping voltammetry of the dissolved heavy metal components. The corresponding metal ions yielded well-resolved stripping peaks at $-1.12\,V$ (Zn), $-0.68\,V$ (Cd), and $-0.53\,V$ (Pb) that corresponded to particular DNA targets. Because the height of each peak reflected the concentration of the corresponding DNA target, the method provided a convenient approach for multiplex electrochemical detection.

The multiplex method has been extended to detect unknown SNPs by using monobase-modified semiconductor NPs [50]. In this application, each mutation site of a duplex DNA strand immobilized on a magnetic bead surface captured a different nanocrystal–mononucleotide conjugate by base pairing. The diverse population of inorganic nanocrystal tags attached to the DNA strand reflected a distinct chemical fingerprint and yielded a distinct voltammetric signature for all of the eight unknown SNPs. For this purpose, ZnS, CdS, PbS, and CuS NPs were linked (using phosphoramidite chemistry through a cysteamine linker) to adenosine, cytidine, guanosine, and thymidine mononucleotides, respectively. Sequential introduction of the monobase-conjugated nanocrystals to the hybrid-coated magnetic-bead solution led to their specific binding to different complementary mismatched sites as well as to previously linked conjugates. Each mutation captured a different proportion of nanocrystal–mononucleotide conjugates, thus enabling bioelectronic coding of all eight possible single-base mismatches in a single voltammetric run.

3.2.2.2 QD Labels for Electrochemical Immunoassay

The concept of applying QDs for electrochemical immunoassay was first demonstrated by Liu et al. [51]. A multiplex electrochemical immunoassay protocol was developed for the simultaneous measurements of multiple protein targets based on the use of different semiconductor nanoparticle tracers (CdS, ZnS, and PbS). A carbamate linkage was used to conjugate the hydroxyl-terminated semiconductor nanoparticles with the secondary antibodies and a sandwich protocol was used to link the nanocrystal tags and magnetic beads modified with primary antibodies. Each biorecognition event yielded a distinct voltammetric peak whose position and size reflected the identity and concentration, respectively, of the corresponding antigen. The multiprotein detection capability was enhanced by amplification afforded by electrochemical stripping transduction (to yield fmol detection limits) and by efficient magnetic separation (to minimize nonspecific adsorption effects). The concept was demonstrated for a simultaneous immunoassay of 2-microglobulin, IgG, bovine serum albumin, and C-reactive protein in connection with ZnS, CdS, PbS, and CuS colloidal crystals, respectively. Wu et al. have reported a similar approach using commercial QD (ZnS@CdS) labels for electrochemical immunoassay of a cancer-related protein biomarker, interleukin-1α (IL-1α) [52]. The concentration of IL-1α was determined by electrochemical stripping analysis of the cadmium component of the captured QD labels after an acid dissolution step. The voltammetric response was highly linear over the range 0.5–50 ng ml^{-1} IL-1α, and the limit of detection was estimated to be 0.3 ng ml^{-1} (18 pM).

Hansen et al. have reported using a QD/aptamer-based ultrasensitive electrochemical biosensor to detect multiple protein targets [53]. The method was based on a simple single-step displacement assay. This involved co-immobilization of several thiolated aptamers, along with binding of the corresponding QD tagged proteins on a gold surface, addition of the protein sample, and quantification of the displacement through electrochemical detection of the remaining nanocrystals. Such electronic transduction of aptamer–protein interactions is extremely attractive for meeting the low power, size, and cost requirements of decentralized diagnostic systems. Unlike two-step sandwich assays used in QD-based electrochemical immunoassays, the new aptamer biosensor protocol relies on a single-step displacement protocol. This biosensor allows sample assays with target concentrations that are three to four orders of magnitude lower, corresponding to a detection limit of 20 ng l^{-1} (0.5 pM) or 54.5 amol (2 pg) in a 100 µL sample.

QDs have also been used as labels for potentiometric immunoassays. Thurer et al. [54] reported the potentiometric bioanalysis of proteins in a microtiter plate format with semiconductor nanocrystal labels. After sandwich immunoreaction on the microtiter plate, the captured CdSe QDs were found to be easily dissolved/oxidized in minutes with hydrogen peroxide. The released Cd^{2+} ions were measured by Cd^{2+}-selective micropipette electrodes. The potentiometric protein immunoassay exhibited a log–linear response ranging from 0.15 to 4.0 pmol of IgG, with a detection limit of <10 fmol in the 150 µL sample wells.

A versatile immunosensor using CdTe quantum dots as electrochemical and fluorescent labels has been developed recently by Zhu et al. for sensitive detection of protein [55]. This sandwich-type sensor was fabricated on an indium tin oxide chip covered with a well-ordered gold nanoparticle monolayer. Gel imaging systems were used successfully to develop a novel highly-efficient optical immunoassay, which could perform simultaneous detection of samples with a series of different concentrations of a target analyte. Similarly, analysis by stripping voltammetry yielded a linear range that was between 0.1 and 500 ng ml^{-1}. The assay sensitivity could be further increased to 0.005 ng ml^{-1} with a linear range from 0.005 to 100 ng ml^{-1} by stripping voltammetric analysis.

Other nanomaterials have also been combined with QDs to design electrochemical immunoassays. For example, Ho et al. have developed a sensitive system for the detection of a protein tumor marker, carcinoembryonic antigen (CEA), that was based on a screen-printed graphite electrode modified with carbon nanoparticles (CNPs), poly(ethylene imine), and anti-CEA antibodies [56]. The signal amplification strategy used CdS nanocrystals as biotracers and CNPs to enhance electron transfer with the electrode and thereby improve both the sensitivity and detection limit for CEA. The calibration curve for CEA concentration was linear in the range 0.032–10 ng ml^{-1} and the detection limit was 32 pg ml^{-1} (equivalent to 160 fg in a 5 µL sample). This method was precise and sensitive for determining urinary CEA – a better marker than serum CEA for the early detection of urothelial carcinoma.

3.2.3
Carbon Nanotubes

Since their discovery in 1991 [57], carbon nanotubes (CNTs) have attracted considerable attention in the field of electroanalytical chemistry. The high surface area, hollow geometry, and the special mechanical properties of these nanomaterials combined with their electronic conductivity and ability to promote electron-transfer reactions have provided new avenues to design sensors for biomolecules and inorganic compounds [58–64]. Compared to traditional carbon-based electrochemical transducers, CNT-based electrochemical transducers offer more sensitive platforms for electrochemical detection of nucleic acids and proteins.

Most early work focused on the use of CNT-based electrochemical transducers for the sensitive detection of nucleic acids and proteins by directly measuring the electrochemistry of nucleic acids and proteins. For example, Erdem et al. have reported a pencil graphite electrode (PGE) modified with CNTs used for the detection of nucleic acids and DNA hybridization based on the voltammetric signal of guanine [65]. The performance of this inexpensive pencil electrode modified with CNTs was favorable compared to CNT-modified glassy carbon electrodes and resulted in more sensitive and reproducible determinations of nucleic acids. Li *et al.* have demonstrated a new electrochemical (EC) platform based on a CNT nanoelectrode array for ultrasensitive detection of DNA [66]. The use of aligned multiwall CNTs provided a new bottom-up scheme for fabricating reliable nanoelectrode arrays. These workers studied the EC characteristics of CNT nanoelectrode arrays with both bulk and immobilized redox species. Combining such a nanoelectrode platform with mediation of guanine oxidation (using tris(2,2′-bipyridine)ruthenium(II), $[Ru(bpy)_3]^{2+}$), a detection limit lower than a few attomoles of the oligonucleotide targets was achieved. The sensitivity was further improved down to thousands of target DNAs after method optimization. Rusling's group has reported a CNT "forest" amperometric immunosensor platform with multi-labeled secondary antibody–CNT bioconjugates for highly-sensitive detection of a cancer biomarker in serum and tissue lysates [67]. Greatly enhanced sensitivity was attained by using bioconjugates featuring HRP labels and secondary antibodies (Ab_2) linked to CNTs at a high HRP/Ab_2 ratio. This approach provided a detection limit of 4 pg ml^{-1} (100 amol ml^{-1}) for prostate-specific antigen (PSA) in 10 µL of undiluted calf serum, a mass detection limit of 40 fg.

CNTs have also been used as labels for extremely sensitive electrochemical detection of DNA and proteins. In this capacity, CNTs have been used as a carrier to load large numbers of tracers, usually enzymes, for monitoring bioaffinity reactions. For example, Wang *et al.* have explored CNTs as labels for ultrasensitive electrochemical detection of DNA and proteins [68]. Enzyme tracers (ALP) were attached to the functionalized CNT surface in the presence of coupling reagents (EDC and NHS (N-hydroxysulfosuccinimide)). An average coverage of around 9600 enzyme molecules per CNT (1 µm length) was obtained. These heavily enzyme-loaded CNT labels were used for the magnetic bead-based electrochemical detection of DNA, resulting in an extremely low detection limit of around 1 fg ml^{-1} (54 aM) – about 820 copies or 1.3 zmol in the 25 µL sample. The authors showed a 100-fold enhancement in the electrochemical signal resulted from the enzyme-CNT label when compared to a system consisting of a single-enzyme label. Further signal amplification was produced by using layer-by-layer (LBL) self-assembly to coat multiple enzyme layers on the CNT tag [69]. The electrostatic LBL self-assembling enzymes onto CNT carriers maximized the ratio of enzyme tags per binding event to offer the greatest amplification factor reported to date and allow detection of as few as 80 copies (5.4 aM) of DNA and 2000 protein molecules (67 aM).

Another effective method for enhancing the electrical detection of DNA hybridization with CNTs was also demonstrated by loading a large number of CdS nanoparticles on the CNT surface [70]. Monolayer-protected CdS NPs were anchored to acetone-activated CNTs by hydrophobic interactions. SEM images indicated that the CdS NPs were attached along the CNT sidewall, with a loading of around 500 particles per CNT. In this case, dual hybridization (i.e., sandwich assay) was performed on a streptavidin-modified 96-well microplate. Analysis was performed after hybridization by ultrasensitive stripping voltammetric detection of the dissolved CdS. A substantial (500-fold) lowering of the detection limit was obtained compared to conventional single particle stripping hybridization assays, reflecting the high CdS loading on the CNT carrier. Even a large (250-fold) excess of non-complementary oligonucleotides had a minimal effect on the response. Such use of CNTs as carriers for multiple electrochemical tags offers great promise for ultrasensitive detection of other biorecognition events.

3.2.4
Apoferritin Nanovehicles

Apoferritin has been used as a template to prepare various nanoparticles [71–78] and the applications of these nanoparticles for electrochemical bioassay have been explored extensively [79–85]. Apoferritin consists of a spherical protein shell, around 12.5 nm in diameter, composed of 24 subunits surrounding an aqueous cavity with a diameter of about 8 nm. The shell can accommodate around 4500 iron atoms [71]. The protein cage of apoferritin can be disassociated into 24 subunits at low pH (2.0) and reconstituted in a high pH (8.5) environment. The structural properties of apoferritin have been used to synthesize size-restricted bioinorganic nanocomposites [71–78]. Small molecules, complexes, and metal ions, such as neutral red, the gadolinium complex (GdHPDO3A), and the uranium ion, have also been captured in the cavities of apoferritin for magnetic resonance imaging, uranium neutron-captured therapy, etc.

These unique properties have been used to load apoferritin with electrochemically-active labels for bioassays [79–83]. Two types of apoferritin nanoparticle labels have been developed: redox probe-loaded and metallic phosphate-loaded (Figure 3.5). The preparation of hexacyanoferrate(III)-loaded apoferritin is illustrated in

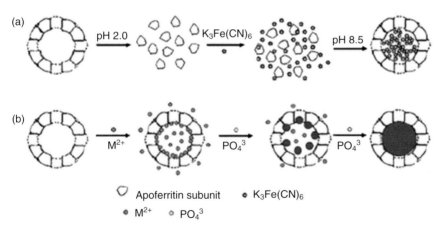

Figure 3.5 (a) Preparation of $K_3Fe(CN)_6$-loaded apoferritin nanoparticle labels and (b) apoferritin template synthesis of a metallic phosphate nanoparticle tag. Reprinted with permission from Reference [80]. Copyright 2007 American Chemical Society.

Figure 3.5a [79]. Briefly, apoferritin was dissociated into subunits at pH 2 and then reconstituted at pH 8.5, thereby trapping hexacyanoferrate(III) in solution within its interior. Around 150 hexacyanoferrate(III) ions were loaded into an apoferritin. Inspired by the use of semiconductor nanoparticle labels in other electrochemical bioassays, Liu et al. developed a simple and facile synthesis based on an apoferritin template to prepare uniform-size, metal phosphate nanoparticle labels for a highly sensitive electrochemical bioassay (Figure 3.5b) [80]. The new metallic phosphate nanoparticle tags are easier to prepare and to functionalize than semiconductor nanoparticle tags. Releasing metal components from the nanoparticles was also more convenient. Release could be performed under the mild condition of a pH 4.6 acetate buffer, rather than the strong acid solution (1 M HNO_3) needed to dissolve semiconductor nanoparticle tags.

Applications of electroactive apoferritin nanoparticles in electrochemical bioassays have typically involved these nanoparticles as labels, which are prepared by conjugating DNA or antibody to the apoferritin protein cage. Such use of apoferritin nano-vehicles offers convenient loading and release of the markers by mild pH control instead of harsher conditions (e.g., strong nitric acid for dissolution of QDs) and avoids the complicated synthesis of semiconductor nanoparticle. For example, Liu et al. have reported a sensitive electrochemical DNA detection bioassay based on hexacyanoferrate(III)-loaded apoferritin tags [79]. An amino-modified DNA probe was conjugated with the marker-loaded apoferritin nanoparticle (MLANP) surface by using the coupling reagent 1-ethyl-3-(dimethylaminopropyl)carbodiimide (EDC) hydrochloride . The biofunctionalized apoferritin NP was thus used as a label for electrochemical DNA detection in connection with a magnetic bead (MB)-based sandwich hybridization assay. The detection limit was estimated to be $3\,ng\,L^{-1}$ (460 fM, based on S/N = 3) with a 60 min hybridization time. This limit corresponded to 150 fg (23 amol) of DNA in the 50 μL hybridization solution.

Metallic phosphate-loaded apoferritin NPs have also been used as labels for electrochemical quantification of SNPs [80]. A cadmium phosphate-loaded apoferritin NP probe, which was modified with nucleotides complementary to the mutant site, formed duplex DNA in the presence of DNA polymerase. Subsequent electrochemical stripping analysis of the cadmium component provided a means to quantify the concentration of mutant DNA. The method was sensitive enough to detect 21.5 amol of mutant DNA, which would enable the quantitative analysis of nucleic acids without polymerase chain reaction preamplification.

Applications of these apoferritin nanoparticles have also been demonstrated in electrochemical immunoassays. In one example, hexacyanoferrate-loaded apoferritin was modified with biotin and used as a label for a magnetic bead-based sandwich immunoassay of IgG [82]. Square-wave voltammetry was used to measure the released hexacyanoferrate tracers from the captured nanoparticle labels. A detection limit of $0.08\,ng\,ml^{-1}$ (0.52 pM) was obtained with a 60 min immunoreaction time. This detection limit corresponded to 26 amol of IgG in the 50 μL sample solution. In another case, metal phosphate-loaded apoferritin nanoparticles were used as a label for the electrochemical immunoassay of tumor necrosis factor α (TNF-α), a protein biomarker, with a detection limit of 2 pg ml^{-1} (77 fM) or 2.33×10^6 TNF-α biomarker molecules (3.9 amol) in a 50 µl sample. This was a significantly lower detection limit than that obtained with a semiconductor NP tag [83]. The ability to simultaneously measure multiple proteins in a single assay was also demonstrated by detecting TNF-α and macrophage chemotactic protein-1 (MCP-1) protein biomarkers with cadmium phosphate and lead phosphate nanoparticle tags [83]. In addition to the preparation of single-metal-component metal phosphate nanoparticle tags, multiple metal phosphate loaded apoferritin nanoparticle tags were also prepared by the apoferritin template [80]. By incorporating different predetermined levels of multiple metal ions, such compositionally encoded nanoparticles can lead to a large number of recognizable voltammetric signatures and hence to a reliable, simultaneous detection of a larger number of protein biomarkers. The apoferritin template synthesis of the encoded nanoparticles provides a simpler, faster approach to prepare electrochemical nanoparticle labels for diverse bio-applications.

3.2.5
Liposomes

Liposomes are vesicles composed of lipids, with their hydrophobic chains forming a bilayer wall and their polar head-groups oriented towards the extravesicular solution and the inner cavity [84]. The size of liposomes varies, ranging from nanometers to several micrometers, depending on the synthesis conditions [85]. Owing to their high surface area, large internal volume, and capability to conjugate with various biorecognition elements, liposomes have been widely used as bioassay labels. Their role has been to encapsulate enzymes, fluorescent dyes, electrochemical and chemiluminescent markers, DNA, RNA, ions, and radioactive isotopes [85, 86]. Applications of liposomes in immunoassays have been reviewed by different research groups [84–86].

The use of liposomes in electrochemical bioassays has enabled highly sensitive detection of DNA [87, 88]. One study employed biotinylated-HRP-functionalized liposomes as probes for the two-step amplification of antigen–antibody interactions and the detection of DNA [89]. The amplified detection of DNA was reported by the application of oligonucleotide-functionalized redox enzymes that acted as bioelectrocatalytic probes for the formation of the double-stranded DNA complex. Alternatively, the DNA detection was reported by the use of dendritic oligonucleotides as a branched sensing interface for the target DNA. Specific anti-double-stranded DNA antibodies or proteins were used as specific biomolecular amplifying probes and allowed to bind to oligonucleotide–DNA assemblies immobilized on the electronic transducers. The selective association of enzymes with the double-stranded oligonucleotide–DNA assembly and subsequent biocatalyzed precipitation of an insoluble product on the transducers was used as an amplification path for the oligonucleotide–DNA recognition event. Charged oligonucleotide-functionalized liposomes or biotin-labeled liposomes were used for the dendritic amplification of oligonucleotide–DNA binding events. Control

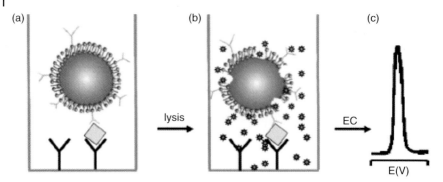

Figure 3.6 Schematic of an electrochemical immunoassay based on markers loaded with liposome labels: (a) immunocapturing liposome labels; (b) lysis to release the markers; (c) electrochemical detection. Reprinted with permission from Reference [7]. Copyright 2007 Elsevier.

of surface properties of the electronic transducers by the liposome micromembranes enabled amplification of the DNA sensing events that yielded a detection limit of 6.5×10^{-13} M.

Applications of liposomes as labels in electrochemical immunoassay and immunosensors have also been explored. Figure 3.6 displays the typical liposome label-based electrochemical immunoassay. Most of the early work in this area was based on ion-encapsulated liposome labels and ion-selective electrodes for serum protein analysis [90–92]. For example, D'Orazio and Rechnitz used trimethylphenylammonium ion (TMPA$^+$)-loaded sheep red-blood-cell ghosts (which acted in the same capacity as liposomes) to quantitate the complement enzymes present in serum samples by using a TMPA$^+$ ion selective electrode [90]. They were later able to adapt this technique to indirectly detect antibodies in bovine serum albumin. Similarly, Shiba et al. have demonstrated the potentiometric determination of tetrapentylammonium ion (TPA$^+$) with a TPA$^+$ ion-selective electrode upon complement disruption of the TPA$^+$-loaded liposomes [92];. Shiba et al. first used potassium ion-loaded liposomes and a potassium ion-selective electrode for monitoring the complement-mediated immune lysis reaction [92]. Liposomes encapsulating electroactive molecules, such as ferrocyanide and ascorbic acid, have also been used as labels for electrochemical immunoassay [93, 94]. Kannuck has reported hexacyanoferrate(II)-encapsulated liposome for selective and sensitive (0.1 nM) determination of immuno-agents in serum matrices [93]. Ferrocyanide was encapsulated in the cavity of liposomes at concentrations of approximately 104 molecules per liposome. The hexacyanoferrate(II) ions were released from within the liposome by either addition of surfactant or complement lysis of the membrane and were monitored by differential-pulse voltammetry on a polymer modified electrode. Liposomes encapsulating ascorbic acid have been used in a competitive assay for the pesticide atrazine using both lateral and horizontal flow formats and amperometric detection following Triton X-100-induced lysis [94].

The signal response was 1–3 min, and the detection limit in tap water was below 1 µg l^{-1}. Lee et al. have reported a disposable liposome immunosensor for theophylline that combined an immunochromatographic membrane and a thick-film electrode [95]. An anti-theophylline antibody was immobilized in an antibody competition zone, and hexacyanoferrate(II)-loaded liposomes were immobilized in a signal generation zone. When a theophylline sample solution was applied to the immunosensor – pre-loaded with theophylline–melittin conjugate in a sample loading zone – the theophylline and theophylline–melittin conjugate migrated through the anti-theophylline antibody zone where competitive binding occurred. Unbound theophylline–melittin conjugate migrated further into the signal generation zone where it disrupted the liposomes to release the electroactive hexacyanoferrate(II). These species were then detected amperometrically. The detection limit for this immunosensor was 5 µg ml^{-1}, which enabled it to monitor theophylline over a clinically relevant range of 10–20 µg ml^{-1}. Additional benefits were the one-step assay and analysis time of only 20 min.

Liposome-encapsulated enzymes, such as HRP, have also been used as labels to develop sensitive electrochemical immunosensors. For example, Haga et al. have reported a liposome immunosensor for the detection of theophylline [96]. The immunosensor was composed of a Clark-type oxygen electrode and HRP encapsulated liposomes. Sample theophylline and theophylline-tagged liposomes competed for antibody sites. Binding to the antibody caused activation of the complement, which lysed the liposomes and resulted in the release of the entrapped horseradish peroxidase. The released HRP catalyzed the conversion of NADH into NAD$^+$, thereby depleting oxygen which was monitored. The authors reported a detection limit for theophylline of 0.72 ng ml^{-1}.

3.2.6
Silica Nanoparticles

Silica nanoparticles have many advantages for use in the development of labels, carriers, and vehicles for various applications. Some of these advantages include ease of fabrication and functionalization as well as high stability in various environments. Silica nanoparticles have been used successfully in bioimaging, optical bioassays, drug delivery, diagnosis, and therapeutic applications [97]. Silica nanoparticles have been employed as labels in electrochemical biosensors and bioassays by doping electroactive components into the silica nanoparticles during synthesis. For example, Zhu et al. have reported a new and sensitive electrochemical DNA hybridization detection protocol based on the voltammetric detection of tris(2,2'-bipyridine)cobalt(III) ([Co(bpy)$_3$]$^{3+}$)-doped inside silica nanoparticles that were used as the oligonucleotide (ODN) labeling tag [98]. Although it is not possible to directly link electroactive [Co(bpy)$_3$]$^{3+}$ with DNA, it can be doped into the silica nanoparticles, which can then be linked to probe DNA with trimethoxysilylpropyldiethylenetriamine (DETA) and glutaraldehyde reagents. The silica-NP DNA probe was used to hybridize with target DNA immobilized on the surface of a glassy carbon electrode. Only the complementary DNA sequence (cDNA) could form a double-stranded DNA complex (dsDNA) with the probe labeled with $[Co(bpy)_3]^{3+}$ and give an obvious electrochemical response. A three-base mismatch sequence and non-complementary sequence had a negligible response. Owing to the large number of $[Co(bpy)_3]^{3+}$ ions doped inside the silica nanoparticles of the DNA probe, the assay showed a high sensitivity and allowed the detection of target oligonucleotides at levels as low as $2.0 \times 10^{-10}\,\text{mol}\,\text{l}^{-1}$.

Wang et al. have developed an "electroactive" silica nanoparticle in which poly(guanine) (poly[G]) was used to functionalize silica nanoparticles. These workers demonstrated that the poly[G]-functionalized silica NPs could serve as a biological label for a sensitive electrochemical immunoassay [99, 100]. The "electroactive" silica nanoparticle labels were prepared by covalently binding poly[G] and avidin to the silica NP surface using the conventional peptide coupling reagents EDC and NHS. It was found that there were approximately 60 strands of poly[G]$_{20}$ per silica nanoparticle. Accordingly, the average surface coverage of poly[G]$_{20}$ on a silica surface was determined to be 8.5×10^{12} molecules cm^{-2}. With such a high concentration of guanine sites, the functionalized silica NPs could be used as a label for an amplified electrochemical immunoassay. Figure 3.7 shows the principles of this electrochemical immunoassay [99]. After a complete sandwich immunoassay, a solution of $[Ru(bpy)_3]^{2+}$ was added and the catalytic current resulting from guanine oxidation was measured. This current is proportional to the amount of guanine near the electrode, which in turn depended on the concentration of target analyte in the original sample. The application was first demonstrated by using IgG as a model protein [99]. It was found that the current was proportional to the logarithm of IgG concentration. The limit of detection for this immunosensor [based on a signal-to-noise ratio (S/N) of 3] was estimated to be about $0.2\,\text{ng}\,\text{ml}^{-1}$ (about 1.3 pM), which corresponded to about 39 amol of mouse IgG in a 30 µl sample solution. The same electroactive silica nanoparticle label was also used to detect a protein biomarker, TNF-α [100]. After optimizing the experimental parameters (e.g., concentration of $[Ru(bpy)_3]^{2+}$, incubation time of TNF-α, etc.), the detection limit for TNF-α was found to be $5.0 \times 10^{-11}\,\text{g}\,\text{ml}^{-1}$ (2.0 pM), which corresponded to 60 amol of TNF-α in a 30 µl of sample.

Another strategy for electrochemical bioassay based on silica nanoparticle labels is to use horseradish peroxidase (HRP)-functionalized silica nanoparticles as a label [101]. The enzyme-functionalized silica nanoparticles were fabricated by co-immobilization of HRP and α-fetoprotein antibody (anti-AFP, the secondary antibody, Ab2), a model protein, onto the surface of SiO_2 nanoparticles using γ-glycidoxypropyltrimethoxysilane (GPMS) as the linkage. Through sandwich-type immunoreactions with AFP, the enzyme-functionalized silica nanoparticle labels were brought close to the surface of gold substrates, as confirmed by scanning electron microscopy (SEM) images. Enhanced detection sensitivity was achieved where the large surface area of SiO_2 nanoparticle carriers increased the amount of HRP bound per sandwich immunoreaction. Electrochemical and chemiluminescence measurements showed 29.5- and 61-fold increases in detection

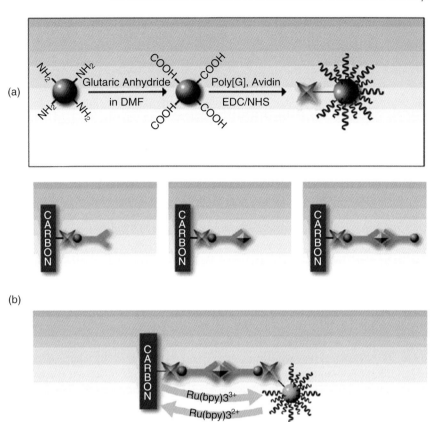

Figure 3.7 (a) Schematic of the procedure for preparation of poly[G]- and avidin-functionalized silica NP conjugate and (b) schematic of the procedure and principle for electrochemical immunoassay based on poly[G]-functionalized silica NPs. Reprinted with permission from Reference [100]. Copyright 2006 American Chemical Society.

signals for AFP, respectively, in comparison with a traditional sandwich immunoassay. An improved particle synthesis using a "seed-particle growth" route yielded particles of narrow size distribution. This characteristic allowed consistent loading of HRP and anti-AFP on each microsphere and ensured subsequent immunosensing possessed of high sensitivity and reproducibility.

3.2.7
Nanowires and Nanorods

One-dimensional nanostructures, such as nanorods and nanowires, have recently attracted attention due to their potential applications in various novel nanodevices [102–104]. Considerable effort has been spent on the synthesis of these nanostructures. The applications of nanowires and nanorods for the bioassay of DNA and proteins have been explored [105–107]. In one example, Wang et al. have reported solid state electrochemical detection of DNA hybridization based on indium nanorods modified with probe DNA [108]. The cylindrical indium nanorods were prepared by a template-directed electrochemical synthesis that involved the plating of indium into the pores of a host membrane. The linear relationship between the charge passed during the preparation and the resulting particle size allowed tailoring of the electrical DNA assay towards higher sensitivity. The resulting micrometer-long rods thus offered a much lower detection limit (250 zmol) than common bioassays that used spherical nanoparticle tags.

In another example, multi-segment CdTe-Au-CdTe nanowires were synthesized using electrochemical deposition into porous alumina templates [109]. After functionalization with thiol-ended ssDNA receptors, the nanowires were immobilized on a field-effect transistor (FET) platform to perform ultrasensitive detection of biomolecules based on modulation of nanowire conductance. Adding ssDNA molecules, which hybridized with complementary receptors at the nanowire surface, significantly changed the nanowire conductance and enabled quantitative detection of target DNA molecules.

Rather than sensing the simultaneous conductance changes of many nanowires, recent efforts have been directed toward building sensors around single nanowires. An example is the single nanowire biosensors developed by Mulchandani et al. [110]. Polypyrrole (Ppy) nanowires were first synthesized by electrochemical polymerization using an alumina template. The single nanowire chemoresistive sensor device was then assembled using AC dielectrophoretic alignment followed by maskless anchoring to a pair of gold electrodes separated by 3 μm. To establish an efficient, covalent method for bio-functionalization of the polymer nanowire, glutaraldehyde and EDC chemistries were compared. EDC was found to be the most effective chemistry and was used to surface-functionalize a single Ppy nanowire with cancer antigen (CA 125) antibody. The resulting nano-immunosensor for CA125 biomarker had excellent sensitivity with a low detection limit of $1 \, U \, ml^{-1}$ and a dynamic range up to $1000 \, U \, ml^{-1}$ in 10 mM phosphate buffer.

3.3
Nanomaterial-Based Electrochemical Devices for Point-of-Care Diagnosis

The studies described above demonstrate the benefits of using bioconjugated nanomaterials for the amplified electrochemical transduction of DNA hybridization events and immunoreactions. They also indicate the broad potential that these technologies may have for the detection of pathogens, genetic mutations, and targets of pharmacogenomic and industrial interest (e.g., transgenic crops) in the near future. Meanwhile, electrochemical methods based on nanomaterials provide a wide variety of relatively simple, accurate and inexpensive senor designs for diagnostic applications. Such methodologies have demonstrated very low detection limits when the electroactive components or analytes are delivered efficiently to the electrode surface and the heterogeneous hybridization reactions proceed with reasonable yields.

However, in the case of DNA detection, most of these established methods cannot function well without prior PCR amplification of the sample or they suffer from matrix affects when clinical samples such as blood, plasma, and urea are employed. The ultimate challenge is to apply these technologies in point-of-care and clinical diagnosis. This prospect was first demonstrated using semiconductor nanoparticle labels for electrochemical protein assays [111, 112]. In the first report, a disposable diagnosis device was developed that integrated the immunochromatographic strip technique with an electrochemical immunoassay using quantum dots (CdS@ZnS) as labels [111]. The device coupled the speed and low cost of the conventional immunochromatographic strip test and the high sensitivity of the nanoparticle-based electrochemical immunoassay. A sandwich immunoreaction was performed on the immunochromatographic strip and captured QD labels on the test zone were determined by highly sensitive stripping voltammetry with a disposable screen-printed electrode (Figure 3.8). The new device, coupled with a portable electrochemical analyzer, shows great promise for in-field and point-of-care quantitative testing of disease-related protein biomarkers. For instance, the response of the optimized device was highly linear over the range $0.1–10\,\text{ng}\,\text{ml}^{-1}$ IgG and the detection limit was $30\,\text{pg}\,\text{ml}^{-1}$ with a 7-min immunoreaction time or $10\,\text{pg}\,\text{ml}^{-1}$ with a 20-min immunoreaction time.

Recently, our group have developed a multiplex electrochemical immunoassay based on the use of gold nanoparticle (Au-NP) probes and immunochromatographic strips (ISs) [113]. The approach takes advantage of the speed and low cost of the conventional IS tests and the high sensitivities of nanoparticle-based electrochemical immunoassays. Rabbit IgG (R-IgG) and human IgM (H-IgM)

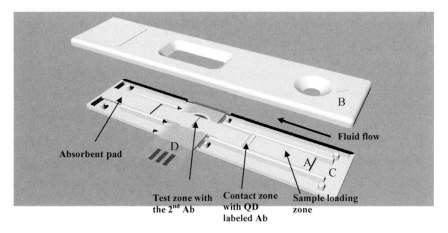

Figure 3.8 Schematic of the immunochromatographic electrochemical biosensor: (A) a test strip, (B) a cover, (C) a bottom, and (D) a screen-printed electrode. Reprinted with permission from Reference [112]. Copyright 2008 Elsevier.

were used as model targets for demonstration. The Au-NP based sandwich immunoreactions were performed on the IS, and the captured gold nanoparticle labels in the test zones were determined by highly sensitive stripping voltammetric measurement of the dissolved gold(III) ions with a carbon paste electrode. The detection limits were 1.0 and 1.5 ng ml^{-1} for R-IgG and H-IgM, respectively, with a linear range of 2.5–250 ng ml^{-1}. The total assay time was around 25 min. Such a multiplex electrochemical immunoassay could allow simultaneous, parallel detection of numerous proteins and it is expected to open up new opportunities for protein diagnostics and biosecurity.

For point-of-care examination, nanowire-based conductometric detection has also been employed in immunochromatographic testing [114]. While gold NPs are widely used labels for antibodies to produce colorimetric signals, this tracer does not produce suitable inter-particle electric conduction due to the presence of protein barriers against electron transfer (e.g., immunoglobulins and blocking agents). To overcome this problem, the authors introduced a conducting polymer, polyaniline. The polymer acted as a conductivity-modulating agent on the gold surface once an antibody specific to human albumin, used as model analyte, was immobilized. This novel modification amplified the conductance signal 4.7 times compared with the plain gold and the signal was also 2.3-fold higher than a photometric system under the same analytical conditions.

3.4
Conclusions

Electrochemical methods based on nanomaterials provide a wide variety of simple, accurate, and inexpensive senor designs for bioassay applications. The enormous signal enhancement associated with the use of nanomaterial labels and with the formation of nanomaterial–biomolecule complexes provides the basis for ultrasensitive electrochemical detection of disease-related gene or protein markers, biothreat agents, or infectious agents. The sensitivities and detection limits of these nanomaterial-based electrochemical biosensors and bioassay are comparable (or better than) with that of traditional bioassays, such as ELISA and PCR. The remarkable sensitivity of the new nanomaterial-label-based sensing protocols opens up the possibility for detecting disease markers, biothreat agents, or infectious agents that cannot be measured by conventional methods. However, the challenges for current research are to determine (i) how well nanomaterial-based electrochemical biosensors and bioassays will work with clinical samples, such as blood and plasma, (ii) whether large-scale multiplex detection is feasible, and (iii) whether this technology can be applied to point-of-care and clinical diagnosis. It is expected that coupling advanced nanomaterial-based electrochemical biosensors and bioassays with other major technological advances, such as lateral-flow, test-strip technology and electronics technology, will result in powerful, easy-to-use, hand-held devices for fast, sensitive, and low-cost biodetection.

Acknowledgments

G. L. acknowledges the financial support from the North Dakota Experimental Program to Stimulate Competitive Research (EPSCoR) and new faculty start-up funds of North Dakota State University and Y. L acknowledges the financial support from grant number U01 NS058161-01 from the National Institutes of Health CounterACT Program through the National Institute of Neurological Disorders and Stroke, partially by CDC/NIOSH Grant RO1 OH008173-01, and partially by grant number U54 ES16015 from the National Institute of Environmental Health Sciences (NIEHS), NIH. The contents of this publication are solely the responsibility of the authors and do not necessarily represent the official views of the Federal Government.

References

1. Eggins, B.R. (ed.) (1997) *Biosensors: An Introduction*, John Wiley & Sons, Inc., New York.
2. Warsinke, A., Benkert, A., and Scheller, F.W. (2000) *Fresenius' J. Anal. Chem.*, **366**, 622.
3. Ghindilis, A.L., Atanasov, P., Wilkinst, M., and Wilkins, E. (1998) *Biosens. Bioelectron.*, **13**, 113.
4. Chen, W. (2008) *J. Nanosci. Nanotechnol.*, **8**, 1019.
5. Chen, W., Zhang, J.Z., and Joly, A.G. (2004) *J. Nanosci. Nanotechnol.*, **4**, 919.
6. Wang, J. (2005) *Analyst*, **130**, 421.
7. Liu, G. and Lin, Y. (2007) *Talanta*, **74**, 308.
8. Kerman, K., Saito, M., Tamiya, E., Yamamura, S., and Takamura, Y. (2008) *Trends Anal. Chem.*, **27**, 585.
9. Wang J. (2006) *Biosens. Bioelectron.*, **21**, 1887.
10. Lucarelli, F., Marrazza, G., Turner, A.P.F., and Mascini, M. (2004) *Biosens. Bioelectron.*, **19**, 515.
11. Minunni, M., Tombelli, S., and Mascini, M. (2007) Analytical applications of QCM-based nucleic acid biosensors, in *Piezoelectric Sensors* (eds. C. Steinem and A. Janshoff), Springer Series on Chemical Sensors and Biosensors, Vol. **5**, Springer, Berlin, p. 211.
12. Tothill, I.E. (2009) *Semin. Cell Dev. Biol.*, **20**, 55.
13. Murphy, L. (2006) *Curr. Opin. Chem. Biol.*, **10**, 177.
14. Erdem, A. (2007) *Talanta*, **74**, 318.
15. Authier, L., Grossiord, C., Brossier, P., and Limoges, B. (2001) *Anal. Chem.*, **73**, 4450.
16. Wang, J., Xu, D., Kawde, A.-N., and Polsky, R. (2001) *Anal. Chem.*, **73**, 5576.
17. Cai, H., Xu, C., He, P., and Fang, Y. (2001) *J. Electroanal. Chem.*, **510**, 78.
18. Wang, J., Polsky, R., and Xu, D. (2001) *Langmuir*, **17**, 5739.
19. Kawde, A.-N. and Wang, J. (2004) *Electroanalysis*, **16**, 101.
20. Cai, H., Wang, Y., He, P., and Fang, Y. (2002) *Anal. Chim. Acta*, **469**, 165.
21. Pumera, M., Teresa, M., Eda, C., Pividori, M.I., Eritja, R., Merkocüi, A., and Alegret, S. (2005) *Langmuir*, **21**, 9625.
22. Kerman, K., Saito, M., Morita, Y., Takamura, Y., Ozsoz, M., and Tamiya, E. (2004) *Anal. Chem.*, **76**, 1877.
23. Cai, H., Xu, Y., Zhu, N.N., He, P.G., and Fang, Y.Z. (2002) *Analyst*, **127**, 803.
24. Fu, X.H. (2008) *Bioprocess Biosyst. Eng.*, **31**, 69.
25. Hwang, S., Kim, E., and Kwak, J. (2005) *Anal. Chem.*, **77**, 579.
26. Rijiravanich, P., Somasundrum, M., and Surareungchai, W. (2008) *Anal. Chem.*, **80**, 3904.
27. Polsky, R., Gill, R., Kaganovsky, L., and Willner, I. (2006) *Anal. Chem.*, **78**, 2268.
28. Liang, K. and Mu, W. (2008) *Chem. Lett.*, **37** (10), 1078.

29 Wu, Z.S., Jiang, J.H., Shen, G.L., and Yu, R.Q. (2007) *Hum. Mutat.*, **28**, 630.
30 Hayatt, M.A. (ed.) (1989) *Colloidal Gold – Principles, Methods and Applications*, Academic Press, San Diego.
31 Gonzalez-Garcia, M.B., Fernandez-Sanchez, C., and Costa-Garcia, A. (2000) *Biosens. Bioelectron.*, **15**, 315.
32 Dequaire, M., Degrand, C., and Limoges, B. (2000) *Anal. Chem.*, **72**, 5521.
33 Liu, G. and Lin, Y. (2005) *J. Nanosci. Nanotechnol.*, **5**, 1060.
34 Chu, X., Fu, X., Chen, K., Shen, G., and Yu, R. (2005) *Biosens. Bioelectron.*, **20**, 1805.
35 Chumbimuni-Torres, K.Y., Dai, Z., Rubinova, N., Xiang, Y., Pretsch, E., Wang, J., and Bakker, E. (2006) *J. Am. Chem. Soc.*, **128**, 13676.
36 Velev, O.D. and Kaler, E.W. (1999) *Langmuir*, **15**, 3693.
37 Mao, X., Jiang, J.H., Chen, J.W., Huang, Y., Shen, G.L., and Yu, R.Q. (2006) *Anal. Chim. Acta*, **557**, 159.
38 Liao, K. and Huang, H. (2005) *Anal. Chim. Acta*, **538**, 159.
39 Das, J., Aziz, M.A., and Yang, H. (2006) *J. Am. Chem. Soc.*, **128**, 16022.
40 Tang, D., Yuan, R., and Chai, Y. (2008) *Anal. Chem.*, **80**, 1582.
41 Chan, W.C.W. and Nie, S. (1998) *Science*, **281**, 2016.
42 Winter, J.O., Liu, T.Y., Korgel, B.A., and Schmidt, C.E. (2001) *Adv. Mater.*, **13**, 1673.
43 Winter, J.O., Gomez, N., Gatzert, S., Schmidt, C.E., and Korgel, B.A. (2005) *Colloids Surf. A*, **254**, 147.
44 Merkoçi, A., Aldavert, M., Marín, S., and Alegret, S. (2005) *Trends Anal. Chem.*, **24**, 341.
45 Sun, C., Liu, B., and Li, J. (2008) *Talanta*, **75**, 447.
46 Wang, J., Liu, G., Polsky, R., and Merkoc, A. (2002) *Electrochem. Commun.*, **4**, 722.
47 Zhu, N.N., Zhang, A.P., He, P.G., and Fang, Y.Z. (2003) *Analyst*, **128**, 260.
48 Zhu, N.N., Zhang, A.P., He, P.G., and Fang, Y.Z. (2004) *Electroanalysis*, **16**, 577.
49 Wang, J., Liu, G., and Merkocüi, A. (2003) *J. Am. Chem. Soc.*, **125**, 3214.
50 Liu, G., Lee, T.M.H., and Wang, J. (2005) *J. Am. Chem. Soc.*, **127**, 38.
51 Liu, G., Wang, J., Kim, J., Jan, M.R., and Collins, G.E. (2004) *Anal. Chem.*, **76**, 7126.
52 Wu, H., Liu, G., Wang, J., and Lin, Y. (2007) *Electrochem. Commun.*, **9**, 1573.
53 Hansen, J.A., Wang, J., Kawde, A., Xiang, Y., Gothelf, K.V., and Collins, G. (2006) *J. Am. Chem. Soc.*, **126**, 2228.
54 Thurer, R., Vigassy, T., Hirayama, M., Wang, J., Bakker, E., and Pretsch, E. (2007) *Anal. Chem.*, **79**, 5107.
55 Cui, R., Pan, H.C., Zhu, J.J., and Chen, H.Y. (2007) *Anal. Chem.*, **79**, 8494.
56 Ho, J.A., Lin, Y., Wang, L., Hwang, K., and Chou, P. (2009) *Anal. Chem.*, **81**, 1340.
57 Iijima, S. (1991) *Nature*, **354**, 56.
58 Dresselhaus, M.S., Dresselhaus, G., and Eklund, P.C. (1996) *Science of Fullerenes and Carbon Nanotubes: Their Properties and Applications*, Academic Press, New York.
59 Li, W.Z., Xie, S.S., Qian, L.X., Chang, B.H., Zou, B.S., Zhou, W.Y., Zhao, R.A., and Wang, G. (1996) *Science*, **274**, 1701.
60 Ren, Z.F., Huang, Z.P., Xu, J.W., Wang, J.H., Bush, P., Siegal, M.P., and Provencio, P.N. (1998) *Science*, **282**, 1105.
61 Murakami, H., Hirakawa, M., Tanaka, C., and Yamakawa, H. (2000) *Appl. Phys. Lett.*, **76**, 1776.
62 Ebbesen, T.W., Lezec, H.J., Hiura, H., Bennet, J.W., Ghaemi, H.F., and Thio, T. (1996) *Nature*, **382**, 54.
63 Wang, J., Kawde, A., and Mustafa, M. (2003) *Analyst*, **128**, 912.
64 Wang, J. (2005) *Electroanalysis*, **17**, 7.
65 Erdem, A., Papakonstantinou, P., and Murphy, H. (2006) *Anal. Chem.*, **78**, 6656.
66 Li, J., Ng, H.T., Cassell, A., Fan, W., Chen, H., Ye, Q., Koehne, J., Han, J., and Meyyappan, M. (2003) *Nano Lett.*, **3**, 597.
67 Yu, X., Munge, B., Patel, V., Jensen, G., Bhirde, A., Gong, J.D., Kim, S.N., Gillespie, J., Gutkind, J.S., Papadimitrakopoulos, F., and Rusling, J.F. (2006) *J. Am. Chem. Soc.*, **128**, 11199.

68 Wang, J., Liu, G., and Jan, M.R. (2004) *J. Am. Chem. Soc.*, **126**, 3010.
69 Munge, B., Liu, G., Collins, G., and Wang, J. (2005) *Anal. Chem.*, **77**, 4662.
70 Wang, J., Liu, G., Rasul, J.M., and Zhu, Q. (2003) *Electrochem. Commun.*, **5**, 1000.
71 Ford, G.C., Harrison, P.M., Rice, D.W., Smith, J.M.A., Treffry, A., and White, Y.J. (1984) *Philos. Trans. R. Soc. London B*, **304**, 551.
72 Meldrum, F.C., Wade, V.J., Nimmo, D.L., Heywood, B.R., and Mann, S. (1991) *Nature*, **349**, 684.
73 Douglas, T., Dickson, D.P.E., Betteridge, S., Charnock, J., Garner, C.D., and Mann, S. (1995) *Science*, **269**, 54.
74 Meldrum, F.C., Heywood, B.R., and Mann, S. (1992) *Science*, **257**, 522.
75 Dominguez-Vera, J.M. and Colacio, E. (2003) *Inorg. Chem.*, **42**, 6983.
76 Webb, B., Frame, J., Zhao, Z., Lee, M.L., and Watt, G.D. (1994) *Arch. Biochem. Biophys.*, **309**, 178.
77 Aime, S., Frullano, L., and Crich, S.G. (2002) *Angew. Chem. Int. Ed.*, **41**, 1017.
78 Hainfeld, J.F. (1992) *Proc. Natl. Acad. Sci. U. S. A.*, **89**, 11064.
79 Liu, G., Wang, J., Lea, S.A., and Lin, Y.H. (2006) *Chembiochem*, **7**, 1315.
80 Liu, G., Wu, H., Dohnalkova, A., and Lin, Y. (2007) *Anal. Chem.*, **79**, 5614.
81 Liu, G. and Lin, Y. (2007) *J. Am. Chem. Soc.*, **129**, 10394.
82 Liu, G., Wang, J., Wu, H., and Lin, Y. (2006) *Anal. Chem.*, **78**, 7417.
83 Liu, G., Wu, H., Wang, J., and Lin, Y. (2006) *Small*, **2**, 1139.
84 Rongen, H.A.H., Bult, A., and Van Bennekom, W.P. (1997) *J. Immunol. Methods*, **204**, 105.
85 Edwards, K.A. and Baeumner, A.J. (2006) *Talanta*, **68**, 1432.
86 Edwards, K.A. and Baeumner, A.J. (2006) *Talanta*, **68**, 1421.
87 Patolsky, F., Lichtenstein, A., and Willner, I. (2000) *J. Am. Chem. Soc.*, **122**, 418.
88 Patolsky, F., Lichtenstein, A., and Willner, I. (2000) *Angew. Chem. Int. Ed.*, **39**, 940.
89 Alfonta, L., Singh, A.K., and Willner, I. (2001) *Anal. Chem.*, **73**, 91.
90 D'Orazlo, P. and Rechnltz, G.A. (1977) *Anal. Chem.*, **49**, 2083.
91 D'Orazlo, P. and Rechnltz, G.A. (1979) *Anal. Chim. Acta*, **109**, 25.
92 Shiba, K., Wanstabe, Y., Ogawa, S., and Fujiwara, S. (1980) *Anal. Chem.*, **52**, 1610.
93 Kannuck, R.M. and Bellama, J.M. (1988) *Anal. Chem.*, **60**, 142.
94 Baumner, A.J. and Schmid, R.D. (1998) *Biosens. Bioelectron.*, **13**, 519.
95 Lee, K.S., Kim, T., Shin, M., Lee, W., and Park, J. (1999) *Anal. Chim. Acta*, **380**, 17.
96 Haga, M., Sugawara, S., and Itagaki, H. (1981) *Anal. Biochem.*, **118**, 286.
97 Smith, J.E., Wang, L., and Tan, W. (2006) *Trends Anal. Chem.*, **25**, 848.
98 Zhu, N., Cai, H., He, P., and Fang, Y. (2003) *Anal. Chim. Acta*, **481**, 181.
99 Wang, J., Liu, G., and Lin, Y. (2006) *Small*, **2**, 1134.
100 Wang, J., Liu, G., Engelhard, M.H., and Lin, Y. (2006) *Anal. Chem.*, **78**, 6974.
101 Wu, Y., Chen, C., and Liu, S. (2009) *Anal. Chem.*, **81**, 1600.
102 Laocharoensuk, R., Bulbarello, A., Hocevar, S.B., Mannino, S., Ogorevc, B., and Wang, J. (2007) *J. Am. Chem. Soc.*, **129**, 7774.
103 Lapierre-Devlin, M.A., Asher, C.L., Taft, B.J., Gasparac, R., Roberts, M.A., and Kelley, S.O. (2005) *Nano Lett.*, **5**, 1051.
104 Jimenez, J., Sheparovych, R., Pita, M., Narvaez Garcia, A., Dominguez, E., Minko, S., and Katz, E. (2008) *J. Phys. Chem. C*, **112**, 7337.
105 Cao, G. and Liu, D. (2008) *Adv. Colloid Interface Sci.*, **136**, 45.
106 Zhu, N., Chang, Z., He, P., and Fang, Y. (2006) *Electrochim. Acta*, **51**, 3758.
107 Andreu, A., Merkert, J.W., Lecaros, L.A., Broglin, B.L., Brazell, J.T., and El-Kouedi, M. (2006) *Sens. Actuators, B*, **114**, 1116.
108 Wang, J., Liu, G., and Zhu, Q. (2003) *Anal. Chem.*, **75**, 6218.
109 Wang, X. and Ozkan, C.S. (2008) *Nano Lett.*, **8**, 398.
110 Bangar, M.A., Shirale, D.J., Chen, W., Myung, N.V., and Mulchandani, A. (2009) *Anal. Chem.*, **81** (6), 2168.

111 Liu, G., Lin, Y., Wang, J., Wu, H., Wai, C.M., and Lin, Y. (2007) *Anal. Chem.*, **79**, 7644.

112 Lin, Y., Wang, J., Liu, G., Wu, H., Wai, C.M., and Lin, Y. (2008) *Biosens. Bioelectron.*, **23**, 1659.

113 Mao, X., Baloda, M., Gurung, A.S., Lin, Y., and Liu, G. (2008) *Electrochem. Commun.*, **10**, 1636.

114 Kim, J.H., Cho, J.H., Cha, G.S., Lee, C.W., Kim, H.B., and Paek, S.H. (2000) *Biosens. Bioelectron.*, **14**, 907.

4
Chemical and Biological Sensing by Electron Transport in Nanomaterials
Jai-Pil Choi

4.1
Introduction

Nanotechnology is an emerging, interdisciplinary field of both science and engineering that combines areas of physics, chemistry, biology, and material science. This technology allows scientists and engineers to design, fabricate, characterize, manipulate, and utilize materials that have well-defined structural features on the nanometer scale (1–100 nm). At this scale, the physical and chemical properties of materials, so-called nanomaterials, can be different from those of bulk or of atoms/molecules having the same chemical composition. For example, when gold (Au) forms clusters consisting of 25 atoms (ca. 1.1 nm diameter), discrete molecular electronic states are developed, producing a HOMO–LUMO (highest occupied molecular orbital–lowest unoccupied molecular orbital) energy gap of 1.3 eV. In bulk Au, however, discrete electronic states merge into continuous bands, and no energy gap exists between an occupied conduction band and unoccupied valence band. This is a well-known example of the quantum-size effect and its physical or chemical effects are profound [1, 2].

Recently a wide variety of nanomaterials have been developed by synthetic methods that can reliably control both size and structure. The range of structure includes nanorods, nanowires, nanotubes, and nanoclusters, which encompasses nanoparticles, nanocrystals, and quantum dots. These nanomaterials are useful for developing chemical and/or biological sensors because of their high surface-to-volume ratio and varied surface-dependent properties. Their large surface area per unit volume is an important factor in miniaturizing devices as well as increasing the performance and throughput of sensors. Their surface-dependent properties can be utilized to develop indirect sensing methods, whereby a change in a surface property induced by an analyte can be converted into a suitable response signal. In contrast, direct sensing methods are based on the intrinsic properties (e.g., spectroscopic, thermal, electrochemical) of the analyte. It is well known that the indirect approach dramatically expands the range of potential analytes and usually improves analytical performance, such as detection limit.

Trace Analysis with Nanomaterials. Edited by David T. Pierce and Julia Xiaojun Zhao
Copyright © 2010 WILEY-VCH Verlag GmbH & Co. KGaA, Weinheim
ISBN: 978-3-527-32350-0

Trace analysis typically means a process that can ultimately quantity or detect a few parts per million (ppm) of analyte in a sample [3, 4]. Chemical or biological sensors designed for trace analysis should ideally have high sensitivity, low detection limit, a wide dynamic range, high selectivity, short response time, good reversibility, and long-term stability. In practice, however, it is difficult to develop a sensor to satisfy all of these features, with selectivity and reversibility being the most difficult features to balance. A compromise usually results because high selectivity requires strong interactions between the sensing material and the analyte, while high reversibility requires the opposite [5].

Like many other fields of scientific research, many developments in chemical and biological sensors based on electron transport are to be found at the micro- and nano-interface. In these sensors, nanomaterials should be supported on a suitable platform, which often provides additional functionalities, such as sample delivery and signal collection. Such platforms are typically designed in micrometer dimensions, and this is another important reason for miniaturization besides portability. Therefore, the capability of assembling nanomaterials is even more important to precisely control nanomaterial architecture and component miniaturization. Unlike optical spectroscopic sensors, sensors based on electron transport have great advantages in miniaturization and cost. These sensors do not require bulky optical components, light sources, or detectors to capture a response. Nowadays, a wide variety of micro- and microarray electrodes and battery-powered minipotentiostats are commercially available at relatively low costs. Recent developments in integrated circuit and lithographic techniques allow further miniaturization of electron transport-based sensors. Another advantage is that these sensors rarely suffer from the high background signals frequently observed in fluorescence-based optical spectroscopic sensors.

In this chapter we first discuss fundamental electron transport mechanisms for solid-state and solution-borne nanomaterials to understand the conceptual working principles of chemical and biological sensors based on electron transport. Then we review the current trends of electron-transport-based sensors for trace analysis, and draw out the motivation for future development and improvement of electron-transport-based sensing with nanomaterials.

4.2
Electron Transport through Metal Nanoparticles

In general, metal nanoparticles consist of a nanometer-sized core of metal atoms and surface-protecting ligands (in most cases as a monolayer), which protect the metal cores against aggregation and decomposition. The ligands are mostly organic molecules such as n-alkane- or aryl thiols, disulfides, amines, phosphines, phosphine oxides, etc. Because the metal cores are surrounded by this organic dielectric medium they can accumulate electrons and holes and thereby behave like capacitors with an extremely small capacitance ($\sim 10^{-18}$ F). In the next three subsections, we discuss the coulomb blockade effect, voltammetry of metal nanoparticles, and electron transport through nanoparticle assemblies to provide some basics and a better understanding of electron transport properties of metal nanoparticles.

4.2.1
Coulomb Blockade Effect and Single Electron Transfer

The coulomb blockade effect [6–11] occurs when the capacitance of a mesoscopic system (e.g., a nanoparticle) is so small that the energy needed to add charge (either an electron or hole) exceeds the available thermal energy. Under this condition, charge cannot be transported through the system at zero bias due to abnormal resistance of the tunnel junctions. For a small spherical particle of capacitance (C), the charging energy (E_c) at low temperature can be expressed as follows [12]:

$$C = 2\pi\varepsilon\varepsilon_0 d \tag{4.1}$$

$$E_c = \frac{e^2}{2C} = \frac{e^2}{4\pi\varepsilon\varepsilon_0 d} \tag{4.2}$$

where

e is the elementary charge,
ε is the dielectric permittivity,
ε_0 is the dielectric permittivity of free space,
d is the diameter of a spherical particle.

To inject or remove a single charge, the thermal energy $k_B T$ (where k_B = the Boltzmann constant and T = absolute temperature) should be larger than E_c. For example, if a metal nanoparticle has a diameter of 10 nm and is protected by an organic monolayer of $\varepsilon = 3$, the E_c of this nanoparticle is anticipated to be 7.69×10^{-21} J (48 meV). By this model, a temperature of ca. 285 °C would be needed to transport charge through the monolayer junction of this nanoparticle at zero bias.

When an external voltage (bias) is applied across the mesoscopic system, the charging energy can be lowered by the amount $|V_{bias}| > e/2C$. The process can be observed from scanning tunneling microscopy (STM) experiments (Figure 4.1a). A plot of applied voltage versus measured current shows discrete steps, called a coulomb staircase [13–19], arising from the incremental increase in tunneling current with greater applied voltage. Here the nanoparticle core has double tunneling junctions, and it is expected that the tunneling current flows at critical voltage bias (V_c) [13]:

$$V_c = \frac{Ze}{C} + \left(\frac{1}{C}\right)\left(Q_0 + \frac{e}{2}\right) \tag{4.3}$$

where

Z is the integral nanoparticle charge,
e is the elementary charge,
C is the capacitance,
Q_0 is a fraction associated with STM tip–substrate work function differences.

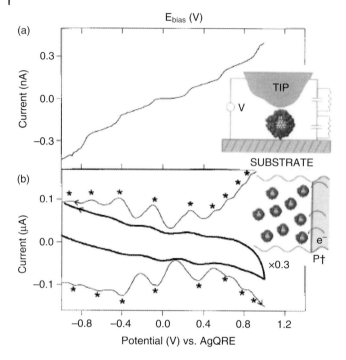

Figure 4.1 (a) Au STM tip addressing a single cluster adsorbed on an Au-on-mica substrate (inset) and Coulomb staircase *I*–*V* curve at 83 K; potential is tip–substrate bias; equivalent circuit of the double tunnel junction gives capacitances $C_{upper} = 0.59\,aF$ and $C_{lower} = 0.48\,aF$; (b) voltammetry (CV—, 100 mV s^{-1}; DPV—, *s are current peaks, 20 mV s^{-1}, 25 mV pulse, top and bottom are negative and positive scans, respectively) of a 0.1 mM 28 kDa cluster solution in 2:1 toluene–acetonitrile/0.05 M H$_4$NClO$_4$ at a $7.9 \times 10^{-3}\,cm^2$ Pt electrode, 298 K, with an Ag wire pseudo-reference electrode. Reprinted with permission from Reference [13]. Copyright 1997 American Chemical Society.

Each increment in current represents a single electron-transfer event. If C is independent of the bias voltage, therefore, coulomb blockades should occur consecutively at a regular voltage-bias spacing (ΔV_c):

$$\Delta V_c = \frac{Ze}{C} + \left(\frac{1}{C}\right)\left(Q_O + \frac{e}{2}\right) - \left\{\frac{(Z-1)e}{C} + \left(\frac{1}{C}\right)\left(Q_O + \frac{e}{2}\right)\right\} = \frac{e}{C} \qquad (4.4)$$

4.2.2
Voltammetry of Metal Nanoparticles in Solutions

The advent of new techniques to stabilize and purify nanoparticles allows voltammetric measurements of nanoparticles directly from the solution. Murray [19] has defined three voltammetric regimes of nanoparticles, based on the ranges of core size: bulk-continuum, quantized double layer charging, and molecules-like voltammetry.

The bulk continuum voltammetric behavior can be observed on large nanoparticles, whose diameters are larger than 3–4 nm. If the double-layer capacitance of an individual nanoparticle (C_{CLU}) is larger than 6.2 aF (aF = 10^{-18} F) at 25° ($k_B T$ = 25.7 meV), successive one-electron charging processes occur continuously in voltammetry. Therefore, no voltammetric peaks (or steps) of current are shown in the voltammogram. Instead the smooth, gradual changes in current (double layer capacitive charging current of nanoparticles) are a function of the electrode potential. This capacitive charging current is controlled by mass transport of nanoparticles, and differs from the double-layer charging current of the working electrode.

Quantized double-layer (QDL) charging behaviors are observed in small nanoparticles of less than ~2 nm diameter. In addition, the double-layer capacitance of a nanoparticle should be less than 6 aF in order to observe QDL charging at room temperature. Successive single-electron charging processes occur stepwise, and the constant potential spacing ($\Delta V = e/C_{CLU}$) between successive charging events is developed. The earliest example of QDL charging voltammetry of nanoparticles was reported by Murray et al. [13]. In their differential pulse voltammetry (DPV) experiments, QDL charging of 28 kDa alkane-thiolate-protected Au nanoparticles (1.64 nm diameter) was observed with a nearly constant potential spacing between the current peaks (Figure 4.1b). This potential spacing was very similar to that observed in the STM i–V curve, indicating that the nanoparticle core is charging at the electrolyte solution/nanoparticle/working electrode junctions.

Molecule-like properties of metal nanoparticles appear when the core consists of a few tens of metal atoms. The nanoparticle core size is typically smaller than those showing QDL double layer charging behavior. In this size regime, nanoparticles exhibit discrete electronic states owing to less dense electronic coupling than in their bulk counterparts. Häkkinen et al. have investigated the electronic structure of $Au_{38}(SCH_3)_{24}$ nanoparticles via density-functional calculations [20]. They reported that the densities of states of $Au_{38}(SCH_3)_{24}$ present a threefold degenerate highest occupied molecular orbital (HOMO) state and a set of 18 empty states, leaving the HOMO–LUMO (lowest unoccupied molecular orbital) gap as 0.9 eV, which agrees with the experimentally measured value of 5-kD Au nanoparticles [18].

The voltammetric wave shapes of molecule-like nanoparticles differ from those of QDL charging, in that the formal potentials of molecule-like nanoparticles are not evenly spaced. Emergence of a HOMO–LUMO energy gap is attributed to a widened formal potential spacing ($\Delta E_{+/-}$) between the first one-electron loss and the first one-electron gain of nanoparticle. $Au_{25}(SC2Ph)_{18}$ (where SC2Ph = phenylethanethiolate) is a well-known example of molecule-like nanoparticles [21]. As shown in Figure 4.2, $\Delta E_{+/-}$ is 1.62 V as the electrochemical energy gap. By considering the potential spacing of the doublet peaks as charging energy (w), $\Delta E_{HOMO-LUMO}$ (= $\Delta E_{+/-} - w$) is estimated as 1.33 eV. The optical HOMO–LUMO gap energy is estimated from the electronic absorption band edge, giving 1.33 eV, in good agreement with $\Delta E_{HOMO-LUMO}$. Because of variations in charging energy, however, such good agreement is not always seen.

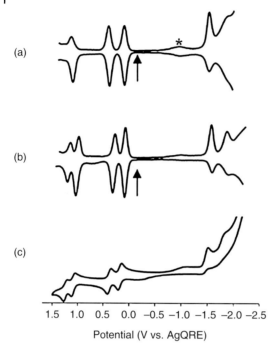

Figure 4.2 (a) 25 °C and (b) −70 °C differential pulse voltammograms at 0.02 V s^{-1}, and (c) −70 °C cyclic voltammogram (0.1 V s^{-1}) of Au$_{25}$(PhC2S)$_{18}$ (corrected from Au$_{38}$(PhC2S)$_{24}$ in Reference [22]) in 0.1 M Bu$_4$NPF$_6$ in degassed CH$_2$Cl$_2$ at a 0.4 mm diameter Pt working, Ag wire quasi-reference (AgQRE), and Pt wire counter electrode. Arrows indicate solution rest potentials, and * indicates wave for incompletely removed O$_2$, which varied from experiment to experiment. Reprinted with permission from Reference [21]. Copyright 2004 American Chemical Society.

$\Delta E_{+/-}$ and the HOMO–LUMO gap energy (both electrochemical and optical) depend on the size of a nanoparticle as a consequence of quantum confinement effect [23, 24].

This effect could be observed within the group of molecule-like Au nanoparticles: Au$_{13}$ [22], Au$_{25}$ [21, 25, 26], Au$_{38}$ [27], Au$_{55}$ [28], and Au$_{75}$ [29]. As shown in Figure 4.3 [19], both electrochemical gap and HOMO–LUMO gap energies increase with decreasing the core size. The emergence of discrete electronic states (molecule-like properties) is anticipated to be found between Au$_{75}$ and Au$_{140}$, and has been detected experimentally at Au$_{75}$ [29]. Further increase in the core size than Au$_{75}$ induces a decrease in electrochemical energy gaps (mainly caused by QDL charging) and disappearance of HOMO–LUMO gaps. This category includes Au$_{140}$ [27, 31] and Au$_{225}$ [30]. Nanoparticles larger than Au$_{225}$ show bulk-continuum voltammetry. Based on these results, the transition between metal-like and molecule-like properties in Au nanoparticles would be found in the size regime of Au$_{75}$ to Au$_{225}$.

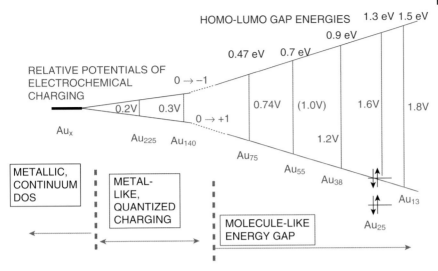

Figure 4.3 Summary of electrochemical behavior by nanoparticle core size, showing estimated energy gaps (HOMO–LUMO gaps) and electrochemical energy gaps (i.e., the spacing between the first oxidation peak and the first reduction current peak for the native nanoparticle). Data for electrochemical energy gaps are as follows: for Au$_x$, $x = 225$ from Reference [30]; $x = 140$ from References [27, 31]; $x = 75$ from Reference [29]; $x = 55$ from Reference [28] and estimate from HOMO–LUMO datum; $x = 38$ from Reference [27]; $x = 25$ from Reference [21]; $x = 13$ from Reference [22]. Reprinted with permission from Reference [19]. Copyright 2008 American Chemical Society.

4.2.3
Electron Transport through Metal Nanoparticle Assemblies

Electron transport through assemblies of metal nanoparticles can be used to develop sensors [32] and electronic devices with nanoscale features [33]. Nanoparticle assemblies are generally prepared as solid-state films having one-, two-, or three-dimensional structures. These films are usually prepared by drop-casting, spin-coating, Langmuir methods, crosslinking precipitation, or stepwise self-assembly. These preparative methods have been well reviewed [34–44]. In general, crosslinking precipitation and stepwise self-assembly methods produce more stable assemblies because the nanoparticles are held together by strong covalent, coordinative, or ionic bonds [44, 45]. With drop-casting, spin-coating, and Langmuir methods, the nanoparticles are only help together by relatively weak van der Waals and dispersion forces.

Electron hopping between nanoparticles is likely to be the main form of electron transport through a nanoparticle film. For sufficiently small nanoparticles, electron hopping in a solid-state film is considered as single-electron exchanges between neighboring donor and acceptor nanoparticles. The hopping transport reaction can be written as in Equation (4.5):

$$\text{NP}_x^Z + \text{NP}_y^{Z+1} \xrightarrow{k_{EX}} \text{NP}_x^{Z+1} + \text{NP}_y^Z \tag{4.5}$$

Here, k_{EX} is the bimolecular self-exchange rate constant ($M^{-1}s^{-1}$), NP denotes an individual nanoparticle, x and y represent neighbor lattice sites, and Z is the charge number. The electron hopping reaction in Equation (4.5) is expected to occur by electron tunneling [46–55] between neighboring nanoparticle cores with surface-protecting ligands serving as an electron tunneling medium and a kind of solvent shell. To better understand this reaction, the role of charge carrier populations (i.e., NP^0 and NP^{1+}) and k_{EX} must be considered. It should be expected that the rate of electron hopping (k_{EX}) is related to the measured electronic conductivity (σ_{EL}) of the nanoparticle film; a faster rate resulting in higher conductivity. Therefore, σ_{EL} becomes highest when equal amounts of NP^Z and NP^{Z+1} are contained in the nanoparticle assembly (second-order reaction) [56, 57]. If metal nanoparticles show QDL or molecule-like charging behavior, it is possible to quantify the populations of charge carriers in the solid state nanoparticle assembly. By comparing the rest potential (E_R) of a nanoparticle solution with the formal potential ($E^{\circ\prime}$) of each charging event, the relative amount ratio of mixed-valent charge carriers can be estimated from:

$$E_R - E^{\circ\prime} = 0.059 \log \frac{[\text{NP}^{Z+1}]}{[\text{NP}^Z]} \tag{4.6}$$

The ratio $[\text{NP}^{Z+1}]/[\text{NP}^Z]$ can be controlled by bulk electrolysis or reducing (or oxidizing) agents. This ratio should be retained in the nanoparticle assembly as long as the charged nanoparticles are stable in the solid state. Controlling $[\text{NP}^{Z+1}]/[\text{NP}^Z]$ may be one way to control the performance factors of electron transport-based sensors (e.g., detection limit, dynamic range, selectivity, and sensitivity), because they are strongly dependent on conductivity.

Conductivity also has an exponential relation with tunneling distance and temperature as shown in Equation (4.7) [56, 58]:

$$\sigma_{EL}(l, T) = \sigma_0 \exp[-l\beta] \exp[-E_A/RT] \tag{4.7}$$

where

l is the tunneling distance,
β is the corresponding electronic coupling term,
E_A is the activation energy of conductivity.

The pre-exponential term, $\sigma_0 e^{-l\beta}$, is the equivalent of an infinite-temperature electronic conductivity. Murray et al. [56] have measured the electronic conductivities of the films of Au nanoparticles stabilized by different thiolate ligands, $-S(CH_2)_n CH_3$ (n = 4–16). An exponential relation between σ_{EL} and l was observed, and ca $0.8\,\text{Å}^{-1}$ for β was obtained at 30 °C. A definitive relationship between k_{EX} and σ_{EL} can be estimated on the basis of the usual (although approximate) cubic lattice model of film structure [56, 57, 59]:

$$k_{EX} = \frac{6RT\sigma_{EL}}{10^{-3} F^2 \delta^2 [\text{NP}^Z][\text{NP}^{Z+1}]} \tag{4.8}$$

where R is the gas constant and F is the Faraday constant. The nanoparticle center-to-center distance (δ) can be estimated by:

$$\delta = 2(r_{core} + l_{app}) = 2 \cdot \sqrt[3]{\frac{0.7}{10^{-3}\left(\frac{4}{3}\right)} \pi C_{FILM} N_A} \qquad (4.9)$$

where

r_{core} is the nanoparticle core radius,
l_{app} is the apparent surface-protecting ligand length,
C_{FILM} is the film concentration,
N_A is Avogadro's number,

and the hexagonal packing fill factor is assumed to be 0.7.

Especially in 3D films, any hypothetical lattice model (i.e., face-centered cubic or hexagonal close packed model) is inadequate to account for a film structure exactly. Therefore, some uncertainties are expected from the use of Equations (4.8) and (4.9).

4.3
Sensing Applications Based on Electron Transport in Nanoparticle Assemblies

Measurements of the electronic conductivities of nanoparticle assemblies not contacted by electrolyte solutions can be utilized for applications in chemical and biological sensors. The electron transport properties of nanoparticle assemblies, reviewed in Section 4.2.3, are affected by nanoparticle size, electron hopping distance, dielectric properties of surface-protecting ligands, the local thermal mobility of a nanoparticle, etc. [56, 57, 59, 60]. Moreover, nanoparticles have a higher surface area-to-volume ratio than the corresponding bulk materials. A large surface area within a small volume may provide advantages in miniaturization of sensor device and increasing sensitivity, because most sensing reactions occur on the surface of sensing materials. These make nanoparticles potentially useful sensing materials. In this section we discuss the usage of nanoparticles for sensors and their working principles.

4.3.1
Chemical Sensors

Charge transport in nanoparticle assemblies can be utilized for chemical analysis in the form of a chemiresistor. These devices use electron or hole transport as a transduction principle, whereby electrical conductivity is changed by the presence of an analyte species and these changes can be correlated to the amount of analyte. Generally, chemiresistors consist of a thin film of sensing material, such as inorganic [61, 62], organic [63], and polymeric compounds [64–67], deposited on a pair of electrodes. Chemiresistors are typically used to detect vapors or gases but are not limited to these types of analytes.

4.3.1.1 Sensors Based on Metal Nanoparticle Films

The first example of a nanoparticle-based chemiresistor was reported by Wohltjen et al. [68]. Their chemiresistor was fabricated with a thin film of Au nanoparticles (2 nm diameter) protected by a monolayer of octanethiols. The sensor allowed fast, reversible detection of organic (toluene, tetrachloroethylene, and 1-propanol) and water vapors (Figure 4.4).

Vapor sorption by the nanoparticle film affected electron tunneling between nanoparticle cores and electron hopping along the atoms of the octanethiol ligand, so that a large effect on the measured current appeared. With toluene vapor, a fast and large response was observed, compared with other vapors, and detection was possible within the range 2.7–11 000 ppmv.

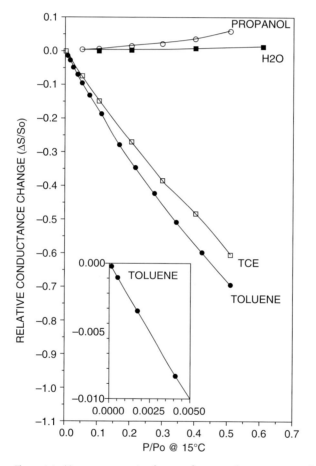

Figure 4.4 Vapor response isotherms of the Au:C8(1:1) MIME sensor to toluene, tetrachloroethylene (TCE), 1-propanol, and water based on 15 °C vapor pressures. Inset: toluene response down to a 2.7 ppmv vapor concentration. Reprinted with permission from Reference [68]. Copyright 1998 American Chemical Society.

Further development of this sensor used chemical functionality on the surface protecting ligands to increase the selectivity of vapor-sorption. Evans et al. [69] have examined the vapor-sensing capabilities of films prepared with Au nanoparticles whose surfaces were protected by p-substituted phenylthiols ($HS–C_6H_4–X$, X = -OH, -COOH, -NH_2, and -CH_3). Characteristics of the functional group, X, were important for controlling the relative strength of inter-particle interactions and vapor sorption at the film interface. Zhong et al. [70] have investigated vapor sorption with different types of nanoparticle films in which nanoparticle cores were linked together by covalent or hydrogen bonding. They observed similar effects of the surface-protecting ligands and also determined that vapor sorption at the film depends on the nanoparticle core size.

Although many groups have observed changes in the conductivity of chemiresistors due to vapor sorption, this effect was not clearly understood. Some studies of unlinked films indicate a decrease in electron-hopping conductivity upon vapor sorption [71–73], while others demonstrate either an increase or a decrease depending on the particular analyte [68, 69, 74]. Generally, unlinked films of large nanoparticles (3–8 nm diameter) do not show an increase in conductivity as analyte vapors are sorbed. Films made from linked nanoparticles rather uniformly display a decrease in conductivity upon vapor sorption [70, 75–78] that does not depend on nanoparticle size. It was commonly accepted in these studies that decreasing conductivity was most likely caused by increased tunneling distance between nanoparticles brought on by film swelling. However, there was no generally accepted explanation for increases in conductivity. One possibility put forward was that the dielectric permittivity of the surface protecting ligands may be affected by vapor sorption and thereby the increase conductivity [69, 74]. However, no experimental evidence was provided.

Recently Murray and coworkers have reported the effect of CO_2 and organic vapor sorption on the electron-hopping conductivity of films of linked and unlinked small nanoparticles (~1.1 nm diameter) [60]. This report provided a better understanding of how conductivity in small nanoparticle films may increase with vapor sorption. When vapor or gas is sorbed, swelling of the film increases the electron tunneling distance (core edge-to-edge distance) and decreases the film conductivity. At the same time, however, local short-range thermal motions of the unlinked nanoparticle cores are enhanced. If these oscillating motions have a sufficient amplitude and frequency to shorten the separation between neighboring nanoparticle cores, electron transport can become accelerated even though the average electron tunneling distance has increased. Note that local thermal mobility implies both physical displacement of the nanoparticle core, altering the instantaneous electron tunneling distance, and associated thermal fluctuation of the intervening surface-protecting ligand shell. In contrast, local thermal motions are suppressed in linked films, leaving only the effect of increased electron tunneling distance to decrease conductivity.

While ligand functional groups can help increase the selectivity of nanoparticle chemiresistor films, the effect is somewhat limited. Another approach is to combine nanoparticle chemiresistor films with a separation method such as gas chromatography (GC). Zellers and coworkers [71, 79–81] have demonstrated that chemiresistors can be employed as a detection method for gas chromatography (GC). A dual-chemiresistor array with two different Au nanoparticle films (n-octanethiolated and benzeneethanethiolated Au nanoparticles) was housed in a 60-μL detector cell [71]. As shown in Figure 4.5, this GC detector exhibited good

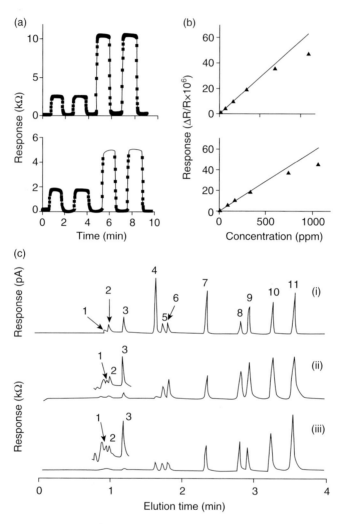

Figure 4.5 (a) Butyl acetate response profiles and (b) calibration curves for chemiresistors coated with C8Au (top) and BC2Au (bottom), where C8 = octanethiolate and BC2 = benzeneethanethiolate; (c) GC detector traces from (i) FID, (ii) BC2Au CR, and (iii) C8Au CR for an eleven-vapor mixture: 1 = ethanol, 2 = acetone, 3 = 2-butanone, 4 = isooctane, 5 = trichloroethylene, 6 = 1,4-dioxane, 7 = toluene, 8 = perchloroethylene, 9 = n-butyl acetate, 10 = chlorobenzene, and 11 = m-xylene. Adapted and reprinted with permission from Reference [71]. Copyright 2002 American Chemical Society.

peak shapes and highly reproducible response. The response was fast and linear with concentrations of 11 organic vapors: ethanol, acetone, 2-butanone, isooctane, trichloroethylene, 1,4-dioxane, toluene, perchloroethylene, n-butyl acetate, chlorobenzene, and *m*-xylene. Limits of detection varied from 0.1 to 24 ppmv depending on vapor and vapor pressures. The nanoparticle-chemiresistor detector was also compared with a conventional flame ionization detector (FID) (Figure 4.5c). The chromatographic peaks recorded from the nanoparticle-chemiresistor array detector were somewhat broader than those from the FID due to (i) a larger dead volume in the chemiresistor array cell, (ii) the low temperature of the chemiresistor array, causing increased possibility of nonspecific adsorption, and (iii) the finite vapor sorption/desorption kinetics of the chemiresistor array coatings. However, like FID, the peak resolution and retention times are still reasonable and interpretable.

4.3.1.2 Sensors Based on Semiconducting Oxide Nanoparticles

Semiconducting oxide nanoparticles, such as SnO_2, WO_3, ZnO, and In_2O_3, are also useful sensing materials to fabricate gas sensitive chemiresistors. Most of these sensors have been developed with n-type metal oxides because the mobility of holes in p-type oxides is much less than the mobility of electrons in n-type oxides. For example, the hole mobility in bulk NiO (a p-type oxide) is ca $0.2\,cm^2\,V^{-1}\,s^{-1}$ [82], whereas electron mobility in bulk SnO_2 (an n-type oxide) is ca $160\,cm^2\,V^{-1}\,s^{-1}$ [82].

As illustrated in Figure 4.6, gas sensing by these by these chemiresistors includes two main functions: recognition of a target gas through a gas–solid interaction, which induces an electronic change of the oxide surface (receptor function), and transduction of the surface change into an electrical signal (transducer function) [82]. These main functions can be strongly influenced by the type of semiconducting oxide, the presence of functionalized surface modifiers, and the size/structure of nanoparticles themselves.

Yamazoe *et al.* [82–84] have reported that the size of semiconducting oxide nanoclusters affect the sensitivity of gas sensors. Sensors for H_2 and CO were

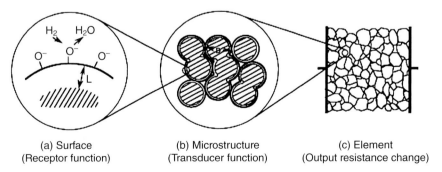

(a) Surface
(Receptor function)

(b) Microstructure
(Transducer function)

(c) Element
(Output resistance change)

Figure 4.6 Receptor and transducer functions of the semiconductor gas sensor. Reprinted with permission from Reference [82]. Copyright 2003 Springer.

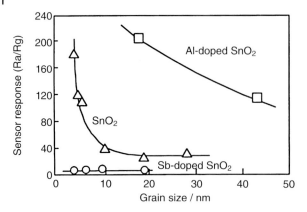

Figure 4.7 Correlation between gas sensitivity at 300 °C and SnO_2 crystallite size of pure and doped SnO_2 elements. Reprinted with permission from Reference [83]. Copyright 1991 Elsevier.

prepared using SnO_2 clusters of various sizes (5–32 nm diameter) [83]. Sensitivity in this work was defined as the ratio of electrical resistance (R_a) measured in air to that (R_s) in the gas sample. Sensors made with smaller clusters (<10 nm in diameter) showed a steep increase in sensitivity for both H_2 and CO as the size of the oxide nanocluster was decreased down to 5 nm. For larger nanoclusters (>10 nm), the sensitivity was almost independent of size (Figure 4.7). This effect of cluster size on sensitivity has been well studied for other semiconducting oxide-based gas sensors [85, 86].

While electron transport is inherent to signal transduction in n-type semiconductor oxide chemiresistors, the mechanism of this transport is fundamentally different from metal nanoparticle assemblies. The adsorption of gas induces an electron-depleted space-charge layer on the SnO_2 cluster and electron transport occurs through this layer. If the cluster size is at least two times larger than the thickness (L) of the electron-depleted space-charge layer, the cluster boundary potential (double Schottky barrier) for electron transfer is virtually constant. Therefore, the sensor resistance becomes independent of the cluster size, resulting in almost no changes in sensitivity in this size regime. When the cluster size is smaller than $2L$, the sensor resistance is affected not only by the resistance at cluster boundary contacts but also by the resistance of a bulk cluster itself. Because charge carrier densities at both surface and bulk cluster decrease with decreasing cluster size, the sensor resistance increases, leading to higher sensitivity.

The sensitivity of semiconducting oxide-based sensors can be significantly affected by surface dopants [82, 85, 87–90], which are different from the n- or p-type dopants used to modify bulk semiconductors. Here doping means the addition of catalytically active sites to the surface of the base sensing material. Surface doping may enhance sensor performance by increasing sensitivity and selectivity to the target gas and decreasing response time. In addition, surface doping may improve the thermal and long-term stability of the sensors. Yamazoe *et al.* have demonstrated [83] the effect with sensors fabricated from ca. 19 nm SnO_2 clusters doped with Al_2O_3 (1 wt%) (Figure 4.7). Compared with sensors fabricated with undoped and Sb_2O_5-doped SnO_2, the Al_2O_3-doped sensors showed at least five-times higher sensitivity for 800 ppmv H_2. Such improvement may have resulted from increased thickness of the electron-depleted space-charge layer (L). The authors suggested that L would increase by a factor of 30 with Al_2O_3-doping, while Sb_2O_5-doping would decrease L [82, 83].

Another important factor that affects the sensitivity of semiconducting oxide-based gas sensors is the microstructure of the sensing layer, such as film thickness and porosity. The diffusion rate of target gas to the sensor surface and the rate of space-charge formation both control the response and the recovery time of these sensors. Therefore, a thin film layer with high porosity is desired to achieve high sensitivity and short response time [91, 92]. Yamazoe *et al.* have reported the effect film thickness on H_2 and H_2S sensitivity by studying thin films of monodispersed SnO_2 clusters (6–16 nm in diameter). Sensitivity was greatly improved by decreasing film thickness and increasing the cluster size [93, 94]. In thin films, the larger cluster size provided higher porosity, causing faster diffusion of target gas. Another way to increase the diffusion rates of target gases is to use one-dimensional nanostructures (e.g., nanowires, nanorods, nanofibers, or nanotubes) [95–97].

4.3.2
Biosensors

Biosensors rely on highly specific recognition processes to detect target analytes. A suitable biosensor platform should facilitate the formation of a probe–target complex that results in a usable signal for electronic readout. In this regard, nanoparticle-based chemiresistors have proven to be particularly suitable platforms. They are extremely sensitive (10^{-15} to 10^{-21} mole range) and well suited to multiple-target detection with different nanoparticles [98]. Surface modification of nanoparticles with molecular monolayers or thin films can introduce functional units that serve as recognition sites to detect biomolecules [99]. In this section we discuss mainly immunosensors and DNA biosensors based on electron transport in DNA–nanoparticle assemblies.

Velev and Kaler reported pioneering work [100] for a miniaturized immunosensor based on the catalytic property of Au nanoparticles to deposit other of metals on their surface. Figure 4.8 shows their experimental configuration. Latex microspheres with adsorbed Protein A were collected in a small gap (7–15 µm) between two gold electrodes. Protein A provided a recognition site for specific binding of the human immunoglobulin (IgG) and subsequent incubation with Au nanoparticles (5 nm in diameter) conjugated to goat antihuman IgG allowed recognition of any captured IgG. The Au nanoparticles were then allowed to deposit silver, which made the electrode gap more conductive. This method allowed the analysis of human IgG with a detection limit of ca 2×10^{-13} M. Note that this detection limit was estimated from the amount of sample used (not adsorbed amount).

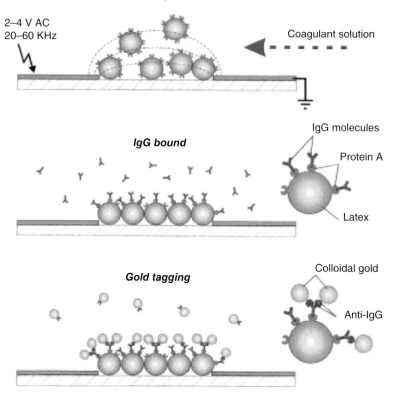
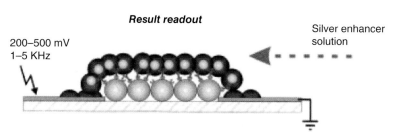

Figure 4.8 Schematic of the sensing mechanism of a miniaturized immunosensor to detect goat antihuman immunoglobulins. The sensing procedure proceeds from top to bottom. IgG denotes human immunoglobulins, and Anti-IgG denotes goat antihuman immunoglobulins. Reprinted with permission from Reference [100]. Copyright 1999 American Chemical Society.

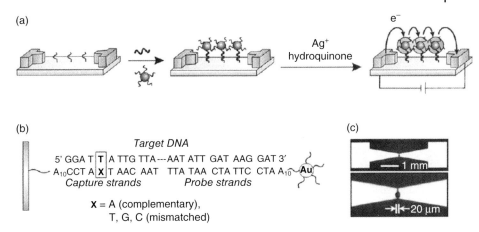

Figure 4.9 (a) Scheme showing concept behind electrical detection of DNA; (b) sequences of capture, target, and probe DNA strands; (c) optical microscope images of the electrodes used in a typical detection experiment. The spot in the electrode gap in the high-magnification image is food dye spotted by a robotic arrayer (GMS 417 Microarrayer, Genetic Microsystems, Woburn, MA). Reprinted with permission from Reference [101]. Copyright 2002 American Association for the Advancement of Science.

A similar approach, shown in Figure 4.9a, has been developed by Mirkin and coworkers [101] for DNA analysis. In this case a small array of microelectrodes with 20 µm gaps was constructed on a Si wafer (1 µm-SiO$_2$ coating) and capture oligonucleotide strands were immobilized on the substrate (SiO$_2$) between the gaps (Figure 4.9c). Using a three-component sandwich assay, 27-base oligonucleotide target strands were hybridized with the capture strands and subsequently with oligonucleotide probe strands tagged with a Au nanoparticle (Figure 4.9b). In the presence of Ag$^+$, the captured Au nanoparticles facilitated hydroquinone-mediated reduction of Ag$^+$ and the deposited silver lowered the resistance between the electrodes. Target DNA was detected at concentrations as low as 5×10^{-13} M using this method. In addition, a high mutation selectivity of ca 100 000 : 1 was achieved by an unusual hybridization behavior that was dependent on electrolyte concentration. When the array was washed with 0.3 M phosphate-buffered saline (pH 7) after DNA-hybridization, the complementary capture strands decreased the resistance (500 Ω) between electrodes, but the one-base mismatched capture strands increased resistance (200 MΩ). These results suggest that highly sensitive DNA analysis may be achieved without polymerase chain reaction or comparable target-amplification.

Immobilizing DNA strands on the nanoparticle surface can change the electrical properties of nanoparticles, because of changes to electron hopping paths and the dielectric permittivity of the surface-protecting layer. Using this concept, Chen and coworkers have developed a nanoparticle film that manifested a significant difference in electrical conductance between immobilized single-stranded (ss) and

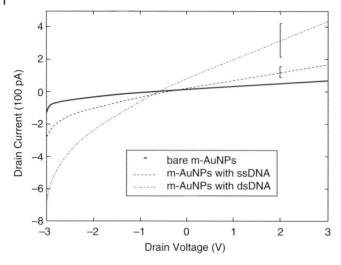

Figure 4.10 Measured *I–V* curves of m-AuNPs at a scan rate of 10 mV s^{-1}. Compared with the s-AuNPs cases, the denser AuNP spots in m-AuNPs lead to much higher current responses. In addition, DNA strands contribute to electron tunneling significantly in these configurations. Reprinted with permission from Reference [102]. Copyright 2006 American Institute of Physics.

double-stranded (ds) DNAs [102]. This DNA biosensor was fabricated with multilayer Au nanoparticles connected together with 1,6-hexanedithiol and decorated with a ss probe DNA oligonucleotide (5′-SH-A10-GTGTGGCAAAGATGTC-CAATCGATTGG-3′). In the presence of the complementary single-stranded oligonucleotide (3′-CACACCGTTTCTACAGGTTAGCTAACG-5′) a specific binding reaction occurred and double-stranded DNA segments were formed on the nanoparticle surfaces. As shown in Figure 4.10, a significant difference in conductance (current per unit voltage) was measured, depending on the surface states of Au nanoparticles: no DNA-immobilized (bare m-AuNPs), probe DNA-immobilized (m-AuNPs with ssDNA), and probe/target-immobilized (m-AuNPs with dsDNA) surface. The conductance change was clearly measurable at a concentration of 1×10^{-13} M of target DNA, which suggests that the direct, electrical detection of ssDNA is possible without any current enhancement agents, such as Ag.

4.4
Concluding Remarks

Chemical and biological sensors that use nanomaterials are attractive for many applications because the properties of nanomaterials – unlike their bulk counterparts – are tunable and surface dependent. It is usually easier to miniaturize chemical and biological sensors using nanomaterials because of their inherently high surface-to-volume ratio. Progress in reliably modifying nanoparticles with molecular and biomolecular ligands suggests that chemical and biological sensors be tailored by functionalizing their surfaces. For example, functionalization of a nanoparticle surface with predesigned receptor moieties can be used to improve sensor sensitivity and selectivity.

The sensor applications of functionalized nanoparticles based on electron transport have rested so far on the collective properties of the nanoparticle assemblies associated with a conductive surface (e.g., electrodes). In these sensors, the interaction between analytes and functionalized nanoparticles perturb the overall conductivity of a nanoparticle assembly. Almost all electron-transport-based sensors have been developed with this principle. However, another possible approach is to utilize the quantized voltammetric properties of small-sized nanoparticles suspended in solution. As described in Section 4.2.2, voltammetric peak potential spacing (ΔV) is strongly affected by the surface dielectric medium (ligand shell) of the nanoparticles. Surface interactions with analytes may induce changes in the surface dielectric, resulting in changes in nanoparticle capacitance and ΔV. This approach may allow sample detection in the solution phase and *in situ* detection of biomolecules.

Presently, the use of single functionalized nanoparticles for electronic sensing is a challenge for the future development of sensors. Compared with sensors prepared with nanoparticle assemblies, a single nanoparticle sensing device may provide extremely sensitive measurements, with the hope of supersensitive sensors. As discussed in Section 4.2.1, a single nanoparticle has two electron-tunneling junctions (inset of Figure 4.1a) and electron transport (tunneling current) is completely governed by capacitance of these two junctions. This single nanoparticle capacitance is in the range of $\sim 10^{-18}$ F [18], and can change greatly in the presence of analytes. While such a supersensitive device may be prototyped by using an STM platform, construction is still problematic with more conventional electronic design methods.

While the basic transduction principles employed in chemical and biological sensors are now well established and have remained unchanged over the years, there have been significant recent advances in both sensing materials and strategies for performance enhancement. For the development of high performance, rapid response, and long lasting sensors, future advancements should be directed to understanding of nanomaterial chemistry and physical behavior more deeply and to developing new techniques that will afford further miniaturization, low power-consumption, and multiplex detection in a single device.

Acknowledgments

Support from the New Faculty Start-Up Fund and Research Corporation (CCSA 10523) is gratefully acknowledged.

References

1 Halperin, W.P. (1986) *Rev. Mod. Phys.*, **58**, 533–606.
2 Cottey, A.A. (1971) *J. Phys. C Solid State Phys.*, **4**, 1734–1736.
3 Brown, R.J.C. and Milton, M.J.T. (2005) *Trends Anal. Chem.*, **24**, 266–274.
4 Dulski, T.R. (1999) *Trace Element Analysis of Metals, Methods and Techniques*, Marcel Dekker, New York.
5 Hierlemann, A. and Gutierrez-Osuna, R. (2008) *Chem. Rev.*, **108**, 563–613.
6 Gorter, C.J. (1951) *Physica*, **17**, 777–780.
7 Giaever, I. and Zeller, H.R. (1968) *Phys. Rev. Lett.*, **20**, 1504–1507.
8 Zeller, H.R. and Giaever, I. (1969) *Phys. Rev.*, **181**, 789–799.
9 Lambe, J. and Jaklevic, R.C. (1969) *Phys. Rev. Lett.*, **22**, 1371–1375.
10 Fulton, T.A. and Dolan, G.J. (1987) *Phys. Rev. Lett.*, **59**, 109–112.
11 Scott-Thomas, J.H.F., Field, S.B., Kastner, M.A., Smith, H.I., and Antoniadis, D.A. (1989) *Phys. Rev. Lett.*, **62**, 583–586.
12 Parak, W.J., Manna, L., and Christian, F. (2004) Quantum dots, in *Nanoparticles: From Theory to Application* (ed. G. Schmid), Wiley-VCH Verlag GmbH, Weinheim, pp. 4–49.
13 Ingram, R.S., Hostetler, M.J., Murray, R.W., Schaaff, T.G., Khoury, J.T., Whetten, R.L., Bigioni, T.P., Guthrie, D.K., and First, P.N. (1997) *J. Am. Chem. Soc.*, **119**, 9279–9280.
14 Andres, R.P., Bein, T., Dorogi, M., Feng, S., Jason, I.H., Kubiak, C.P., Mahoney, W., Osifchin, R.G., and Reifenberger, R. (1996) *Science*, **272**, 1323–1325.
15 Fan, F.-R.F. and Bard, A.J. (1997) *Science*, **277**, 1791–1793.
16 Bockrath, M., Cobden, D.H., McEuen, P.L., Chopra, N.G., Zettl, A., Thess, A., and Smalley, R.E. (1997) *Science*, **275**, 1922–1925.
17 Guo, L., Leobandung, E., and Chou, S.Y. (1997) *Science*, **275**, 649–651.
18 Chen, S., Ingram, R.S., Hostetler, M.J., Pietron, J.J., Murray, R.W., Schaaff, T.G., Khoury, J.T., Alvarez, M.M., and Whetten, R.L. (1998) *Science*, **280**, 2098–2101.
19 Murray, R.W. (2008) *Chem. Rev.*, **108**, 2688–2720.
20 Häkkinen, H., Barnett, R.N., and Landman, U. (1999) *Phys. Rev. Lett.*, **82**, 3264–3267.
21 Lee, D., Donkers, R.L., Wang, G., Harper, A.S., and Murray, R.W. (2004) *J. Am. Chem. Soc.*, **126**, 6193–6199.
22 Menard, L.D., Gao, S.-P., Xu, H., Twesten, R.D., Harper, A.S., Song, Y., Wang, G., Douglas, A.D., Yang, J.C., Frenkel, A.I., Nuzzo, R.G., and Murray, R.W. (2006) *J. Phys. Chem. B*, **110**, 12874–12883.
23 Brus, L.E. (1984) *J. Chem. Phys.*, **80**, 4403–4409.
24 Vahala, K.J. and Sercel, P.C. (1990) *Phys. Rev. Lett.*, **65**, 239–242.
25 Jimenez, V.L., Georganopoulou, D.G., White, R.J., Harper, A.S., Mills, A.J., Lee, D., and Murray, R.W. (2004) *Langmuir*, **20**, 6864–6870.
26 Lee, D., Donkers, R.L., DeSimone, J.M., and Murray, R.W. (2003) *J. Am. Chem. Soc.*, **125**, 1182–1183.
27 Quinn, B.M., Liljeroth, P., Ruiz, V., Laaksonen, T., and Kontturi, K. (2003) *J. Am. Chem. Soc.*, **125**, 6644–6645.
28 Chaki, N.K., Tsunoyama, H., Negishi, Y., and Tsukuda, T. (2006) Isolation and electrochemical studies of Au55:SR clusters, Presented at Symposium on nanoparticles, electrons, and photons. 210th Annual Meeting of the Electrochemical Society, Cancun, Mexico.
29 Balasubramanian, R., Guo, R., Mills, A.J., and Murray, R.W. (2005) *J. Am. Chem. Soc.*, **127**, 8126–8132.
30 Wolfe, R.L. and Murray, R.W. (2006) *Anal. Chem.*, **78**, 1167–1173.
31 Hicks, J.F., Miles, D.T., and Murray, R.W. (2002) *J. Am. Chem. Soc.*, **124**, 13322–13328.
32 Zamborini, F.P., Leopold, M.C., Hicks, J.F., Kulesza, P.J., Malik, M.A., and Murray, R.W. (2002) *J. Am. Chem. Soc.*, **124**, 8958–8964.
33 Joachim, C., Gimzewski, J.K., and Aviram, A. (2000) *Nature*, **408**, 541–548.
34 Murray, C.B., Kagan, C.R., and Bawendi, M.G. (2000) *Annu. Rev. Mater. Sci.*, **30**, 545–610.

35 Fendler, J.H. and Meldrum, F.C. (1995) *Adv. Mater.*, **7**, 607–632.
36 Collier, C.P., Vossmeyer, T., and Heath, J.R. (1998) *Annu. Rev. Phys. Chem.*, **49**, 371–404.
37 Wang, Z.L. (1998) *Adv. Mater.*, **10**, 13–30.
38 Shipway, A.N., Katz, E., and Willner, I. (2000) *ChemPhysChem*, **1**, 18–52.
39 Rao, C.N.R., Kulkarni, G.U., Thomas, P.J., and Edwards, P.P. (2000) *Chem. Soc. Rev.*, **29**, 27–35.
40 Chakraborty, A.K. and Golumbfskie, A.J. (2001) *Annu. Rev. Phys. Chem.*, **52**, 537–573.
41 Pileni, M.P. (2001) *J. Phys. Chem. B*, **105**, 3358–3371.
42 Brust, M. and Kiely, C.J. (2002) *Colloids Surf. A*, **202**, 175–186.
43 Crespo-Biel, O., Ravoo, B.J., Reinhoudt, D.N., and Huskens, J. (2006) *J. Mater. Chem.*, **16**, 3997–4021.
44 Zabet-Khosousi, A. and Dhirani, A.-A. (2008) *Chem. Rev.*, **108**, 4072–4124.
45 Fendler, J.H. (1996) *Chem. Mater.*, **8**, 1616–1624.
46 Marcus, R.J., Zwolinski, B.J., and Eyring, H. (1954) *J. Phys. Chem.*, **58**, 432–437.
47 Hush, N.S. (1958) *J. Chem. Phys.*, **28**, 962–972.
48 Conway, B.E. (1959) *Can. J. Chem.*, **37**, 178–189.
49 Marcus, R.A. (1964) *Annu. Rev. Phys. Chem.*, **15**, 155–196.
50 Salomon, M. and Conway, B.E. (1965) *Discuss. Faraday Soc.*, **39**, 223–238.
51 Gerischer, H. (1966) *J. Electrochem. Soc.*, **113**, 1174–1182.
52 McConnell, H.M. (1961) *J. Chem. Phys.*, **35**, 508–515.
53 Glaeser, R.M. and Berry, R.S. (1966) *J. Chem. Phys.*, **44**, 3797–3810.
54 Schmidt, P.P. (1972) *J. Chem. Phys.*, **57**, 3749–3762.
55 Libby, W.F. (1977) *Annu. Rev. Phys. Chem.*, **28**, 105–110.
56 Wuelfing, W.P., Green, S.J., Pietron, J.J., Cliffel, D.E., and Murray, R.W. (2000) *J. Am. Chem. Soc.*, **122**, 11465–11472.
57 Choi, J.-P. and Murray, R.W. (2006) *J. Am. Chem. Soc.*, **128**, 10496–10502.
58 Terrill, R.H., Postlethwaite, T.A., Chen, C.-H., Poon, C.-D., Terzis, A., Chen, A., Hutchison, J.E., Clark, M.R., Wignall, G., Londono, J.D., Superfine, R., Falvo, M., Johnson, C.S. Jr., Samulski, E.T., and Murray, R.W. (1995) *J. Am. Chem. Soc.*, **117**, 12537–12548.
59 Wuelfing, W.P. and Murray, R.W. (2002) *J. Phys. Chem. B*, **106**, 3139–3145.
60 Choi, J.-P., Coble, M.M., Branham, M.R., DeSimone, J.M., and Murray, R.W. (2007) *J. Phys. Chem. C*, **111**, 3778–3785.
61 Seiyama, T., Kato, A., Fujiishi, K., and Nagatani, M. (1962) *Anal. Chem.*, **34**, 1502–1503.
62 Seiyama, T. and Kagawa, S. (1966) *Anal. Chem.*, **38**, 1069–1073.
63 Janata, J. and Bezegh, A. (1988) *Anal. Chem.*, **60**, 62–74.
64 Lundberg, B. and Sundqvist, B. (1986) *J. Appl. Phys.*, **60**, 1074–1079.
65 Ruschau, G.R., Newnham, R.E., Runt, J., and Smith, B.E. (1989) *Sens. Actuators*, **20**, 269–275.
66 Talik, P., Zabkowska-Wacławek, M., and Wacławek, W. (1992) *J. Mater. Sci.*, **27**, 6807–6810.
67 Lonergan, M.C., Severin, E.J., Doleman, B.J., Beaber, S.A., Grubbs, R.H., and Lewis, N.S. (1996) *Chem. Mater.*, **8**, 2298–2312.
68 Wohltjen, H. and Snow, A.W. (1998) *Anal. Chem.*, **70**, 2856–2859.
69 Evans, S.D., Johnson, S.R., Cheng, Y.L., and Shen, T. (2000) *J. Mater. Chem.*, **10**, 183–188.
70 Han, L., Daniel, D.R., Maye, M.M., and Zhong, C.-J. (2001) *Anal. Chem.*, **73**, 4441–4449.
71 Cai, Q.-Y. and Zellers, E.T. (2002) *Anal. Chem.*, **74**, 3533–3539.
72 Pang, P., Guo, Z., and Cai, Q. (2005) *Talanta*, **65**, 1343–1348.
73 Yang, C.-Y., Li, C.-L., and Lu, C.-J. (2006) *Anal. Chim. Acta*, **565**, 17–26.
74 Foos, E.E., Snow, A.W., Twigg, M.E., and Ancona, M.G. (2002) *Chem. Mater.*, **14**, 2401–2408.
75 Krasteva, N., Besnard, I., Guse, B., Bauer, R.E., Mullen, K., Yasuda, A., and Vossmeyer, T. (2002) *Nano Lett.*, **2**, 551–555.
76 Harnack, O., Raible, I., Yasuda, A., and Vossmeyer, T. (2005) *Appl. Phys. Lett.*, **86**, 034108.

77 Joseph, Y., Besnard, I., Rosenberger, M., Guse, B., Nothofer, H.-G., Wessels, J.M., Wild, U., Knop-Gericke, A., Su, D., Schlogl, R., Yasuda, A., and Vossmeyer, T. (2003) *J. Phys. Chem. B*, **107**, 7406–7413.

78 Joseph, Y., Krasteva, N., Besnard, I., Guse, B., Rosenberger, M., Wild, U., Knop-Gericke, A., Schlögl, R., Krustev, R., Yasuda, A., and Vossmeyer, T. (2004) *Faraday Discuss.*, **125**, 77–97.

79 Rowe, M.P., Steinecker, W.H., and Zellers, E.T. (2007) *Anal. Chem.*, **79**, 1164–1172.

80 Steinecker, W.H., Rowe, M.P., and Zellers, E.T. (2007) *Anal. Chem.*, **79**, 4977–4986.

81 Zhong, Q., Steinecker, W.H., and Zellers, E.T. (2009) *Analyst*, **134**, 283–293.

82 Yamazoe, N., Sakai, G., and Shimanoe, K. (2003) *Catal. Surv. Asia*, **7**, 63–75.

83 Xu, C., Tamaki, J., Miura, N., and Yamazoe, N. (1991) *Sens. Actuators B*, **3**, 147–155.

84 Tamaki, J., Zhang, Z., Fujimori, K., Akiyama, M., Harada, T., Miura, N., and Yamazoe, N. (1994) *J. Electrochem. Soc.*, **141**, 2207–2210.

85 Göpel, W. and Schierbaum, K.D. (1995) *Sens. Actuators, B*, **26**, 1–12.

86 Barsan, N. and Weimar, U. (2001) *J. Electroceram.*, **7**, 143–167.

87 Morrison, S.R. (1987) *Sens. Actuators*, **12**, 425–440.

88 Kohl, D. (1990) *Sens. Actuators, B*, **1**, 158–165.

89 Yamazoe, N. (1991) *Sens. Actuators, B*, **5**, 7–19.

90 Franke, M.E., Koplin, T.J., and Simon, U. (2006) *Small*, **2**, 36–50.

91 Sakai, G., Matsunaga, N., Shimanoe, K., and Yamazoe, N. (2001) *Sens. Actuators, B*, **80**, 125–131.

92 Matsunaga, N., Sakai, G., Shimanoe, K., and Yamazoe, N. (2003) *Sens. Actuators, B*, **96**, 226–233.

93 Vuong, D.D., Sakai, G., Shimanoe, K., and Yamazoe, N. (2004) *Sens. Actuators, B*, **103**, 386–391.

94 Vuong, D.D., Sakai, G., Shimanoe, K., and Yamazoe, N. (2005) *Sens. Actuators, B*, **105**, 437–442.

95 Favier, F., Walter, E.C., Zach, M.P., Benter, T., and Penner, R.M. (2001) *Science*, **293**, 2227–2231.

96 Wang, Y., Jiang, X., and Xia, Y. (2003) *J. Am. Chem. Soc.*, **125**, 16176–16177.

97 Kolmakov, A., Zhang, Y., Cheng, G., and Moskovits, M. (2003) *Adv. Mater.*, **15**, 997–1000.

98 Drummond, T.G., Hill, M.G., and Barton, J.K. (2003) *Nat. Biotechnol.*, **21**, 1192–1199.

99 Templeton, A.C., Wuelfing, W.P., and Murray, R.W. (2000) *Acc. Chem. Res.*, **33**, 27–36.

100 Velev, O.D. and Kaler, E.W. (1999) *Langmuir*, **15**, 3693–3698.

101 Park, S.-J., Taton, T.A., and Mirkin, C.A. (2002) *Science*, **295**, 1503–1506.

102 Tsai, C.-Y., Chang, T.-L., Kuo, L.-S., and Chen, P.-H. (2006) *Appl. Phys. Lett.*, **89**, 203902.

5
Micro- and Nanofluidic Systems for Trace Analysis of Biological Samples
Debashis Dutta

5.1
Introduction

Over the past few decades, developments in micro- and nanoscale science/technology have significantly advanced our ability to analyze biological samples, such as, nucleic acids, proteins, and biological cells, both faster as well as with greater accuracy. For example, simple miniaturization of analysis columns has allowed significant improvements in the efficiency of DNA and protein separations due to fast heat and mass transfer in the system [1]. In addition, our ability to confine these molecules within micro-/nanofabricated structures today permits us to work with smaller sample volumes that in turn have reduced the cost of such analyses. Most importantly, however, these developments have opened doors to new experimental paradigms that were inaccessible earlier by allowing us to probe biological samples at the single molecule and/or cell level [2, 3].

One platform where many of these benefits of miniaturization have been demonstrated already is micro- and nanofluidics [4–7]. A microfluidic device typically consists of a network of micrometer-scale channels (microchannels) connecting different circular wells created on a glass or a plastic [e.g., poly(methyl methacrylate) or poly(dimethylsiloxane)] plate. Depending on the choice for the substrate material, these devices are produced using fabrication techniques borrowed from the semiconductor or the plastic industry [8, 9]. While chemical and biological analyses are performed with such devices by appropriately moving fluid and analyte samples through the closed conduits, the circular wells in these systems act as control ports for guiding the transport processes. In addition, the architecture of micro-/nanofluidic devices also allows easier automation and integration of multiple analysis procedures, e.g. sample preparation, reaction, sample preconcentration, and separation, on a single unit. Such automation and integration not only minimizes the number of manual sample handling steps, improving the accuracy and the reproducibility of the assays, but at the same time reduces operator exposure to potential biohazardous materials. Finally, because these devices can be manufactured using inexpensive methods, they can be customized for a specific biological assay in a time and cost-effective way.

In this chapter we present a detailed review on the emergence of micro- and nanofluidic devices for trace analysis of biological samples since their conception about two decades ago. The material presented here has been organized in three major sections depending on the type of biological sample under consideration. While the first two sections focus on the use of micro-/nanofluidic systems for analyzing nucleic acids and proteins, respectively, the final one discusses scientific developments on the application of fluidic microchips for probing biological cells.

5.2
Nucleic Acid Analysis

The experimental steps that are often essential in the trace analysis of nucleic acid samples are DNA extraction and purification, polymerase chain reaction (PCR), and DNA detection based either on hybridization or separation methods. We begin this section by reviewing the advances in miniaturization of PCR devices followed by describing the application of micro-/nanofluidic technology to enhance DNA hybridization and separation processes. Finally, a brief discussion is presented on the development of novel technologies that can allow fast and inexpensive analyses of DNA samples with resolution down to a single base-pair.

5.2.1
Miniaturization of PCR Devices

Miniaturization of PCR devices is often desirable due to its potential to shorten amplification times, increase sample throughputs, and minimize human/world-to-PCR intervention and contamination [10]. To this end, there has been a significant effort towards miniaturizing this amplification technique on the microfluidic platform. To date, PCR microfluidics has been developed mainly in four formats: chamber stationary PCR, flow-through PCR, thermal convection-driven PCR, and droplet-based PCR.

Chamber stationary PCR devices work very similarly to conventional PCR systems, where the PCR solution is kept stationary and the temperature of the PCR reaction chamber is cycled between three different values [11, 12]. After completion of the PCR reaction, the amplification products are recovered from the chamber for off-line or on-line detection. The chamber stationary PCR format can be implemented on a microchip in either single or multiple chambers. While single chamber micro-PCR devices offer better thermal control, their throughput is often limited. As a result, many sequential PCR experiments need to be performed in these systems for amplifying multiple DNA samples, which increases the analysis time. Moreover, sample contamination due to carry-over from experiment to experiment tends to deteriorate the performance of single chamber PCR units.

To improve PCR throughput, as well as reduce the labor required in PCR assays (e.g., in cleaning PCR chambers), multi-chamber stationary PCR been explored on a single microchip [13, 14]. In these systems, however, special care must be taken in the thermal design of the chamber array to acquire temperature uniformity between chambers. Without a careful design to assure temperature uniformity and the elimination of any possible differences in amplification, the repeatability, sensitivity, efficiency, and specificity of PCR amplification across the different chambers may be compromised.

Another important configuration of PCR microfluidics that is ingeniously realized with a "time–space conversion" concept is flow-through PCR. Instead of being stationary in a chamber, the PCR solution in this case is continuously and repeatedly flown through three different temperature zones necessary for the amplification process. The attractive features of this type of PCR microfluidics include rapid heat transfer and high thermal cycling, allowing for total run times of the order of minutes. In addition, this design has a low possibility of cross-contamination between samples, allowing for very specific amplification and a high potential for further development of integrated PCR devices by incorporating several other functionalities. A significant limitation of this approach, however, is the fixed number of PCR cycles that is dictated by the channel layout in the system.

To date, flow-through PCR microfluidics has been implemented using two main designs. The first involves a serpentine channel layout passing repeatedly through three temperature zones to realize the DNA denaturation, annealing, and extension steps [15]. One attractive feature of this arrangement is the smooth temperature gradient that is established on a monolithic chip, without the need of a forced cooling process. A drawback of this architecture is that the melted single-stranded DNA sample can hybridize with the template strands or their complementary fragments when passing through the extension zone, resulting in decreased amplification efficiency. To circumvent this problem, a novel "circular" arrangement of the three temperature zones has been exploited to realize the sequence of DNA denaturation, annealing, and extension [16]. For this set-up to work well, however, a proper insulation of the thermal zones is often crucial [17]. A spiral channel configuration has also been proposed for flow-through PCR with a circular arrangement of the three temperature zones, allowing for a compact footprint and requirement of minimal number of heaters for temperature control [18, 19].

The third format of microchip PCR, convection-driven PCR microfluidics, although only in its infancy, has been demonstrated by multiple research groups to be capable of rapid DNA amplification. Rayleigh–Bénard convection, for example, has been exploited by Krishnan *et al.* to perform PCR amplification inside a 35-µL cylindrical cavity [20]. The temperature cycling in this design was achieved as the flow continuously shuttled fluid vertically between the two temperature zones for annealing/extension (Figure 5.1a, top, 61.8 °C) and denaturation (Figure 5.1a, bottom, 97.8 °C). A laminar convection PCR device was subsequently designed by Braun *et al.* in which the PCR solution was heated in the center of the chamber by an infrared source [21]. Braun and coworkers' design not only simplified the fabrication of the PCR chamber over Krishnan *et al.*'s device but also allowed for

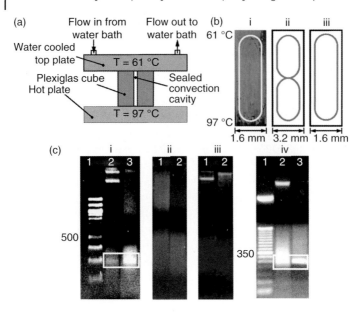

Figure 5.1 (a) Schematic of a Rayleigh–Bénard PCR device [20]; (b) influence of channel geometry on the Rayleigh–Bénard convection; (c) DNA amplification in a Rayleigh–Bénard PCR cell. Reprinted with permission from the American Association for the Advancement of Science.

faster temperature cycling as it did not require heating and cooling of the chamber walls and fittings in each cycle. In a different study, Wheeler et al. have reported a closed PCR microfluidic heating pipe with a "0" shape where the fluid was heated on one side (94.8 °C) and cooled on the other (57.8 °C) [22]. The PCR solution in this design was placed in a polypropylene thin bag sandwiched between two circuit boards. One desirable feature of this architecture was that the entire fluid passed through both the temperature zones, avoiding any dead volumes not being amplified. However, the non-uniform velocity distribution of the fluid flow in this device resulted in uneven convection times, leading to variability in amplification times within the same volume.

The most recently developed format of microchip PCR, droplet-based PCR microfluidics, relies on the thermo-cycling of nanoliter sized droplets containing the PCR solution to realize amplification of DNA samples. The smaller reaction volume in this format allows for faster and more efficient thermo-cycling, further reducing the amplification time in the PCR process. Moreover, because the amplified DNA sample in this design is contained within a smaller volume of liquid, greater detection sensitivity is accomplished. Droplet-based PCR has been demonstrated on the microfluidic platform in aqueous droplets dispersed in an immiscible oil-phase. While initial studies relied on moving the PCR droplets through different temperature zones [23–26], a more recent study has demonstrated temperature cycling of the droplets in this technique using a non-contact heating source [27]. Because the latter method does not require the droplets to be moved through different temperature zones, it has the potential to significantly reduce cross-contamination between the different droplets as well as allow for realization of higher throughputs.

5.2.2
Integration of PCR with Separation, DNA Hybridization, and Sample Preparation

The integration of the PCR technique with capillary electrophoresis (CE) or capillary gel electrophoresis (CGE) combines the advantages of fast DNA amplification and subsequent separation/detection on microfluidic devices without manual sample transfer [28]. Moreover, simple scaling laws suggest an increase in the efficiency of the DNA separations with miniaturization of the separation system, which tends to reduce the overall DNA analysis time. In addition, higher throughputs can be realized on these integrated devices through parallel DNA amplification [29] or multiplex DNA amplification [30] followed by CE separation and detection. More impressively, the sensitivity of these integrated systems has been enhanced to the level so as to be able to detect a single DNA copy, which significantly facilitates our ability to study genetic expressions from individual cells and genetic heterogeneities [31]. Other functional components have been also incorporated into PCR-CE systems—such as sample loading by valves and hydrophobic vents [31], cell lysis [30], and on-chip DNA concentration prior to analysis [32]—to create more "complete" microfluidic devices for DNA analysis. Interestingly, many of these devices are stand-alone units requiring no external lenses, heaters, or mechanical pumps for complete processing and analysis of the DNA samples [33]. While DNA analysis on the microfluidic platform was initially demonstrated on glass microdevices, polymeric materials are becoming an increasingly popular choice for designing PCR-CE/CGE integrated microchips, taking advantage of their optical and electrical insulation properties, microchannel surface charge and chemistry, and lower fabrication cost [34].

Advantages associated with the miniaturization of DNA hybridization assays were first demonstrated using microarray technology that allowed thousands of specific DNA/RNA sequences to be detected simultaneously on silicon, glass, or polymer wafers. However, these DNA microarrays suffer from several limitations, including difficulties in further scaling-down the array-size and the nucleotide densities, diffusion limited kinetics, and the requirement for manual liquid transfer. Moreover, despite the ability of these systems to detect multiple genes in a single assay, detection at the cellular level without any gene sequence amplifications remains limited. To circumvent many of these problems, there has recently been significant interest in developing microfluidics-based DNA hybridization assays. Such devices have not only allowed the integration of DNA extraction, purification, amplification, and detection steps with the hybridization module, but have also significantly reduced DNA hybridization times due to fast mass transfer rates and large surface area to volume ratios in microfluidic channels. To date, highly integrated microfluidic devices have been developed on polycarbonate wafers smaller than a credit card, that are capable of extracting and concentrating nucleic acids from milliliter sized aqueous samples and performing chemical reactions, metering, mixing, and nucleic acid hybridization at the microliter scale [35]. Liu et al. have further enhanced these analytical capabilities by demonstrating a monolithic device that integrated target cell capture, sample preparation, PCR, and DNA hybridization on a single microchip [36]. To realize higher throughputs in microfluidics-based DNA hybridization devices, Trau et al. have incorporated multiple PCR microreactors with integrated DNA microarrays on a single silicon chip that required no buffer exchange or sample transfer during the entire analysis [37]. Flow-through PCR micro-devices have also been integrated with bio-electronic DNA microarrays for genotyping on different substrate materials such as ceramics and plastics [38].

5.2.3
Novel Micro- and Nanofluidic Tools for DNA Analysis

In the past decade, nanofabricated structures have emerged as promising tools for enhancing the resolution and throughput of DNA assays as well as reducing the cost of these analyses. For example, nanoscale constrictions in nanofilter columns [39], assembled colloidal arrays [40], and nanopillar arrays [41] have all been employed successfully as effective matrices for separation of DNA molecules based on the Ogston sieving mechanism. Note that these nanofluidic devices not only offer a greater control over the structure of the sieving matrix to fine tune a desired separation process but also can operate under a wider range of separation conditions, tend to have longer lifetimes, and can be produced at a lower cost. Moreover, theoretical calculations have recently shown that a nanofilter column may also allow significant improvements in the speed of DNA separations when operated under high electric field conditions [42].

Han and Craighead showed that if the size of the nanoscale constriction in a nanofilter column is further miniaturized and made smaller than the diameter of a DNA molecule, novel separations based on an entropic trapping mechanism can be designed (Figure 5.2) [43]. This phenomenon was exploited by the authors to separate mega-base-pair sized DNA strands, about ten times faster than that reported with pulsed field gel electrophoresis. The deeper regions in this design entrapped shorter DNA molecules more effectively, resulting in increased elution times with a decrease in the length of the DNA strand (a trend opposite to that observed in Ogston sieving). More recently, Pennathur et al. have demonstrated novel DNA separations that exploited the strong interaction of the analyte molecules with the channel walls and the Debye layer around them in open nanofluidic channels [44]. To allow the recovery of larger sample volumes of the separated DNA strands (e.g., for further analysis), nanofluidic separation of DNA samples has also been demonstrated in continuous-flow fashion based on length-dependent backtracking [45], rectification of Brownian motion [46], ratcheting in a quasi-two-dimensional sieve [47], and asymmetric bifurcation of laminar flow [48]. Furthermore, the properties of an anisotropic nanofluidic array have been also exploited by Fu et al. to separate short DNA molecules [49].

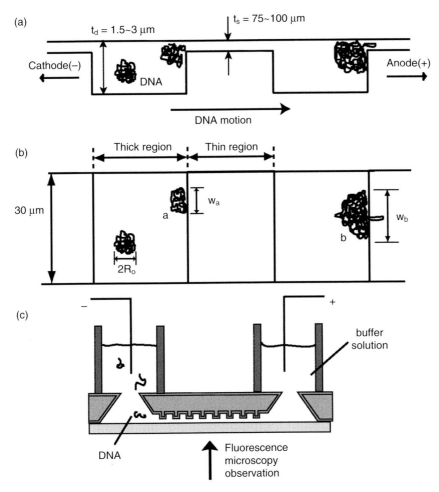

Figure 5.2 A nanofluidic separation device with many entropic traps [43]. (a) Cross-sectional schematic diagram of the device – electrophoresed DNA molecules are trapped in this design whenever they meet a thin region due to steric interactions; (b) top view of the device in operation; (c) experimental set-up used in the study. Reprinted with permission from the American Association for the Advancement of Science.

In addition to providing a promising platform for developing novel DNA separations, nanofluidic devices also offer unique opportunities for fast and inexpensive DNA sequencing with single base-pair resolution through electronic sensing. Kasianowicz *et al.* have demonstrated that a DNA molecule can be detected as a transient decrease in the ionic current when it passes through a nanopore, and the passage duration allows for the determination of the polymer length [50]. If each base in the DNA molecule modulates the signal in a specific and measurable way during translocation, then the base sequence could be determined at rates between 1000 and 10 000 bases per second [51]. It has been proposed that the required measurement sensitivity to distinguish individual bases might be achievable with embedded electrodes in the walls along the axis of a synthetic nanopore [52]. However, the transverse conductance across DNA fragments between electrodes depends critically on geometric rather than electronic structural properties [53]. To enhance the sensitivity of transverse conductance measurements in nanopores with embedded electrodes, other approaches are also being tested. For example, it has been postulated that derivatization of the nanopore surface with functional molecules that preferentially interact with certain nucleotide bases may amplify differences in the conductance measurements for the different nucleotides.

5.3
Protein Analysis

The traditional approach to analyzing protein mixtures primarily consists of three steps: (i) extracting the protein molecules from biological cells, (ii) separating the different proteins using a one- or two-dimensional separation method, and (iii) identifying these components by mass spectrometry (MS). We begin this section by reviewing the miniaturization of common protein separation techniques on microchips, followed by a discussion on the pre-concentration methods that have been demonstrated to enhance the sensitivity of microfluidic protein analysis. The latter part of this section reviews the integration of various sample preparation procedures as well as mass-spectrometric techniques to microfluidic devices.

5.3.1
Protein Separations

Microfluidic systems offer tremendous opportunities for realizing high-speed, high-efficiency protein separations not only because of their fast heat and mass transfer characteristics but also due to their ability to integrate sample preparation, sample pre-concentration and multidimensional separations on a single device. Bousse et al., for example, have demonstrated simultaneous dynamic labeling with fluorescent dyes and sizing of protein mixtures in a sodium dodecyl sulfate (SDS) based polymer solution, yielding a sizing accuracy of 5% or better and high detection sensitivity (30 nM for carbonic anhydrase) [54]. In this work, both on-chip staining and SDS dilution steps occurred within 100 ms, which was about 10^4 times faster than that reported in a SDS-polyacrylamide gel electrophoresis (SDS-PAGE) system. In a different study, researchers from the Sandia National Laboratories were able to separate proteins using an *in situ* UV-initiated crosslinked polyacrylamide (PAAm) gel and a covalent protein labeling protocol instead of dynamic labeling [55].

Tan and coworkers showed that isoelectric focusing (IEF) of protein samples could be realized on the microchip platform with efficiencies of 1.5×10^5 plates (for lysozyme) over a focusing distance of 4.7 cm [56]. The detection limit in this study was observed to be as low as 50 fM. To further improve the resolving power of this micro-IEF separation technique, Cui and coworkers demonstrated that several stages of IEF could be coupled in series on a single microchip device [57]. Lee *et al.* were able to increase the sample loading in such analyses and therefore the concentrations of focused analytes using a dynamic approach based on electrokinetic injection of proteins/peptides from solution reservoirs [58]. This allowed for 10–100-fold enhancement in sample loading after 30 min of electrokinetic injection with a 500 V cm^{-1} electric field. Using simple microfabrication techniques, Huang and Pawliszyn have constructed tapered microchannels to realize a thermally generated pH gradient in the analysis column, which was then used to demonstrate IEF of hemoglobin from dog, cat, and humans [59]. Yager *et al.* showed that protein samples could also be concentrated and fractionated in a continuous fashion using transverse IEF in microfluidic channels [60].

In addition to the studies described above, protein and peptide samples have been successfully analyzed on microchips using the capillary zone electrophoresis (CZE) technique. Hauser's group, for example, was able to separate cytochrome c, HAS, myoglobin, and IgG in less than 1 min. with plate numbers of 1000, 4500, 5400, and 7600, respectively [61]. Recently, Shadpour *et al.* have implemented multiple simultaneous CZE protein/peptide separations on a microchip using 16 identical analysis channels to enhance the throughput of the device [62]. Identification of proteins based on their hydrophobicities has also been realized on fluidic microchips using the micellar electrokinetic chromatographic (MEKC) technique. Moreover, to improve the separation performance and the sensitivity of micro-MEKC devices several types of injectors have been introduced. Furthermore, gradient elution MEKC has been demonstrated on microchips by Kutter *et al.* using both mixing-T and cross-channel injectors [63].

Capillary electrochromatographic (CEC) analysis of tryptic peptides using microfabricated columns based on an array of C18-modified collocate monolith support structures (COMOSSs) has been reported [64]. Separation of peptides and proteins using the CEC technique on the above-described microchip was comparable to that in HPLC. Moreover, the possibility of using COMOSS chips made from polydimethylsiloxane (PDMS) has been demonstrated to reduce the cost of these separation devices [65]. Furthermore, 2-acrylamido-2-methylpropane sulfonic acid (AMPS), poly(4-styrenesulfonic acid), poly(acrylic acid), poly(vinylsulfonic acid), and poly(stearyl methacrylate-*co*-AMPS) have been used for successful modification of the PDMS surface by cerium(IV)-catalyzed polymerization on COMOSS microchips, which provided highly efficient separations of peptide mixtures [66]. Conventional granular packings have been used to create a fritless CEC column in a microchannel for the analysis of biomolecules [67]. In addition, rapid reverse phase HPLC was demonstrated using a prototype fused-silica microchip that was integrated with a polymer plug-based valve injector and a monolithic stationary phase [68]. In this HPLC system, proteins and peptides were injected using a pressure-switchable fluoropolymer valve, separated on a C18 porous polymer monolith, and detected by a miniaturized fluorescence detector. Integration of an on-chip pressure generation unit with a microfluidic chromatographic channel has also been demonstrated to minimize dead volumes in microchip-HPLC systems [69].

While two-dimensional (2D) gel electrophoretic devices still remain the main workhorse for proteomic analyses, they suffer from some major drawbacks such as limited automation, long cycle time, and low sample throughput. Therefore, significant effort is being made to accomplish microchip based 2D protein analysis. Mimicking 2D gel electrophoretic devices, multidimensional separations can be established on a microchip by first focusing protein samples using IEF in a main channel, and then sizing the IEF sections in parallel cross channels [70]. This prototype device has been demonstrated using a PDMS microchip by Chen and coworkers [71]. In this design, IEF was performed in an isolated channel, and the focused samples were then introduced into an array of orthogonal microchannels for performing CGE using a 3D microfluidic network. Li *et al.* have improved this system by implementing both IEF and micro-CGE on a planar plastic chip [72]. Analysis in this device was completed within 10 min with an overall peak capacity of ~1700. Han and coworkers have coupled IEF with micro-CGE via active microvalve control, in which the protein plug after IEF could be isolated with a pneumatic valve and subjected to microchip CGE [73]. Thus, the interruption between running buffers of CGE and IEF could be inhibited to a large extent.

Some other promising 2D microchip systems have also been demonstrated in which the sample plug after the first separation was transferred into the second dimension sequentially at regular intervals. In 2000, Ramsey and coworkers first coupled MEKC (the first dimension) with CZE (the second dimension) on a glass microchip with a peak capacity of 500–1000 [74]. Gated injection was implemented for introducing the samples into the CZE dimension by opening an electric valve for about 0.3 s every 3–4 s. Later, CEC was used instead of MEKC as the first dimension and coupled with CZE [75]. In this design, a 25-cm spiral separation channel (instead of a serpentine layout) modified with octadecylsilane was used for CEC to minimize band broadening while a 1.2 cm straight separation channel segment was employed for CZE. The sample plug from CEC was loaded into the CZE channel for about 0.2 s with a cycle time of 3.2 s. The peak capacity for this design was estimated to be 150. Subsequently, an optimized chip MEKC-CZE was introduced with 1 Hz sampling frequency for the CZE dimension [76]. BSA tryptic digest was separated within 10 min in this device with a peak capacity of 4200. In a different effort, Soper's group presented a tandem PMMA 2D microchip device [77], in which CGE and MEKC were used as the first and the second dimensions, respectively. The sample plug from micro-CGE was loaded into MEKC for 0.5 s with a cycle time of 10.5 s. This system was demonstrated with ten model proteins, and provided a theoretical peak capacity of ~1000.

Notable advances have also been made in developing affinity-based electrophoretic assays on the microfluidic platform. In one such study, Chiem and Harrison have demonstrated a direct assay for monoclonal mouse IgG in mouse ascites fluid [78]. Koutny *et al.* have reported on serum cortisol determination over a range of clinical interest (1–60 $\mu g\,dl^{-1}$) via a competitive assay [79] using a fused silica microchip. Recently, Bharadwaj *et al.* have implemented a kinetic affinity-based CE in a microfluidic device for the detection of Lens culinaris agglutin (LCA)-reactive AFP (L3), a specific marker for hepatocellular carcinoma [80]. The first microchip based MEKC immunoassay was presented by von Heeren et al., who analyzed serum theophylline in human urine and serum samples [81]. To realize higher throughputs, Cheng et al. reported on microfluidic devices that were capable of independently performing six concurrent affinity electrophoresis based immunoassays [82]. Dishinger and Kennedy in a different study have developed a microchip containing four individual channel networks, each capable of performing immunoaffinity-based electrophoretic analysis of perfusate from insulin-producing cells found in the pancreas known as islets of Langerhans [83].

5.3.2
On-Chip Protein Pre-concentration

Real protein samples are usually very dilute, and the amount of sample injection in a microfluidic chip is usually limited. As a result, online pre-concentration of protein is highly desirable on microchips in most cases. To date, two methods have been primarily adopted on microfluidic devices for protein pre-concentration: filtering and electric field gradient focusing (EFGF). Ramsey and coworkers have reported a convenient filtering method using a silicate-based nanoporous membrane that was fabricated by spin-coating a silicate precursor between two glass microchannels and then providing suitable heat treatment [84]. In this design, proteins were concentrated electrokinetically on the porous silica layer prior to sample loading. This system functioned well for both coated and uncoated open channels, and yielded about a 600-fold enhancement in the assay sensitivity. Han's group have demonstrated that similar nanofluidic protein concentrators can also be designed on a PDMS microchip, by electrical breakdown through a gap junction between two PDMS microchannels [85].

Zwitterionic polymer membranes have been produced by Song and coworkers through laser-induced polymerization of the precursor material at the junction of a cross channel in a microchip [86]. In this work, the local and spatially averaged analyte concentration was increased by four and two orders of magnitude, respectively, in about 100 s using moderate voltages (70–150 V) for proteins >5.7 kDa. The degree of concentration in this system was reported to be limited only by the solubility of the proteins. In a different study, another organic membrane was fabricated in situ using photopolymerization of crosslinked PAAm to concentrate proteins (>10 kDa) by over a factor of 1000 in less than 5 min prior to a micro-CGE analysis [87]. The same functionality has also been accomplished by Long et al. using a multilayer device that had a membrane (with 10 nm pores) sandwiched between two layers of PDMS substrates with embedded microchannels [88].

Isotachophoresis (ITP) has been employed by Lin and coworkers to concentrate biomolecules prior to micro-CGE. In this work, a 5 mm SDS–protein sample plug was focused with 50 mM Tris, 0.5% SDS, 2% dithiothreitol (pH 6.8) leading electrolyte, and a 192 mM glycine, 25 mM Tris (pH 8.3) trailing electrolyte. The focused sample was then sized in a polymer solution containing 100 mM Tris-NaH$_2$PO$_4$,

0.1% SDS, 10% glycerol, and 10% dextran (pH 8.3), all of which could be completed within 300 s with an enhancement factor of about 40 [89]. Recently, concentration of HSA and its immunocomplex has been reported using transient ITP on a PMMA microchip, yielding an 800-fold signal enhancement [90]. In separation science, electric field gradient focusing (EFGF) is also known to be a powerful tool for both concentrating and separating protein samples simultaneously. This method has been illustrated on a microchip by Woolley and coworkers through *in situ* fabrication of a conductive membrane using phase-changing sacrificial layers [91]. With this device, about a three-fold improvement in separation resolution and a 10 000-fold enhancement in detection sensitivity were obtained within 40 min. As an alternative, Lee and coworkers have used a weir structure to fabricate a polymer membrane between a separation channel and an electric-field gradient generating channel [92]. With this EFGF device, the authors were able to concentrate green fluorescent protein samples by a factor of ca 4000.

5.3.3
Integrated Microfluidic Devices for Protein Analysis

Microfluidic platforms provide an unrivaled capability to integrate and automate preparatory and analytical functions in a single instrument using a lab-on-a-chip approach. For example, Gao *et al.* have developed a PDMS based device enabling protein digestion, peptide separation, and subsequent protein identification [93]. This device consisted of a capillary tube embedded in a PDMS substrate that contained a micro-PVDF membrane reactor with adsorbed trypsin to catalyze the protein digestion. The peptide products were then concentrated and resolved by electrophoretic separations prior to electrospray ionization for mass spectrometric analysis. Pressure-driven flow was used to drive the protein solution through the reactor and the extent of digestion was regulated by manipulating the dwell time. Another flow-through protein digestion device has been presented by Wang *et al.* that consisted of integrated beads of immobilized trypsin in a microchannel [94].

In addition, significant effort has gone into coupling microfluidic technologies with protein arrays [95] and mass spectrometric systems [96]. An example of the former is the work by Pawlak *et al.* [97], who described the integration of a Zeptosens protein array with a microfluidic delivery system. This device had a high sensitivity and signal-to-noise ratio primarily due to an integrated, planar waveguide detection system. As discussed by Figeys and Pinto, the widespread integration of microfluidic devices with mass spectrometric systems has mainly relied on the electrospray ionization method [98]. Examples of this type of integration include the ESI emitter and sheath gas approach [99], the PDMS devices of Chen *et al.* [100] and Chiou *et al.* [101], and the user-friendly device presented by Pinto *et al.* [102].

Ekström *et al.* [103] have presented a silicon micro-extraction chip (SMEC) with an integrated weir structure for sample clean-up and trace enrichment of peptides. This structure was used to trap a reversed-phase chromatography media (POROS R2 beads), and facilitated sample purification and enzymatic digestion of proteins by trapping beads immobilized with trypsin. Improvements in the weir design have been suggested by Bergkvist et al. [104]. A glass microchip developed by Bousse et al. [54] has integrated the separation, staining, virtual destaining, and detection steps for a protein sizing assay. An example of an assay with integrated cell culture with affinity-based electrophoresis of secreted proteins has been demonstrated by Roper et al. [105]. To perform on-line monitoring of insulin (Ins) secretion from islet cells in this study, the authors designed microfluidic devices allowing for mixing of effluent from an islet with FITC-labeled insulin (Ins*) and insulin antibody (Ab) in a 4 cm long reaction channel via EOF. The mixing time was controlled by adjusting the applied electric potential. For formation of Ab–Ins* complexes, on-chip mixing required 50 s, which was comparable to off-chip mixing time and was augmented by thin film heaters that raised the temperature in the mixing chamber to 38 °C, as a means of producing favorable binding kinetics. Following mixing in the reaction channel, samples were electrokinetically injected into a 1.5 cm long electrophoresis channel where the Ins* and Ab–Ins* complex were separated in 5 s. The integrated on-chip mixing and fast electrophoretic separations allowed for Ins* measurements at 15 s intervals, thus enabling continuous monitoring of Ins secretion with detection limits of 3 nM (or 18 ng ml^{-1}). The method was reported to resolve secretory profiles of both first- and second-phase Ins secretions upon addition of glucose to the islets. The work highlights the use of microchip immunoaffinity assays for high temporal resolution monitoring of cellular secretions of soluble analytes.

5.4
Microfluidic Devices for Single-Cell Analysis

There has been significant interest in employing microfluidic devices and systems for cellomics, the knowledge of different cellular phenotypes and function, for various reasons. For example, because the size of a mammalian cell is close to the dimension of typical microchannels, combining with non-mechanical fluidic micropumps or microvalves allows for precise and automated manipulation of singe cells in cellomic assays. Moreover, the ability to accurately control flow and reagent concentrations around a single or a small cell population permits a better understanding of the effect of drugs and/or external stimuli on cell behavior. Furthermore, the microfluidic platform allows for the integration of various tasks such as reagent delivery, cell culture, sorting, manipulation, lysis, and separation all on a single unit, which enables rapid, highly efficient single-cell analysis to be performed. Owing to these unique advantages, microfluidic techniques have recently emerged as the most potential platform for analyzing single or a small population of cells. In this section we review some of the scientific developments made towards performing fast and more accurate single-cell studies using microfluidic systems.

Integration of cell sampling, single-cell loading, docking, lysing, and CE separation with laser-induced fluorescence (LIF) detection on a single microchip has been demonstrated by Gao et al. [106]. In this device, cellular transport and single-cell loading into the separation channel were controlled by hydrostatic pressure and electrophoretic force, respectively, while single-cell docking was realized by repeatedly connecting and disconnecting a set of low potentials ("switching the high voltage on and off"). The docked cell was then directly lysed by applying a potential of 1.4 kV (280 V cm^{-1}) for 40 ms, and the cell lysate was analyzed to determine the NDA-derivatized glutathione (GSH) content in single human erythrocytes. To obtain better reproducibility in cellomic assays, precise trapping of single cells at a fixed location within a microchannel is essential. Ros' group has realized such cell localization by designing a cell trap composed of microstructured obstacles at a crossing channel. Using this architecture, single cells were trapped, injected, steered, and deposited by means of optical tweezers in a PDMS microfluidic device and consecutively lysed at positions defined by the microstructured obstacles. A green fluorescent protein-construct (T31N-GFP) in single Sf9 insect cells (*Spodoptera frugiperda*) was separated and its level quantitated in this work [107, 108]. Sun and Yin later fabricated a novel microfluidic chip with a weir structure for cell docking and lysis. In this design, individual cells were electrophoretically loaded into a separation channel where they were stopped by the weir and precisely positioned within the separation channel. The trapped cell was then lysed and the lysate material separated to determine the amount of reduced glutathione (GSH) and reactive oxygen species (ROS) in single human carcinoma cells [109].

Dilution of the intracellular contents during derivatization and lysis processes inside the microchannel is a critical issue as the process restricts the determination of the intracellular compounds present in low concentration. Intracellular derivatization can minimize the dilution during derivatization, but only derivatizing reagents that penetrate the cell membrane can be used with this approach. For those derivatizing reagents that cannot penetrate the cell membrane, Sun et al. have developed a method by encapsulating the fluorescent dye fluorescein isothiocyanate (FITC) into liposomes with an average diameter of 100 nm and intracellularly delivering the liposomes to label the intracellular species [110]. In a different effort, Zare's group has developed valved microfluidic devices that can isolate a single cell in a chamber having a volume less than 100 pl, where it can then be lysed and its contents derivatized with minimal dilution prior to analysis (Figure 5.3) [111]. With this tool, the researchers have been able to quantitate levels of amino acids as well as low-copy number proteins within single cells [112].

Single-cell gene expression analysis holds great promise for studying diverse biological systems, but it is challenging to process these precious samples in a reproducible, quantitative, and parallel fashion using conventional methods. Quake's group has developed a microfluidic device with integrated micromechanical valves for single-cell RNA and DNA analysis. All steps including cell capture, cell lysis, mRNA purification, cDNA synthesis, and cDNA purification were implemented in this device. In addition, single NIH/3T3 cell mRNA isolation and cDNA synthesis were demonstrated with quantitative calibration for each step in the process, and gene expression in individual cells was measured [113]. More recently,

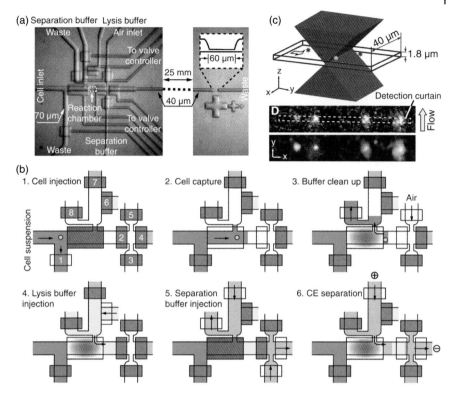

Figure 5.3 A single-cell analysis microchip proposed by Zare's group [112]. (a) Layout of the single-cell chip, showing the cell-manipulation section on the left and the molecule-counting section on the right; (b) analysis procedure for a mammalian or insect cell; (c) schematic illustration of the excitation laser focused by the microscope objective and the dimensions of the molecule-counting channel; (d) one frame from the CCD images of fluorescent molecules flowing across the molecule-counting section (upper panel) and the identification results (lower panel). Reprinted with permission from the American Association for the Advancement of Science.

a higher throughput microfluidic digital PCR device was demonstrated to amplify and analyze multiple genes obtained from single bacterial cells. In this design, several parallel reaction chambers (12 samples × 1176 chambers/sample) were created on single microchip, using micromechanical valves, to act as independent PCR reactors [114].

Information on how cells respond to changes in their external and internal environments is critical to understanding cellular functions. Eriksson *et al.* have realized such experimental conditions by creating an environmental gradient between two media in a microchip. The researchers then moved a trapped cell repeatedly between the different environments using optical tweezers and observed the rapid cytological responses [115].

Direct cell-to-cell communication between adjacent cells is vital to multiple physiological functions, including neurotransmission, immune response, transmitting action potential in cardiac myocytes, and proper organ development. However, current techniques are difficult to scale up for high-throughput screening of cell–cell communication in an array format. Microfluidic devices provide a powerful platform for such studies due to their ability to accurately control the physiological environment around a single cell and also precisely manipulate the position of the other cells adjacent to the target. Lee's group has described a microfluidic device to monitor cell–cell communication via gap junctions between individual cells. This monitoring unit contained a main channel that was 50 μm deep, 20 μm wide and had 37 pairs of cell trapping conduits with a cross-section of 2×2 μm located at the floor. This allowed trapping of multiple cell-pairs for the simultaneous optical observation of functional gap junction intercellular communication. The device operation was verified in this work by observation of dye transfer between mouse fibroblasts (NIH3T3) placed in membrane contact [116]. In a different work, Cooper's group have described the regional microfluidic and electrical manipulation of two cardiac myocytes connected through the intercalated discs. The extracellular space in this work was partitioned into three pools, namely two aqueous reservoirs for the cell ends and the central sealing gap containing the region of the intercalated discs. A lithographically defined partition, combined with a nanopipetting system, was then used to induce Ca^{2+} waves in a single cell and the propagation of these waves was monitored across the intercalated discs. The results demonstrated that Ca^{2+} waves traveled unimpaired along the longitudinal axis of the stimulated cell; however, under physiological conditions, they stopped in front of the intercalated discs. Under non-physiological conditions, when the adjoining cell was either damaged or challenged by a drug, conditions that resulted in Ca^{2+} overload in this cell, propagation of the Ca^{2+} wave across the cell junction was observed [117].

Patch-clamp recording is a significant research tool for electrophysiology. Traditional patch-clamp recording is accomplished by using a micromanipulator to position the tip of a glass pipette against the membrane of a cell. A carefully applied negative pressure through the pipette tip then causes the membrane to invaginate into the pipette and form a giga-ohm seal between the pipette and the cell. Though traditional patch-clamp techniques have been quite successful in electrophysiological research, these techniques usually require complicated and expensive set-ups. Microchip based patch clamp recordings have been accomplished using an array of microchannels, each microchannel in which can be used as a micropipette to draw individual cells. This architecture offers the capability to perform patch-clamp experiments in a high throughput and automated fashion. Lee's group has developed a 12-channel patch-clamp array using microfluidic junctions between a main chamber and lateral recording capillaries, all fabricated by micromolding of PDMS. In this design, the microfluidic integration allowed recording capillaries to be arrayed 20 μm apart, for a total chamber volume of

<0.5 nl, and incorporated partial cure bonding, yielding robust seals on individually selected mammalian cells under optical observation. The device was used to record activation of the voltage-gated potassium channel Kv2.1, which was stably expressed in mammalian CHO cells. Seals were established without the need of a vibration-isolation equipment and the Kv2.1 channel activation data corresponded well with data measured by the traditional pipette-based technique, using the same reagents and protocols [116]. However, this device showed substandard seal quality with cells, with seal resistances measured at around 150–250 MΩ. Moreover, the trapping sites that were contacting the bottom plane of the chip in this design caused the patched cell an uncommon deformation. In addition, the enclosed micro-scale fluid chamber posed problems when efficient fluid exchange and minimal seal quality perturbation was required. The authors later improved their design by creating a patch-clamp array with the cell trapping locations "raised" above the bottom plane of the chip to better resemble traditional glass micropipette openings. In this modified design, the main fluidic chamber was open to the air, providing an easy-to-use platform for fluidic exchange. The seal resistances were characterized with mammalian HeLa cells and the optimum patch aperture dimensions were determined. Extensive whole cell patch-clamp measurements were performed with CHO cells expressing Kv2.1 ion channels, including ion-channel I–V correlation, drug dose–response and drug binding activity characterizations. The results showed that this microfluidic system was suitable for high-throughput electrophysiology research and development, and provided an easy transition from the traditional patch-clamp approach [119].

5.5 Conclusion

Research over the past couple of decades has demonstrated the great potential of micro- and nanofluidic systems for analyzing nucleic acids, proteins/peptides, and biological cells. Genomic assays based on the PCR technique show significant improvements with miniaturization both in terms of speed as well as accuracy. Moreover, the integration of separation and pre-concentration methods on microfluidic devices help in the automation of these assays, significantly reducing the time and cost involved in such experimentation. The ability to manipulate protein/peptide samples in micro-/nanofluidic systems with greater control has also permitted the development of powerful tools that can significantly simplify proteomic studies. To date, most traditional proteomic approaches have been demonstrated on the microchip platform with substantial improvements in assay time and effort. Finally, micro- and nanofluidic devices today allow us to precisely control the physical and chemical environments around a single or small number of biological cell(s), which has already proven to provide analytical capabilities that can significantly improve our understanding of these systems.

References

1 Janasek, D., Franzke, J., and Manz, A. (2006) *Nature*, **442**, 374–380.
2 Mohamadi, M.R., Mahmoudian, L., Kaji, N., Tokeshi, M., et al. (2006) *Nano Today*, **1**, 38–45.
3 Schoch, R.B., Han, J., and Renaud, P. (2008) *Rev. Mod. Phys.*, **80**, 839–883.
4 West, J., Becker, M., Tombrink, S., and Manz, A. (2008) *Anal. Chem.*, **80**, 4403–4419.
5 Dittrich, P.S., Tachikawa, K., and Manz, A. (2006) *Anal. Chem.*, **78**, 3887–3907.
6 Vilkner, T., Janasek, D., and Manz, A. (2004) *Anal. Chem.*, **76**, 3373–3385.
7 Auroux, P.A., Iossifidis, D., Reyes, D.R., and Manz, A. (2002) *Anal. Chem.*, **74**, 2637–2652.
8 Reyes, D.R., Iossifidis, D., Auroux, P.A., and Manz, A. (2002) *Anal. Chem.*, **74**, 2623–2636.
9 Perry, J.L. and Kandlikar, S.G. (2006) *Microfluid Nanofluid*, **2**, 185–193.
10 Auroux, P.A., Koc, Y., de Mello, A., Manz, A., and Day, P.J.R. (2004) *Lab Chip*, **4**, 534–546.
11 Northrup, M.A., Ching, M.T., White, R.M., and Watson, R.T. (1993) Transducers '93, Seventh International Conference on Solid State Sensors & Actuators, Yokohama, Japan, pp. 924–926.
12 Wilding, P., Shoffner, M.A., and Kricka, L.J. (1994) *Clin. Chem.*, **40**, 1815–1818.
13 Poser, S., Schulz, T., Dillner, U., Baier, V., et al. (1997) *Sens. Actuators, A*, **62**, 672–675.
14 Taylor, T.B., Harvey, S.E., Albin, M., Lebak, L., et al. (1998) *Biomed. Microdevices*, **1**, 65–70.
15 Kopp, M.U., de Mello, A.J., and Manz, A. (1998) *Science*, **280**, 1046–1048.
16 Liu, J., Enzelberger, M., and Quake, S. (2002) *Electrophoresis*, **23**, 1531–1536.
17 Chou, C.F., Changrani, R., Roberts, P., Sadler, D., et al. (2002) *Microelectron. Eng.*, **61–62**, 921–925.
18 Hupert, M.L., Witek, M.A., Wang, Y., Mitchell, M.W., et al. (2003) *SPIE Proc.*, **4982**, 52–64.
19 Mitchell, M.W., Liu, X., Bejat, Y., Nikitopoulos, D.E., et al. (2003) *SPIE Proc.*, **4982**, 83–98.
20 Krishnan, M., Ugaz, V.M., and Burns, M.A. (2002) *Science*, **298**, 793–793.
21 Braun, D., Goddard, N.L., and Libchaber, A. (2003) *Phys. Rev. Lett.*, **91**, 158103.
22 Wheeler, E.K., Benett, W., Stratton, P., Richards, J., et al. (2004) *Anal. Chem.*, **76**, 4011–4016.
23 Ohashi, T., Kuyama, H., Hanafusa, N., and Togawa, Y. (2007) *Biomed. Microdevices*, **9**, 695–702.
24 Beer, N.R., Hindson, B.J., Wheeler, E.K., Hall, S.B., et al. (2007) *Anal. Chem.*, **79**, 8471–8475.
25 Pipper, J., Zhang, Y., Neuzil, P., and Hsieh, T.-M. (2008) *Angew. Chem. Int. Ed.*, **47**, 3900–3904.
26 Schaerli, Y., Wootton, R.C., Robinson, T., Stein, V., et al. (2009) *Anal. Chem.*, **81**, 302–306.
27 Kim, H., Dixit, S., Green, C.J., and Faris, G.W. (2009) *Opt. Express*, **17**, 218–227.
28 Woolley, A.T., Hadley, D., Landre, P., de Mello, A.J., et al. (1996) *Anal. Chem.*, **68**, 4081–4086.
29 Waters, L.C., Jacobson, S.C., Kroutchinina, N.K., Khandurina, J., et al. (1998) *Anal. Chem.*, **70**, 5172–5176.
30 Waters, L.C., Jacobson, S.C., Kroutchinina, N.K., Khandurina, J., et al. (1998) *Anal. Chem.*, **70**, 158–162.
31 Lagally, E.T., Medintz, I., and Mathies, R.A. (2001) *Anal. Chem.*, **73**, 565–570.
32 Khandurina, J., McKnight, T.E., Jacobson, S.C., Waters, L.C., et al. (2000) *Anal. Chem.*, **72**, 2995–3000.
33 Burns, M.A., Johnson, B.N., Brahmasandra, S.N., Handique, K., et al. (1998) *Science*, **282**, 484–487.
34 Horsman, K.M., Bienvenue, J.M., Blasier, K.R., and Landers, J.P. (2007) *J. Forensic Sci.*, **52**, 784–799.
35 Anderson, R.C., Su, X., Bogdan, G.J., and Fenton, J. (2000) *Nucleic Acids Res.*, **28**, e60.
36 Liu, R.H., Yang, J., Lenigk, R., Bonanno, J., and Grodzinski, P. (2004) *Anal. Chem.*, **76**, 1824–1831.

37 Trau, D., Lee, T.M.H., Lao, A.I.K., Lenigk, R., et al. (2002) *Anal. Chem.*, **74**, 3168–3173.

38 Sadler, D.J., Changrani, R., Roberts, P., Chou, C.F., and Zenhausern, F. (2003) *IEEE Trans. Compon. Packag. Technol.*, **26**, 309–316.

39 Fu, J., Mao, P., and Han, J. (2005) *Appl. Phys. Lett.*, **87**, 263902.

40 Zeng, Y. and Harrison, D.J. (2007) *Anal. Chem.*, **79**, 2289–2295.

41 Kaji, N., Tezuka, Y., Takamura, Y., Ueda, M., et al. (2004) *Anal. Chem.*, **76**, 15–22.

42 Laachi, N., Declet, C., Matson, C., and Dorfman, K.D. (2007) *Phys. Rev. Lett.*, **98**, 098106.

43 Han, J. and Craighead, H.G. (2000) *Science*, **288**, 1026–1029.

44 Pennathur, S., Baldessari, F., Santiago, J.G., Kattah, M.G., et al. (2007) *Anal. Chem.*, **79**, 8316–8322.

45 Huang, L.R., Tegenfeld, J.O., Kraeft, J.J., Sturm, J.C., Austin, R.H., and Cox, E.C. (2002) *Nat. Biotechnol.*, **20**, 1048–1051.

46 Chou, C.F., Bakajin, O., Turner, S.W.P., and Duke, T.A.J. (1999) *Proc. Natl. Acad. Sci. U. S. A.*, **96**, 13762–13765.

47 Duke, T.A.J. and Austin, R.H. (1998) *Phys. Rev. Lett.*, **80**, 1552–1555.

48 Huang, L.R., Cox, E.C., Austin, R.H., and Sturm, J.C. (2004) *Science*, **304**, 987–990.

49 Fu, J., Schoch, R.B., Stevens, A.L., Tannenbaum, S.R., and Han, J. (2007) *Nat. Nanotechnol.*, **2**, 121.

50 Kasianowicz, J.J., Brandin, E., Branton, D., and Deamer, D.W. (1996) *Proc. Natl. Acad. Sci. U. S. A.*, **93**, 13770–13773.

51 Deamer, D.W. and Akeson, M. (2000) *Trends Biotechnol.*, **18**, 147–151.

52 Lagerqvist, J., Zwolak, M., and Di Ventra, M. (2006) *Nano Lett.*, **6**, 779–782.

53 Zhang, X.G., Krstic, P.S., Zikic, R., Wells, J.C., and Fuentes-Cabrera, M. (2006) *Biophys. J.*, **91**, L4–L6.

54 Bousse, L., Mouradian, S., Minalla, A., Yee, H., Williams, K., and Dubrow, R. (2001) *Anal. Chem.*, **73**, 1207–1212.

55 Renzi, R.F., Stamps, J., Horn, B.A., Ferko, S., et al. (2005) *Anal. Chem.*, **77**, 435.

56 Tan, W., Fan, Z.H., Qiu, C.X., Ricco, A.J., and Gibbons, I. (2002) *Electrophoresis*, **23**, 3638.

57 Cui, H.C., Horiuchi, K., Dutta, P., and Ivory, C.F. (2005) *Anal. Chem.*, **77**, 7878.

58 Li, Y., DeVoe, D.L., and Lee, C.S. (2003) *Electrophoresis*, **24**, 193.

59 Huang, T. and Pawliszyn, J. (2002) *Electrophoresis*, **22**, 3504–3510.

60 Macounova, K., Cabrera, C.R., and Yager, P. (2001) *Anal. Chem.*, **73**, 1627–1633.

61 Abad-Villar, E.M., Tanyanyiwa, J., Fernadez-Abedul, M.T., Costa-Garcia, A., and Hauser, P.C. (2004) *Anal. Chem.*, **76**, 1282–1288.

62 Shadpour, H., Hupert, M.L., Patterson, D., Liu, C., et al. (2007) *Anal. Chem.*, **79**, 870–878.

63 Kutter, J.P., Jacobson, S.C., and Ramsey, J.M. (1997) *Anal. Chem.*, **69**, 5165–5171.

64 Wu, J.T., Huang, P.Q., Li, M.X., and Lubman, D.M. (1997) *Anal. Chem.*, **69**, 2908–2913.

65 Slentz, B.E., Penner, N.A., Lugowska, E., and Regnier, F.E. (2001) *Electrophoresis*, **22**, 3736–3743.

66 Slentz, B.E., Penner, N.A., and Regnier, F.E. (2003) *J. Chromatogr. A*, **984**, 97–107.

67 Ceriotti, L., de Rooij, N.F., and Verpoorte, E. (2002) *Anal. Chem.*, **74**, 639–647.

68 Reichmuth, D.S., Shepodd, T.J., and Kirby, B.J. (2005) *Anal. Chem.*, **77**, 2997–3000.

69 Fuentes, H.V. and Woolley, A.T. (2007) *Lab Chip*, **7**, 1524–1531.

70 Herr, A.E., Molho, J.I., Drouvalakis, K.A., Mikkelsen, J.C., et al. (2003) *Anal. Chem.*, **75**, 1180–1187.

71 Chen, X.X., Wu, H.K., Mao, C.D., and Whitesides, G.M. (2002) *Anal. Chem.*, **74**, 1772–1778.

72 Li, Y., Buch, J.S., Rosenberger, F., DeVoe, D.L., and Lee, C.S. (2004) *Anal. Chem.*, **76**, 742–748.

73 Wang, Y.C., Choi, M.N., and Han, J.Y. (2004) *Anal. Chem.*, **76**, 4426–4431.

74 Rocklin, R.D., Ramsey, R.S., and Ramsey, J.M. (2000) *Anal. Chem.*, **72**, 5244–5249.
75 Gottschlich, N., Jacobson, S.C., Culbertson, C.T., and Ramsey, J.M. (2001) *Anal. Chem.*, **73**, 2669–2674.
76 Ramsey, J.D., Jacobson, S.C., Culbertson, C.T., and Ramsey, J.M. (2003) *Anal. Chem.*, **75**, 3758–3764.
77 Shadpour, H. and Soper, S.A. (2006) *Anal. Chem.*, **78**, 3519–3527.
78 Chiem, N. and Harrison, D.J. (1997) *Anal. Chem.*, **69**, 373–378.
79 Koutny, L.B., Schmalzing, D., Taylor, T.A., and Fuchs, M. (1996) *Anal. Chem.*, **68**, 18–22.
80 Bharadwaj, R., Park, C.C., Kazakova, I., Xu, H., and Paschkewitz, J.S. (2008) *Anal. Chem.*, **80**, 129–134.
81 von Heeren, F., Verpoorte, E., Manz, A., and Thormann, W. (1996) *Anal. Chem.*, **68**, 2044–2053.
82 Cheng, S.B., Skinner, C.D., Taylor, J., Attiya, S., et al. (2001) *Anal. Chem.*, **73**, 1472–1479.
83 Dishinger, J.F. and Kennedy, R.T. (2007) *Anal. Chem.*, **79**, 947–954.
84 Foote, R.S., Khandurina, J., Jacobson, S.C., and Ramsey, J.M. (2005) *Anal. Chem.*, **77**, 57–63.
85 Lee, J.H., Chung, S., Kim, S.J., and Han, J.Y. (2007) *Anal. Chem.*, **79**, 6868–6873.
86 Song, S., Singh, A.K., and Kirby, B.J. (2004) *Anal. Chem.*, **76**, 4589.
87 Hatch, A.V., Herr, A.E., Throckmorton, D.J., Brennan, J.S., and Singh, A.K. (2006) *Anal. Chem.*, **78**, 4976–4984.
88 Long, Z.C., Liu, D.Y., Ye, N.N., Qin, J.H., and Lin, B.C. (2006) *Electrophoresis*, **27**, 4927–4934.
89 Huang, H.Q., Xu, F., Dai, Z.P., and Lin, B.C. (2005) *Electrophoresis*, **26**, 2254–2260.
90 Mohamadi, M.R., Kaji, N., Tokeshi, M., and Baba, Y. (2007) *Anal. Chem.*, **79**, 3667–3672.
91 Kelly, R.T., Li, Y., and Woolley, A.T. (2006) *Anal. Chem.*, **78**, 2565–2570.
92 Liu, J.K., Sun, X.F., Farnsworth, P.B., and Lee, M.L. (2006) *Anal. Chem.*, **78**, 4654–4662.
93 Gao, J., Xu, J., Locascio, L.E., and Lee, C.S. (2001) *Anal. Chem.*, **73**, 2648–2655.
94 Wang, C., Oleschuk, R., Ouchen, F., Li, J.J., et al. (2000) *Rapid Commun. Mass Spectrom.*, **14**, 1377–1383.
95 Figeys, D. (2002) *Proteomics*, **2**, 373–382.
96 Figeys, D. and Aebersold, R. (1998) *Electrophoresis*, **19**, 885–892.
97 Pawlak, M., Schick, E., Bopp, M.A., Schneider, M.J., et al. (2002) *Proteomics*, **2**, 383–393.
98 Figeys, D. and Pinto, D. (2001) *Electrophoresis*, **22**, 208–216.
99 Wen, J., Lin, Y., Xiang, F., Matson, D.W., et al. (2000) *Electrophoresis*, **21**, 191–197.
100 Chen, S.H., Sung, W.C., Lee, G.B., Lin, Z.Y., et al. (2001) *Electrophoresis*, **22**, 3972–3977.
101 Chiou, C.H., Lee, G.B., Hsu, H.T., Chen, P.W., and Liao, P.C. (2002) *Sens. Actuators, B*, **86**, 280–286.
102 Pinto, D.M., Ning, Y.B., and Figeys, D. (2000) *Electrophoresis*, **21**, 181–190.
103 Ekström, S., Malmström, J., Wallman, L., Löfgren, M., et al. (2002) *Proteomics*, **2**, 413–421.
104 Bergkvist, J., Ekström, S., Wallman, L., Löfgren, M., et al. (2002) *Proteomics*, **2**, 422–429.
105 Roper, M.G., Shackman, J.G., Dahlgren, G.M., and Kennedy, R.T. (2003) *Anal. Chem.*, **75**, 4711–4717.
106 Gao, J., Yin, X.F., and Fang, Z.L. (2004) *Lab Chip*, **4**, 47–52.
107 Hellmich, W., Pelargus, C., Leffhalm, K., Ros, A., and Anselmetti, D. (2005) *Electrophoresis*, **26**, 3689–3696.
108 Ros, A., Hellmich, W., Regtmeier, J., Duong, T.T., and Anselmetti, D. (2006) *Electrophoresis*, **27**, 2651–2658.
109 Sun, Y. and Yin, X.F. (2006) *J. Chromatogr. A*, **1117**, 228–233.
110 Sun, Y., Lu, M., Yin, X.F., and Gong, X.G. (2006) *J. Chromatogr. A*, **1135**, 109–114.
111 Wu, H.K., Wheeler, A., and Zare, A.N. (2004) *Proc. Natl. Acad. Sci. U. S. A.*, **101**, 12809–12813.
112 Huang, B., Wu, H., Bhaya, D., Grossman, A., et al. (2007) *Science*, **315**, 81–84.
113 Marcus, J.S., Anderson, W.F., and Quake, S.R. (2006) *Anal. Chem.*, **78**, 3084–3089.

114 Ottesen, E.A., Hong, J.W., Quake, S.R., and Leadbetter, J.R. (2006) *Science*, **314**, 1464–1467.
115 Eriksson, E., Enger, J., Nordlander, B., Erjavec, N., *et al.* (2007) *Lab Chip*, **7**, 71–76.
116 Lee, P.J., Hung, P.J., Shaw, R., Jan, L., and Lee, L.P. (2005) *Appl. Phys. Lett.*, **86**, 223902.
117 Klauke, N., Smith, G., and Cooper, J.M. (2007) *Lab Chip*, **7**, 731–739.
118 Ionescu-Zanetti, C., Shaw, R.M., Seo, J.G., Jan, Y.N., *et al.* (2005) *Proc. Natl. Acad. Sci. U. S. A.*, **102**, 9112–9117.
119 Lau, A.Y., Hung, P.J., Wu, A.R., and Lee, L.P. (2006) *Lab Chip*, **6**, 1510–1515.

Part 2
Environmental Analysis

6
Molecularly Imprinted Polymer Submicron Particles Tailored for Extraction of Trace Estrogens in Water

Edward Lai, Anastasiya Dzhun, and Zack De Maleki

6.1
Introduction

In an era focused on environmental health, the water industry is faced with the challenge of ensuring a safe supply of drinking water from sustained sources of varying quality [1]. The wide variety of chemicals that are released daily into the environment has attracted great attention worldwide. Of particular concern is the potential of adverse effects on human health through consumption of drinking water. Water contaminants include toxic metals, carcinogenic organic compounds, synthetic chemicals, pharmaceuticals, illicit drugs, cosmetics, personal care products, and food supplements, together with their respective metabolites and transformation products [2]. Some of these chemicals have been found to disrupt the endocrine system of fish, wildlife, and humans [1]. They are called endocrine-disrupting chemicals (EDCs) and constitute a very large group of natural and synthetic compounds with a broad range of biological activities. The main sources of EDCs in the rivers and lakes of North America and Europe are sewage effluent and agricultural runoff.

More emphasis needs to be placed on one group of EDCs, the estrogens and androgens. Naturally produced estrogens such as 17β-estradiol (E2), estrone (E1), and their metabolites are excreted into the environment. Previous studies have shown that women, on average, excrete 2–3 μg of E2 and 7–8 μg of E1 per day [3]. In the US, livestock animals excrete a combined total of 10–30 kg of E2 and 80 kg of 17α-estradiol on a daily basis [4]. These two estrogens are differentiated from other estrogens by the presence of an aromatic ring, and differ from each other in the location, number, and nature of the oxygen functions. The E2 molecule has a 3-hydroxyl group (Figure 6.1).

The synthetic female hormone, 17α-ethynylestradiol (EE2), is used as oral contraceptive or for hormonal therapy. It is excreted un-metabolized from the human body. Many estrogens are not completely broken down during wastewater treatment and are discharged into waterways. The major estrogens in effluent water are E1, E2, and EE2 [5]. These compounds are very active at low concentrations and can affect the hormonal systems of both animals and humans. It has been

Trace Analysis with Nanomaterials. Edited by David T. Pierce and Julia Xiaojun Zhao
Copyright © 2010 WILEY-VCH Verlag GmbH & Co. KGaA, Weinheim
ISBN: 978-3-527-32350-0

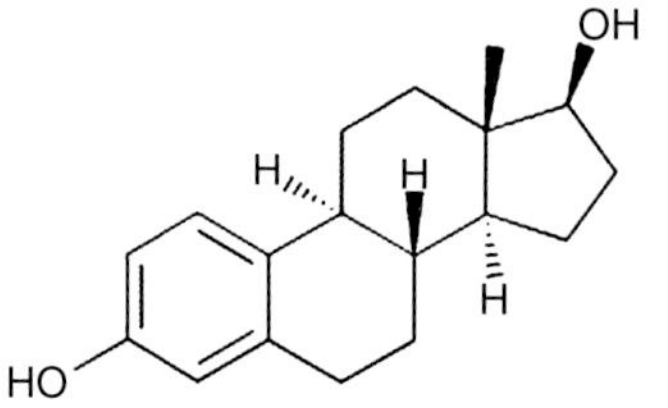

Figure 6.1 Molecular structure of 17β-estradiol (E2).

shown that estrogens interfere with the growth, development, and reproduction of aquatic organisms. Research on their effects in wildlife has demonstrated male fish feminization, such as abnormal changes to their reproductive organs and development of ovaries. In humans, estrogens lead to a decrease of sperm production and development of testicular and prostate cancers. Unfortunately, the concentration of estrogens in one river has been measured to be 100-times higher than the level that can have significant impact on endocrine systems [6].

It is likely to be a decade before human epidemiological studies can play a major role in setting guidelines for safe levels of human exposure. In the meantime, the water industry needs to identify how best to maintain a sustainable supply of safe drinking water, which requires the detection and removal of potentially harmful contaminants. Extensive monitoring of intake and discharge waters is becoming routine in water treatment plants. Trace analysis for EDCs is commonly performed by the use of chromatographic methods. High-performance liquid chromatography (HPLC) with fluorescence detection (FD) is often preferred for the quantitative analysis of estrogens. It has several advantages over other instrumental methods, including direct sample injection and no further derivatization [7]. However, the very similar physical and chemical properties of estrogens make their separation challenging and difficult. Techniques in water analysis have to be developed for preconcentration of trace estrogens to exceed the detection limits of HPLC-FD.

Molecularly imprinted polymers (MIPs) are crosslinked polymeric materials that exhibit high binding capacity and selectivity towards a target molecule, purposely present as a template during the synthesis process. Current research in the field of molecularly imprinted materials is focusing on applications for selective extraction of environmentally harmful chemicals. MIPs have been applied in solid-phase extraction (SPE), liquid chromatography (LC), capillary electrochromatography (CEC), and binding assays. MIPs have also been used for selective preconcentration and removal of the target molecules [8]. By molecular recognition, the MIP binds the target chemical from water with high affinity and specificity [9]. Recently a selective MIP has been synthesized for isoxicam preconcentration, followed by its spectrophotometric determination based on hydrogen bonding interactions between the drug and alizarin yellow GG [10]. This method was able to evaluate isoxicam in the range $1\,\text{ng}\cdot\text{ml}^{-1}$ to $20\,\mu\text{g}\cdot\text{ml}^{-1}$, with a limit of determination of $1\,\text{ng}\cdot\text{ml}^{-1}$. The retention capacity and preconcentration factor of prepared sorbent were $18.5\,\text{mg}\cdot\text{g}^{-1}$ and 200, respectively; the prepared MIPs could be reused at least five times. The MIP capability for isoxicam selection and extraction from the solution was higher than non-imprinted polymer (NIP). Under optimum conditions, this procedure can be successfully applied to assay trace amounts of isoxicam (with anti-inflammatory properties) in pharmaceutical and biological samples.

MIPs are considered one of the most promising separation methods for removal phenolic compounds in wastewater treatment. In a recent work, a MIP was prepared by bulk polymerization in acetonitrile using 2,4-dinitrophenol as template, acrylamide as functional monomer, ethylene glycol dimethacrylate as crosslinker, and benzoyl peroxide as initiator [11]. An adsorption process for removal of nitrophenol using the fabricated MIP was evaluated under various pH and time conditions. The maximum adsorption of nitrophenol by the fabricated MIP was found to be $3.50\,mg\cdot g^{-1}$ whereas the adsorption of 2,4-dinitrophenol was found effective at pH 6.0. A kinetic study showed that nitrophenol adsorption followed a second-order adsorption rate and the adsorption isotherm data were explained well by the Langmuir model.

As E2 is a primary contaminant and the most active estrogen in wastewater, an E2 selective MIP has been synthesized by a thermo-polymerization method using methacrylic acid (MAA) as functional monomer, ethylene glycol dimethacrylate (EGDMA) as crosslinker, acetonitrile as porogenic solvent, and E2 as template [12]. The MIP showed obvious affinity for E2 in acetonitrile solution, which was confirmed by adsorption experiments. After optimizing the conditions for molecularly imprinted solid-phase extraction (MISPE), two structurally related estrogenic compounds (estriol and diethylstilbestrol) were used to evaluate the selectivity of the MIP cartridges. The MIP cartridges exhibited high selectivity for E2, as the recoveries were 85 ± 7% for MIPs and 19 ± 2% for non-imprinted polymer (NIP) cartridges. The detection and quantification limits correspond to 0.023 and $0.076\,\mu g\cdot ml^{-1}$, respectively. Furthermore, the MISPE methods were used to selectively extract E2 from fish and prawn tissue prior to HPLC analysis. This MISPE-HPLC procedure could eliminate all matrix interference simultaneously and had good recoveries of 81 ± 3%.

One goal of our own research over the last two years has been to evaluate the binding properties of MIP submicron particles that were imprinted specifically for E2. Binding of this target compound was monitored against time for comparison with a NIP control and a different MIP imprinted for EE2. The large surface area-to-volume ratio of submicron particles guarantees high binding capacity and recovery. Thus, MIP submicron particles can be applied as a final treatment for effluent water after solids are removed from raw wastewater. After preconcentration of trace E2 in water, desorption of E2 from the MIP submicron particles was studied to assess the feasibility of rapid regeneration for reuse. Of course, the trace amount of desorbed E2 can be eluted by a small volume of solvent, possibly containing a strong organic base or acid modifier, for HPLC-FD or LC-MS analysis. Under optimal conditions, recoveries can be 100 ± 3%, precision can be 4 ± 2%, and quantification can be achieved by matrix-matched calibration, as reported for a dispersive solid-phase microextraction method in the analysis of four tetracyclines in water samples by high-performance liquid chromatography [13].

6.2
Principle of Molecular Recognition by Imprinting

Living organisms interact at the molecular level through recognition of the chemical, physical, and biological information transfer [14]. The noncovalent chemical interactions are based on the formation of weak forces between molecules, such as hydrogen bonding, ion pairing (via electrostatic interactions), hydrophobic attractions, van der Waals forces, π–π stacking interactions, and shape complementarity that minimizes cross reactivity [15]. When these interactions occur at the same time, complexes with high stability are formed. These concepts are useful for the synthesis of three-dimensional polymeric structures capable of recognizing template molecules with high selectivity [16]. Template-specific polymers can be applied in analytical chemistry for preconcentration/removal of target analytes [17]. Scatchard plot analysis typically reveals that the template–polymer system shows two-site binding behavior with dissociation constants of 0.5 ± 0.2 and $10 \pm 6\,\mu\text{mol}\cdot\text{l}^{-1}$ [18]. Steric factors, electrostatic factors, large surface area, and multiple interaction sites all influence the molecular recognition of analyte molecules in solution by imprinted cavities [19]. For example, a new MIP has been prepared with styrene and MAA as functional co-monomers [20]. Strong π–π interactions occurring between phenyl groups of styrene and tamoxifen promote rebinding of the analyte by the specific sites. The enhanced hydrophobic character of the imprinted polymer enables the direct percolation of urine through MIP-SPE and the easy elimination of endogenous salts from urine with only one aqueous washing step. HPLC-UV analysis has confirmed high extraction recoveries (85%) for tamoxifen and its metabolite with an enrichment factor of 8.

There are two approaches to molecular imprinting. The covalent approach, developed by Gunter Wulff and his coworkers, uses covalent bonding to assemble template-monomer complexes in solution prior to polymerization [21]. Molecular recognition is dependent on the reversible formation and cleavage of these bonds. Another approach, developed by Klaus Mosbach and his coworkers, uses functional monomers that are allowed to prearrange around the template through noncovalent interactions [22]. For noncovalent imprinting, a complex is formed through interactions of the template with monomers that have complementary functional groups (Figure 6.2). Crosslinkers are added to fix the arrangement of functional groups during polymerization. Afterwards the template molecules are removed, leaving cavities that resemble the steric and chemical properties of the template. These imprinted cavities can rebind molecules that are structurally identical or analogous to the template. Increasing the number of noncovalent interactions can produce exceptionally high affinity for analyte. However, it is difficult to control weak interactions between the template and monomers during polymerization. This can lead to an increased heterogeneity among the binding cavities. A combination of the two approaches is semi-covalent, in which covalent bonds are formed during the imprinting process and recognition is through noncovalent interactions between the analyte and the polymer [23]. Semi-covalent and covalent methods do make good recognition sites, but are limited in the choice of

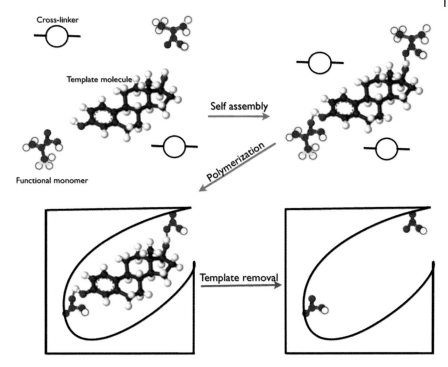

Figure 6.2 Preparation of MIP for SPE of 17β-estradiol.

templates. Generally, it is more difficult to make recognition sites for bigger templates. The bigger the template molecule, the harder it is to extract all template molecules from the recognition sites, yielding a low number of vacant sites for the rebinding process.

6.2.1
Monomers, Crosslinkers, and Porogen Solvents

Functional monomers are chosen to complement the chemical functionality of a template. Basic groups on the template are usually allowed to interact with acidic groups on the monomers, or vice versa. The most common monomer used is methacrylic acid (MAA) because it contains a carboxyl group that can be a hydrogen bond acceptor and donor. MAA transfers its proton to basic functional groups on the template, which creates electrostatic interactions [24]. Similarly, poly(acrylic acid) (PAA) can be obtained by the radical polymerization of AA. The selectivities of MIP and NIP particles were recently evaluated by Feás et al. in binding experiments of the four synthesized polymeric materials (MIP_{MAA}, MIP_{AA}, NIP_{MAA}, and NIP_{AA}) [25]. The effects of monomers on the surface morphology, binding capacity, and swelling properties of imprinted and non-imprinted polymers were studied. Morphology of the polymeric materials was assessed with scanning electron microscopy (SEM). These studies revealed differences in monomer and polymer functions when polymerization occurred in the presence of template. Non-specific retention of the template to NIP_s was higher for NIP_s-PAA than for NIP_s-PMAA. In terms of specific binding ($\Delta Q = Q_{MIP} - Q_{NIP}$), MIP_{MAA} showed a greater value (53 ± 1%) than MIP_{AA} (50 ± 1%).

The noncovalent imprinting process is commonly performed by thermal decomposition of, at 60 °C, 2,2′-azobisisobutyronitrile (AIBN) as the initiator in dichloromethane. Ethylene glycol dimethacrylate (EGDMA) is typically used as the crosslinker although trimethylolpropane triacrylate (TRIM) is another common crosslinker [26]. Crosslinker monomers join functional monomers to create binding cavities and build a three-dimensional polymer network. Thus, the crosslinker has to be in high proportion. Polymerization typically requires 60–70% crosslinking of monomers because the morphology, rigidity, selectivity, and binding affinity of MIP strongly depend on it. The porogen, the solvent used for polymerization, has to dissolve all the template, functional monomers, and crosslinkers to achieve good interactions. It fills out the space in order to increase the porosity of the three-dimensional polymer structure and should be non-polar so as not to interfere with ionic or van der Waals forces. Typical solvents used are acetonitrile, chloroform, and toluene [27].

MIPs can be used as selective sorbents for the SPE of target analytes from sample matrices. MIPs are often called synthetic antibodies in comparison with immuno-based sorbents, and they offer several advantages over antibodies, including rapid preparation and high chemical stability. A review by Pichon in 2007 described the use of MIPs in SPE with emphasis on their synthesis, the various parameters affecting the selectivity of MISPE, their potential to selectively extract analytes from complex aqueous samples or organic extracts, their on-line coupling with LC, and their potential in miniaturized devices [28]. The use of MIPs packed in HPLC columns and the preparation of imprinted fibers for solid-phase microextraction (SPME) have been discussed in another review by Martin-Esteban and coworkers [29]. Thus, this chapter does not pretend to be a collection of MISPE-related papers but gives an overview of the significant attempts carried out during the last two years to improve the performance of MIPs in solid-phase extraction using submicron particles.

6.2.2
Rebinding of Target Analytes

Newly synthesized MIP submicron particles need to be evaluated for their binding characteristics, particularly their affinity and efficiency. Farrington and Reagan have investigated the physical characteristics of MIP submicron particles, including the particle size distribution on exposure of the MIP to different solvents [30]. This was related to the ability of the MIP to rebind ibuprofen under the same conditions, which further enhanced their understanding of the nature of MIP recognition. In addition to particle size analysis by dynamic light scattering, the MIP submicron particles can be assessed by batch binding with different analyte concentrations until equilibrium is reached. After equilibration, the remaining analyte in solution can be measured by HPLC with fluorescence detection (FD) or liquid chromatography–mass spectrometry (LC-MS). By difference, the amount of analyte bound to the MIP can be determined. This indirect measurement of binding percentage gives information on the binding affinity of the new MIP, especially in a Scatchard plot. To assess the new MIP for efficient imprinting by the known amount of template, its binding percentage is compared to that measured for the non-imprinted polymer (NIP) as a control. The difference is correlated with the number of selective binding cavities in the MIP [31]. In addition, a second imprinted polymer (MIP2) is prepared using a template that is structurally similar to the first template. The reason for making MIP2, other than the NIP, is that the NIP surface morphology, steric properties, and physical characteristics can be significantly different from those of the MIP. MIP2 helps to confirm the specific imprinting effect of the MIP.

6.2.3
Computational Modeling

MIPs binding with phenoxyacetic acid (PA) as a dummy template molecule have been synthesized via thermal initiation in aqueous medium [32]. The retention behaviors of benzoic acid, PA, 2-methyl-4-chlorophenoxyacetic acid, 4-chlorophenoxyacetic acid, and 2,4-dichlorophenoxyacetic acid (2,4-D) on this MIP column indicated that the MIP could selectively retain phenoxyacetic herbicides. To investigate these recognition mechanisms, the interactions between the functional monomer 4-vinylpyridine (4-VP) and PA or 2,4-D were investigated by computational modeling. ^{1}H NMR spectroscopy of 2,4-D titrated by 4-VP was also recorded. The chemical shift of the 2,4-D acidic proton (12.15–14.32 ppm) showed the existence of the ion-pair interaction. This kind of dummy MIP could be useful as a solid phase to extract 2,4-D, 4-CPA or MCPA and avoid leakage of a trace amount of target analyte remaining in the MIP column.

A MIP for the recognition of the methyl-carbamate pesticide, carbaryl, in water has been synthesized using a molecular modeling approach to select the monomers [33]. The functional monomers with the highest binding energy to the template were used for the synthesis of MIPs. A flow-injection system for the selective and sensitive monitoring of carbaryl in water has been constructed, based on its native fluorescence for emission detection. This flow-injection system demonstrated high specificity for carbaryl against other methyl-carbamate pesticides (e.g., bendiocarb and carbofuran). A detection limit for the target molecule of $0.27 \mu g \cdot l^{-1}$ (3-ml sample injections) was found, with throughput of seven analyses per hour. The polymer was easily regenerated for subsequent sample injections (up to 200 cycles tested) and showed good stability for at least 4 months after preparation.

MIPs using dimethoate (a widely used organophosphate insecticide) as the template molecule have been prepared for selective recognition and enrichment of dimethoate from tea leaves [34]. Six functional monomers were examined, which showed that the MIP prepared by methyl methacrylate (MMA) had the largest imprinting factor of 7.9 for dimethoate. Molecular dynamics (MDs) simulations were carried out for the six different molecular systems to predict the interaction energies, the closest approach distances, and the active site groups. The dynamic adsorption of dimethoate on the MMA-based-MIP was in accord with the Langmuir isotherm. This MIP had a marked selectivity for dimethoate compared to structurally related organophosphorus pesticides. MISPE attained a large selective enrichment factor of 100 for dimethoate from tea leaves. It was a useful tool in the trace detection of environmental contaminants.

The selective adsorption properties of dimethoate MIPs were further studied by computational and conformational methods at the molecular level through a MD simulation [35]. The MD simulation was confirmed through chromatographic evaluation of dimethoate on the MIP and NIP columns. Both MD modeling and chromatographic evaluations showed that the MIP based on the butyl methacrylate (BMA) functional monomer had the best selective recognition for dimethoate compared to other functional monomers, including methyl methacrylate (MMA) and ethyl methacrylate (EMA). Two parameters, the imprinting factor indicator and the competitive factor indicator, provided an insight into the imprinting selectivity of the MIPs for dimethoate versus other structurally related organophosphorus pesticides. The influence of the rebinding solvents on the adsorption properties of MIPs was also investigated. It was found that a good rebinding solvent should have less affinity with both template and polymer and a good solubility; a large self-association among solvent molecules could enhance the affinity between template and MIP. The results of MD simulation were in good agreement with those of chromatographic evaluation, and indicated that MD simulation could be an effective tool for the design of new MIPs.

Molecular recognition can be accomplished by imprinting its structure on the surface or in the bulk material. A surface molecular imprinting technique was recently combined with a simple sol–gel process to synthesize a new imprinted amino-functionalized silica gel material for solid-phase extraction–high performance liquid chromatography (SPE–HPLC) determination of diethylstilbestrol (DES) [36]. Moreover, Li et al. have used molecular dynamics simulations and computational screening to identify functional monomers capable of interacting with sulfadimidine (SM_2), which is an inexpensive veterinary drug [37]. A library of 15 kinds of common functional monomers for preparing MIPs was built and their interactions with SM_2 in acetonitrile were calculated using the molecular dynamics software GROMACS 3.3. According to the theoretical calculation results, surface molecularly imprinted silica (MIP-silica) with SM_2 as template was prepared by surface-imprinting technique using MAA as functional monomer and divinylbenzene as crosslinker in acetonitrile. The surface composition of the MIP-silica was determined by Fourier-transform infrared (FT-IR) spectroscopy and energy-dispersive X-ray spectrometer (EDS). Scanning electron microscopy (SEM) was used to characterize the morphological properties of the MIP-silica. The synthesized MIP-silica was then tested by an equilibrium-adsorption method, and the MIP-silica demonstrated high binding specificity to the SM_2. The molecular recognition of SM_2 was analyzed in detail by using molecular modeling software Gaussian 03. Furthermore, Koohpaei et al. have used central composite design (CCD) to increase the precision and accuracy of synthesis and optimization of MIP for ametryn and other similar analogues [27].

A rational design approach was taken by Farrington et al. to the planning and synthesis of a MIP capable of extracting caffeine from a standard solution and from food samples containing caffeine [38]. Data from NMR titration experiments, in conjunction with molecular modeling, were used to predict the relative ratios of template to functional monomer and furthermore to determine both the choice and amount of porogen used for the MIP preparation. In addition, the molecular modeling program yielded information regarding the thermodynamic stability of the pre-polymerization complex. Post-polymerization analysis of the MIP by BET yielded significant information regarding the nature of the size and distribution of the pores within the polymer matrix. Both the physical characteristics of MIP and analysis of pre-polymerization complex could yield vital information to predict how well a given MIP would perform.

6.3
Analytical Application of MIPs for Biopharmaceuticals and Toxins

The analysis of alkylphosphonic acids, degradation products of V and G nerve agents (such as VX, sarin, or soman), has been an important task for compliance to the Chemical Weapons Convention. Detection of these contaminants at low concentration levels was difficult with complex matrices due to the amount of interfering substances. MISPE has allowed selective extraction of these compounds from complex samples, thus making their detection easier [39]. MIPs of cholesterol have been prepared by UV-initiated polymerization and used as SPE sorbents for direct extraction of cholesterol from different biological samples (human serum, cow milk, yolk, shrimp, pork and beef) [40]. Over the concentration range 10–80 $\mu g\,ml^{-1}$, recoveries ranged from 80 to 93%. Compared with conventional C18 SPE, almost all of the matrix interferences were removed after MISPE and higher selectivity was achieved.

MIPs have also been applied for therapeutic drug monitoring, for instance, of mycophenolic acid in patient plasma [41]. The determination of trace biopharmaceutical residues in food and contaminants in feedstuffs has been a growing concern over the past few years [42]. Residual antibacterials in food constitute a risk to human health, especially because they can contribute to the transmission of antibiotic-resistant pathogenic bacteria through the food chain. Therefore, to ensure food safety EU and USA regulatory agencies have established lists of forbidden or banned substances and tolerance levels for authorized veterinary drugs (e.g., antibacterials). During the past decade, the use of powerful mass spectrometric detectors in combination with innovative chromatographic technologies has solved many problems related to sensitivity and selectivity of LC-MS analysis. However, in determining polar pollutants such as antibiotics and biopharmaceuticals for environmental investigation, the analyst has often faced two major challenges – poor detectability of analytes and highly variable matrix interferences – that compromise quantification [43]. Sample preparation remains the bottleneck step in these methods regarding analysis time and sources of error.

Owing to the complexity of new pharmaceutical compounds, current research in medical, food, and environmental analyses has focused on the development of efficient methods for sample preconcentration and clean-up [44]. Traces of pharmaceuticals are continuously introduced into the aquatic environment mainly by sewage treatment plant effluents [45]. Contemporary analytical methods have the sensitivity required for contaminant detection and quantification, but direct application of these methods on food samples can rarely be performed. In fact, the matrix introduces severe disturbances and analysis can only be performed after some clean-up and preconcentration steps. It is important to develop SPE sorbents for simplification of HPLC analysis, for instance, to attain high selectivity and sensitivity in the detection of polar environmental pollutants. Conventional SPE has limitations in the extraction of polar pharmaceuticals selectively from complex mixtures [46]. Recent years have seen a significant increase of the MISPE technique in food contaminant analysis. In fact, this technique seems to be particularly suitable for extraction applications where analyte selectivity in the presence of a very complex and structured matrix is the main problem [47]. MIP has been coupled with electrochemical fluorimetry detection for the efficient determination of methotrexate in serum and urine [48]. Methotrexate (generally used to treat cancers of the breast, skin, head, neck, or lung) was preconcentrated in the MISPE micro-column, and then eluted. The eluate was detected by fluorescence spectrophotometer after electrochemical oxidation. Under the selected experimental conditions, the detection limit was $0.8\,\text{ng}\cdot\text{ml}^{-1}$. The selectivity and sensitivity of fluorimetry was greatly improved by the MISPE method.

The development of MIP submicron particles for SPE would eliminate the potential loss of trace compounds during preparation of unknown samples and would provide rapid preconcentration of trace analytes in the samples. MIPs have high resistance to degradation processes that can be caused by mechanical stress, high temperature, high pressure, acid, base, metal ions, and organic solvents. In addition, MIP can be stored for long periods and can be used repeatedly without loss of the imprinting effect. These properties are desired in the field of analytical chemistry, as well as for environmental applications. There are several additional advantages with MIP sorbents when used for extraction in water, such as simplified sample preparation, no clean up of humic and fulvic acids, no acidic hydrolysis of analytes, and no protonation of basic analytes [49]. Widstrand and coworkers were able to imprint the SPE sorbent for different pharmaceutical compounds [50]. MIP material was packed into an SPE micro-column, with 80% recoveries to achieve a detection limit of $1\,\text{ng}\cdot\text{l}^{-1}$ in water. Each MIP exhibited high selectivity toward the target analyte in raw wastewater samples. Sample matrix components from the MIP micro-column can be eluted with a proper choice of organic solvents. Thus, high-affinity binding to the MIP is an important factor that can influence the % recovery of target analyte.

The use of custom-made MIPs with high selectivities for target molecules in SPE is becoming an increasingly important sample preparation technique in biopharmaceutical analysis. However, the potential risk of leakage of the template molecules during desorption has limited many applications. The use of a mimicking template, called a dummy molecular imprinting polymer (DMIP), that bears the structure of a related molecule and acts as a putative template molecule may provide a useful solution to this problem. MIPs with selective recognition properties for zearalenone (ZON), an estrogenic mycotoxin, have been prepared by Uracca et al. using cyclododecyl 2,4-dihydroxybenzoate (CDHB) that exhibits resemblance to ZON in terms of size, shape, and functionality as template instead of the natural toxin [51]. They could be excellent for clean-up and preconcentration of the mycotoxin in contaminated food samples. Kubo et al. have prepared MIPs for an amnesic shellfish poison, domoic acid [52]. They tested several commercial aromatic dicarboxylic compounds (isomers of phthalic acid) for use as dummy templates. The highest selective recognition ability was found when o-phthalic acid was used. The ability was due to the similarity in shape around the carboxylic acids of domoic acid and o-phthalic acid. The effective chromatographic separation of domoic acid in the extract from blue mussels was achieved with a LC column packed with the DMIP.

A confirmatory method was described for the determination of the illegal antibiotic chloramphenicol using a specifically developed MIP as the sample clean-up technique [53]. The MIP was produced using an analog to chloramphenicol as the template molecule. Using an analog of the analyte as the template avoided the traditional drawback associated with MIPs of residual template leeching or bleeding. The MIP described was used for the solid-phase extraction of chloramphenicol from various sample matrices, including honey, urine, milk, and plasma. A full analytical method with quantification by LC-MS/MS was fully validated, according to the European Union (EU) criteria for the analysis of veterinary drug residues. Zhang et al. have reported that 2-methylphenoxyacetic acid (2-MPA, which is similar in shape, size, and functionality to phenoxyacetic herbicides) was suitable to be used as a dummy template to prepare MIPs for retaining phenoxyacetic herbicides [54]. They employed computational molecular modeling to study the ion-pair interactions between template molecules and functional monomer 4-vinylpiridine (4-VP). The data indicated that the cross-selectivities of MIPs for phenoxyacetic acid herbicides depended on the binding energies of complexes. An antihistaminic/anticholinergic compound, cyproheptadine (CPH), and azatadine (AZA) were used recently as templates in the development of MIPs and DMIPs [55]. The results indicated that the DMIPs exhibited equal recognition of CPH, thereby avoiding the problem of leakage of the original template during desorption, relative to MIPs synthesized in the presence of CPH as the print molecule. Examination of the surface structures of the DMIPs and MIPs by SEM showed appreciable differences in structural morphology, even though molecular modeling of CPH and AZA had suggested that both substrates were similar in shape and volume. These results were well supplemented by data obtained for swelling ratios and solvent uptake.

6.4
Preparation of MIP Submicron Particles

A novel MIP material has been prepared using quercetin as the template molecule, methacrylic acid as a functional monomer, and macroporous chitosan beads as a functional matrix [56]. The MIP exhibited higher selectivity for quercetin in a mixture of flavonoids than the NIP control. Langmuir and Freundlich adsorption models were successfully applied to describe the equilibrium isotherms. Direct injection, enrichment and HPLC analysis of ultra-trace bisphenol A (BPA) in water samples using molecularly imprinted polymeric microspheres (MIPMs) in a column has been developed by Jiang et al. [57]. Under optimal conditions, the MIPM column could be used to simultaneously extract, enrich, separate, and determine ultra-trace BPA in one analysis. The enrichment factor for BPA was 10 000 when 40 ml of water sample was directly injected and analyzed. The limit of quantification was $0.1\,nmol\cdot l^{-1}$ and the recoveries were 99 ± 3%.

Preparation of molecularly imprinted polymer microspheres (MIPMs) for chloramphenicol (CAP) by aqueous suspension polymerization was first reported by Shi et al. [58]. The resulting MIPMs had the ability to specifically adsorb CAP, and MISPE based on these MIPMs was shown to be applicable for clean-up and preconcentration of trace CAP in milk and shrimp samples with high recoveries of 92 ± 1% and 85 ± 1%, respectively. Combined with MISPE, the conventional HPLC-UV analysis sensitivity for CAP in foods was significantly increased. Uniform-sized molecularly imprinted polymer (MIP) beads for metsulfuron-methyl (MSM) were firstly prepared by a one-step swelling and polymerization method using 4-vinylpyridine (4-VPY) and ethylene glycol dimethacrylate (EDMA) as functional monomer and crosslinker, respectively [59]. The chromatographic behavior of MSM and structurally related sulfonylureas (SUs) on the resultant MIP column were evaluated. The imprinted polymer revealed specific affinity to the template and fair resolution of SUs was also obtained. Furthermore, the MSM-MIP was used as the solid-phase extraction (SPE) material to enrich MSM in real water samples before reversed-phase HPLC analysis. The recovery of MSM from 100 ml of drinking water at a $50\,ng\cdot l^{-1}$ spike level was 99 ± 1% with a RSD of 1.1%. The detection limit was about $6.0\,ng\cdot l^{-1}$ of MSM when enriching a 100 ml water sample. A simple, sensitive, and specific method has been developed by Wang and Zhang for high-throughput detection of dipyridamole [60]. The method is based on chemiluminescence (CL) imaging assay combined with MIP recognition. MIP microspheres were prepared using precipitation polymerization with MAA as functional monomer, TRIM as the crosslinker, and dipyridamole as the template. The microspheres were coated in 96-microtiter well plates with poly(vinyl alcohol) (PVA) as glue. The amount of polymer-bound dipyridamole was determined based on the dipyridamole peroxyoxalate chemiluminescence reaction. The emitted light was measured with a high-resolution charge-coupled device (CCD). The proposed method exhibited high selectivity and sensitivity to dipyridamole. Under optimum conditions, the relative CL imaging intensity was proportional to the concentration of dipyridamole, ranging from 0.02 to $10\,\mu g\cdot ml^{-1}$. The detection limit was

0.006 µg·ml^{-1}. The method could perform 96 independent measurements simultaneously in 30 min. These results show that the MIP-based CL imaging can become a useful analytical technology for quick detection of dipyridamole in real-world samples.

Uniformly-sized MIPs for atrazine, ametryn, and irgarol have been prepared by Sambe et al. using a multistep swelling and polymerization method [61]. The MIP for atrazine prepared using MAA showed good molecular recognition abilities for chlorotriazine herbicides, while the MIPs for ametryn and irgarol prepared using 2-(trifluoromethyl) acrylic acid (TFMAA) showed excellent molecular recognition abilities for methylthiotriazine herbicides. A restricted access media-molecularly imprinted polymer (RAM-MIP) for irgarol was prepared followed by *in situ* hydrophilic surface modification using glycerol dimethacrylate and glycerol monomethacrylate as hydrophilic monomers. The RAM-MIP was applied to selective enrichment of methylthiotriazine herbicides (simetryn, ametryn, and prometryn) in river water, followed by their separation and UV detection via column-switching HPLC. With a 100-ml loading of river water sample, the detection limits were 25 pg·ml^{-1}. The recoveries of simetryn, ametryn, and prometryn at 50 pg·ml^{-1} were 101%, 95.6%, and 95.1%, respectively. Haginaka has recently reviewed all the preparation methods for spherical and monodispersed MIPs in micrometer sizes [62]. Those methods include suspension polymerization in water, liquid perfluorocarbon, and mineral oil, seed polymerization and dispersion/precipitation polymerization. The other methods are the use of beaded materials such as a spherical silica or organic polymer for grafting MIP phases onto the surfaces of porous materials or filling the pores of silica with MIPs followed by dissolution of the silica. Furthermore, applications of MIP microspheres as affinity-based chromatography media, HPLC stationary phases, and solid-phase extraction media will be useful for pharmaceutical, biomedical, and environmental analysis.

MIP submicron particles have been synthesized in the authors' laboratory for selective SPE of E2. MAA and ethylene glycol dimethacrylate (EGDMA) or TRIM were added to a solution of E2 in acetone/acetonitrile (1:3 v/v) inside a screw-cap vial. AIBN (2 wt% of MAA and EGDMA or TRIM) was added to the mixture. After sonication for 1 min and deoxygenation with nitrogen for 5 min, the vial was sealed and placed in a thermostated water bath at 60 °C to polymerize over 24 h. More details on the preparation of submicron particles are available from a previous report [63]. The proper functional monomer (MAA) was selected and its concentration was optimized to obtain the best MIP particles. This ensured that the imprinting effect would be at its maximum strength, by forming stable template-monomer complexes at the pre-polymerization stage, and it produced high-affinity imprinted cavities throughout the MIP matrix. The newly synthesized MIP submicron particles can extract E2 from water samples. This ability to recognize E2 in aqueous solution is an important property. The choice of porogen solvent as acetone–acetonitrile (1:3 v/v) was to improve the noncovalent imprinting effect and to yield particles in a submicron size range.

The newly synthesized MIP submicron particles were evenly dispersed throughout the porogen solvents to appear as a colloid. The polymerized colloid turned milky white in color, yielding submicron particles in abundance [64]. After drying, electrostatic properties of the MIP submicron particles were noticeable, resulting in some mass loss via dust clouds when the vial was opened for washing with 1% triethylamine (TEA) in methanol several times until the E2 template was no longer detected in the supernatant. NIP submicron particles, and MIP2 submicron particles imprinted with 17 α-ethynylestradiol (EE2), have been prepared similarly for use as a control and a reference, respectively.

6.5
Binding Properties of MIP Submicron Particles with E2

The MIP submicron particles (prepared using TRIM as the crosslinker) have been tested for their rebinding efficiency with E2. At the end of incubation, the concentration of remaining E2 in the supernatant is determined by HPLC-FD. This is subtracted from the original concentration of E2 to calculate the amount of E2 bound with the MIP submicron particles. The results have shown that 20 mg of particles absorb $99 \pm 2\%$ of the target molecule E2, which is practically quantitative and complete. Not all of the imprinted cavities (based on the number of moles of E2 used as template in the preparation of MIP submicron particles) are needed for binding the analyte molecules during incubation. Recently, Le Noir and his coworkers imprinted MIP particles with E2 and found that the % recovery of E2 was only 77% [65]. The morphology of MIP submicron particles was examined by scanning electron microscopy (SEM). The average particle size was estimated to be 300 ± 30 nm (Figure 6.3).

Although these MIP submicron particles are not perfect in their spherical shape (in comparison with those obtained using EGDMA as the crosslinker), their binding characteristics are impressive. Interestingly, the NIP submicron particles also bound E2 in aqueous solution at a significant efficiency of $93 \pm 2\%$ due to

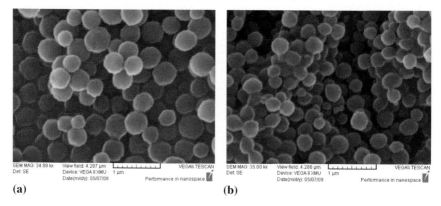

Figure 6.3 Scanning electron microscopy images of (a) molecularly imprinted polymer (MIP) submicron particles and (b) non-imprinted polymer (NIP) submicron particles.

non-specific binding. The difference in MIP and NIP binding results proved that the MIP particles have only a small number of specific cavities to bind 3% more of E2 strongly (as well as low-affinity sites for non-specific binding of E2). An important observation is that the mixing of MIP submicron particles with E2 solutions is instant, resulting in a homogeneous dispersion. When NIP submicron particles are mixed with E2 solutions, sonication for 5–10 s is required to obtain a homogeneous dispersion. The accessibility of binding sites depends on how porous the MIP and NIP submicron particles are. It is possible that the specific surface area of MIP particles is larger than that of NIP particles.

Polymeric particles were recently imprinted with E2 using methacrylic acid as functional monomer and divinylbenzene as crosslinker [66]. Binding studies of these MIP particles (5 ± 1 μm) showed a recovery of 88% for E2 in deionized water and 81% in surface water. The corresponding NIP particles attained 78% recovery. Wei and Mizaikoff found that strong binding sites were mostly positioned at the surface of the MIP particles. As non-imprinted polymers had smaller surface areas, they would have smaller pore volumes [67].

6.5.1
Models of E2 Binding with MIP Submicron Particles

The characterization of MIP binding has been mostly phenomenological, like Langmuir–Freundlich isotherms [68], binding site models, chromatographic k and α values, etc., as they relate to different applications [69]. Determination of the number and strength of binding sites on MIP particles requires an understanding of different binding models. There are two common binding models: homogeneous and heterogeneous. The heterogeneous binding model suggests the possibility of having high-affinity and low-affinity sites. Previous studies on MIP binding sites have proven that MIPs typically contain a higher number of non-selective binding sites in comparison to the needed high-affinity sites [70]. The heterogeneity occurs because of noncovalent interactions, which allow only a fraction of the functional monomers to be arranged in a complementary geometry around the template molecules, while the rest are randomly located throughout the polymer matrix. At low concentrations of E2, analyte molecules are attracted by the low-affinity sites that are found on the MIP particle surface and resemble the non-specific binding sites of NIP [71]. At high concentrations of E2, the high-affinity sites are activated, possibly beyond the diffusion energy barrier, to bind the target analytes. Apparently, the increased amount of E2 in the sample solution can increase the chance of E2 molecules contacting these binding sites inside MIP particles.

Previous studies by Yu *et al.* in our laboratory indicated that a steric exclusion effect, for larger molecules than the target analyte, would reduce the binding frequencies of the analyte with the MIP [7]. The exposure of MIP to two compounds in one mixture, E2 and EE2, showed that MIP was selective towards E2 but the recovery of targeted E2 experienced a decrease in the presence of EE2 (which is a larger molecule than E2) when compared to binding of E2 alone in the sample solution. Moreover, molecules that are smaller than the target analyte have decreased selectivity even though their size did not block the entrance of the analyte to the binding sites. This test provided evidence for shape selectivity of MIPs and antagonistic versus agonistic properties of other chemicals on the binding sites.

The selectivity of heterogeneous MIP binding sites can be enhanced by site-selective chemical modification. Umpleby et al. were able to selectively eliminate low-affinity binding sites by esterification with diazomethane, which selectively inactivates hydrogen bonding on carboxylic acid groups [70]. Thus the final MIP contained only high-affinity binding sites, and the binding specificity of MIP would be optimized. A promising result for enhancing the selectivity of MIP has been reported by McNiven et al. [72]. They treated the polymer with methyl iodide in the presence of the template molecules. Consequently, the non-specific binding sites were blocked with methyl iodide while the high-affinity sites were protected by the template. This afforded strong binding of, and specificity to, the target analyte.

6.5.2
Kinetics of MIP Binding with E2

Binding of E2 molecules with MIP particles would require time for diffusion into the macroporous polymer matrix of imprinted cavities. The authors' own results showed that MIP can bind up to $64 \pm 3\%$ E2 at the optimal reaction time of 3 min. Previously, Jiang et al. studied uptake kinetics of NIP using a simple imprinted amino-functionalized silica gel [36]. The results showed that the maximum binding of $47 \pm 3\%$ was achieved after the reaction time is above or equal to 60 min.

Ultrasonication has an effect on the binding efficiency for it can speed up the binding of E2 with MIP submicron particles. The authors' own results showed that the E2 binding efficiency was increased to $98 \pm 2\%$ when 2-min of ultrasonication was applied. Apparently, ultrasonication is an effective way to agitate the water sample, thereby dispersing the submicron particles evenly throughout the entire volume of water. Ultrasonication helped E2 molecules penetrate the macroporous polymer matrix, to enter the binding cavities that were otherwise not as easily accessible.

6.6
Trace Analysis of E2 in Wastewater Treatment

A new MIP for trace analysis of diclofenac in environmental water samples was developed [73]. Diclofenac is a non-steroidal anti-inflammatory drug (NSAID) which belong to the most frequently detected pharmaceuticals in the water-cycle in Europe. The MIP was synthesized using 2-vinylpyridine (2-VP) as a functional monomer and EGDMA as a crosslinker in a bulk thermal polymerization method. Scatchard plot analysis revealed that two classes of binding sites were formed with dissociation constants of $55 \pm 1\,\mu\text{mol} \cdot l^{-1}$ and $1.4 \pm 0.1\,\text{mmol} \cdot l^{-1}$, respectively. This resulted in an MISPE-LC/DAD method allowing the direct extraction of the analyte with a selective wash using dichloromethane–acetonitrile (94:6, v/v) followed by elution with dichloromethane/methanol (85:15, v/v). The recovery of diclofenac was 96%, with good precision (RSD = 3.3%, n = 3). The MISPE method was applicable to raw influent and final effluent wastewater samples from sewage treatment plant. Analysis results were in good agreement with the corresponding LC/TIS/MS/MS data obtained by an independent laboratory.

It is important, for public health, to clean up wastewater from contamination by natural and synthetic estrogens. Existing techniques to degrade EDCs include activated carbon absorption, chlorination, membrane filtration, ozonation, reverse osmosis and UV radiation. All of these treatments are expensive with harmful byproducts. They lack selectivity and are impacted by the presence of other organic substances like pharmaceuticals, surfactants and humic/fulvic acids. Moreover, 5–95% of EDC residual levels in the effluent water can persist to affect the environment. New remedies for trace EDCs would need to be developed in government and university laboratories worldwide [74]. For instance, iron tetra-amido macrocyclic ligand (a synthetic catalyst which speeds oxidation) combined with hydrogen peroxide showed potential for degradation of EDCs in sewage effluent [75]. Some species of the genus *Rhodococcus* could have the enzyme that degrades the steroidal skeleton of E2 [76]. However, biological treatment of trace E2 is barely adequate because microorganisms metabolize other EDCs too present in effluent water [77]. The inability of conventional wastewater treatment processes, to be selective towards removal of trace E2, gives room for development of more selective extraction methods using MIPs.

The efficiency of using bisphenol A-molecularly imprinted polymeric microspheres (MIPMs) to remove phenolic estrogens from different sources of water were evaluated by Lin *et al.* [78]. The highest removal efficiency was observed at pH 5. MIPMs were more suitable to remove trace estrogens in a large volume than concentrated estrogens in a small volume. The removal efficiency of spiked tap water, lake water and river water were better than that of distilled water. MIPMs had higher removal selectivity and efficiency than activated carbons. Moreover, they can be re-used for 30 times without losing any efficiency.

There are several issues when applying MIP submicron particles for the cleanup of wastewater after secondary treatment in a plant. The processing of large volumes of wastewater for treatment with particles is one challenge to engineering. Another challenge is how best to eliminate all submicron particles before the treated wastewater is released into a river or lake. The clogging of membrane filters is a potential problem with submicron particles. This problem can be alleviated if ultrasonication is applied to dislodge the particles clogging any pores of a membrane that is mounted horizontally and facing down.

For cost savings, the reuse of MIP or NIP submicron particles is an important feature that warrants further research [79]. Removing E2 from the binding cavities is a necessary step The authors' lab confirmed that 10% trifluoroacetic acid (TFA) in water did not have the chemical ability needed for desorption of E2 from MIP/NIP submicron particles. Methanol alone was better for desorbing E2 than TFA, with a desorption efficiency of 46 ± 4%. After much research, 3% of TEA solution in methanol–water (1:40 v/v) was determined to be the best desorption agent for E2. It was an optimal concentration which would not damage the HPLC column but strong enough to break the hydrogen bonding, van der Waals, ionic and hydrophobic interactions between E2 and the carboxyl groups inside each binding cavity. One precaution is that some molecules of TEA may be left behind in the binding cavities. They can interfere with the binding of new E2 molecules during the next extraction [80]. Therefore, an additional wash with methanol or other volatile solvents to remove any residual TEA can be very beneficial.

6.7
Current Progress

The great affinity of MIP particles for the target analyte makes them a good tool for trace analysis. Both MIP and NIP particles have high potential for applications in the removal of E2 from wastewater. Syringe filters containing MIP or NIP particles are efficient in trapping E_2. Maximum binding for MIP submicron particles is $30 \pm 10\,\mu\text{mol}$ (= $8 \pm 2\,\text{mg}$) of E_2 per gram of MIP particles, which is similar to typical binding capacities of MIP micro-particles ($2 \pm 1\,\mu\text{m}$ in diameter) [81]. However, saturation binding would not take as long as 50–100 h. Removal of trace E2 has been achieved with $80 \pm 10\%$ efficiency for E2 at low concentrations ($\mu\text{g} \cdot \text{ml}^{-1}$). This efficiency can be improved by either pre-loading more NIP particles in one syringe filter or using several NIP pre-loaded syringe filters in series to filter the water. The ease of polymerization and relatively low cost of preparation add to the attraction of using NIP particles. These particles are re-useable after removal of the bound analyte, which is especially cost-effective in their application for cleanup of wastewater. Binding site competition (and poisoning) could potentially be caused by chemicals (with similar hydrophobicity to E2) and dissolved organic matter in wastewater. This would affect the extraction efficiency of NIP particles more than MIP particles.

Other naturally occurring steroids such as progesterone and testosterone can be analyzed by similar methods [82]. These bio-identical molecules paradoxically can be either beneficial or toxic at elevated levels. Owing to difficulty in monitoring these hormones at trace quantities in biological matrices, MIPs were used for preconcentration and clean-up in the sample preparation step. Two- and three-dimensional MIP nanosized structures can also be used for the fabrication of future sensors and arrays [83]. In a new study, Wang *et al.* have synthesized estrone-MIP coated Fe_3O_4 magnetic hybrid nanoparticles with controlled size using a semi-covalent imprinting strategy. The estrone–silica monomer complex was synthesized by the reaction of 3-(triethoxysilyl)propyl isocyanate with estrone, where the template was linked to the silica coating on the iron oxide core *via* a thermally reversible bond. Removal of the template by a simple thermal reaction produced specific estrone recognition sites on the surface of silica shell. The resulting magnetic nanoparticles exhibited a much higher specific recognition and saturation magnetization for biochemical separation of estrone [84].

MIPs were also considered to be one of the most promising selective separation methods for phenolic compound in wastewater treatment [11]. Bisphenol A (BPA)-imprinted polysulfone particles were prepared by a phase inversion technique, with a porosity of about 70% [85]. The binding amounts increased very rapidly at the beginning, and reached a maximum at 30 h. New restricted access materials (RAMs) combined to MIPs with a hydrophilic external layer have been prepared, for recognition of *p*-acetaminophenol (AMP) [86]. Glycidil methacrylate (GMA) epoxide ring opening with perchloric acid was performed for hydrophilic modification of the polymeric surface. With RAM-MIPs, non-specific hydrophobic interaction between templates (or analogues) and polymeric matrices was drastically reduced. A dual-phase solvent system for improved extraction capability of magnetic MIP beads in aqueous sample has been proposed by Hu *et al.* [87]. Their method integrated MIP extraction and micro-liquid–liquid extraction (micro-LLE) into one step. Magnetic MIP beads were applied to aqueous media by adding a micro-volume of *n*-hexane. The magnetic MIP beads preferred to suspend in the organic phase. The target analyte in the water sample was extracted into the organic phase by micro-LLE and then bound to the magnetic MIP beads. The specificity was significantly improved with the imprinting efficiency increasing to 4.4, from 0.5 in pure aqueous media. The method is selective, low in organic solvent consumption, and has potential to broaden the range of MIP applications in environmental analysis. A novel protein imprinted polymer for recognition of lysozyme has been reported by Zhang *et al.* [88]. Acryloyl-β-cyclodextrin, which offered a hydrophilic exterior and a hydrophobic cavity, was allowed to self-assemble with the template protein through hydrogen interaction and hydrophobic interaction. Polymerization in the presence of acrylamide (as an assistant monomer) resulted in a new type of protein imprinted polymer. A column packed with the lysozyme imprinted beads for HPLC could effectively separate lysozyme from a mixture of lysozyme–cytochrome *c*, lysozyme–bovine serum albumin, lysozyme–avidin and lysozyme–methylated bovine serum albumin, which demonstrated its high selectivity for biochemical analysis.

6.8
Recent Advances in MIP Technology for Continuing Development

A new piezoelectric quartz crystal (PQC) sensor using MIP particles as sensing material has been developed for fast and onsite determination of pirimicarb in contaminated vegetables [89]. Three kinds of MIP particles were prepared by conventional bulk polymerization (MIP-B) and precipitation polymerization in either acetonitrile (MIP-P1) or chloroform (MIP-P2). MIP-P2, with a uniform spherical shape and mean diameter of 50 nm, showed the best performance. It achieved a steady-state response within 5 min and exhibited low response to those pesticides with similar structures to pirimicarb (such as atrazine, carbaryl, carbofuran, and aldicarb). For onsite determination of pirimicarb in aqueous extract from contaminated vegetables, satisfactory recoveries from 96 to 103% were attained at pirimicarb concentrations ranging from 8×10^{-6} to 2×10^{-4} mol·l^{-1}. A screening method for haloacetic acid (HAA) disinfection by-products in drinking water was based on the use of a piezoelectric quartz crystal microbalance (QCM) transducing system, where the electrode was coated with a trichloracetic acid (TCAA) MIP [90]. This MIP consisted of poly(ethylene glycol dimethacrylate-co-4-vinylpyridine). The coated QCM was able to specifically detect the analytes in water samples in terms of the mass change in relation to acid–base interactions with the analytes. The TCAA-MIP coated QCM showed high specificity for the determination of TCAA in aqueous solutions. The achieved limit of detection was below the guidelines for maximum permissible levels (60 µg·l^{-1} for mixed HAAs).

An MIP-based optrode has been reported by Navarro-Villoslada for zearalenone (ZON) mycotoxin analysis [91]. The automated flow-through assay is based on the displacement of highly fluorescent tracers by the analyte from a MIP prepared by UV irradiation of a mixture of cyclododecyl 2,4-dihydroxybenzoate (template, ZON mimic), 1-allyl piperazine (functional monomer), TRIM (crosslinker), and AIBN in acetonitrile (porogen). Three fluorescent analogues of ZON were molecularly engineered for the assay development. These pyrene-containing tracers also inform on the characteristics of the microenvironment of the MIP binding sites. The ZON displacement fluorosensor showed a detection limit of 2.5×10^{-5} M in acetonitrile. A positive cross-reactivity was found for β-zearalenol, but not for resorcinol, resorcylic acid, E2, estrone, or bisphenol A.

A thin film of MIP with selective binding sites for dimethoate has been developed by Du et al. [92]. This film was cast on a gold electrode by electrochemical polymerization of o-phenylenediamine and template dimethoate via cyclic voltammetry scans and further deposition of Ag nanoparticles. Surface plasmon resonance and cyclic voltammetric signals were recorded simultaneously during the electropolymerization, controlling the thickness of the polymer film to be 25 nm. Recognition for the target molecule was observed by measuring the amperometric response of an oxidation–reduction probe, $K_3Fe(CN)_6$. The peak current was proportional to the concentration of dimethoate from 1.0 to 1000 ng·mL^{-1}, with a detection limit of 0.5 ng·mL^{-1}.

Most potentiometric sensing electrodes offered detection limits on the order of 1 µmol·l^{-1}, rarely stretching down to 0.1 µmol·l^{-1}. In 2007, a biomimetic potentiometric sensor was developed by dispersing the atrazine MIP particles in di-n-octyl phthalate plasticizer and then embedding in a poly(vinyl chloride) matrix [93]. The sensor responds to atrazine over a wide working range of 0.0001–10 mM, with a detection limit of 0.5 µmol·l^{-1} (or 0.1 µg.ml^{-1}). This sensor had a response time of 2 min. Recent advances have made it possible to bring these levels down to 0.01–10 nmol·l^{-1} for some inorganic ions. MIPs or plastic antibodies, which rely on a lock and key mechanism, can in principle selectively rebind and sense a particular analyte in a host of analogous species of similar size, shape, and geometry. Thus, the integration of MIPs with potentiometric transducers has immense potential in the fabrication of commercial sensing devices [94].

The principle of molecular imprinting has been proven successful in creating binding sites of specific recognition within polymers. After almost three decades of development, some evidence of large molecule imprinting has finally appeared. Maury et al. synthesized and evaluated new lipomonosaccharide-imprinted polymers in 2007 [95]. To bring the molecular imprinting community up-to-date, Hillberg and Tabrizian have described some of the new and innovative work that endeavored to take molecular imprinting away from its chromatographic and synthetic past to new, exciting, and developing fields such as drug delivery, biotechnology, biosensors, protein recognition, and novel materials [96]. Various different two- and three-dimensional approaches have been developed recently for the recognition of proteins, viruses, and cells. Traditional imprinting methods have been adapted to suit the mass transfer requirements of these large biological templates.

Molecular imprinting is continuing as a developed methodology that provides molecular assemblies of desired structures and properties. It is being increasingly used for applications in separation processes, micro-reactors, immunoassays, antibody mimics, catalysis, artificial enzymes, chemosensors, and biosensors. A new concept for the preparation of selective sorbents with high flow path properties was presented in 2007 by Le Noir et al. on embedding MIPs into various macroporous gels (MGs) [77]. A MIP was first synthesized with E2 as template for the selective adsorption of this endocrine disrupter. The composite macroporous gel/MIP (MG/MIP) monoliths were then prepared at subzero temperatures. Complete recovery of E2 from a $2\,\mu g \cdot l^{-1}$ aqueous solution was achieved using the poly(vinyl alcohol) (PVA) MG/MIP monoliths whereas only 49–74% was removed with NIPs. The PVA MG/MIP monolith columns were operated at an almost ten times higher flow rate ($50\,ml \cdot min^{-1}$) than the MIP columns ($1-5\,ml \cdot min^{-1}$). The possibility for processing the particulate-containing wastewater effluents at high flow rates with selectivity on E2 removal, as well as the easy preparation of the monoliths, made the macroporous MG/MIP systems attractive sorbents for the preconcentration of ultratrace E2 in water for quantitative analysis.

The ambient processing conditions and versatility make a sol–gel glassy matrix suitable for molecular imprinting. The progress of sol–gel based MIPs for various new applications can be observed from the growing number of publications [97]. Combining sol–gel process with molecular imprinting enables the production of sensors with greater sensitivity and selectivity. Considerable attention has been drawn to recent developments like the use of organically modified silane (ORMOSILS) precursor for the synthesis of hybrid molecular imprinted polymers (HMIPs) and the application of surface sol–gel processes for molecular imprinting. The development of molecular imprinted sol–gel nanotubes for biochemical separation and bio-imprinting is a new advancement under progress. Templated xerogels and molecularly imprinted sol–gel films provide a good platform for various priority sensor applications.

Molecularly imprinted nanoparticles have been successfully encapsulated into polymer nanofibers with a simple electrospinning method by Yoshimatsu et al. [98]. The composite nanofibers form non-woven mats that can be used as an affinity membrane to greatly simplify solid-phase extraction of drug residues. Upward of 100% of propranolol-imprinted nanoparticles can be easily encapsulated into poly(ethylene terephthalate) nanofibers, ensuring that the composite materials have a high specific binding capacity. As confirmed by radioligand binding analysis, specific binding sites in these composite materials remain easily accessible and are chiral-selective. Using the new composite nanofiber mats as SPE materials, trace amount of propranolol ($1\,\text{ng}\cdot\text{ml}^{-1}$) in tap water can be easily detected. There is no problem of template leakage from the composite nanofibers. Without the SPE, the existence of propranolol residues in water cannot be confirmed with even tandem HPLC–MS/MS analysis.

Acknowledgments

Financial support of the Natural Sciences and Engineering Research Council (NSERC) Canada is gratefully acknowledged (31554). Special thanks go to Audrey Murray, Asten Huang, Toby Cheung, Woomee Cho, Yiyan Li, Lerato Magosi, and Shuyi Wu for their technical assistance in various experiments and for many helpful scientific discussions.

References

1 Falconer, I.R., Chapman, H.F., Moore, M.R., and Ranmuthugala, G. (2006) *Environ. Toxicol.*, **21**, 181–191.
2 Besse, J.P. and Garric, J. (2008) *Toxicol. Lett.*, **176**, 104–123.
3 Liu, R., Zhou, J.L., and Wilding, A. (2004) *J. Chromatogr. A*, **1022**, 179–189.
4 Meng, Z., Chen, W., and Mulchandani, A. (2005) *Environ. Sci. Technol.*, **39**, 8958–8962.
5 Labadie, P., Cundy, A.B., Stone, K., Andrews, M., Valbonesi, S., and Hill, E.M. (2007) *Environ. Sci. Technol.*, **41**, 4299–4304.
6 Canadian Broadcasting Corporation (2008) Estrogen levels skyrocket in river around Montreal, http://www.cbc.ca/canada/montreal/story/2008/09/17/estrogen-stlawrence.html (accessed 18 March 2010).
7 Yu, J.C.C., Hrdina, A., Mancini, C., and Lai, E.P.C. (2007) *J. Nanotechnol.*, **7**, 3095–3103.
8 Andersson, L.I. (2000) *J. Chromatogr. B*, **745**, 3–13.
9 Li, Y., Li, X., Li, Y., Qi, J., Bian, J., and Yuan, Y. (2009) *Environ. Pollut.*, **157**, 1879–1885.
10 Rezaei, B., Mallakpour, S., and Majidi, N. (2009) *Talanta*, **78**, 418–423.
11 Zakaria, N.D., Yusof, N.A., Haron, J., and Abdullah, A.H. (2009) *Int. J. Mol. Sci.*, **10**, 354–365.
12 Jiang, T., Zhao, L., Chu, B., Feng, Q., Yan, W., and Lin, J.M. (2009) *Talanta*, **78**, 442–447.
13 Tsai, W.H., Huang, T.C., Huang, J.J., Hsue, Y.H., and Chuang, H.Y. (2009) *J. Chromatogr. A*, **1216**, 2263–2269.
14 Piletsky, S. and Turner, A. (2006) *Molecular Imprinting of Polymers*, Landes Bioscience, Texas, USA.
15 Farrington, K. and Regan, F. (2009) *Talanta*, **78**, 653–659.
16 Yan, M. and Ramstrom, O. (2005) *Molecularly Imprinted Materials: Science and Technology*, Marcel Dekker, New York.
17 Long, C., Mai, Z., Yang, Y., Zhu, B., Xu, X., Lu, L., and Zou, X. (2009) *J. Chromatogr. A*, **1216**, 2275–2281.
18 Li, Y.H., Yang, T., Qi, X.L., Qiao, Y.W., and Deng, A.P. (2008) *Anal. Chim. Acta*, **624**, 317–325.
19 Lehn, J.M. (1995) *Supramolecular Chemistry: Concepts and Perspectives*, VCH, Weinheim.
20 Claude, B., Morin, P., Bayoudh, S., and De Ceaurriz, J. (2008) *J. Chromatogr. A*, **1196**, 81–88.

21 Wulff, G. (1993) *Forum*, **11**, 85–87.
22 Mosbach, K. (1994) *Techniques*, **19**, 9–14.
23 Komiyama, M., Takeuchi, T., Mukawa, T., and Asanuma, H. (2003) *Molecular Imprinting: From Fundamentals to Applications*, Wiley-VCH Verlag GmbH, Weinheim.
24 Skreenivasan, K. (2000) *J. Appl. Polym. Sci.*, **82**, 889–893.
25 Feás, X., Fente, C.A., Hosseini, S.V., Seijas, J.A., Vázquez, B.I., Franco, C.M., and Cepeda, A. (2009) *Mat. Sci. Eng. C*, **29**, 398–404.
26 Tang, K., Chen, S., Gu, X., Wang, H., Dai, J., and Tang, J. (2008) *Anal. Chim. Acta*, **614**, 112–118.
27 Koohpaei, A.R., Shahtaheri, S.J., Ganjali, M.R., Forushani, A.R., and Golbabaei, F. (2008) *Talanta*, **75**, 978–986.
28 Pichon, V. (2007) *J. Chromatogr. A*, **1152**, 41–53.
29 Tamayo, F.G., Turiel, E., and Martín-Esteban, A. (2007) *J. Chromatogr. A*, **1152**, 32–40.
30 Farrington, K. and Regan, F. (2007) *Biosens. Bioelectron.*, **22**, 1138–1146.
31 Pichon, V. and Chapuis-Hugon, F. (2008) *Anal. Chim. Acta*, **622**, 48–61.
32 Zhang, H., Song, T., Zhang, W., Hua, W., and Pan, C. (2007) *Bioorg. Med. Chem.*, **15**, 6089–6095.
33 Sánchez-Barragán, I., Karim, K., Costa-Fernández, J.M., Piletsky, S.A., and Sanz-Medel, A. (2007) *Sens. Actuators, B*, **123**, 798–804.
34 Lv, Y., Lin, Z., Feng, W., Zhou, X., and Tan, T. (2007) *Biochem. Eng. J.*, **36**, 221–229.
35 Lv, Y., Lin, Z., Tan, T., Feng, W., Qin, P., and Li, C. (2008) *Sens. Actuators, B*, **133**, 15–23.
36 Jiang, X., Zhao, C., Jiang, N., Zhang, H., and Liu, M. (2008) *Food Chem.*, **108**, 1061–1067.
37 Li, Y., Li, X., Li, Y., Dong, C., Jin, P., and Qi, J. (2009) *Biomaterials*, **30**, 3205–3211.
38 Farrington, K., Magner, E., and Regan, F. (2006) *Anal. Chim. Acta*, **566**, 60–68.
39 Le Moullec, S., Bégos, A., Pichon, V., and Bellier, B. (2006) *J. Chromatogr. A*, **1108**, 7–13.
40 Shi, Y., Zhang, J.H., Shi, D., Jiang, M., Zhu, Y.X., Mei, S.R., Zhou, Y.K., Dai, K., and Lu, B. (2006) *J. Pharm. Biomed. Anal.*, **42**, 549–555.
41 Yin, J., Wang, S., Yang, G., and Chen, Y. (2006) *J. Chromatogr. B*, **844**, 142–147.
42 Marazuela, M.D. and Bogialli, S. (2009) *Anal. Chim. Acta*, **645**, 5–17.
43 O'Connor, S. and Aga, D.S. (2007) *Trends Anal. Chem.*, **26**, 456–465.
44 Haupt, K. (2001) *Analyst*, **126**, 747–756.
45 Buchberger, W.W. (2007) *Anal. Chim. Acta*, **593**, 129–139.
46 Zhang, Z.L. and Zhou, J.L. (2007) *J. Chromatogr.*, **1154**, 205–213.
47 Baggiani, C., Anfossi, L., and Giovannoli, C. (2007) *Anal. Chim. Acta*, **591**, 29–39.
48 Chen, S. and Zhang, Z. (2008) *Spectrochim. Acta A*, **70**, 36–41.
49 Weigel, S., Kallenborn, R., and Huhnerfuss, H. (2004) *J. Chromatogr. A*, **1023**, 183–195.
50 Widstrand, C., Yilmaz, E., Boyd, B., Billing, J., and Rees, A. (2006) *Am. Lab.*, **38**, 12–14.
51 Urraca, J.L., Marazuela, M.D., Merino, E.R., Orellana, G., and Moreno-Bondi, M.C. (2006) *J. Chromatogr. A*, **1116**, 127–134.
52 Kubo, T., Nomachi, M., Nemoto, K., Sano, T., Hosoya, K., Tanaka, N., and Kaya, K. (2006) *Anal. Chim. Acta*, **577**, 1–7.
53 Boyd, B., Björk, H., Billing, J., Shimelis, O., Axelsson, S., Leonora, M., and Yilmaz, E. (2007) *J. Chromatogr. A*, **1174**, 63–71.
54 Zhang, H., Song, T., Zong, F., Chen, T., and Pan, C. (2008) *Int. J. Mol. Sci.*, **9**, 98–106.
55 Feás, X., Seijas, J.A., Vázquez-Tato, M.P., Regal, P., Cepeda, A., and Fente, C. (2009) *Anal. Chim. Acta*, **631**, 237–244.
56 Xia, Y.Q., Guo, T.Y., Song, M.D., Zhang, B.H., and Zhang, B.L. (2006) *React. Funct. Polym.*, **66**, 1734–1740.
57 Jiang, M., Zhang, J.H., Mei, S.R., Shi, Y., Zou, L.J., Zhu, Y.X., Dai, K., and Lu, B. (2006) *J. Chromatogr. A*, **1110**, 27–34.

58 Shi, X., Wu, A., Zheng, S., Li, R., and Zhang, D. (2007) *J. Chromatogr. B*, **850**, 24–30.
59 Liu, X., Chen, Z., Zhao, R., Shangguan, D., Liu, G., and Chen, Y. (2007) *Talanta*, **71**, 1205–1210.
60 Wang, L. and Zhang, Z. (2008) *Sens. Actuators, B*, **133**, 40–45.
61 Sambe, H., Hoshina, K., and Haginaka, J. (2007) *J. Chromatogr. A*, **1152**, 130–137.
62 Haginaka, J. (2008) *J. Chromatogr. B*, **866**, 3–13.
63 Wei, S., Molinelli, A., and Mizaikoff, B. (2006) *Biosens. Bioelectron.*, **21**, 1943–1951.
64 Lai, E.P.C., De Maleki, Z., and Wu, S. (2010) *J. Appl. Polym. Sci.*, **116**, 1499–1508.
65 Le Noir, M., Lepeuple, A., Guieysse, B., and Mattiasson, B. (2007) *Water Res.*, **41**, 2825–2831.
66 Celiz, M.D., Aga, D.S., and Colón, L.A. (2009) *Microchem. J.*, **92**, 174–179.
67 Wei, S., and Mizaikoff, B. (2007) *Bioelectronics*, **23**, 201–209.
68 Cacho, C., Turiel, E., Martín-Esteban, A., Ayala, D., and Pérez-Conde, C. (2006) *J. Chromatogr. A*, **1114**, 255–262.
69 Toth, B., Pap, T., Horvath, V., and Horvai, G. (2007) *Anal. Chim. Acta*, **591**, 17–21.
70 Umpleby, R.J., Rushton, G.T., Shah, R.N., Rampey, A.M., Bradshaw, J.C., Berch, J.K., and Shimizu, K.D. (2001) *Macromolecules*, **34**, 8446–8452.
71 García-Calzón, J.A. and Díaz-García, M.E. (2007) *Sens. Actuators, B*, **123**, 1180–1194.
72 McNiven, S., Yokobayashi, Y., Cheong, S.H., and Karube, I. (1997) *Chem. Lett.*, **26**, 1297–1298.
73 Sun, Z., Schüssler, W., Sengl, M., Niessner, R., and Knopp, D. (2008) *Anal. Chim. Acta*, **620**, 73–81.
74 Health Canada (2004) Effects of a Synthetic estrogen on aquatic populations: a whole ecosystem study, http://www.hc-sc.gc.ca/sr-sr/finance/tsri-irst/proj/endocrin/tsri-94-eng.php (accessed 18 March 2010).
75 Black, H. (2008) *Environ. Health Perspect.*, **116**, A159.
76 Yoshimito, T., Nagai, F., Fujimoto, J., Watanabe, K., Mizukoshi, H., Makino, T., Kimura, K., Saino, H., Sawada, H., and Omura, H. (2004) *Appl. Environ. Microbiol.*, **70**, 5283–5289.
77 Le Noir, M., Plieva, F., Hey, T., Guieysse, B., and Mattiasson, B. (2007) *J. Chromatogr. A*, **1154**, 158–164.
78 Lin, Y., Shi, Y., Jiang, M., Jin, Y., Peng, Y., Lu, B., and Dai, K. (2008) *Environ. Pollut.*, **153**, 483–491.
79 Mahony, J.O., Nolan, K., Smyth, M.R., and Mizaikoff, B. (2005) *Anal. Chim. Acta*, **534**, 31–39.
80 Theodoridis, G., Kantifes, A., Manesiotis, P., Raikos, N., and Tsoukali-Papdopoulou, H. (2003) *J. Chromatogr. A*, **987**, 103–109.
81 Yang, K., Li, B., Zhou, H., Ma, J., Bai, P., and Zhao, C. (2007) *J. Appl. Polym. Sci.*, **106**, 2791–2799.
82 Gadzała-Kopciuch, R., Ricanyová, J., and Buszewski, B. (2009) *J. Chromatogr. B*, **877**, 1177–1184.
83 Tokonami, S., Shiigi, H., and Nagaoka, T. (2009) *Anal. Chim. Acta*, **641**, 7–13.
84 Wang, X., Wang, L., He, X., Zhang, Y., and Chen, L. (2009) *Talanta*, **78**, 327–332.
85 Yang, K., Ma, J., Zhou, H., Li, B., Yu, B., and Zhao, C. (2009) *Desalination*, **245**, 232–245.
86 Puoci, F., Iemma, F., Cirillo, G., Curcio, M., Parisi, O.I., Spizzirri, U.G., and Picci, N. (2009) *Eur. Polym. J.*, **45**, 1634–1640.
87 Hu, Y., Liu, R., Zhang, Y., and Li, G. (2009) *Talanta*, **79**, 576–582.
88 Zhang, W., Qin, L., He, X.W., Li, W.Y., and Zhang, Y.K. (2009) *J. Chromatogr. A*, **1216**, 4560–4567.
89 Sun, H. and Fung, Y. (2006) *Anal. Chim. Acta*, **576**, 67–76.
90 Suedee, R., Intakong, W., and Dickert, F.L. (2006) *Talanta*, **70**, 194–201.
91 Navarro-Villoslada, F., Urraca, J.L., Moreno-Bondi, M.C., and Orellana, G. (2007) *Sens. Actuators, B*, **121**, 67–73.
92 Du, D., Chen, S., Cai, J., Tao, Y., Tu, H., and Zhang, A. (2008) *Electrochim. Acta*, **53**, 6589–6595.
93 Prasad, K., Prathish, K.P., Gladis, J.M., Naidu, G.R.K., and Rao, T.P. (2007) *Sens. Actuators, B*, **123**, 65–70.
94 Rao, T.P. and Kala, R. (2008) *Talanta*, **76**, 485–496.

95 Maury, D., Couderc, F., Garrigues, J.C., and Poinsot, V. (2007) *Talanta*, **73**, 340–345.

96 Hillberg, A.L. and Tabrizian, M. (2008) *IRBM*, **29**, 89–104.

97 Gupta, R. and Kumar, A. (2008) *Biotechnol. Adv.*, **26**, 533–547.

98 Yoshimatsu, K., Ye, L., Lindberg, J., and Chronakis, I.S. (2008) *Biosens. Bioelectron.*, **23**, 1208–1215.

7
Trace Detection of High Explosives with Nanomaterials
Wujian Miao, Cunwang Ge, Suman Parajuli, Jian Shi, and Xiaohui Jing

7.1
Introduction

Detection and quantification of high explosives and related compounds have attracted much attention in recent years due to the pressing needs associated with global security and growing concerns with the environment and human health [1–5]. Such detection is necessary in various complex environments, including mine fields, munitions storage facilities, ground and seawater, transportation areas, and blast sites. In each of these settings, sensitive and timely detection of explosive materials is required to ensure the safety and security of the surrounding area.

Explosive detection techniques can be broadly classified into two categories: bulk and trace detection [1, 2]. In bulk detection, a macroscopic mass of the explosive material is detected directly, usually by viewing images made by X-ray scanners or similar equipment such as millimeter-wave and terahertz imaging spectrometers [3]. Other recently developed bulk detection techniques include neutron techniques, nuclear quadrupole resonance (NQR), and laser techniques. In trace detection, the explosive is detected by chemical identification of microscopic residues, usually in the form of a vapor or a particulate. Table 7.1 summarizes commonly used trace detection methods and their features.

Additionally, two or more detection methods from the same or different types (e.g., two different bulk detection technologies or a bulk detection plus a trace detection technique) can be combined [15]. This integration can be achieved by using simultaneous detection or by a two-step detection method. For example, simultaneous operation of NQR and X-ray imaging or two-step operation of X-ray computed tomography (CT) [63] and IMS can be performed. The strength of one technique may thus compensate the weakness of the other, or the vulnerability of one detection device to a potential countermeasure could be compensated by another detection device.

Chemical explosives commonly used by the military and terrorists can be categorized into three groups (Table 7.2). The first group contains nitrated explosives that are generally used by military. On the basis of chemical structural and

Trace Analysis with Nanomaterials. Edited by David T. Pierce and Julia Xiaojun Zhao
Copyright © 2010 WILEY-VCH Verlag GmbH & Co. KGaA, Weinheim
ISBN: 978-3-527-32350-0

Table 7.1 Commonly used trace detection methods for high explosives and their features.

Method	Feature	References
Mass spectrometry (MS) coupled with gas or high-performance liquid chromatography (GC/MS, HPLC/MS)	Sensitive but expensive, not portable	[5–18]
Ion mobility spectrometry (IMS)	Sensitive but with matrix effect	[19–25]
Micro-mechanical sensors (e.g., micro-cantilevers)	Sensitive, low power consumption and real-time operation, but mainly for vapors, with moisture effect, less selective	[26–30]
Electrochemical methods	Inexpensive, fast, portable, less sensitive	[31–36]
Chemiluminescence (CL)	High sensitivity but poor selectivity	[37–42]
Colorimetric tests	Simple, inexpensive, fast but less sensitive and low specificity	[43, 44]
Fluorescence (FL) spectroscopy	Sensitive with interferences effect	[45–52]
Surface plasmon resonance	Label-free, sensitive but subject to external contamination	[53–57]
Surface-enhanced Raman scattering spectroscopy (SERS)	Sensitive, but complicated technique and difficult to operate	[58–62]

functional properties, this group can be divided into three subgroups: (i) nitroaromatics (NACs), (ii) nitroamines, and (iii) nitrate esters. The second group of high explosives is peroxide based, which includes hexamethylene triperoxide diamine (HMTD) and triacetone triperoxide (TATP). HMTD and TATP have become popular materials used by terrorists because they are easily prepared from readily obtained ingredients, although the synthesis is fraught with danger. For example, TATP was used in terrorist bombings of the London subway system in 2005 and by the infamous shoe bomber on a trans-Atlantic flight in 2001 [64]. As HMTD and TATP contain three peroxide linkages per molecule, their explosive output is much higher than most organic peroxides. HMTD is estimated to have 60% and TATP 88% blast strength of TNT [2]. Plastic explosives form the third group of explosives, in which one or more of the first group explosives are plasticized to make a moldable material, such as C-4 and Semtex H. To retain the best explosive output, the inert plasticizers are usually added as less than 10–15% of the overall weight. Plastic explosives were originally developed for convenient use in military demolitions but have since been widely used in terrorist bombs. Figure 7.1 shows the structures and abbreviations of explosives listed in Table 7.2.

Compared with other organic compounds, explosives have exceptionally high density (e.g., military explosives generally have a density greater than $1.6 \,g\, cm^{-3}$,

Table 7.2 Commonly used high explosives and their chemical properties.

Explosive	Name/contents	Formula	Density (g cm^{-3})	N (%)	O (%)
1. Nitrated explosives					
(i) Nitroaromatics					
TNT	2,4,6-Trinitrotoluene	$C_7H_5N_3O_6$	1.65	18.5	42.3
Picric acid	2,4,6-Trinitro-1-phenol	$C_6H_3N_3O_7$	1.77	18.3	48.9
Tetryl	N-Methyl-N,2,4,6-tetranitroaniline	$C_7H_5N_5O_8$	1.73	24.4	44.6
(ii) Nitramines					
RDX	1,3,5-Trinitro-1,3,5-triazacyclohexane	$C_3H_6N_6O_6$	1.82	37.8	43.2
HMX	1,3,5,7-Tetranitro-1,3,5,7-tetrazocane	$C_4H_8N_8O_8$	1.96	37.8	43.2
CL 20	Hexanitrohexaazaisowurtzitane	$C_6H_6N_{12}O_{12}$		38.4	43.8
(iii) Nitrate esters					
PETN	Pentaerythritol tetranitrate	$C_5H_8N_4O_{12}$	1.76	17.7	60.7
Nitrocellulose	Cellulose nitrate	$C_6H_7N_3O_{11}$	1.2	14.1	59.2
2. Peroxide-based explosives					
HMTD	Hexamethylene triperoxide diamine	$C_6H_{12}N_2O_6$	1.6	13.5	46.1
TATP	Triacetone triperoxide	$C_9H_{18}O_6$	1.2	0	43.2
3. Plastic explosives					
C-4	RDX + plasticizer				
Semtex H	RDX + PETN + plasticizer				
Detasheet	PETN + plasticizer				

Table 7.2). All explosives have very high oxygen and/or nitrogen contents that cause dramatic volume changes (from solid to gas) when an explosion occurs. Many bulk detection methods are in fact based on the properties of high density and high oxygen and nitrogen contents of explosives.

Although a wide variety of explosive detection technologies are currently available, this chapter will mainly focus on trace detection methods involving nanomaterials. Such methods generally possess many advantages, which include high sensitivity, good selectivity, fast response, portability, and low cost. These features are clearly desirable for all analytical systems, and are essential in combating explosives-based terrorism.

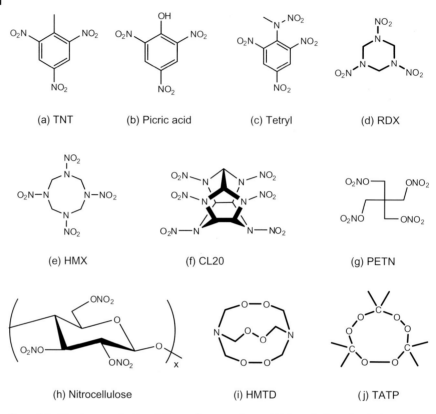

Figure 7.1 Structures and abbreviations of commonly used high explosives: (a–c) nitroaromatics; (d–f) nitramines; (g,h) nitrate esters; and (i,j) peroxide-based explosive compounds.

7.2
Techniques for Trace Detection of High Explosives

7.2.1
Electrochemical Sensors

The inherent redox properties of nitrated and peroxide-based explosives make them ideal candidates for electrochemical monitoring [34, 36]. Electrochemical sensors (ESs) for explosive detection provide several advantages over spectroscopic and spectrometric techniques such as SERS, MS, and IMS. They are characterized by a reasonable sensitivity, low cost, and can be easily used as field detectors and remote control devices [65–69]. Various ESs for the detection of nitroaromatic compounds (NACs, Figure 7.1) have been reported using different sensing materials, including bare carbon and Au as well as boron-doped diamond electrodes [65–68, 70]. A polyphenol-coated screen-printed carbon electrode was also used for highly sensitive voltammetric measurements of TNT in the presence of surface-active substances [71]. Electrochemical responses of several NACs have been compared at glassy carbon (GC), Pt, Ni, Au, and Ag electrodes, revealing that Au and Ag were suitable in capillary electrophoresis (CE) amperometric detection. A bimetal electrode, prepared by depositing Ag on Au, offered a superior performance by exploiting the sensitivity of Au while suppressing its response toward acetonitrile (MeCN). In this case a ten-fold lower detection limit than UV measurement was achieved for the explosive compounds, corresponding to 70–110 ppb (parts per billion) [72].

7.2.1.1 Nanomaterial Modified Electrodes

Metallic and metal-oxide nanoparticles (NPs) can increase the rate of many chemical reactions due to their high ratio of surface atoms with free valences. In addition to their high surface area-to-volume ratio, NP derivatized materials provide size controllability, chemical stability, and surface tenability. As such, these nanomaterials provide an ideal platform for sensing/biosensing and catalytic applications. The use of electrodes modified with NPs of transition and precious metals, which have distinct properties compared to that of the bulk materials, have provided new opportunities for the development of ESs [73, 74].

The modification of electrode surfaces with redox-active metal NPs has led to various ESs. Filnovsky *et al.* have found that modification of carbon with noble metal NPs is a promising approach for obtaining electrodes with highly catalytic activity for the detection of trace aromatic compounds [75]. Modification of the electrode was performed with composites of nanometer-sized, mesoporous TiO_2 containing inserted or deposited NPs of Ru, Pt, or Au. Cyclic voltammetry indicated that TNT can be reduced on carbon-paper electrodes modified with these composites at potentials around –0.5 V (vs Ag/AgCl/Cl⁻) in aqueous solutions. Remarkable electrochemical activity of the electrode toward the reduction of TNT was observed, suggesting that the composite material may play a specific role in facilitating the TNT reduction process. Modified electrodes based on mesoporous SiO_2-MCM-41 coatings have also been shown recently to enhance sensitivity through adsorptive accumulation of the target NACs explosives [76].

Carbon nanotubes (CNTs) are often used as modifying materials for improving the electrochemical detection of explosives. GC electrodes modified with multi-walled CNTs (MWCNTs) offer a significant improvement in electrochemical sensitivity toward TNT from seawater [77]. Metal NPs (Pt, Au, or Cu) together with MWCNTs and single-walled CNTs (SWCNTs) have also been used to form nanocomposites to improve electroactivity and selectivity for TNT and several other NACs [78]. Among the various combinations tested, a synergistic effect was observed for the nanocomposite containing Cu NPs and SWCNTs solubilized in Nafion and this combination provided the best sensitivity for detecting TNT and other NACs. Adsorptive stripping voltammetry of TNT with the electrode modifier resulted in a detection limit of 1 ppb and linearity up to three orders of magnitude. Selectivity towards the number and position of the nitro groups in different NACs was also found to be reproducible and distinct. The Cu-SWCNT-modified GC electrode was demonstrated for analysis of TNT in tap water, river water, and contaminated soil.

7.2.1.2 "Artificial Peroxidase"-Modified Electrodes Based on Prussian Blue

The peroxide-based explosives TATP and HMTD are easy to synthesize from readily available precursor chemicals. However, their detection is challenging since they lack electrochemically reducible nitro groups, they do not fluoresce, and they exhibit minimal UV absorption [79].

While TATP and HMTD can be determined using expensive instruments such as chemical-ionization mass spectrometers or IR spectrometers [14], these bulky instruments are not suitable for field screening or trace analysis of peroxide-based explosives. Accordingly, there is an urgent need for developing highly sensitive and yet small, easy-to-use, field deployable devices for on-site testing of peroxide explosives. Activity in this direction has focused primarily on peroxidase-based optical (fluorescent or colorimetric) assays of the hydrogen peroxide produced from UV- or acid-treated [79] peroxide explosives. However, enzymatic assays often suffer from limited stability and high cost of the biocatalyst. Surprisingly, little attention has been given to the development of electrochemical devices for monitoring peroxide explosives [80], despite the fact that these devices are uniquely qualified for meeting the size, cost, and low power requirements of field detection of TATP and HMTD.

Electrodes modified with polycrystalline Prussian Blue (PB) offer highly selective, low potential, and stable electrocatalytic detection of H_2O_2 [81, 82]. A highly sensitive electrochemical assay of TATP and HMTD at such an electrode has been reported [32]. The method involves UV light degradation of the peroxide explosives and a low potential (0.0 V vs Ag/AgCl [3 M KCl]) amperometric sensing of H_2O_2 generated at the PB modified electrode. Nanomolar detection limits have been obtained following short (15 s) irradiation times. Electrochemical detection based on direct reduction of peroxide explosives cannot be carried out because the reduction of -O–O- peroxide groups is ineffective. Although the PB modified electrode is specific toward H_2O_2 reduction at a low potential where unwanted reactions of co-existing compounds are negligible [81–83], selective detection of TATP and HMTD is difficult. The high catalytic activity of PB leads also to a very high sensitivity towards H_2O_2. The behavior of PB-coated electrodes resembles that of peroxidase-based enzyme electrodes, and hence PB has often been denoted as "artificial enzyme peroxidase" [84].

When needed, the PB film can be covered with a permselective (size-exclusion) coating that can further enhance the sensor selectivity, stability, and overall performance [85]. In addition, relevant samples may be treated with catalase to remove H_2O_2 that may originate from cleaning agents [86]. The electrochemical route can be further developed into disposable microsensors in connection to single-use screen-printed electrode strips and a hand-held meter (similar to those used for self-testing of blood glucose). Preliminary data with such PB-coated screen-printed electrodes are promising. The PB-transducer can be readily adapted for gas-phase electrochemical detection of trace TATP and HMTD when combined with an appropriate solid electrolyte coating [32, 33].

7.2.2
Electrogenerated Chemiluminescence

Electrogenerated chemiluminescence (also known as electrochemiluminescence, ECL) is a process whereby light is generated from electrochemical reactions [87–90]. ECL has been used for various analytical applications, including immunoassay and DNA detection [91–93]. ECL has distinct advantages over other spectroscopic-based detection systems. Unlike FL methods, ECL does not require a light source, which effectively frees the method from scattered-light and from interferences of luminescent impurities. Moreover, specificity of the ECL luminophore (e.g., $[Ru(bpy)_3]^{2+}$, where bpy = 2,2′-bipyridine) and the co-reactant species [94] decreases side effects such as self-quenching. Surprisingly, there are only few studies in the literature on ECL detection of high explosives [95–99]. Wilson et al. have reported ECL-enzyme immunoassays for TNT and PETN, where the amount of analyte in the sample was determined by measuring light emission when H_2O_2 was generated electrochemically in the presence of luminol and its ECL enhancer p-iodophenol [95–97]. Haptens corresponding to these explosives were covalently attached to high-affinity dextran-coated paramagnetic beads (MBs). The beads were mixed with the corresponding Fab fragments and the sample. After adding a second antispecies-specific antibody labeled with horseradish peroxidase (HRP), the mixture was measured for ECL [96, 97] once the beads were magnetically collected on the surface of the working electrode. Detection limits of 0.11 and 19.8 ppb were obtained for TNT and PETN, respectively. The increase in sensitivity obtained when Fab fragments were used instead of whole antibodies was attributed to the higher affinity of the divalent whole antibodies, which could affect the related chemical equilibria. Figure 7.2 schematically shows the HRP-ECL system used for the determination of TNT and PETN. HRP label catalyzes the chemiluminescent oxidation of luminol in the presence of electrochemically generated H_2O_2 [100]. Although the enzyme catalyzes the chemiluminescent oxidation of luminol directly, addition of an enhancer such as p-iodophenol increases the efficiency of the reaction by up to three orders of magnitude [101].

The luminol/H_2O_2 system generally exhibits substantial non-specific background ECL and requires the use of a strong alkaline reaction medium for sufficient light emission [88]. To reduce the detection limit, an ECL system with very little background and high amplification is preferred. Bard's group previously achieved such a system for determination of DNA and C-reactive protein by using polystyrene microspheres containing $[Ru(bpy)_3]^{2+}$ [92, 93]. Based on this work, Miao's group have recently described the ultrasensitive detection of TNT extracted from soil and water samples collected from a military firing range and a nearby creek [99]. The method is illustrated in Figure 7.3. Magnetic beads (MBs, 1 μm in diameter) coated with TNT antibodies were mixed with the TNT antibody coated-polystyrene beads (PSBs, 10 μm in diameter). The PSBs were also preloaded

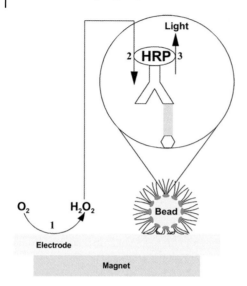

Figure 7.2 HRP-ECL detection system. (1) H_2O_2 is generated by reducing dissolved oxygen electrochemically; (2) H_2O_2 oxidizes HRP (horseradish peroxidase); (3) HRP oxidizes p-iodophenol, which mediates the chemiluminescent oxidation of luminol. Reprinted with permission from Reference [101]. Copyright 1986 Academic Press.

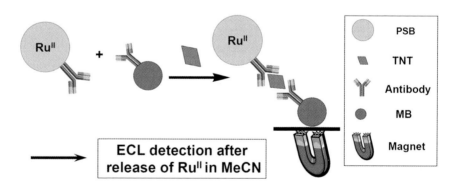

Figure 7.3 Schematic diagram showing the general principle of ultrasensitive ECL detection of TNT at high amplification with $Ru(bpy)_3^{2+}$-containing polystyrene beads. Reprinted with permission from Reference [99]. Copyright 2009 Elsevier.

with a large number of $[Ru(bpy)_3]^{2+}$ ECL labels. In the presence of TNT, a sandwich-type MB-TNT-PSB aggregate formed and was separated magnetically from the reaction medium. The captured aggregates were then transferred into a MeCN solution where the PSB dissolved, the ECL label was released, and an ECL signal was measured using tri-n-propylamine (TPrA) as an ECL co-reactant with an anodic electrode potential. Unlike traditional ECL immunoassays, where only one or a few ECL labels can be attached to one antibody, the PSB strategy allows tens of millions ECL labels to bind to one or a few antibodies.

Using this improved method, the integrated ECL intensity was found to be linearly proportional to the standard TNT concentration over the range 0.10–1000 ppt (parts per trillion, or pg ml^{-1}). The ultimate detection limit is believed to be 0.10 ± 0.01 ppt (absolute detection limit in a mass of ~0.1 pg) or less; this is approximately 1100-, 1000-, and 600-times lower than that obtained from luminol-H_2O_2-enzyme ECL [97], FL-based liquid array displacement immunoassays [102], and surface plasmon resonance [103], respectively.

Very recently, Miao's group have reported the sensitive ECL detection and quantification of HMTD explosives with a detection limit of 50 μM [98]. HMTD was used as an ECL co-reactant and light emission was measured in a MeCN solution at an anodic Pt electrode when $[Ru(bpy)_3]^{2+}$ and the ECL enhancing agent $AgNO_3$ were present. ECL enhancement of the HMTD/$[Ru(bpy)_3]^{2+}$ system by $AgNO_3$ (up to 27 times) was primarily ascribed to the chemical oxidation of HMTD by electrogenerated $NO_3^{\bullet}$ and Ag(II) species (strong oxidizing agents), which resulted in an increase in [HMTD$^{\bullet+}$] and hence the ECL intensity.

7.2.3
Fluorescence-Based Sensors

Fluorescence (FL) sensors have been widely used for the detection of nitrated explosives. The sensors are generally developed on the basis of either FL quenching or competitive FL immunoassays and both types will be reviewed.

The mechanism of FL-quenching used for detecting vaporous explosives, such as dinitrotoluene (DNT) and TNT, is mainly based on electron transfer from the electron-rich, fluorescent organic materials to those electron-deficient NACs. This electron transfer leads to fluorescence quenching of the organic materials in proportion to the amount of NAC present. Both the exciton diffusion length and surface area of the sensing films have a strong influence on sensitivity [52, 104].

FL-quenching-based detection represents one of the most sensitive and convenient methods for identification of explosives [105]. Only the chromophore molecules that interact directly with the analyte molecule are quenched; the remaining chromophore molecules continue to fluoresce. The basic sensor design [15, 106] (Figure 7.4) consists of a FL excitation source, such as a blue light-emitting diode. Light passes through a lens and filter, allowing a narrow wavelength band (e.g., 430 nm) to impinge on a nanocomposite polymer film. A pump pulls in air samples across film for a given period of time, after which fluorescence spectra are measured at a particular excitation wavelength. If the air sample contains explosive vapors, the PMT detector will sense a FL quenching in light intensity and trigger an alarm.

7.2.3.1 Quenching Sensors Based on Fluorescent Polymer Porous Films
Among various photoluminescent materials, conjugated polymers have been the most extensively explored as chemosensory materials for the FL detection of

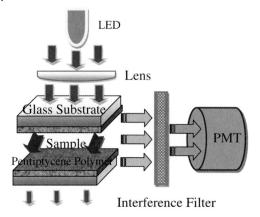

Figure 7.4 Fluorescent polymer sensor design. Modified from Reference [106].

electron-deficient analytes such as NACs [107–111]. Some conjugated polymers exhibit a high sensitivity to NACs explosives, resulting in strong quenching of their FL emissions [111–113]. Swager and coworkers have reported an amplified response to analyte binding in aggregated systems and solid films of conjugated polymers by intermolecular exciton migration [52, 107]. Multiphoton FL quenching has been observed for the real-time detection of TNT [109–111, 114]. A recent report indicated that molecular imprinting of conjugated polymer matrices can greatly improve their chemosensory selectivity to NACs [115]. Other photoluminescent materials such as polytetraphenylsilole, polytetraphenylgermole, photoluminescent silica films [112, 113, 116, 117], and silica microspheres with physisorbed dyes [118] have also exhibited high FL response to solution- and vapor-borne NACs explosives at low concentrations.

Porous silicon (PSi) sensors have been intensively studied [119–132] because of the high surface area of PSi and the variety of optical transduction mechanisms upon exposure to different analytes [105]. However, most of these unhybridized sensors demonstrate no specificity for target molecules and require high analyte vapor pressures ($\sim$1–100 $\times$ 10^{-5} mmHg) to detect changes in the PSi reflectance or luminescence [105]. Hybridization of PSi platforms with polymers that exhibit high affinity for NAC explosives has shown promise toward improving these characteristics. Entrapping polymers inside the PSi microcavity (MC) also has other beneficial effects, such as increasing initial FL quantum yield (above that of PSi's self-luminescence), amplifying FL quenching due to more effective energy migration [52], and producing fine spectral patterning of the polymer FL band due to MC structure. Polymer hybridization causes sufficient changes in the MC reflectivity that sensitive detection of explosive vapors can also be performed by these measurements. Levitsky *et al.* [105] have demonstrated recently these properties with fluorescent polymer-PSi MC devices, in which a conjugated chemosensitive polymer entrapped in PSi MC allowed detection of vapors of explosive NACs via a modulation of both FL and reflectance signals. The MC resonant peak in the reflectance spectra shifted upon vapor exposure and the broad polymer FL band showed patterning that was sensitive to NAC vapor exposure.

Structure–property studies have demonstrated many potential applications of nanostructured materials fabricated on different length scales, particularly for trace analysis. An important development in this area has been the fabrication of materials with hierarchical porous structures, which combine the multiple advantages of different pore sizes [133–143]. For instance, a material with interwoven meso- and macroporous structures can provide a high specific surface area and more interaction sites via small pores, whereas the presence of additional macropores can offer increased mass transport and easier accessibility to the active sites through the material. These features make such materials highly suitable for applications in catalysis, separations, and sensor devices, especially if specific action sites or recognition units are attached to these materials. For example, a series of porphyrin or metalloporphyrin-doped silica films with bimodal porous structures have been utilized to detect trace amounts of vapors of explosives such as TNT, DNT, and nitrobenzene (NB) [141–143]. The results have demonstrated that an appropriate combination of macropores and mesopores can achieve high molecule permeability and high density of interaction sites. As a result, silica films with bimodal porous structures exhibit much more efficient FL quenching than single modal porous films. For example, close to 55% of the initial FL was quenched after 10 s of exposure to 10 ppbv TNT vapor; a result that was nearly double that of conjugated polymer based sensor materials reported previously [71, 109, 112, 117]. Besides their remarkable TNT sensitivity, these hybrid films have several additional advantages over other FL-based sensory materials. These include relatively simple preparation, low material costs, broad recognition of different NACs, and more stabile organic sensing elements (because of the inorganic matrix). Reversibility is also preserved, since the sensory properties of the constructed films can be easily recovered by washing with toluene.

7.2.3.2 Quenching Sensors Based on Fluorescent Nanofibril Films

One-dimensional crystalline structures of organic molecules on the nanometer scale are good candidates for explosives detection. This is because of their long exciton diffusion length [144] and their intrinsic large surface-to-volume ratio. Such one-dimensional (1D) nano- and supernanostructures, when self-assembled with extended planar molecular surfaces, enable effective 1D π–π stacking favorable for exciton migration by cofacial intermolecular electronic coupling [144] and flexible tuning of morphologies on the nano- or microscopic scale.

A fluorescent nanofibril film, fabricated from the alkoxycarbonyl- substituted, carbazole-cornered, arylene-ethynylene tetracycle (ACTC), has been reported to be an efficient sensing film for detection of explosives [50]. The incorporation of carbazole enhanced the electron-donating power of the molecule and thus increased the efficiency of FL quenching by oxidative explosives (e.g., NACs). The quenching response of these films was significantly faster than for other organic materials [107, 145]. This behavior was consistent with the fibril porous structure of the film, which facilitates both gaseous adsorption and exciton migration across the film. The quenching efficiency obtained for the ACTC films was also higher than other explosive sensing materials of the same thickness [146]. Both the porous film morphology and the extended one-dimensional π–π stacking facilitated access of quencher molecules to the excited states, thereby enhancing quenching. Another important characteristic of these films is the relative independence of their FL quenching response with film thickness. This behavior is in contrast to other organic film sensors for which quenching efficiency is inversely proportional to the film thickness, owing to diffusion-limited transport of excitons and gaseous adsorbates.

In another example, porphyrin-doped nanofibrous membranes have been fabricated without the addition of polymers and have been used as novel FL-quenching chemosensors for the rapid vapor detection of trace levels (10 ppbv) of TNT [51]. The films were prepared by sol–gel chemistry and the electrospinning technique. Owing to their large surface area and good gas permeability, these fluorescent nanofibrous membranes exhibit remarkable sensitivity to trace TNT vapor compared to tightly crosslinked silica films. However, their sensitivity was strongly dependent on the morphology and phase aggregation of the nanofibers. Reducing the nanofiber diameter and introducing a pore structure considerably enhanced the sensitivity of the resulting materials. Because of the strong hydrogen bonding between imino hydrogens of the porphyrin molecules and nitro groups of NACs, as well as their π-stacking, the porphyrin units had a relatively high affinity for NACs molecules, which provided a strong driving force for fast FL quenching.

Several effects are considered responsible for the remarkable sensitivity of electrospun nanocomposite fibers towards trace TNT vapor [141]. First, their unique bimodal porous structure provides a necessary condition for the facile diffusion of analytes to sensing elements, while the large surface area increases the number interaction sites between analyte molecules and sensing elements. Second, their strong binding and energy level matching to the TNT molecule ensure high quenching efficiency. This matching has been demonstrated in theoretical studies, which indicate that the FL quenching per unit time is affected by various factors, including the vapor pressure of analyte, the exergonicity of electron transfer, and the binding strength between sensing elements and analytes.

7.2.3.3 Quenching Sensors Based on Quantum Dots

Colloidal semiconductive nanocrystals (also called quantum dots, QDs) are spherical particles in a size regime dominated by strong quantum confinement of charge carriers. This confinement lifts the degeneracy of the carrier states within the conduction and valence bands, and increases the effective band gap energy significantly with decreasing particle size. An important consequence of this confinement effect is an inherent size dependence of optical properties, such as absorption and photoluminescence [147, 148].

Luminescent QDs have the potential to circumvent some of the functional limitations encountered by organic dyes in biotechnological applications. Recently, QDs with high quantum yields have found wider application as FL sensors [149–

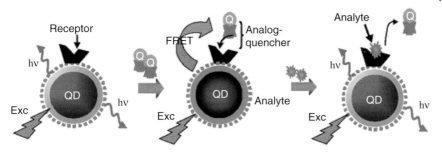

Figure 7.5 Schematic of a hybrid QD–antibody fragment FRET-based TNT sensor. Reprinted with permission from Reference [155]. Copyright 2005 American Chemical Society.

153]. The photoluminescence of QDs is readily tunable within a large spectral range by changing the QD size or introducing dopant ions. The dopants can potentially stimulate a spectral response toward a particular target analyte [152]. More importantly, QDs can be chemically modified to attach recognition receptors on their surfaces, thereby enhancing selectivity toward target species [151, 153]. Such FL chemosensors form the basis of the "lab-on-QDs" concept and have a remarkable advantage over other detection schemes in terms of sensitivity and selectivity [154]. For example, Goldman and coworkers [155, 156] have demonstrated QDs-based chemosensors based on CdSe QDs hybridized with antibody segments and dye molecules. Their specific detection toward TNT was achieved through fluorescence resonance energy transfer (FRET) between the QDs and dye. As shown in Figure 7.5, the hybrid sensor consisted of anti-TNT specific antibody fragments (receptors) attached to a hydrophilic QD via metal-affinity coordination. A dye-labeled TNT analog (analog-quencher) pre-bound in the antibody binding site quenched the QD PL via proximity-induced FRET. Addition of soluble TNT analyte displaced the dye-labeled analog, eliminating FRET and resulting in a concentration-dependent recovery of QD PL.

A general strategy for FRET-based biosensor design and construction employing multifunctional surface-tethered components from the above research team has been proposed (Figure 7.6a) [157, 158] and used in the detection of TNT and related compounds [158]. The modular biosensor consists of two modules: the biorecognition module and the modular arm. Both modules are specifically attached to a surface in a particular orientation. Choices for surface attachment include biotin–avidin chemistry, metal-affinity coordination, thiol bonding, hydrophobic interactions, DNA-directed immobilization, etc [157, 159]. The biorecognition module can consist of proteins (enzymes, receptors, bacterial periplasmic binding proteins [bPBPs], antibody fragments, peptides), aptamers, carbohydrates, DNA, PNA, RNA, etc. This module is site-specifically dye-labeled in the current configuration. The modular arm may consist of flexible moieties such as DNA, PNA, RNA, peptides, polymers, etc. The modular arm is also site-specifically dye-labeled. An analog of the primary analyte is attached to the distal end of the flexible arm to act as the recognition analog. Binding of this recognition element in the

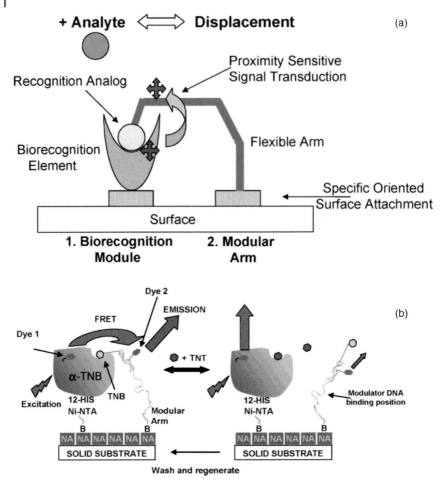

Figure 7.6 (a) Schematic of the modular biosensor consisting of two modules: the biorecognition module and the modular arm (see main text for detailed description); (b) schematic of the TNT targeting biosensor. The dye-labeled anti-TNT scFv fragment (1, the biorecognition module) is attached to the surface with Bio-X-NTA (Bio = biotin; X = aminomethoxy spacer; NTA = nitrilotriacetic acid chelator) coordinating the twelve histidines (12-HIS) and orienting the protein on the NeutrAvidin (NA). The dye-labeled TNB (1,3,5-trinitrobenzene, a TNT analog) DNA arm (2, modular arm) is attached to the NA via complementary hybridization to a biotinylated (B) flexible DNA linker. Both are added in equimolar amounts. ScFv binding of the TNB analog brings the protein located dye and DNA located dye into proximity, establishing FRET. Addition of TNT displaces the TNB analog and DNA arm, disrupting FRET in a concentration-dependent manner. Reprinted with permission from Reference [158]. Copyright 2005 American Chemical Society.

binding pocket of the biorecognition element assembles the sensor into the ground state by bringing both dyes into proximity, which establishes FRET. Addition of analyte competitively displaces the analog and signal transduction is designed to be sensitive to this displacement. FRET donor/acceptor can be placed on either module. Mechanisms of controlling binding affinity include stiffening the flexible arm or switching in of different affinity biorecognition elements. As shown in Figure 7.6b, the sensor consists of a dye-labeled anti-TNT antibody fragment (scFv) that interacts with a co-functional surface-tethered DNA arm. The arm consists of a flexible biotinylated DNA oligonucleotide specifically modified with a dye and terminated in a TNB (an analog of TNT, 1,3,5-trinitrobenzene) recognition element. Both of these elements are tethered to a NeutrAvidin (NA) surface with the TNB recognition element bound in the antibody fragment binding site, bringing the two dyes into proximity and establishing a baseline level of FRET. Addition of TNT or related explosive compounds (e.g., RDX and DNT) to the sensor environment alters FRET in a concentration-dependent manner. The sensor can be regenerated repeatedly by washing away of analyte and specific reformation of the sensor assembly. Sensor dynamic range can also be altered through the addition of DNA oligonucleotides that hybridize to a portion of the cofunctional arm. Although the authors have used quenching of organic dyes for biosensor signal generation, the use of optical components such as QDs should be also possible.

Although QDs-based chemosensors have been demonstrated extensively, the use of ion doping in QDs for obtaining spectral sensitivity to the analytes of interest has rarely been explored [160]. ZnS is a particularly suitable host material for a large variety of dopants because of its wide band gap (3.6 eV at 300 K) [161, 162]. When ZnS QDs were doped with Mn^{2+}, where a few Mn atoms were substituted for Zn atoms, an efficient visible orange luminescence at about 590 nm was observed [163]. This FL has been recently found to quench upon the binding of TNT, leading to the development of a new type of explosive sensor [154]. ZnS/Mn^{2+} QDs capped with organic amine were used to bind TNT species from solution and atmosphere. Binding was promoted by acid–base pairing interactions between the electron-rich amino ligands and the electron-deficient aromatic rings of TNT. The FL quenching mechanism was associated with electron transfer from the conductive band of ZnS to the lowest unoccupied molecular orbital (LUMO) of TNT anions. The amino ligands provide a two- to five-fold increase in quenching response to binding of NACs. Moreover, a large difference in quenching efficiency was observed for different types of NACs analytes, which was dependent on the electron-accepting abilities of the nitro analytes. Use of the amine-capped QDs enabled the detection of TNT in solution down to 1 nM and of TNT vapor to several ppbv.

7.2.3.4 Quenching Sensors Based on Organic Supernanostructures

Organic 1D nanostructures, such as nanofibers, nanowires, nanoribbons, and nanotubes, are usually obtained from π-conjugated materials [164–166] and have attracted considerable attention because of their potential applications in the field of chemical sensors. Because the optoelectronic properties of organic semiconductors in the solid state are strongly correlated with their structural hierarchy [165, 166], morphology control plays a key role in optimizing their optoelectronic performance. Several well-controllable organic nanostructures, including 1D microbelts and 3D flower-shaped supernanostructures, have been used for NACs sensing. These nanostructures can be easily self-assembled by a simple drop casting of the oligoarene solutions in different solvents [164]; generally the process yields high purity of the self-assembled nano/microstructures. The oligoarene derivative typically used is a planar condensed benzothiophene compound. Application of these structures in FL-quenching sensors for explosives has improved the detection speed for DNT and TNT more than 700-fold.

7.2.3.5 Fluoroimmunoassays Using QD-Antibody Conjugates

Goldman and coworkers [148, 167, 168] have developed a strategy for trace analysis that is based on the use of antibody-conjugated QDs in plate-based competitive immunoassays. The method, shown in Figure 7.7, was used for detection of ng quantities of the TNT-surrogate, TNB fluorescein, and RDX in aqueous samples. The QD–antibody conjugates were formed by using either a molecular adapter protein or using avidin. The analytes of interest competed with the surface-confined antigen for QD–antibody conjugates and a FL signal was measured from the plate after a washing step. An inverse relationship between the measured FL intensity and the analyte concentration was followed.

7.2.3.6 Displacement Immunosensors

The displacement immunosensor is a faster, more efficient detection system than a traditional immunoassay [169, 170]. It may be categorized as a type of chromatographic immunoassay [171]. With a displacement immunosensor, immunoassays are conducted with a device that is set under continuous buffer flow (Figure 7.8). The key components of the system are antibodies (Ab) immobilized on a solid support (e.g., micro-sized beads [172], agarose gel [173], or membranes

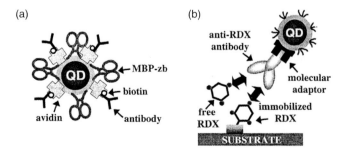

Figure 7.7 QD–antibody conjugates prepared using molecular bridges: (a) mixed surface conjugate after purification by crosslinked amylose affinity chromatography; (b) schematic of competitive assay for the explosive RDX dissolved in water. Reprinted with permission from Reference [168]. Copyright 2002 Elsevier.

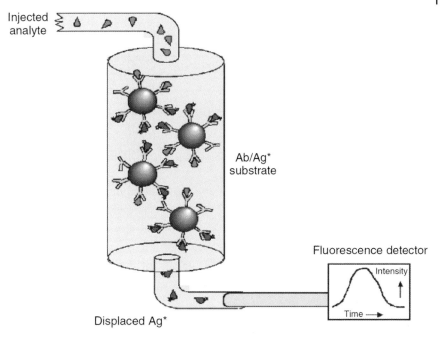

Figure 7.8 Schematic of the displacement immunosensor method. Step 1: monoclonal antibodies are covalently attached to a solid substrate and incubated with labeled antigen analog (Ab/Ag* substrate). Step 2: sample is injected into the flow stream and the labeled antigen is displaced. Step 3: displaced labeled antigen and excess sample move downstream, where a detector measures the label. Reprinted with permission from Reference [169]. Copyright 2008 Elsevier.

[174]), antigen analogs (Ag*) that are labeled with a reporter molecule (e.g., a fluorophore), and the associated hardware needed to establish a controlled flow system. Generally, monoclonal antibodies and reporter molecules are first linked to a prepared surface and a target antigen/analyte, respectively. The Ag* is then allowed to react with the immobilized Ab until equilibrium is reached (e.g., 2–15 h). To perform an assay, the solid support coated with the Ab/Ag* complex is placed in a buffer flow. When a sample containing the analyte of interest is injected into the flow stream, the Ag* molecules are displaced into the buffer and measured downstream using a FL detector. The measured FL intensity is proportional to the concentration of analyte molecules injected, within a predetermined linear range for each antibody.

The entire analytical process only takes 30–90 s, and multiple sample injections can be made on a single prepared surface. However, high analyte concentrations (>10–20% of surface capacity) could degrade the sensor response very rapidly [169]. Table 7.3 lists explosive analytes and formats reported for the displacement immunosensor.

Table 7.3 Explosive analytes and formats reported for the displacement immunosensor. Modified from Reference [169].

Analyte	Format used	Sample sources	References
2,4-Dinitrophenol (DNP)	Beads	Lab standards	[172]
TNT	Agarose gel	Lab standards	[173]
	Membranes	Groundwater	[174, 175]
	Capillary	Lab standards, seawater	[176–178]
	Sol–gel on poly(methyl methacrylate)	Groundwater	[179, 180]
		Lab standards	[181]
RDX	Agarose gel	Groundwater	[174, 182]
	Membranes	Groundwater	[179]
	Capillary	Lab standards	[183]
PETN	Copolymer beads	Lab standards	[184]

7.2.4 Microcantilever Sensors

Microcantilever sensors are a type of surface-effect sensor that rely on a substrate, often a microstructure, with an adsorption affinity for specific molecules [27]. They have been attracting attention as potential detectors for explosive vapors due to their advantages such as miniature size, array-based detection, high sensitivity, real-time operation, and low power consumption [26]. They are based on micromachined silicon. The cantilevers are typically 100 μm long, 40 μm wide, and 1 μm thick and are coated with sensory molecules or materials. Microcantilever sensors can be operated in static mode where adsorption induced surface stress is monitored as bending of a cantilever or in dynamic mode where mass loading due to molecular adsorption is monitored as a variation in resonance frequency of the cantilever. In the static mode, adsorption of molecules on one side of the cantilever makes the cantilever bend [185, 186]. Differential adsorption on the surfaces results in differential stress. Low-frequency, low-spring constant cantilevers result in large bending due to adsorption-induced forces. In the dynamic mode, the resonance frequency of the cantilever varies sensitively as a function of mass loading. The sensitivity of the dynamic mode of operation is related directly to the frequency of the cantilever; the higher the frequency, the higher the sensitivity. Figure 7.9 shows a schematic diagram of cantilever bending due to differential molecular adsorption.

Since individual cantilevers lack the ability to selectively identify adsorbed targets, chemically selective layers (receptors) are often coated on the cantilever surface.

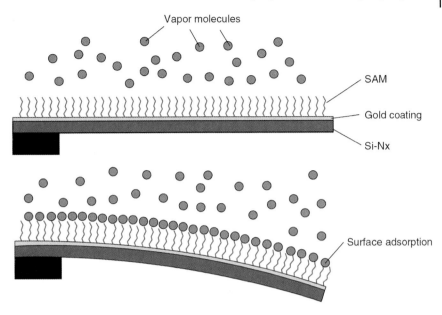

Figure 7.9 Schematic diagram of a cantilever bending due to differential adsorption of analyte molecules. Reprinted with permission from Reference [27]. Copyright 2007 Elsevier.

This allows some form of adsorbed molecular speciation when used in conjunction with cantilever arrays. As the vapor pressures of most explosives are extremely small at room temperature, their vapors are often produced by maintaining the sources at higher temperatures. Scanning the temperature of the cantilever allows detection of various explosives according to their temperature of deflagration.

A photothermal microcantilever technique for standoff detection of explosive residues has been reported [187]. Figure 7.10 shows the experimental setup. The cantilevers were deposited with a thin film of 300 nm gold (on one side), with a 3 nm thick using Cr film serving as an adhesion layer. The standoff photothermal spectroscopy was demonstrated as a variation of photothermal deflection spectroscopy. In photothermal deflection spectroscopy, sequential exposure of an adsorbate-covered bimaterial cantilever to infrared (IR) radiation causes the cantilever to bend more when the adsorbed molecules absorb IR irradiation. Deflection of the cantilever in the cited work was measured by the light (635 nm diode laser) reflected from the apex of the cantilever into a position sensitive detector (PSD). In general, a plot of the cantilever bending as a function of illuminating IR wavelength resembles the traditional IR absorption spectrum of the adsorbate. The sensitivity of the method is related directly to the thermal sensitivity of the bimaterial cantilever. The typical sensitivity of a bimaterial cantilever, $100\,\text{ng}\,\text{cm}^{-2}$, is sufficient to detect the explosive contamination generally found on explosive devices. The sensitivity of the technique could be further improved by optimizing the bimaterial cantilever and by using higher intensity infrared sources.

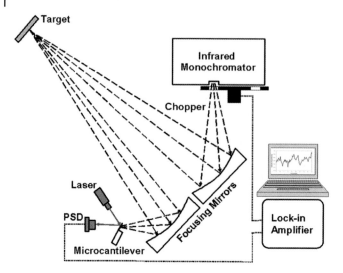

Figure 7.10 Experimental setup of a photothermal microcantilever-based technique for standoff detection of explosive. IR light is reflected off the target and focused onto the microcantilever sensor. The photothermal induced bending of the cantilever is measured by using the same laser reflection technique found in AFM. Reprinted with permission from Reference [187]. Copyright 2008 American Institute of Physics.

7.2.5
Metal Oxide Semiconductor (MOS) Nanoparticle Gas Sensors

The metal oxide semiconductor (MOS) gas sensor [188] is a common detector for volatile organic compounds because of its sensitivity, low-cost, and easy manufacture. Several reports on its use for explosive detection have appeared. In the first study, the sensing of military grade TNT in various substrates (air, sand and soil) was investigated by using TiO_2 thin film sensors with static headspace sampling [189]. It was found that MOS sensors can feasibly detect solid explosives. However, the low vapor concentration of solid explosives makes their detection an extremely difficult and challenging task. To solve this problem, some researchers have suggested using Pt or Pd catalysts in the carrier gas line to increase the sensitivity of the MOS sensors. Such a change could allow the sampling of solids, liquids, and gases with a gas sensor [117]. In another study, several explosives (e.g., DNT, NH_4NO_3, and picric acid) were investigated by using ZnO-doped nanoparticle sensors with additives of Sb_2O_3, TiO_2, V_2O_5, and WO_3 [190].

7.2.6
Surface-Enhanced Ramam Scattering Spectroscopy

Surface-enhanced Raman scattering spectroscopy (SERS) typically yields higher analyte sensitivity than FL spectroscopy due to enhanced absorption cross-sections. It also provides detailed vibrational spectra of adsorbed species, thereby yielding greater selectivity than FL methods. As such, SERS is one of the most incisive analytical methods for chemical and biochemical detection and analysis [59, 191]. The metallic NPs (e.g., Au and Ag NPs [192]) that make SERS possible are of fundamental interest since they possess unique size-dependent properties. The use of SERS for trace explosive detection was first investigated during the late 1990s [192, 193], and SERS detection of 2,4-DNT vapor to ~1 ppbv was demonstrated [194]. Subsequently [195], a field-portable unit demonstrated a limit of detection of 5 ppbv vapor DNT and the ability to locate buried land mines. More recently [196], nano-engineered SERS substrates have been employed, and sensitivity to ppbv levels for some nerve agent and explosive simulants has been demonstrated.

SERS has also been used to selectively detect functionalized TNT. Additional signal enhancement due to the resonance Raman effect has resulted in detection limits much better than 1 nM in solution [197]. With similar ideas, detection of explosive RDX via reduction and subsequent functionalization has been reported [198].

7.3
Conclusions

Six types of techniques involving nanomaterials for trace detection and quantification of high explosives have been reviewed. These techniques are based on electrochemistry, ECL, FL, microcantilevers, metal oxide semiconductive NPs, and surface enhanced scattering spectroscopy. A wide variety of nanomaterials have been used, which include metallic and metallic oxide NPs, CNTs, polycrystalline metal complexes, conjugated conductive polymer nanostructures, and QDs. Use of nanomaterials generally can increase the sensitivity, lower the detection limit, and improve the selectivity of explosive detection. This is because nanomaterials possess several unique properties, including an electrocatalytic effect, large surface area, and size-dependent tunability. Although various techniques are available for trace detection of high explosives, each technique has its own advantages and limitations in terms of cost, portability, sensitivity, reliability, and selectivity, etc. For example, electrochemical methods are generally applicable to electroactive explosives in solution, whereas FL-quenching-based sensors are sensitive to nitroaromatics vapors but not to peroxide-based explosives. ECL-based analytical methods offer many distinct features. However, at present studies on ECL-based trace detection of explosives are still limited. It is foreseeable that ECL, with the combination of QDs-based immunoassays, could be used in simultaneous multiplexing detection of explosives. In addition, sensitive detection and determination of the entire range of nitrated explosives may be realized on the basis of the ECL quenching or inhibition of an ECL standard such as the $[Ru(bpy)_3]^{2+}$/TPrA system after introduction of a target analyte into the system. Most of the studies so far have focused on detection of nitroaromatics and a few are related to peroxide explosives. Nanotechnology will continue to play an important role in developing new explosive sensors that not only have better sensitivity and higher selectivity toward analytes of interest but also offer greater field portability, faster response, cheaper maintenance, and easier operation.

Acknowledgments

The partial support of this work by grants from the National Science Foundation MRSEC (NSF-DMR 0213883) via the University of Southern Mississippi and the NFS CAREER award (NFS-0955878) (WJM), the Overseas Project of Jingsu Province (China) (WJM and CWG), and the Jiangsu Government Scholarship for Overseas Studies (China) (JS and XHJ) is gratefully acknowledged.

References

1. Yinon, J. (ed.) (2007) *Counterterrorist Detection Techniques of Explosives*, Elsevier, B. V., Amsterdam.
2. Marshall, M. and Oxley, J.C. (eds) (2009) *Aspects of Explosive Detection*, Elsevier B. V., Amsterdam.
3. Davies, A.G., Burnett, A.D., Fan, W., Linfield, E.H., and Cunningham, J.E. (2008) *Mater. Today*, **11** (3), 18–26.
4. Jacoby, M. (2009) *Chem. Eng. News*, **87** (22), 10–13.
5. Armitt, D., Zimmermann, P., and Ellis-Steinborner, S. (2008) *Rapid Commun. Mass Spectrom.*, **22** (7), 950–958.
6. Cotte-Rodriguez, I., Hernandez-Soto, H., Chen, H., and Cooks, R.G. (2008) *Anal. Chem.*, **80** (5), 1512–1519.
7. Crowson, A. and Beardah, M.S. (2001) *Analyst*, **126** (10), 1689–1693.
8. Kojima, K., Sakairi, M., Takada, Y., and Nakamura, J. (2000) *J. Mass Spectrom. Soc. Jpn.*, **48** (5), 360–362.
9. McLuckey, S.A., Goeringer, D.E., Asano, K.G., Vaidyanathan, G., and Stephenson, J.L., Jr. (1996) *Rapid Commun. Mass Spectrom.*, **10** (3), 287–298.
10. Pacheco-Londono, L., Primera, O.M., Ramirez, M., Ruiz, O., and Hernandez-Rivera, S. (2005) *Proc. SPIE-Int. Soc. Opt. Eng.*, **5778** (Pt. 1, Sensors, and Command, Control, Communications, and Intelligence (C3I) Technologies for Homeland Defense IV), 317–326.
11. Stambouli, A., El Bouri, A., Bouayoun, T., and Bellimam, M.A. (2004) *Forensic Sci. Int.*, **146** (Suppl.), S191–S194.
12. Tamiri, T. and Zitrin, S. (1986) *J. Energ. Mater.*, **4** (1–4), 215–237.
13. Widmer, L., Watson, S., Schlatter, K., and Crowson, A. (2002) *Analyst*, **127** (12), 1627–1632.
14. Xu, X., van de Craats, A.M., Kok, E.M., and de Bruyn, P.C.A.M. (2004) *J. Forensic Sci.*, **49** (6), 1230–1236.
15. Yinon, J. (2006) *Am. Lab.*, **38** (12), 18, 20–23.
16. Yinon, J., Boettger, H.G., and Weber, W.P. (1972) *Anal. Chem.*, **44** (13), 2235–2237.
17. Zhao, X. and Yinon, J. (2002) *J. Chromatogr. A*, **977** (1), 59–68.
18. Yinon, J. (2007) Detection of explosives by mass spectrometry, in *Counterterrorist Detection Techniques of Explosives* (ed. J. Yinon), Elsevier B. V., Amsterdam, Ch. 2, pp. 41–60.
19. Buttigieg, G.A., Knight, A.K., Denson, S., Pommier, C., and Bonner Denton, M. (2003) *Forensic Sci. Int.*, **135** (1), 53–59.
20. Lokhnauth, J.K. and Snow, N.H. (2006) *J. Chromatogr. A*, **1105** (1–2), 33–38.
21. Marr, A.J. and Groves, D.M. (2003) *Int. J. Ion Mobil. Spectrom.*, **6** (2), 59–62.

22. Matz, L.M., Tornatore, P.S., and Hill, H.H. (2001) *Talanta*, **54** (1), 171–179.
23. Oxley, J.C., Smith, J.L., Kirschenbaum, L.J., Marimganti, S., and Vadlamannati, S. (2008) *J. Forensic Sci.*, **53** (3), 690–693.
24. Wilson, R. and Brittain, A. (1997) *Explosives in the Service of Man: The Nobel Heritage (Special Publication)* (eds J.E. Dolan and S.S. Langer), RSC Special Publication 203, Royal Society of Chemistry, London, pp. 92–101.
25. Eiceman, G.A. and Schmidt, H. (2009) Advances in ion mobility spectrometry of explosives, in *Aspects of Explosives Detection* (eds M. Marshall and J.C. Oxley), Elsevier B. V., Amsterdam, Ch. 9, pp. 171–202.
26. Datskos, P.G., Lavrik, N.V., and Sepaniak, M.J. (2003) *Sensor Lett.*, **1** (1), 25–32.
27. Senesac, L. and Thundat, T. (2007) Explosive vapor detection using microcantilever sensors, in *Counterterrorist Detection Techniques of Explosives* (ed. J. Yinon), Elsevier B. V., Amsterdam, Ch. 5, pp. 109–130.
28. Pinnaduwage, L.A., Boiadjiev, V., Hawk, J.E., and Thundat, T. (2003) *Appl. Phys. Lett.*, **83** (7), 1471–1473.
29. Pinnaduwage, L.A., Wig, A., Hedden, D.L., Gehl, A., Yi, D., Thundat, T., and Lareau, R.T. (2004) *J. Appl. Phys.*, **95** (10), 5871–5875.
30. Senesac, L. and Thundat, T.G. (2008) *Mater. Today*, **11** (3), 28–36.
31. Fu, X., Benson, R.F., Wang, J., and Fries, D. (2005) *Sens. Actuators, B*, **106** (1), 296–301.
32. Lu, D., Cagan, A., Munoz, R.A.A., Tangkuaram, T., and Wang, J. (2006) *Analyst*, **131** (12), 1279–1281.
33. Munoz, R.A.A., Lu, D., Cagan, A., and Wang, J. (2007) *Analyst*, **132** (6), 560–565.
34. Wang, J. (2004) *NATO Sci. Ser. II Math. Phys. Chem.*, **159** (Electronic Noses & Sensors for the Detection of Explosives), 131–142.
35. Wang, J., Bhada, R.K., Lu, J., and MacDonald, D. (1998) *Anal. Chim. Acta*, **361** (1–2), 85–91.
36. Wang, J. (2007) Electrochemical sensing of explosives, in *Counterterrorist Detection Techniques of Explosives* (ed. J. Yinon), Elsevier B. V., Amsterdam, Ch. 4, pp. 91–108.
37. Bruno, J.G. and Cornette, J.C. (1997) *Microchem. J.*, **56** (3), 305–314.
38. Jimenez, A.M. and Navas, M.J. (2004) *J. Hazard. Mater.*, **106** (1), 1–5.
39. Meaney, M.S. and McGuffin, V.L. (2008) *Anal. Bioanal. Chem.*, **391** (7), 2557–2576.
40. Monterola, M.P.P., Smith, B.W., Omenetto, N., and Winefordner, J.D. (2008) *Anal. Bioanal. Chem.*, **391** (7), 2617–2626.
41. Nguyen, D.H., Locquiao, S., Huynh, P., Zhong, Q., He, W., Christensen, D., Zhang, L., and Bilkhu, B. (2004) *NATO Sci. Ser. II Math. Phys. Chem.*, **159** (Electronic Noses & Sensors for the Detection of Explosives), 71–80.
42. Jimenez, A.M. and Navas, M.J. (2007) Detection of explosives by chemiluminescence, in *Counterterrorist Detection Techniques of Explosives* (ed. J. Yinon), Elsevier B. V., Amsterdam, Ch. 1, pp. 1–40.
43. Pamula, V.K. (2004) *NATO Sci. Ser. II Math. Phys. Chem.*, **159** (Electronic Noses & Sensors for the Detection of Explosives), 279–288.
44. Almog, J. and Zitrin, S. (2009) Colorimetric detection of explosives, in *Aspects of Explosives Detection* (eds M. Marshall and J.C. Oxley), Elsevier B. V., Amsterdam, Ch. 4, pp. 51–59.
45. Charles, P.T., Bart, J.C., Judd, L.L., Gauger, P.R., Ligler, F.S., and Kusterbeck, A.W. (1997) *Proc. SPIE-Int. Soc. Opt. Eng.*, **3105** (Chemical, Biochemical, and Environmental Fiber Sensors IX), 80–87.
46. Germain, M.E. and Knapp, M.J. (2008) *Inorg. Chem.*, **47** (21), 9748–9750.
47. Goodpaster, J.V. and McGuffin, V.L. (2001) *Anal. Chem.*, **73** (9), 2004–2011.
48. la Grone, M.J., Cumming, C.J., Fisher, M.E., Fox, M.J., Jacob, S., Reust, D., Rockley, M.G., and Towers, E. (2000) *Proc. SPIE-Int. Soc. Opt. Eng.*, **4038** (Pt. 1, Detection and Remediation Technologies for Mines and Minelike Targets V), 553–562.
49. Malashikhin, S. and Finney, N.S. (2008) *J. Am. Chem. Soc.*, **130** (39), 12846–12847.

50. Naddo, T., Che, Y., Zhang, W., Balakrishnan, K., Yang, X., Yen, M., Zhao, J., Moore, J.S., and Zang, L. (2007) *J. Am. Chem. Soc.*, **129** (22), 6978–6979.
51. Tao, S., Li, G., and Yin, J. (2007) *J. Mater. Chem.*, **17** (26), 2730–2736.
52. Thomas, S.W., III and Swager, T.M. (2009) Detection of explosives using amplified fluorescent polymers, in *Aspects of Explosives Detection* (eds M. Marshall and J.C. Oxley), Elsevier B. V., Amsterdam, Ch. 10, pp. 203–222.
53. Sarkar, D. and Somasundaran, P. (2002) *Langmuir*, **18** (22), 8271–8277.
54. Larsson, A., Angbrant, J., Ekeroth, J., Mansson, P., and Liedberg, B. (2006) *Sens. Actuators, B*, **B113** (2), 730–748.
55. Miura, N., Shankaran, D.R., Kawaguchi, T., Matsumoto, K., and Toko, K. (2007) *Electrochemistry*, **75** (1), 13–22.
56. Shankaran, D.R., Kawaguchi, T., Kim, S.J., Matsumoto, K., Toko, K., and Miura, N. (2007) *Int. J. Environ. Anal. Chem.*, **87** (10–11), 771–781.
57. Shankaran, D.R., Matsumoto, K., Toko, K., and Miura, N. (2005) *Chem. Sensors*, **21** (Suppl. A), 166–168.
58. Baker, G.A. and Moore, D.S. (2005) *Anal. Bioanal. Chem.*, **382** (8), 1751–1770.
59. Smith, W.E. (2008) *Chem. Soc. Rev.*, **37** (5), 955–964.
60. De La Cruz-Montoya, E., Jerez, J.I., Balaguera-Gelves, M., Luna-Pineda, T., Castro, M.E., and Hernandez-Rivera, S.P. (2006) Enhanced Raman spectroscopy of 2,4,6-TNT in anatase and rutile titania nanocrystals, in *Optics and Photonics in Global Homeland Security II* (ed. T.T. Saito and D. Lehrfeld), Proceedings of the SPIE, Vol. **6203**, SPIE, Bellingham, WA, p. 62030X-7.
61. Ruffin, P.B., Brantley, C., and Edwards, E. (2008) Innovative smart micro sensors for army weaponry applications, in *Nanosensors and Microsensors for Bio-Systems 2008* (ed. V.K. Varadin), Proceedings of the SPIE, Vol. **6931**, SPIE, Bellingham, WA, pp. 693102–693112.
62. Ko, H., Chang, S., and Tsukruk, V.V. (2009) *ACS Nano*, **3** (1), 181–188.
63. Smith, R.C. and Connelly, J.M. (2009) CT technologies, in *Aspects of Explosives Detection* (eds M. Marshall and J.C. Oxley), Elsevier B. V., Amsterdam, Ch. 7, pp. 131–146.
64. Everts, S. (2008) *Chem. Eng. News*, **86** (39), 34.
65. Wang, J., Chen, G., Chatrathi, M.P., Fujishima, A., Tryk, D.A., and Shin, D. (2003) *Anal. Chem.*, **75** (4), 935–939.
66. Wang, J. and Thongngamdee, S. (2003) *Anal. Chim. Acta*, **485** (2), 139–144.
67. Krausa, M. (2004) *NATO Sci. Ser. II Math. Phys. Chem.*, **167** (Vapour and Trace Detection of Explosives for Anti-Terrorism Purposes), 1–9.
68. Krausa, M. and Schorb, K. (1999) *J. Electroanal. Chem.*, **461** (1, 2), 10–13.
69. Filanovsky, B., Tur'yan, Y.I., Kuselman, I., Burenko, T.y., and Shenhar, A. (1998) *Anal. Chim. Acta*, **364** (1–3), 181–188.
70. Kim, J.-W., JKim.-H., and Tung, S. (2008) Nanoscale flagellar-motor based MEMS biosensor for explosive detection. Nano/Micro Engineered and Molecular Systems, 2008. NEMS 2008. 3rd IEEE International Conference on, 2008, pp. 630–632.
71. Wang, J., Thongngamdee, S., and Kumar, A. (2004) *Electroanalysis*, **16** (15), 1232–1235.
72. Hilmi, A., Luong, J.H.T., and Nguyen, A.-L. (1999) *Anal. Chem.*, **71** (4), 873–878.
73. Weng, J., Xue, J., Wang, J., Ye, J.-S., Cui, H., Sheu, F.-S., and Zhang, Q. (2005) *Adv. Funct. Mater.*, **15** (4), 639–647.
74. Pardo-Yissar, V., Gabai, R., Shipway, A.N., Bourenko, T., and Willner, I. (2001) *Adv. Mater.*, **13** (17), 1320–1323.
75. Filanovsky, B., Markovsky, B., Bourenko, T., Perkas, N., Persky, R., Gedanken, A., and Aurbach, D. (2007) *Adv. Funct. Mater.*, **17** (9), 1487–1492.
76. Zhang, H.-X., Cao, A.-M., Hu, J.-S., Wan, L.-J., and Lee, S.-T. (2006) *Anal. Chem.*, **78** (6), 1967–1971.

77. Wang, J., Hocevar, S.B., and Ogorevc, B. (2004) *Electrochem. Commun.*, **6** (2), 176–179.
78. Hrapovic, S., Majid, E., Liu, Y., Male, K., and Luong, J.H.T. (2006) *Anal. Chem.*, **78** (15), 5504–5512.
79. Schulte-Ladbeck, R., Kolla, P., and Karst, U. (2003) *Anal. Chem.*, **75** (4), 731–735.
80. Schulte-Ladbeck, R. and Karst, U. (2003) *Chromatographia*, **57** (Suppl.), S/61–S/66.
81. Borisova, A.V., Karyakina, E.E., Cosnier, S., and Karyakin, A.A. (2009) *Electroanalysis*, **21** (3–5), 409–414.
82. Karyakin, A.A. (2001) *Electroanalysis*, **13** (10), 813–819.
83. Zhang, X., Wang, J., Ogorevc, B., and Spichiger, U.E. (1999) *Electroanalysis*, **11** (13), 945–949.
84. Karyakin, A.A., Karyakina, E.E., and Gorton, L. (2000) *Anal. Chem.*, **72** (7), 1720–1723.
85. Lukachova, L.V., Kotel'nikova, E.A., D'Ottavi, D., Shkerin, E.A., Karyakina, E.E., Moscone, D., Palleschi, G., Curulli, A., and Karyakin, A.A. (2002) *Bioelectrochemistry*, **55** (1–2), 145–148.
86. Schulte-Ladbeck, R., Kolla, P., and Karst, U. (2002) *Analyst*, **127** (9), 1152–1154.
87. Miao, W. (2007) *Handbook of Electrochemistry* (ed. C.G. Zoski), Elsevier, Amsterdam.
88. Miao, W. (2008) *Chem. Rev.*, **108** (7), 2506–2553.
89. Bard, A.J. (ed.)(2004) *Electrogenerated Chemiluminescence*, Marcel Dekker, Inc., New York.
90. Richter, M.M. (2004) *Chem. Rev.*, **104** (6), 3003–3036.
91. Miao, W. and Bard, A.J. (2003) *Anal. Chem.*, **75** (21), 5825–5834.
92. Miao, W. and Bard, A.J. (2004) *Anal. Chem.*, **76** (23), 7109–7113.
93. Miao, W. and Bard, A.J. (2004) *Anal. Chem.*, **76** (18), 5379–5386.
94. Miao, W. and Choi, J.-P. (2004) Coreactants, in *Electrogenerated Chemiluminescence* (ed. A.J. Bard), Marcel Dekker, Inc., New York, Ch. 5, pp. 213–272.
95. Wilson, R., Clavering, C., and Hutchinson, A. (2003) *Analyst*, **128** (5), 480–485.
96. Wilson, R., Clavering, C., and Hutchinson, A. (2003) *J. Electroanal. Chem.*, **557**, 109–118.
97. Wilson, R., Clavering, C., and Hutchinson, A. (2003) *Anal. Chem.*, **75** (16), 4244–4249.
98. Parajuli, S. and Miao, W. (2009) *Anal. Chem.*, **81** (13), 5267.
99. Pittman, T.L., Thomson, B., and Miao, W. (2009) *Anal. Chim. Acta*, **632** (2), 197–202.
100. Zhou, H., Kasai, S., and Matsue, T. (2001) *Anal. Biochem.*, **290** (1), 83–88.
101. Thorpe, G.H.G. and Kricka, L.J. (1986) Enhanced chemiluminescent reactions catalyzed by horseradish peroxidase, *Bioluminescence and Chemiluminescence, Part B* (eds M. DeLuca and W.D. McElroy), Methods in Enzymology, Vol. **133**, Academic Press, London, pp. 331–353.
102. Anderson, G.P., Moreira, S.C., Charles, P.T., Medintz, I.L., Goldman, E.R., Zeinali, M., and Taitt, C.R. (2006) *Anal. Chem.*, **78** (7), 2279–2285.
103. Shankaran, D.R., Kawaguchi, T., Kim, S.J., Matsumoto, K., Toko, K., and Miura, N. (2006) *Anal. Bioanal. Chem.*, **386** (5), 1313–1320.
104. Thomas, S.W., Joly, G.D., and Swager, T.M. (2007) *Chem. Rev.*, **107** (4), 1339–1386.
105. Levitsky, I.A., Euler, W.B., Tokranova, N., and Rose, A. (2007) *Appl. Phys. Lett.*, **90** (4), 041904-3.
106. Cumming, C.J., Aker, C., Fisher, M., Fok, M., la Grone, M.J., Reust, D., Rockley, M.G., Swager, T.M., Towers, E., and Williams, V. (2001) *IEEE Trans. Geosci. Remote Sens.*, **39** (6), 1119–1128.
107. Rose, A., Zhu, Z., Madigan, C.F., Swager, T.M., and Bulovic, V. (2005) *Nature*, **434** (7035), 876–879.
108. Zahn, S. and Swager, T.M. (2002) *Angew. Chem. Int. Ed.*, **41** (22), 4225–4230.
109. Yang, J.-S. and Swager, T.M. (1998) *J. Am. Chem. Soc.*, **120** (46), 11864–11873.

110. Yamaguchi, S. and Swager, T.M. (2001) *J. Am. Chem. Soc.*, **123** (48), 12087–12088.
111. Narayanan, A., Varnavski, O.P., Swager, T.M., and Goodson, T., III (2008) *J. Phys. Chem. C*, **112** (4), 881–884.
112. Sohn, H., Calhoun, R.M., Sailor, M.J., and Trogler, W.C. (2001) *Angew. Chem. Int. Ed.*, **40** (11), 2104–2105.
113. Sohn, H., Sailor, M.J., Magde, D., and Trogler, W.C. (2003) *J. Am. Chem. Soc.*, **125** (13), 3821–3830.
114. Narayanan, A., Varnavski, O.P., Swager, T.M., and Goodson, T., III (2007) *PMSE Preprints*, **96**, 678.
115. Li, J., Kendig, C.E., and Nesterov, E.E. (2007) *J. Am. Chem. Soc.*, **129** (51), 15911–15918.
116. Toal, S.J., Magde, D., and Trogler, W.C. (2005) *Chem. Commun.*, **43**, 5465–5467.
117. Content, S., Trogler, W.C., and Sailor, M.J. (2000) *Chem. Eur. J.*, **6** (12), 2205–2213.
118. Albert, K.J. and Walt, D.R. (2000) *Anal. Chem.*, **72** (9), 1947–1955.
119. Chan, S., Horner, S.R., Miller, B.L., and Fauchet, P.M. (2001) *Mater. Res. Soc. Symp. Proc.*, **638** (Microcrystalline and Nanocrystalline Semiconductors – 2000), F10 4 1–F10 4 6.
120. Chvojka, T., Holec, T., Jelinek, I., Nemec, I., Jindrich, J., Lorenc, M., Koutnikova, J., Kral, V., and Dian, J. (2003) *Proc. SPIE-Int. Soc. Opt. Eng.*, **5036** (Photonics, Devices, and Systems II), 51–56.
121. De Stefano, L., Alfieri, D., Rea, I., Rotiroti, L., Malecki, K., Moretti, L., Della Corte, F.G., and Rendina, I. (2007) *Phys. Status Solidi A*, **204** (5), 1459–1463.
122. De Stefano, L., Moretti, L., Rendina, I., and Rotiroti, L. (2005) *Sens. Actuators, B*, **111-112**, 522–525.
123. De Stefano, L., Rendina, I., Moretti, L., Rossi, A.M., Lamberti, A., Longo, O., and Arcari, P. (2003) *Proc. SPIE-Int. Soc. Opt. Eng.*, **5118** (Nanotechnology), 305–309.
124. De Stefano, L., Rendina, I., Moretti, L., Tundo, S., and Rossi, A.M. (2004) *Appl. Opt.*, **43** (1), 167–172.
125. Furbert, P., Lu, C., Winograd, N., and DeLouise, L. (2008) *Langmuir*, **24** (6), 2908–2915.
126. Mahmoudi, B., Gabouze, N., Guerbous, L., Haddadi, M., Cheraga, H., and Beldjilali, K. (2007) *Mater. Sci. Eng. B.*, **138** (3), 293–297.
127. Min, N.K., Lee, C.W., Jeong, W.S., and Kim, D.I. (1999) *J. Korean Electrochem. Soc.*, **2** (1), 17–22.
128. Mulloni, V. and Pavesi, L. (2000) *Appl. Phys. Lett.*, **76** (18), 2523–2525.
129. Rea, I., Iodice, M., Coppola, G., Rendina, I., Marino, A., and De Stefano, L. (2009) *Sens. Actuators, B*, **139** (1), 39–43.
130. Rotiroti, L., De Stefano, L., Rendina, I., Moretti, L., Rossi, A.M., and Piccolo, A. (2005) *Biosens. Bioelectron.*, **20** (10), 2136–2139.
131. Saarinen, J.J., Weiss, S.M., Fauchet, P.M., and Sipe, J.E. (2005) *Opt. Express*, **13** (10), 3754–3764.
132. Zangooie, S., Bjorklund, R., and Arwin, H. (1997) *Sens. Actuators, B*, **43** (1–3), 168–174.
133. Coppens, M.O., Sun, J., and Maschmeyer, T. (2001) *Catal. Today*, **69** (1–4), 331–335.
134. Falcaro, P., Malfatti, L., Kidchob, T., Giannini, G., Falqui, A., Casula, M.F., Amenitsch, H., Marmiroli, B., Grenci, G., and Innocenzi, P. (2009) *Chem. Mater.*, **21** (10), 2055–2061.
135. Fujita, S., Nakano, H., Ishii, M., Nakamura, H., and Inagaki, S. (2006) *Microporous Mesoporous Mater.*, **96** (1–3), 205–209.
136. Kuang, D., Brezesinski, T., and Smarsly, B. (2004) *J. Am. Chem. Soc.*, **126** (34), 10534–10535.
137. Li, Y., Cai, W., Cao, B., Duan, G., Sun, F., Li, C., and Jia, L. (2006) *Nanotechnology*, **17** (1), 238–243.
138. Loiola, A.R., da Silva, L.R.D., Cubillas, P., and Anderson, M.W. (2008) *J. Mater. Chem.*, **18** (41), 4985–4993.
139. Yacou, C., Fontaine, M.-L., Ayral, A., Lacroix-Desmazes, P., Albouy, P.-A., and Julbe, A. (2008) *J. Mater. Chem.*, **18** (36), 4274–4279.
140. Yamauchi, Y., Gupta, P., Fukata, N., and Sato, K. (2009) *Chem. Lett.*, **38** (1), 78–79.

141. Tao, S. and Li, G. (2007) *Colloid. Polym. Sci.*, **285** (7), 721–728.
142. Tao, S., Shi, Z., Li, G., and Li, P. (2006) *ChemPhysChem*, **7** (9), 1902–1905.
143. Tao, S., Yin, J., and Li, G. (2008) *J. Mater. Chem.*, **18** (40), 4872–4878.
144. Datar, A., Balakrishnan, K., Yang, X., Zuo, X., Huang, J., Oitker, R., Yen, M., Zhao, J., Tiede, D.M., and Zang, L. (2006) *J. Phys. Chem. B*, **110** (25), 12327–12332.
145. Yang, J.-S. and Swager, T.M. (1998) *J. Am. Chem. Soc.*, **120** (21), 5321–5322.
146. Toal, S.J. and Trogler, W.C. (2006) *J. Mater. Chem.*, **16** (28), 2871–2883.
147. Mattoussi, H., Radzilowski, L.H., Dabbousi, B.O., Thomas, E.L., Bawendi, M.G., and Rubner, M.F. (1998) *J. Appl. Phys.*, **83** (12), 7965–7974.
148. Goldman, E.R., Balighian, E.D., Kuno, M.K., Labrenz, S., Tran, P.T., Anderson, G.P., Mauro, J.M., and Mattoussi, H. (2002) *Phys. Status Solidi B*, **229** (1), 407–414.
149. Medintz, I.L., Uyeda, H.T., Goldman, E.R., and Mattoussi, H. (2005) *Nat. Mater.*, **4** (6), 435–446.
150. Han, M., Gao, X., Su, J.Z., and Nie, S. (2001) *Nat. Biotechnol.*, **19** (7), 631–635.
151. Medintz, I.L., Clapp, A.R., Brunel, F.M., Tiefenbrunn, T., Tetsuo Uyeda, H., Chang, E.L., Deschamps, J.R., Dawson, P.E., and Mattoussi, H. (2006) *Nat. Mater.*, **5** (7), 581–589.
152. Peng, H., Zhang, L., Kjallman, T.H.M., and Soeller, C. (2007) *J. Am. Chem. Soc.*, **129** (11), 3048–3049.
153. Dayal, S. and Burda, C. (2007) *J. Am. Chem. Soc.*, **129** (25), 7977–7981.
154. Tu, R., Liu, B., Wang, Z., Gao, D., Wang, F., Fang, Q., and Zhang, Z. (2008) *Anal. Chem.*, **80** (9), 3458–3465.
155. Goldman, E.R., Medintz, I.L., Whitley, J.L., Hayhurst, A., Clapp, A.R., Uyeda, H.T., Deschamps, J.R., Lassman, M.E., and Mattoussi, H. (2005) *J. Am. Chem. Soc.*, **127** (18), 6744–6751.
156. Medintz, I.L., Goldman, E.R., Clapp, A.R., Uyeda, H.T., Lassman, M.E., Hayhurst, A., and Mattoussi, H. (2005) *Proc. SPIE-Int. Soc. Opt. Eng.*, **5705** (Nanobiophotonics and Biomedical Applications II), 166–174.
157. Medintz, I.L., Anderson, G.P., Lassman, M.E., Goldman, E.R., Bettencourt, L.A., and Mauro, J.M. (2004) *Anal. Chem.*, **76** (19), 5620–5629.
158. Medintz, I.L., Goldman, E.R., Lassman, M.E., Hayhurst, A., Kusterbeck, A.W., and Deschamps, J.R. (2005) *Anal. Chem.*, **77** (2), 365–372.
159. Wada, A., Mie, M., Aizawa, M., Lahoud, P., Cass, A.E.G., and Kobatake, E. (2003) *J. Am. Chem. Soc.*, **125** (52), 16228–16234.
160. Santra, S., Yang, H., Holloway, P.H., Stanley, J.T., and Mericle, R.A. (2005) *J. Am. Chem. Soc.*, **127** (6), 1656–1657.
161. Denzler, D., Olschewski, M., and Sattler, K. (1998) *J. Appl. Phys.*, **84** (5), 2841–2845.
162. Rossetti, R., Hull, R., Gibson, J.M., and Brus, L.E. (1985) *J. Chem. Phys.*, **82** (1), 552–559.
163. Bhargava, R.N., Gallagher, D., Hong, X., and Nurmikko, A. (1994) *Phys. Rev. Lett.*, **72** (3), 416.
164. Wang, L., Zhou, Y., Yan, J., Wang, J., Pei, J., and Cao, Y. (2009) *Langmuir*, **25** (3), 1306–1310.
165. Hoeben, F.J.M., Jonkheijm, P., Meijer, E.W., and Schenning, A.P.H.J. (2005) *Chem. Rev.*, **105** (4), 1491–1546.
166. Grimsdale, A.C. and Mullen, K. (2005) *Angew. Chem. Int. Ed.*, **44** (35), 5592–5629.
167. Goldman, E.R., Anderson, G.P., Tran, P.T., Mattoussi, H., Charles, P.T., and Mauro, J.M. (2002) *Anal. Chem.*, **74** (4), 841–847.
168. Mattoussi, H., Kenneth, K., Goldman, E.R., Anderson, G.P., and Mauro, J.M. (2002) Colloidal semiconductor quantum dot conjugates in biosensing, in *Optical Biosensors: Present and Future*, 1st edn (eds F.S. Ligler and C.R. Taitt), Elseveir B. V., Amsterdam, Ch. 17, pp. 537–569.
169. Kusterbeck, A.W. and Blake, D.A. (2008) Flow immunosensors, in *Optical Biosensors: Today and Tomorrow*, 2nd edn (eds F.S. Ligler and C.R. Taitt), Elsevier B. V., Amsterdam, Ch. 5, pp. 243–286.
170. Ghindilis, A.L., Atanasov, P., Wilkins, M., and Wilkins, E. (1998) *Biosens. Bioelectron.*, **13** (1), 113–131.

171. Hage, D.S. and Nelson, M.A. (2001) *Anal. Chem.*, **73** (7), 198 A–205 A.
172. Kusterbeck, A.W., Wemhoff, G.A., Charles, P.T., Yeager, D.A., Bredehorst, R., Vogel, C.W., and Ligler, F.S. (1990) *J. Immunol. Methods*, **135** (1–2), 191–197.
173. Whelan, J.P., Kusterbeck, A.W., Wemhoff, G.A., Bredehorst, R., and Ligler, F.S. (1993) *Anal. Chem.*, **65** (24), 3561–3565.
174. Bart, J.C., Judd, L.L., Hoffman, K.E., Wilkins, A.M., and Kusterbeck, A.W. (1997) *Environ. Sci. Technol.*, **31** (5), 1505–1511.
175. Rabbany, S.Y., Marganski, W.A., Kusterbeck, A.W., and Ligler, F.S. (1998) *Biosens. Bioelectron.*, **13** (9), 939–944.
176. Narang, U., Gauger, P.R., and Ligler, F.S. (1997) *Anal. Chem.*, **69** (14), 2779–2785.
177. Narang, U., Gauger, P.R., and Ligler, F.S. (1997) *Anal. Chem.*, **69** (10), 1961–1964.
178. Charles, P.T., Rangasammy, J.G., Anderson, G.P., Romanoski, T.C., and Kusterbeck, A.W. (2004) *Anal. Chim. Acta*, **525** (2), 199–204.
179. Charles, P.T., Gauger, P.R., Patterson, C.H., Jr., and Kusterbeck, A.W. (2000) *Environ. Sci. Technol.*, **34** (21), 4641–4650.
180. Charles, P.T., Kusterbeck, A.W., Patterson, C.H., Jr., Lundgren, J.S., Holt, D.B., and Shriver-Lake, L.C. (2001) *J. Proc. Anal. Chem.*, **6** (4), 140–145.
181. Holt, D.B., Gauger, P.R., Kusterbeck, A.W., and Ligler, F.S. (2002) *Biosens. Bioelectron.*, **17** (1–2), 95–103.
182. Bart, J.C., Judd, L.L., and Kusterbeck, A.W. (1997) *Sens. Actuators, B*, **39** (1–3), 411–418.
183. Charles, P.T. and Kusterbeck, A.W. (1999) *Biosens. Bioelectron.*, **14** (4), 387–396.
184. Judd, L.L., Kusterbeck, A.W., Conrad, D.W., Yu, H., Myles, H.L., Jr., and Ligler, F.S. (1995) *Proc. SPIE-Int. Soc. Opt. Eng.*, **2388** (Advances in Fluorescence Sensing Technology II), 198–204.
185. Thundat, T., Finot, E., Hu, Z., Ritchie, R.H., Wu, G., and Majumdar, A. (2000) *Appl. Phys. Lett.*, **77** (24), 4061–4063.
186. Berger, R., Delamarche, E., Lang, H.P., Gerber, C., Gimzewski, J.K., Meyer, E., and Guntherodt, H.-J. (1997) *Science*, **276** (5321), 2021–2024.
187. Van Neste, C.W., Senesac, L.R., Yi, D., and Thundat, T. (2008) *Appl. Phys. Lett.*, **92** (13), 134102–134105.
188. Gardner, J.W. (2004) *NATO Sci. Ser. II Math. Phys. Chem.*, **159** (Electronic Noses & Sensors for the Detection of Explosives), 1–28.
189. Pardo, M., Benussi, G.P., Niederjaufner, G., Faglia, G., and Sberveglieri, G. (2001) Detection of TNT vapour with the Pico-1-nose, in *Electronic Noses and Olfaction 2000* (ed. J.W. Gardner and K.C. Persaud), Sensors (series ed. B.E. Jones), Institute of Physics, Bristol, UK, pp. 67–74.
190. Gui, Y., Xie, C., Xu, J., and Wang, G. (2009) *J. Hazard. Mater.*, **164** (2–3), 1030–1035.
191. Jerez Rozo, J.I., Chamoun, A.M., Pena, S.L., and Hernandez-Rivera, S.P. (2007) *Proc. SPIE-Int. Soc. Opt. Eng.*, **6538** (Sensors, and Command, Control, Communications, and Intelligence (C3I) Technologies for Homeland Security Defense VI), 653824/1–653824/12.
192. Kneipp, K., Wang, Y., Kneipp, H., Dasari, R.R., and Feld, M.S. (1995) *Exper. Tech. Phys. (Lemgo, Germany)*, **41** (2), 225–234.
193. Haas, J.W., III, Sylvia, J.M., Spencer, K.M., Johnston, T.M., and Clauson, S.L. (1998) *Proc. SPIE-Int. Soc. Opt. Eng.*, **3392** (Pt. 1, Detection and Remediation Technologies for Mines and Minelike Targets III), 469–476.
194. Spencer, K.M., Sylvia, J.M., Janni, J.A., and Klein, J.D. (1999) *Proc. SPIE-Int. Soc. Opt. Eng.*, **3710** (Pt. 1, Detection and Remediation Technologies for Mines and Minelike Targets IV), 373–379.
195. Sylvia, J.M., Janni, J.A., Klein, J.D., and Spencer, K.M. (2000) *Anal. Chem.*, **72** (23), 5834–5840.

196. Bertone, J.F., Cordeiro, K.L., Sylvia, J.M., and Spencer, K.M. (2004) *Proc. SPIE-Int. Soc. Opt. Eng.*, **5403** (Pt. 2, Sensors, and Command, Control, Communications, and Intelligence (C31) Technologies for Homeland Security and Homeland Defense III), 387–394.

197. McHugh, C.J., Keir, R., Graham, D., and Smith, W.E. (2002) *Chem. Commun.*, **6**, 580–581.

198. McHugh, C.J., Smith, W.E., Lacey, R., and Graham, D. (2002) *Chem. Commun.*, **21**, 2514–2515.

8
Nanostructured Materials for Selective Collection of Trace-Level Metals from Aqueous Systems

Sean A. Fontenot, Timothy G. Carter, Darren W. Johnson, R. Shane Addleman, Marvin G. Warner, Wassana Yantasee, Cynthia L. Warner, Glen E. Fryxell, and John T. Bays

8.1
Introduction

Several separations and filtration techniques are available for the collection of trace-level species and each method must be matched to a specific application. For trace-level metals in solution arguably the best and most widely used methods involve solid phase sorbent materials that provide effective capture of desired metal species. Effectiveness of the solid phase sorbent for any given application is determined by availability, cost, and performance. Activated carbon and ion exchangers are widely available and relatively cheap but lack, in most cases, the performance necessary for many analytical applications. Activated carbon and ion exchangers generally fail to have the selectivity and affinity needed for trace analyte collection from actual environmental matrices.

To understand and respond to situations involving toxic materials it is critical to quickly identify the toxic material(s) involved and the extent of contamination. This is a key issue for circumstances ranging from responding to terrorist attacks to monitoring the effects of environmental remediation. Unfortunately, analytical technology does not presently exist to meet these needs. Instruments powerful enough to meet the required speed, sensitivity, and selectivity requirements often do not function well outside of rigorously controlled laboratory conditions and are usually very complex and expensive. Simple screening methods that provide immediate results in the field enable on-site, near real-time decisions. These field screening methods are typically less costly and more rapid than formal laboratory analysis; this is significant since site testing and monitoring typically involves extensive sampling. To meet this need, a wide range of field screening methods for identifying chemical, biological, and nuclear materials is presently being marketed and used. Unfortunately, existing field assay methods are typically inadequate because they lack the selectivity and sensitivity needed to provide reliable information. The degree and type of improvement needed vary with the application but sensitivity improvements of greater than 1000× are typically required, and much larger enhancements would usually be preferred. This large

Trace Analysis with Nanomaterials. Edited by David T. Pierce and Julia Xiaojun Zhao
Copyright © 2010 WILEY-VCH Verlag GmbH & Co. KGaA, Weinheim
ISBN: 978-3-527-32350-0

leap in analytical performance is very unlikely to be achieved with incremental improvements in measurement procedure, instrument design, or improved electronics. A new analytical approach is required.

In many circumstances the deficiencies in selectivity and sensitivity could be addressed with high-performance sorbent materials that selectively concentrate target analytes. In addition to concentrating target analytes the sorbent can exclude interfering species and provide a uniform, well-defined sample matrix for analysis. Sorbents coupled with instrumentation could be used for either real-time analysis of the signature species or as a rapid screening method to flag those samples that require more detailed analysis. Sorbents coupled with rugged, compact instrumentation could provide portable, yet highly sensitive, field analyzers that could be quickly reconfigured for new analytes simply by changing the sorbent material. Sorbents used in these systems could be designed to be reusable, renewable, or disposable depending upon the instrument configuration desired and material chemistries involved. These same sorbents could be used to improve the performance of traditional laboratory techniques with more effective sample clean up.

This chapter is a discussion and review of various advanced nanostructured materials applicable to the selective collection of trace-level analytes from aqueous systems for sensing and separation applications. For consistency and comparison, when possible, materials expressing thiol surface chemistry are used as examples. While a plethora of surface chemistries exist, and many have relevance to environmental challenges, thiol surface chemistry is highly effective for the capture of many toxic heavy metals from aqueous systems, and serves as a useful baseline to compare materials performance. Further, thiol surface chemistry has been demonstrated on a range of nanostructures and therefore provides continuity and a common platform for nanomaterial comparison. This chapter is organized into sections by material type. Discussions are broken down into the materials science and application of:

- functionalized nanoporous silica material;
- functionalized magnetic nanomaterials;
- carbon-based nanostructured materials;
- other materials such as zeolites, and imprinted polymers;
- concluding thoughts on economics and the future of nanostructured materials in trace analysis.

8.2
Sorbents for Trace-Metal Collection and Analysis: Relevant Figures of Merit

Solid-phase sorbent materials provide effective capture and preconcentration of trace-level metal species from aqueous matrices. Effective solid-phase extraction (SPE) sorbents allow separation for remediation purposes, but they are also ideal for analytical applications since preconcentration of analytes prior to assay separates the analyte(s) from the sample matrix and concentrates them into a smaller volume prior to measurement. Once preconcentration is complete, the analyte can be stripped from the SPE material by an appropriate method (e.g., via acid, thermalization, etc.) and then quantified by analysis. Alternatively, the SPE material can be assayed directly for some applications [1]. This allows one to improve the sensitivity, selectivity, and speed of an analytical process.

Sorbents may be evaluated in terms of affinity and capacity for target species as well as how quickly they can extract target species (kinetics). Most of the sorbents discussed here consist of support materials having surfaces functionalized with organic metal-binding groups. The surface chemistry of the SPE material dictates the binding affinity of the material for a given analyte, and adjusting the surface chemistry of the SPE material allows it to be applied to different classes of analytes or function in different matrices. Capacity is a function of the number of binding units on the sorbent, which, in turn, depends on the surface area of the support and the degree of functionalization. Kinetics depend mostly on mass transport through the sorbent material and the extent to which the material may be dispersed in the sample matrix.

Binding affinities of sorbents are typically expressed using the distribution coefficient (K_d); K_d, (ml g^{-1}) is simply a mass-weighted partition coefficient between the solid phase and liquid supernatant phase as represented in Equation (8.1):

$$K_d = \frac{(C_o - C_f)}{C_f} \times \frac{V}{M} \qquad (8.1)$$

where

C_o and C_f are the initial and final concentrations, respectively, in the solution of the target species,
V is the solution volume in ml,
M is the mass of the sorbent.

The V/M term is simply the liquid to solid (L/S) ratio of the sample solution to the sorbent. In other words, V/M is the ratio of the volume of sample analyzed to the amount of sorbent used and provides a measure of the amount of sorbent needed for a given analysis. The higher the K_d, the more effective the sorbent material is at capturing and holding the target species. In practical terms, a sorbent with a high K_d for a given target will allow uptake of the target even at very low concentrations. In general, K_ds of ~10^3 ml g^{-1} are considered good and those above 10^4 ml g^{-1} are outstanding [2].

For applications involving trace analysis, capacity is less important than affinity since saturation of the sorbent is typically not an issue and the amount of analyte extracted by the SPE material is limited by the magnitude of the analyte distribution coefficient (K_d). By employing high surface area, dispersible, and specifically functionalized sorbents one can drive the interaction of the sorbents with the analytes in the sample, effectively facilitating both separation and preconcentration. The remainder of this chapter uses K_d as the figure of merit for demonstrating analyte selectivity.

8.3
Thiol-Functionalized Ordered Mesoporous Silica for Heavy Metal Collection

Removal of heavy metals, such as mercury (Hg), lead (Pb), thallium (Tl), cadmium (Cd), and arsenic (As), from natural waters has attracted considerable attention because of environmental contamination and their adverse effects on environmental and human health. Introduction of sorbent functionalities into nanoporous structures has significantly improved the performance of those sorbent materials in metal removal when compared to conventional sorbent beds [3–5]. This section reviews the ability of thiol-functionalized ordered mesoporous silica, termed thiol self-assembled monolayer on a mesoporous support (SH-SAMMS™), to capture heavy metals. This material is superior to commercial sorbents, such as Duolite® GT-73 resins, and Darco® KB-B activated carbon. SH-SAMMS also presents new uses beyond environmental applications. These include chelation therapies and biomonitoring of heavy metals.

8.3.1
Performance Comparisons of Sorption Materials for Environmental Samples

The ability of a sorbent material to capture a metal ion depends on the pK_a of its primary ligating functional group, the stability constant of the metal–ligand complex, the presence of competitive ligands in solution, the pH of the solution [6], and the metal ion's capacity to undergo hydrolysis [7]. Figure 8.1 shows the binding affinity (K_d) of SH-SAMMS to various metal ions in HNO_3 spiked filtered river water. As anticipated from Pearson's hard-soft acid-base theory (HSAB) [8], soft ligands like thiol groups prefer to bind soft metals like Hg and Ag, rather than a relatively harder metal such as Co. Based on the K_ds, SH-SAMMS is an outstanding sorbent for Hg, Ag, Pb, Cu, and Cd ($K_d > 50\,000\,\text{ml}\,\text{g}^{-1}$) and a good sorbent for As in river water at neutral pH. SH-SAMMS could effectively bind Hg, Ag, and As for the entire pH range (pH 0 to 8.5), Cu from pH 2 to 8.5, and Cd and Pb from 6 to 8.5. Work in our laboratory reveals that Hg is strongly bound to SH-SAMMS even in acid concentrations as high as 5 M (HCl and HNO_3). Above pH 7, a noticeable drop in K_d of Ag, Hg, Cd, and Pb (all still above 10^5) may be a result of strong association between the metal ions and native anions in the water (e.g., between Ag^+ and reduced sulfur groups [9] or Ag^+ and Hg^{2+} and chloride anions [10, 11]).

Ionic strength can also vary in most pH-adjusted samples (e.g., pH 0–8.5) and may affect the metal binding affinity of a sorbent. Table 8.1 shows that ionic strength does not significantly affect the SH-SAMMS material until the concentration reaches 1 M sodium (as acetate) concentration. In the three natural waters, the affinity of SH-SAMMS for all metals tested remained virtually unaffected.

Table 8.2 summarizes K_d of metal ions from filtered groundwater (pH 6.8–6.9), measured for different sorbents. Note that thallium was added into the solutions as Tl^+ and arsenic as As^{3+}. In terms of K_d, SH-SAMMS was significantly superior to the commercial GT-73 and activated carbon (Darco KB-B) for capturing Hg, Cd,

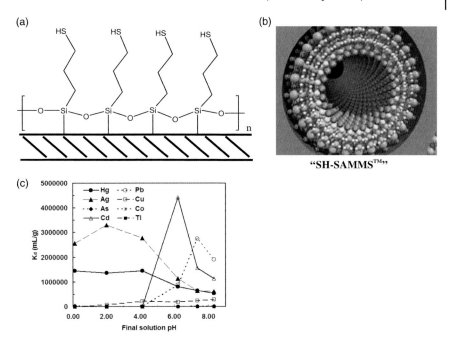

Figure 8.1 Diagram of SH-SAMMS showing a propylthiol monolayer (a) and the monolayer-lined pores (b); (c) effect of pH on the K_d of SH-SAMMS, measured in HNO_3 spiked filtered river water with a liquid to solid (L/S) ratio of 5000:1.

Table 8.1 Matrix effect on K_d (ml g^{-1}) of various metals extracted by SH-SAMMS.[a] Data reprinted with permission from Reference [12]. Copyright 2008 Elsevier.

Matrix	pH	Cu	As	Ag	Cd	Hg	Pb
Columbia River water, WA[b]	7.66	9.5×10^5	7.2×10^4	7.1×10^6	1.0×10^7	3.6×10^5	5.3×10^6
Hanford groundwater, WA[b]	8.00	1.3×10^6	5.7×10^4	1.1×10^7	1.1×10^7	5.9×10^5	5.6×10^6
Sequim Bay sea water, WA[b]	7.74	1.3×10^6	9.2×10^4	7.5×10^6	1.5×10^7	2.5×10^6	3.4×10^6
0.001 M CH_3COONa	7.14	–	4900	–	1.5×10^6	3.8×10^5	3.1×10^5
0.01 M CH_3COONa	7.19	–	2900	–	3.3×10^6	3.0×10^5	8.5×10^5
0.1 M CH_3COONa	7.21	–	4100	–	3.9×10^6	6.6×10^5	6.6×10^5
1.0 M CH_3COONa	7.28	–	2000	–	7.5×10^5	5.7×10^5	9.2×10^4

a) Reported as average value of three replicates, L/S of 5000 ml g^{-1}.
b) 0.45 micron filtered.

Table 8.2 K_d (ml g^{-1}) of metal ions extracted by selected sorbents from Hanford groundwater. [a] Data reprinted with permission from Reference [13]. Copyright 2007 American Chemical Society.

Sorbent	Final pH	Co	Cu	As	Ag	Cd	Hg	Tl	Pb
DMSA-Fe$_3$O$_4$	6.91	3000	2.7×10^5	5400	3.6×10^6	1.0×10^4	9.2×10^4	1.4×10^4	2.3×10^6
Fe$_3$O$_4$	6.93	1600	7400	5800	1.3×10^4	2400	1.6×10^4	4000	7.8×10^4
SH-SAMMS	6.80	430	1.7×10^6	950	6.7×10^7	6.6×10^4	1.1×10^6	1.5×10^4	3.5×10^5
Thiol resin[b]	6.76	890	6300	1200	1.6×10^4	1500	1.0×10^4	2200	4.1×10^4
Activated carbon[c]	6.90	790	2.6×10^4	750	2.7×10^4	1300	3.1×10^4	21	1.9×10^5

a) Reported as average value of three replicates, L/S = 10 000 ml g^{-1}, in 0.45 micron filtered groundwater.
b) GT-173 by Rohm & Hoss.
c) Darco KB-B.

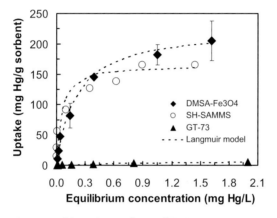

Figure 8.2 Adsorption isotherm of Hg in groundwater (pH 8.1) on DMSA-Fe$_3$O$_4$ (L/S, 5×10^5), SH-SAMMS (L/S, 5×10^5), and GT-73 (L/S, 7000); dashed lines represent Langmuir modeling of the data. Reprinted with permission from Reference [13]. Copyright 2007 The American Chemical Society.

Ag, Pb, and Tl. The superior affinity results from the multidentate chelation ability and the suitable substrate of SH-SAMMS, characteristics that are not found in the commercial resins.

Figure 8.2 shows the adsorption isotherms of Hg on SH-SAMMS, DMSA-Fe$_3$O$_4$, and GT-73 in filtered groundwater (pH 8.1) [13]. The isotherm curves present Hg uptake as a function of the equilibrium Hg solution concentration at room temperature. The saturation adsorption capacity of Hg on SH-SAMMS was found to be 167 mg g^{-1} in filtered groundwater (pH 8.1) while that of GT-73 was only 8 mg g^{-1} in the same matrix. In acidified river water (pH 1.99) a Hg adsorption capacity of 400 mg g^{-1} of SH-SAMMS has been achieved. The large surface areas of the DMSA-Fe$_3$O$_4$ (114 m^2 g^{-1}, see Section 8.4 for more details) and SH-SAMMS (740–680 m^2 g^{-1}) afforded a high number of ligand sites on these materials, leading to large ion loading capacities. Although the capacity of GT-73 for Hg in DI water (pH 4–6) was previously reported to be 600 mg g^{-1} [14], the measured Hg capacity in groundwater reported in the highlighted study was only 8 mg g^{-1}, which suggests very poor selectivity of the GT-73's binding sites in groundwater.

In addition to covalently linking functional head-groups to mesoporous silica, an alternative connective approach utilizes weak interactions through π–π stacking between an "anchored" aromatic monolayer and chemisorbed functionalized ligands. One example studied various mono and bis-functionalized mercaptomethyl benzenes adsorbed onto mesoporous silica that had been prepared with a covalently attached phenyl monolayer (phenyl-SAMMS) [15]. Comparisons between these materials and SH-SAMMS show similar uptake capacities and kinetics. For instance, K_d values for phenyl-SAMMS chemisorbed with 1,4-bis(mercaptomethyl)benzene show similar capture levels with the covalently attached SH-SAMMS in Hanford well water matrix spiked with 500 ppb Hg^{2+}, Pb^{2+}, Cd^{2+}, and Ag$^+$ ions. Therefore, the metal affinity levels of the chemisorbed phenyl-SAMMS are near equal to that of the covalently bound SH-SAMMS. Because of the weak yet stable interaction imparted by the π–π stacking, these materials have shown the ability to be stripped of their metal-loaded ligands by simple organic washes, thus leading to possible regeneration of the base phenyl-SAMMS material. The ability to regenerate sorbents is an attractive feature for both water purification and preconcentration-based technologies. Regeneration of the base material with a simple infield wash can greatly reduce the overall usage cost; an essential criterion for sorbent materials.

In addition to sorption affinity, capacity, and selectivity to the target metal ions, it is important that a sorbent material offer rapid sorption to minimize the contact time required for removal metal ions. Figure 8.3 shows the uptake rate of Pb on various sorbent materials measured in filtered groundwater (pH 7.7). Lead could be captured by SH-SAMMS at a much faster rate than the two commercial sorbents, Chelex-100 (an EDTA-based resin) and GT-73 (a thiol functionalized resin). Specifically, after one minute of contact time, over 99 wt.% of 1 mg l^{-1} of Pb was removed by SH-SAMMS, while only 48 wt.% and 9 wt.% of Pb were removed using Chelex-100 and GT-73, respectively. It took over 10 min for Chelex-100 and 2 h for GT-73 to remove over 96% of Pb [14]. GT-73 and Chelex-100 are synthesized by attaching chelating ligands to porous polymer resins, which dominate their physical properties. Thus GT-73 and Chelex-100 are subject to solvent swelling and have dendritic porosities. The SH-SAMMS has a rigid silica support and open pore structure that allow for rapid diffusion of analytes into the binding sites, resulting in extremely fast sorption kinetics.

8.3.2
Performance Comparisons of Sorption Materials for Biological Samples

Sorbents that can capture heavy metals in biological media such as blood, gastrointestinal fluids, and urine are highly desirable for several reasons. In addition

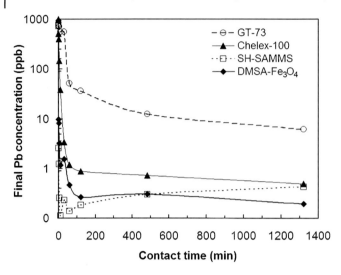

Figure 8.3 Kinetics of adsorption of 1000 ppb of Pb in filtered groundwater, pH 7.77, all with L/S of 1000, except DMSA-Fe$_3$O$_4$ with L/S of 2000. Reprinted with permission from Reference [13]. Copyright 2007 American Chemical Society.

to enhanced detection (discussed in Section 8.2), the decorporation of toxic metals from blood and gastric intestinal fluids is anticipated to provide a breakthrough in chelation therapy.

Since the 1940s, *in vivo* toxic metal immobilization has involved the use of intravenous ethylenediamine-tetraacetate (EDTA) treatment and oral or intravenous dimercaptosuccinic acid (DMSA) treatment following metal exposures. Solid sorbents are potentially better than their liquid counterparts because, as oral drugs, they can minimize the gut absorption of ingested chemicals harmful to the human body. In addition, when used in or hemoperfusion devices, they can remove the chemicals in blood that have been absorbed systemically from all routes of exposure (oral, dermal, and inhalation). This decreases the burden on the kidneys for clearing the toxic metal-bound liquid chelating agents.

SH-SAMMS captures a large percentage (~90%) of As, Cd, Hg, and a moderate percentage of Pb, presented in human urine and blood at relevant exposure levels (50 µg l^{-1}). Not only can SH-SAMMS remove inorganic Hg^{2+} (Table 8.3) but it can also remove methyl Hg^{2+} (CH$_3$Hg$^+$, which is a much more problematic form of Hg); for example, at L/S of 200, 87% of methyl Hg^{2+} was removed from human plasma containing 100 µg l^{-1} of the analyte. Figure 8.4 also shows the K_d of the four metals on SH-SAMMS measured in synthetic gastrointestinal fluids [16], with a pH similar to what might be encountered within the various regions of the gastrointestinal tract (pH 1–3 in stomach, pH 5.5–7 in large intestine, pH 6–6.5 in duodenum, and pH 7–8 in jejunum and ileum) [17].

In addition to their use in chelation therapies, sorbent materials that can effectively preconcentrate heavy metals from complex matrices like blood and urine

Table 8.3 Percent removal[a] of heavy metals from biological matrices using SH-SAMMS.

Matrix	L/S (ml g^{-1})	Initial metal concentration (µg L^{-1})	As (%)	Cd (%)	Hg (%)	Pb (%)
Human blood	1000	50	90	93	92	15
Human urine	1000	50	93	87	89	33

a) Reported as average values of three replicates, SD < 5%.

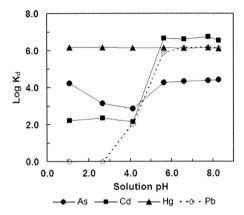

Figure 8.4 K_d of As, Cd, Hg, and Pb, measured on SH-SAMMS in synthetic gastrointestinal fluids prepared by adjusting synthetic gastric fluid (contained 0.03 M NaCl, 0.085 M HCl, pH 1.11) with 0.2 M NaHCO$_3$ to the desired pH; initial metal ion concentrations of 50 µg L^{-1} and L/S of 5000.

will substantially improve biomonitoring of exposure to these species. Direct preconcentration of toxic metals from biological samples without prior acid digestion has been a challenging task for two main reasons. The proteins and macromolecules in complex biological samples can compete with the sorbent materials for binding metal ions, resulting in low capture extent (e.g., Pb example in Table 8.3). Additionally, these biomolecules tend to foul sensors, resulting in rapid degradation in response. This is especially evident with electrochemical sensors where proteins adsorb on electrode surfaces and form an insulating layer. By employing Nafion as an antifouling layer and SH-SAMMS as the metal preconcentrator, we have successfully used a SH-SAMMS-Nafion composite in electrochemical sensors for low ppb (µg l^{-1}) detection of heavy metals (e.g., Cd and Pb) in urine without sample pretreatment or protein fouling [12]. The resulting sensor offers detection limits similar to state-of-the-art ICP-MS methods but, unlike ICP-MS, the sensor is portable and will facilitate rapid biomonitoring of exposure to toxic metals. SH-SAMMS nanomaterials have recently emerged as highly effective sensors for trace metal detection. Their high efficacy for metal concentration has led to their exploration to enhance a range of analytical methods, including ICP-MS, radiological measurements, and X-ray fluorescence.

8.4
Surface-Functionalized Magnetic Nanoparticles for Heavy Metal Capture and Detection

Recently, the use of functionalized superparamagnetic nanoparticles in environmentally relevant applications such as selective capture and preconcentration of specific analytes from complex samples for sensitive detection has been reported in an increasingly large number of publications and reviews [13, 18–33]. In this section we focus on the application of engineered magnetic nanoparticles (diameter < 100 nm in most cases) that contain a thiol surface functionality for the purpose of separating and detecting a wide variety of analytes from complex environmental matrices.

Preconcentration is an ideal application for functionalized magnetic nanoparticles since they provide a controllable sorbent material for solid-phase extraction (SPE). Once dispersed in solution they can rapidly contact high volumes of the sample matrix, selectively capture target analytes, and then can be recovered and manipulated by the application of a relatively strong (often > 1 T), but easily generated, magnetic field. It has been demonstrated that the intrinsically high surface area arising from the nanoscale dimensions of these nanomaterials and the ability to impart specific surface functionalization make them very effective for SPE [1]. The K_d values for various different SPE materials, including functionalized superparamagnetic nanoparticles, in filtered groundwater is shown above in Table 8.2.

Magnetic nanoparticles have the potential to be extremely effective SPE materials for the preconcentration, removal, and detection of environmental contaminants. The nanoparticle surface can be tailored to target a wide range of analytes in much the same manner as the SAMMS materials discussed above. This is accomplished by incorporation onto the surface small organic molecules that contain two sets of functional groups – one with an affinity toward the iron oxide surface (i.e., carboxylic acid and/or silane) and another with an affinity toward the target metal analyte of interest (i.e., thiol).

Complementing the ability to modify the surface chemistry with analyte-selective ligands, these materials demonstrate superparamagnetism that arises from their nanoscale single magnetic domain structures [25]. Superparamagnetic behavior manifests itself in nanoparticles smaller than a critical diameter that is both material and temperature dependent. Throughout this section, we predominantly address iron oxide nanoparticles with diameters ranging from ca. 5 to 20 nm, which falls within the established critical diameter for this material (ca. 15–20 nm) [25]. From a practical standpoint, a superparamagnetic nanoparticle has little to no remnant magnetization after exposure to a magnetic field and low to no coercivity (the field required to bring the magnetization to zero), meaning that they will not magnetically agglomerate at room temperature [25]. This is significant for applications where it is desirable to have the nanoparticles well dispersed in the sample matrix and easily manipulated by an external magnetic field.

By exploiting the ability to remove the magnetic nanoparticles from solution with an external field and the ability to tailor the surface functionality of the nanoparticles through synthetic means, it is possible to both separate and detect with great sensitivity a wide range of heavy metal analytes. We focus our discussion on the attachment of small molecules for the purposes of both separating the target analyte from complex samples containing interferents and detecting them once separation is complete. In doing so we hope to demonstrate the efficacy and future potential of magnetic nanomaterials for the effective preconcentration and sensing of environmentally relevant heavy metal analytes from complex sample matrices (e.g., river, ground-, and ocean water).

Iron oxide nanoparticles are most commonly used for these applications since they can be made cheaply, in large quantities, and methods for their surface functionalization are well established [20, 23–25, 30, 34–37]. The iron oxide core can be made using various methods, depending on the desired size, dispersity, and magnetic characteristics. The surface can be further modified to contain the thiol functionality necessary for the intended application. In some cases applications may require exposure of the nanoparticles to harsh chemical environments, necessitating encasement of the nanoparticles in an inert shell, typically silica. Silica encasement of iron oxide nanoparticles is a common treatment to render the material more robust in low pH or biological environments and has the advantage of the silanol surface chemistry for silane ligand modification [38–42]. Noble metals are also used for encasement of the iron oxide core, depending on the application. For example, gold or silver encasement allows one to take advantage of both the magnetic and optical characteristics, such as plasmon resonance, of the core–shell materials when concentration and detection of the materials is desired [43–55].

In our work we have demonstrated that functionalized superparamagnetic nanoparticles can effectively disperse in aqueous environmental samples and sequester a wide variety of analytes including heavy metals [13, 33]. Specifically, we have employed thiol-modified Fe_3O_4 nanoparticles that are approximately 6 nm in diameter to remove Hg, Ag, Pb, Cd, and Tl from natural waters (i.e., river, ground-, and ocean water) [56]. Figure 8.5 illustrates the behavior of these materials.

The magnetic nanoparticles used in this study were highly dispersible in aqueous media, but were removed with relative ease by exposing the sample to a magnetic field. In this case the field strength was ca 1.2 T generated by a NdFeB rare earth magnet [13]. Using this setup the nanoparticles removed over 99 wt% of $1\,mg\,l^{-1}$ Pb within 1 min of contact time and had a Hg capacity of over $227\,mg\,g^{-1}$. This capacity is nearly 30-fold larger than that of conventional resin-based sorbents [13]. To determine the efficacy of extraction of heavy metals by the magnetic nanoparticles, various measurements to determine the distribution coefficient (K_d) were made, as summarized in Table 8.2. Figure 8.6 illustrates that at near neutral pH in river water, the thiol-modified magnetic nanoparticles are outstanding

Figure 8.5 Schematic of (a) DMSA-modified Fe₃O₄ nanoparticles, and (b) removal of the nanoparticles from the liquid phase using NdFeB magnets; initial solution (left-hand panel), after 10 s with the magnet (middle panel), and when the magnet was moved to a distant position (right-hand panel). Reprinted with permission from Reference [56]. Copyright 2008 The American Chemical Society.

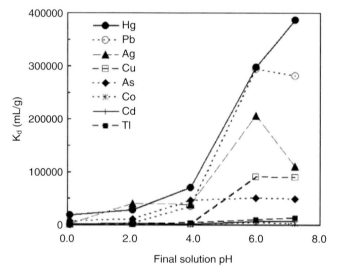

Figure 8.6 Effect of pH on K_d, measured in HNO₃ spiked unfiltered river water (L/S, $10^5:1$). Reprinted with permission from Reference [13]. Copyright 2007 The American Chemical Society.

sorbent materials for soft metals such as Hg, Ag, Pb, Cu, and As ($K_d > 50\,000$). In addition, the materials were demonstrated to be a good sorbent for harder metals such as Cd, Co, and Tl [13]. Once the metals were extracted, the trace detection of the heavy metal analyte was carried out using ICP-MS after contact with the magnetic nanoparticles [13].

Materials such as magnetic nanoparticles [13] and polymer/nanoparticle composites [57–60] offer the unique capability for magnetically directed separation and sensing processes. Studies from other groups have shown similar characteristics of functionalized magnetic nanoparticles and microparticles modified with a wide

8.4 Surface-Functionalized Magnetic Nanoparticles for Heavy Metal Capture and Detection

variety of affinity ligands for the extraction of heavy metals from environmental samples [61–64]. This field still remains relatively undeveloped when one is discussing the use of superparamagnetic nanoparticles between 5 and 20 nm in diameter. For instance, the behavior of nanomaterials dispersed in the environment is challenging to fully understand due to their tendencies toward aggregation and/or decomposition as well as their mobility. Many of the above reports address materials constructed from nanoparticle/polymer composites, which fall well outside of the size range of what is traditionally considered a nanomaterial (i.e., they have sizes >100 nm) [62].

Our group has demonstrated the use of both magnetic and nonmagnetic [33] high surface area sorbent materials to enhance the electrochemical detection of toxic heavy metals from natural waters [56]. Both functionalized magnetic nanoparticles and mesoporous silica were modified with a wide range of thiol-containing organic molecules that possess a high affinity toward soft heavy metals and were placed or collected at an electrode surface (Figure 8.7) [33, 56]. For instance, superparamagnetic Fe_3O_4 nanoparticles functionalized with dimercaptosuccinic acid (DMSA) were used to first bind the heavy metal contaminants from complex samples and then subsequently carry them to the surface of a magnetic electrode (Figure 8.7) [56].

By using an applied magnetic field to remove the target analytes from solution prior to electrochemical analysis, they are effectively isolated from the huge number of potential interferents present in complex sample matrixes. Using this system we have successfully overcome, at least to some extent, two of the biggest problems that prevent widespread adoption of electrochemical sensors for the analysis of metal ions in biological samples: (i) the binding of the target metals to

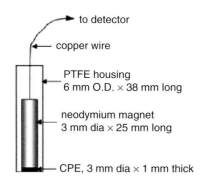

(a) Magnetic electrode

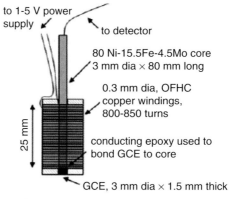

(b) Electromagnetic electrode

Figure 8.7 Schematic diagrams of (a) magnetic electrode and (b) electromagnetic electrode that preconcentrate metal ions using superparamagnetic nanoparticles. Reprinted with permission from Reference [56]. Copyright 2008 The Royal Society of Chemistry.

proteins present in the sample matrix, leading to a lowered signal response, and (ii) electrode fouling caused by proteins. As can be seen from Figure 8.8, we have successfully measured concentrations of Pb in urine as low as 10 ppb with as little as 20 s of preconcentration (after the optimal 90 s of preconcentration the detection limit dropped to 2.5 ppb Pb).

Furthermore, Figure 8.9 shows that the magnetic nanoparticles can also enable the detection of multiple heavy metals (i.e., Cd, Pb, Cu, and Ag) from various natural river and ocean waters with only ~2.5 min of preconcentration time.

In addition to this work several other groups have reported the use of magnetic nanoparticles in the electrochemical analysis of other environmentally relevant targets (e.g., proteins and nucleic acids) besides heavy metals [21, 22, 65]. The reader is directed to two recent reviews for the application of high surface area magnetic nanomaterials for the detection of biological analytes [20, 66]. Importantly, even though the bulk of the work that has been performed in this area was aimed at clinical applications, the detection of biological species is of paramount importance to environmental sensing due to the biological origin of many common environmental contaminants [67]. A recent example that employs magnetic nanomaterials for the detection of a protein biomarker to pesticide

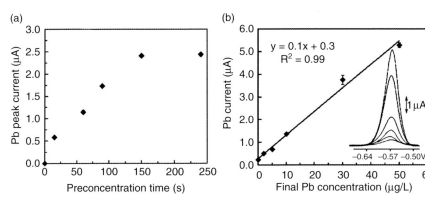

Figure 8.8 (a) Signals of 10 ppb Pb measured at DMSA–Fe$_3$O$_4$-magnetic sensors in samples containing 25 vol.% rat urine with varied preconcentration time; (b) linear Pb calibration curve measured at DMSA–Fe$_3$O$_4$ magnetic sensors in Pb-spiked samples containing 25 vol.% rat urine. Reprinted with permission from Reference [56]. Copyright 2008 The Royal Society of Chemistry.

Figure 8.9 Sensor measurements of (a) background metal ions in seawater (dashed line) and river water (solid line) and (b) background metal ions (thin line) and metals spiked (thick line) in seawater, after 150 s of preconcentration time. Inset: metal concentrations, measured with ICP-MS, and the distribution coefficients of multiple metal ions (S/L of 0.01 g l^{-1} of DMSA–Fe$_3$O$_4$, initial metal conc. of 500 ppb each, pH of 7.20 for river water and 7.64 for seawater). Reprinted with permission from Reference [56]. Copyright 2008 The Royal Society of Chemistry.

8.4 Surface-Functionalized Magnetic Nanoparticles for Heavy Metal Capture and Detection

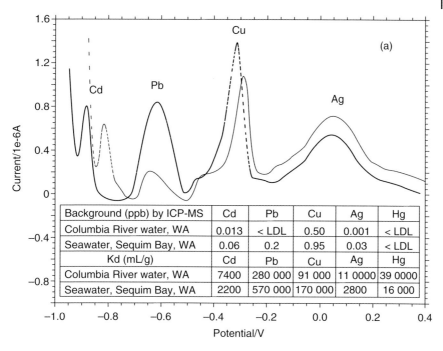

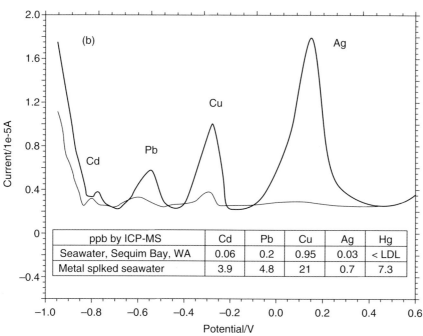

exposure utilizes a similar magnetic electrode to that described above and shown in Figure 8.7 [68]. In this work, the magnetic particles have been bound with gold nanoparticles to provide an extremely responsive material for electrochemical analysis [68, 69]. These composite nanomaterials have been used to bind and separate protein biomarker targets from solution followed by detection, obviating the need for the amplification used in typical protein detection [68, 69].

Although the bulk of reported detection schemes focus on the use of magnetic nanoparticles in electrochemical assays, the use of magnetic nanoparticles in trace analyte optical detection scenarios has received a great deal of attention. Recent reviews by Corr et al. and Katz et al. go into great depth on the formation of nanomaterial composites for biological detection and biomedical applications [18, 21]. Even when one considers all of the potential benefits associated with using a fluorescent nanomaterial that also is magnetic, many complications can arise. Primarily, the use of materials such as magnetic nanoparticles in an optical detection platform can scatter, absorb, or even quench the optical signal from the fluorescent reporter, which leads to a decrease in signal output. Despite this limitation, magnetic/fluorescent materials have great promise in environmental sensing because they will enable separation/preconcentration of target analytes from a complex sample prior to analysis, preventing unwanted optical noise from background interferents. Concurrently, these materials will optically label the target analyte upon binding, allowing for rapid sample analysis using traditional optical methods. We believe that as the materials production methods continue to mature and more magnetic/fluorescent composite nanomaterials become available there will be an explosion in the use of these types of materials in environmental sensing applications.

8.5
Nanoporous Carbon Based Sorbent Materials

Activated carbon is arguably the oldest and most widely utilized sorbent in human history [70]. Activated carbon has been around since antiquity and has grown into a major industry today, enjoying sales of many hundreds of millions of dollars each year. Activated carbon has several features that are attractive for preconcentration of analytes from aqueous systems – it is affordable, widely available, has a high surface area, an open pore structure, is stable towards hydrolysis, has good chemical stability, and excellent thermal stability. Activated carbon is widely used in the removal of various contaminants from water in municipal drinking water purification (e.g., chlorocarbons arising as a by-product of the chlorination process [71]). The utility of activated carbon as a sorbent material is centered around its ability to capture a wide variety of chemical species. Such non-specific adsorption is not necessarily desirable for analytical preconcentration as it entrains many other species that are not of interest and wastes valuable capture capacity doing so. Chemical selectivity can be very valuable for analytical preconcentration.

There have been numerous efforts over the years to chemically modify activated carbon in an effort to enhance its capture efficiency for specific analytes. Many of these efforts have involved some sort of controlled oxidation, or acid treatment, as a way to increase the degree of oxygenation of the carbon backbone. For example, activated carbon treated with sulfuric acid has been used to adsorb pollutants from wastewater [72]. Activated carbon has also been treated with various oxidants to enhance its adsorption capacity. Treatment of granular activated carbon with potassium bromate has been found to enhance the sorption of Ni ions [73].

8.5.1
Chemically-Modified Activated Carbons

Numerous studies have used chemically modified activated carbons to capture a wide variety of metal ions from aqueous environments. Various activated carbons have been used to capture chromate from wastewater [74]. Similarly, various activated carbons' capture efficiency for Ni^{2+} has been systematically compared [75]. Selective sorption of Pt^{2+} from a mixture of metals in solution was also studied using chemically modified activated carbon [76]. Likewise, toxic metals like Pb^{2+} [77], Cd^{2+} [78] and Cu [79] have all been concentrated from aqueous media using carbon-based sorbents. Activated carbons have even had polymer chains intercalated into their porous architectures in an effort to enhance their ability to bind toxic heavy metals [80].

Chemically modified activated carbons have been used for the preconcentration of specific analytes for water quality analysis. For example, various trace-level toxic elements have been concentrated from water samples for analysis by neutron activation [81, 82]. This area has been reviewed recently [83].

Clearly, activated carbon is a broad-scale sorbent, capable of sorbing a wide variety of analytes. When sampling natural waters, this can lead to undesirable fouling and competition issues. Therefore, it would be desirable to attach specific ligands (as opposed to generic "activation") inside the carbon scaffold so that additional chemical selectivity might be imparted.

Simple ligands have been attached on the carbon backbone to enhance metal binding affinity. For example, activated carbon has been sulfonated to produce a sulfonic acid moieties (Scheme 8.1) that are known to be effective for ion exchange [84]. Activated carbon has also been nitrated, and the nitro groups subsequently reduced to amines (Scheme 8.2), which were then used to capture various transition metals and lanthanide ions from aqueous media [85].

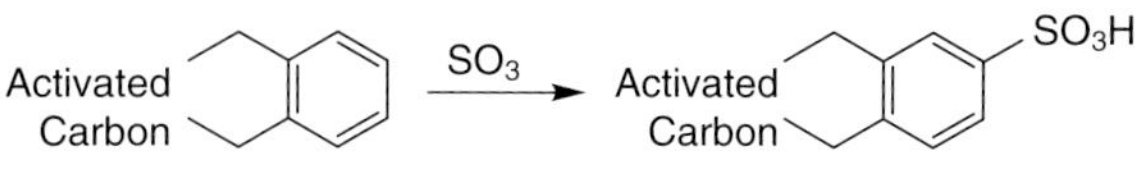

Scheme 8.1

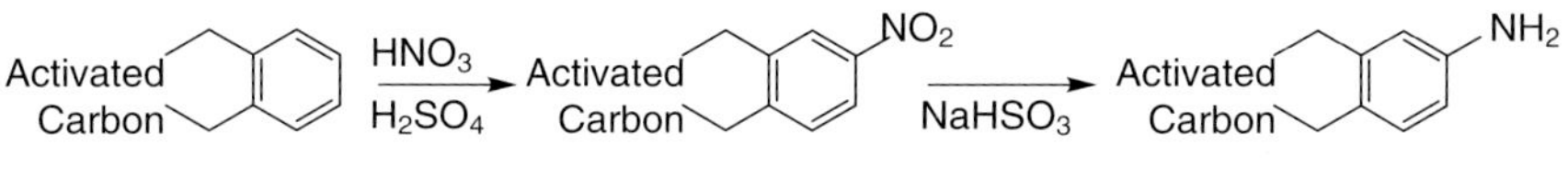

Scheme 8.2

More recently, activated carbon has been chloromethylated, analogous to the synthesis of the polymeric system known as Merrifield's resin [86]. Chloromethylation allows for the easy introduction of a wide variety of chemical functionality through simple substitution reactions (Scheme 8.3). In this case, the chloride was displaced by a sulfur-containing nucleophile, and the resulting thiolated activated carbon was shown to be an effective, and selective, heavy metal sorbent [87]. Table 8.4 compares the performance of thiolated nanoporous carbon with conventional activated carbon. The thiol functionality improves the affinity of the material for softer heavy metals. Notably, a portion of the functionality appears to be located inside micropores (an inherent limitation of activated carbon) and therefore has limited chemical accessibility.

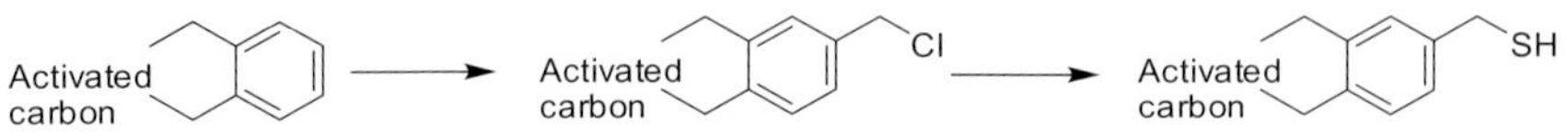

Scheme 8.3

More sophisticated chelating ligands, such as N,N'-bis(salicylidene)-1,2-phenylenediamine, have also been immobilized on activated carbon. This sorbent material has been used to capture ultra-trace levels of copper from aquatic media prior to analysis [88]. Moving to a more sophisticated ligand design allows for greater discrimination in the binding chemistry.

Chemically modifying activated carbon has the advantages of being simple and direct. This substrate also has a high surface area and is readily available in bulk. However, this approach is limited due to the microporosity inherent to activated carbon, as well as the latent functionality of the activated carbon backbone (e.g.,

Table 8.4 K_d (ml g^{-1} sorbent) values based metal sorption experiments using thiolated activated carbon (AC-CH$_2$SH). All experiments were performed in triplicate and averaged. Data reprinted with permissions from Reference [87]. Copyright 2010 Elsevier.

Sorbent	Final pH	Average K_d							
		Co^{2+}	Cu^{2+}	As^{3+}	Ag$^+$	Cd^{2+}	Hg^{2+}	Tl$^+$	Pb^{2+}
AC-CH$_2$SH	0.17	280	260	180	1700	0	1.6×10^6	96	91
	2.02	160	260	78	1400	83	1.1×10^6	19	120
	4.31	120	2100	0	5800	270	1.8×10^6	110	1500
	6.37	1100	5.5×10^4	160	6.2×10^4	1400	2.2×10^6	560	8.6×10^4
	7.33	1900	1.0×10^5	0	3.4×10^5	5000	6.1×10^6	1500	1.2×10^5
	8.49	2100	8.8×10^4	0	4.1×10^5	4300	2.0×10^7	1700	1.1×10^5
Activated carbon	2.12	0	55	0	220	0	2600	73	170
	4.22	110	5400	0	820	170	4800	250	6600
	7.61	1300	5.3×10^4	23	3400	2900	9700	1800	6.7×10^4

carboxylic acids, phenols, ketones, etc.). Both factors lead to non-selective metal ion (and organic) sorption, and therefore to a significant amount of undesirable competing adsorption. Clearly, it would be advantageous to work with a carbon scaffold having large pores that were constructed around specifically tailored functionality.

8.5.2
Templated Mesoporous Carbons

In recent years a great deal of effort has gone into the study of templated mesoporous carbons [89–100]. This approach generally uses a templated mesoporous silica as scaffold upon which some suitable organic precursor is arrayed, and subsequently polymerized and carbonized (typically at 800–1000 °C). After the carbonization stage, the silica template is generally removed by digestion with either HF or NaOH, leaving a free-standing nanoporous carbon scaffold that is structurally related to the original silica template.

Because of the high temperatures involved in the carbonization stage, this synthetic strategy tends to have very little flexibility in terms of functional "handles" that can be used to bind metal ions, or other analytes. A clever solution to this problem has been reported by Mokaya and coworkers in their syntheses of N-doped mesoporous carbons [97–99]. In this work a stream of acetonitrile (CH_3CN) vapor was entrained in an inert atmosphere and passed through a tube furnace (generally at 900–1100 °C) containing a sample of the silica template. The acetonitrile was carbonized on the silica surface, and the resulting mesoporous carbon was found to contain ~8% N, with a surface area of ~$1000\,m^2g^{-1}$, and a pore volume of $0.83\,cm^3g^{-1}$. XPS analysis of this material suggests that the N functionality is a mixture of "pyridine-like N" and quaternary ammonium salts. Use of these materials as a sorbent to capture metal ions and other analytes has not yet been demonstrated but pyridine ligands are well known to bind metal ions and quaternary ammonium salts can be used for anion exchange.

High surface area N-containing mesoporous carbons have been made using other strategies as well. It is possible to start with the N-containing arene intact and polymerize the heteroaromatic precursor (Scheme 8.4). For example, 1,10-phenanthroline is a diamine that is well known to chelate various transition metal cations [101]. Utilization of this material in the synthesis of templated mesoporous

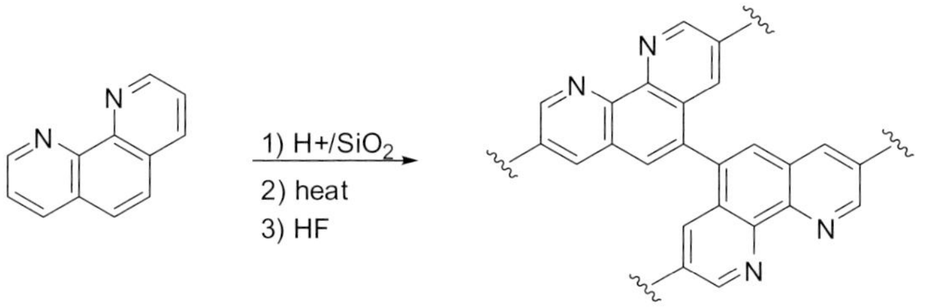

Scheme 8.4

carbon resulted in a product that had a surface area of approximately $870\,m^2\,g^{-1}$ and 30–35 Å pores [102]. Owing to the reluctance of the 1,10-phenanthroline nucleus to undergo electrophilic aromatic substitution, this strategy was found to require temperatures of 700–800 °C for carbonization. As a result, N loss (which takes place above about 600 °C [103]) was able to compete and these materials were found to contain ~5% N. This material was found to have excellent chemical and thermal stability and was shown to be an effective sorbent for transition metal cations (e.g., Ni^{2+}).

High surface area pyridine-based mesoporous carbons have also been made using the cyclotrimerization of diethynylpyridines (Scheme 8.5) by taking advantage of the high reactivity of the acetylene group [104]. The structure of products obtained from this approach depended on the regiochemistry of the acetylene groups on the pyridine precursor. The best results were obtained with the 2,5-diethynylpyridine precursor, which gave a product with a surface area of $1930\,m^2\,g^{-1}$, a pore volume of $2.14\,cm^3\,g^{-1}$, and ~4% N. These high surface area pyridine-based materials have not been evaluated as preconcentration sorbents, but should be useful for capturing transition metal cations, organic acids, and

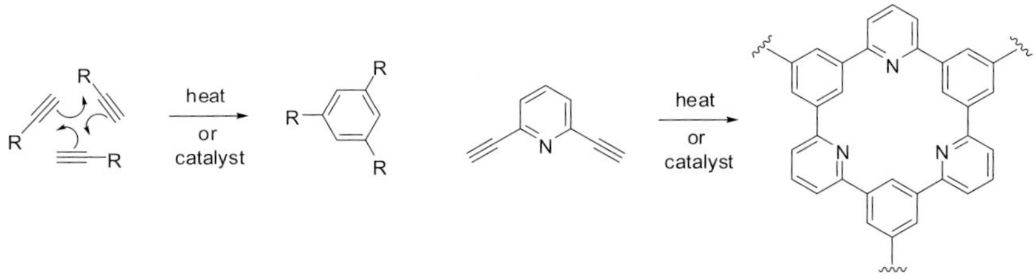

Scheme 8.5

trigonal (or tetrahedral) anions.

The primary limitation of this synthetic strategy is the high temperature required for the polymerization/carbonization process. Above 600 °C, pyridine rings undergo ring–ring fusion reactions and N is lost [103]. Therefore, it would be desirable to use more reactive polymerization chemistry in an effort to lower these temperatures as far as possible so as to preserve as much of the N functionality as possible. $SiCl_4$-catalyzed cyclotrimerization of commercially available 2,6-diacetylpyridine inside an SBA-15 template (Scheme 8.6) has been shown to result in a mesoporous carbon at temperatures as low as 600 °C [105]. The product was found to have a surface area of $1275\,m^2\,g^{-1}$, 35 Å pores, and contain 6.8% N. Again, this pyridine-based nanoporous sorbent should be useful for capturing transition metal cations, organic acids, and (in protonated form) trigonal or tetrahedral anions.

Other heteroaromatic precursors can also be used in this chemistry. For example, acid-catalyzed polymerization of 2-thiophenemethanol inside an SBA-15 template (Scheme 8.7), and subsequent carbonization of the intermediate product at 700–800 °C, was found to create a S-functionalized mesoporous carbon (S-FMC) [106].

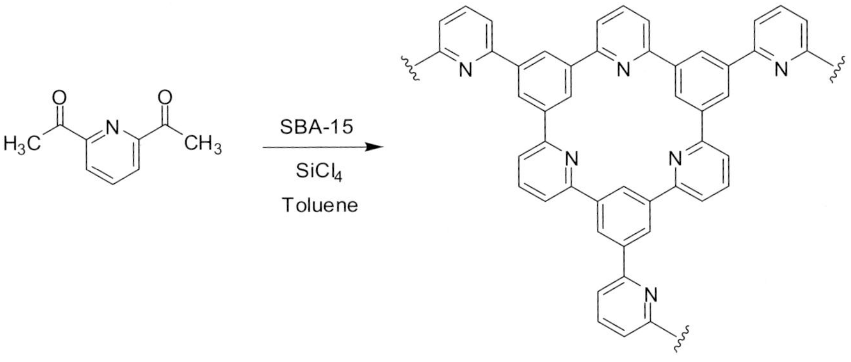

Scheme 8.6

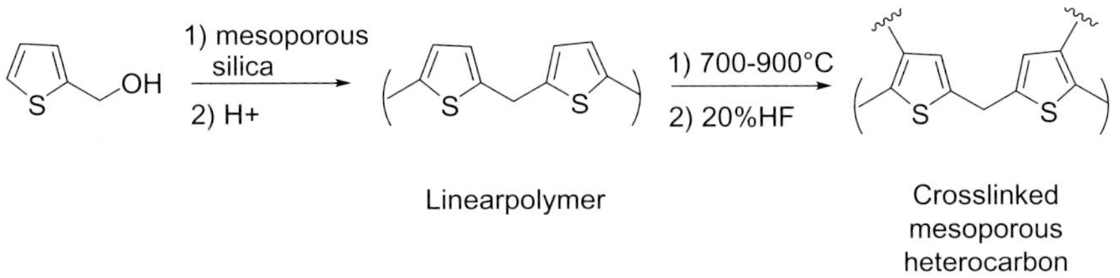

Scheme 8.7

The S content, surface area, and pore volume depended on the carbonization temperature, but between 700 and 800 °C, the S content was 4.9–7.2%, the surface area was 1620–1930 m^2 g^{-1}, and the pore volume was 1.68–2.14 cm^3 g^{-1}. Like all mesoporous carbons, these S-FMCs showed excellent thermal and chemical stability; boiling them in buffers from pH 1 to 13 for 24 h induced no discernible change. Indeed, these S-FMCs were found to be effective heavy metal sorbents. The K_ds for Hg^{2+} were >250 000 over the same broad range of pH (1–13). Very few heavy metal sorbents are capable of effective metal capture over such a wide range of pH.

One of the advantages of carbon-based sorbents is their chemical stability and resistance to chemical degradation. Unfortunately, this chemical stability also means that they are resistant to several commonly used forms of chemical functionalization. Such chemical functionalization is needed to augment their poor selectivity. Chemical selectivity is desirable for analytical preconcentration applications (and also in remediation applications) so as not to consume the sorbent's capacity to bind non-target species. Installation of specific ligand chemistries (e.g., thiols, thiophenes, chelating diamines, etc.) helps to overcome this shortcoming. These solutions preserve the chemical stability of the carbon backbone while adding chemically selective ligands with improved affinity for metal species, and extend the usefulness of carbon-based sorbents beyond that of a general sorbent.

8.6
Other Nanostructured Sorbent Materials

8.6.1
Zeolites

Zeolites are aluminosilicate materials that exhibit well-defined, highly porous structures. They occur naturally and may be prepared by several methods. The general zeolite structure is a three-dimensional network of repeating isomorphous SiO_4 and AlO_4^- tetrahedra linked by oxygen atoms (Figure 8.10). This yields an anionic lattice with acidic (bridging OH groups on Al–O–Si linkages) and basic (Al tetrahedra) sites. Charge balance is maintained by extra-lattice Na^+, K^+, Ca^{2+}, or Mg^{2+} atoms. To date, there are over 40 different naturally occurring zeolites that differ in structure and in Al:Si ratio [107]. Small amounts of Fe are also found in natural zeolites [108].

Zeolites are widely used in both separation and catalysis applications [109]. Like activated carbon materials, zeolites are commonly employed as general sorbents [110]. Their high surface area allows removal of organic species from solution although their effectiveness as sorbents for organic species is limited compared to that of activated carbon materials [110]. Their anionic framework makes zeolites natural cation exchangers while their well-defined pores lend some degree of preference to the ions absorbed. The most widely-used and studied natural zeolite, clinoptilolite, exhibits a general selectivity: Pb^{2+} > Cd^{2+} > Cs^+ > Co^{2+} > Cr^{3+} > Zn^{2+} > Ni^{2+} > Hg^{2+} [111]. Clinoptilolite is also effective in sorption of Sr^{2+} and

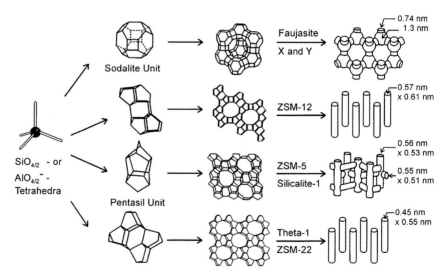

Figure 8.10 Structures of several zeolites as well as their pore shapes and dimensions. Reprinted with permission from Reference [115]. Copyright 2000 Elsevier.

Sb^{2+} cations [112, 113]. Other natural zeolite materials may differ in selectivity. For instance, scolecite follows the series $Cr^{3+} > Mn^{2+} > Cd^{2+} > Ni^{2+}$ [114]. This selectivity is desirable for adsorption of these analytes at trace levels in matrices that contain competing species.

Adsorption of metals by zeolites is governed by pH, temperature, and is made further complex by the different surface sites found within the lattices. The pH affects not only the state of lattice hydroxyl groups and metal speciation but, at very low pH, protonation of surface sites reduces the lattice affinity for positively charged species [116].

Analysis of adsorption isotherms indicates that adsorption of many metal cations occurs via by both ion exchange and chemisoprtion processes. For Pb^{2+} and Zn^{2+} ion exchange occurs quickly and is followed by slower chemisorption [113, 117, 118]. In most cases, sorption favors cations with higher charges and smaller radii [119].

Adsorption occurs most effectively for metals that exist as cations. Because they have little affinity for the anionic lattice, metals that form oxoanions or other anionic species are not as effectively adsorbed. Organic contaminants such as phenol that form complexes with metal ions interfere to various degrees with metal sorption [120]. It is likely that these metal–ligand complexes hinder penetration into pores or form neutral or anionic complexes that have no affinity for the anionic lattice. In some cases, the sorbent may be regenerated with high concentrations of competing cations, such as in solutions of $NaNO_3$. Acid may also be used to strip adsorbed metals, although this has been shown to damage some zeolites [116].

Natural zeolites may be modified to enhance their ability to absorb anionic species. When treated with Fe^{2+} and Fe nanoparticles, these functionalized materials show impressive affinity for arsenate and arsenite anions compared to the native zeolite [121, 122]. Fe^{2+} is also used to treat activated carbon to achieve the same effect [121]. An Al-functionalized zeolite has also been shown to remove arsenate [123]. The small size of natural zeolite pores (typically between 0.4 and 1.2 nm) lends them their selectivity and high surface area but limits their use for adsorption of larger molecules, restricts mass transport through the material, and can result in high back pressure in flow systems [124].

Synthetic zeolites have been designed in an effort to prepare zeolites and zeolite-like materials without these limitations. Two strategies have been explored: (i) making zeolites with larger pores and (ii) inserting larger pores into zeolite materials. These modified zeolites have been reviewed recently [125, 126]. Synthetic mesoporous zeolites may possess different metal selectivity profiles than their natural counterparts. However, little work has been done involving synthetic zeolites as sorbents.

Like activated carbon materials, zeolites are low-cost, high surface area, semi-selective sorbents. Their selectivity can be tailored by modifying lattice constituents, by functionalizing the lattice, or by changing the porous network topology. Like the functionalized nanoporous materials discussed in this section, the high cost of functionalization renders the more sophisticated materials less practical as sorbents for remediation but makes them ideal for detection applications [127].

8.6.2
Ion-Imprinted Polymers

Molecular imprinting is a technique for preparing polymeric matrices that are capable of highly selective solid-phase extraction. Imprinted polymers are prepared by polymerizing functional and often crosslinkable monomers in the presence of an imprint molecule. This preparation of imprinted polymers is typically divided into three steps (Figure 8.11). During an imprint step, functional monomers form a complex with the imprint molecule which either closely resembles or is identical to a target analyte. The monomers are then crosslinked during a polymerization step. This fixes their position and orientation in the network. The imprint molecule is then removed during a leaching step. The polymer network is left with functional monomers pre-organized in a geometry optimum for binding the template molecule.

The first ion-imprinted polymer was developed by Nishide *et al.* in 1976 [129, 130]. These early poly(4-vinlypyridine) resins could be imprinted to preferentially adsorb Cu^{2+}, Ni^{2+}, Hg^{2+}, Zn^{2+}, and Cd^{2+}. Since then, efforts have been made to selectively bind different metals, improve the efficiency of absorption, and develop robust, regenerable imprinted polymers. Many of these efforts have involved incorporating metal chelating systems into polymer matrices. For example, carboxylic acid derivatized monomers have been shown to produce a resin specific to UO_2^{2+} [131]. 5,7-Dichloroquinoline derivatives have been used as the chelating monomers for Dy- and Sm^{3+}-specific polymers [132, 133]. A polymerizable 3-oxapentanediamide derivative was employed in a polymer that could be used to

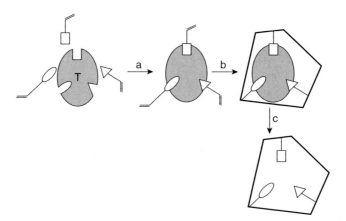

Figure 8.11 Representation of imprinting process: (a) complexation; (b) polymerization; (c) leaching. T = imprint molecule. Reprinted with permission from Reference [128]. Copyright 2004 Elsevier.

separate Ca^{2+} from Mg^{2+} ions and vice versa [134]. Lemaire, *et al.* have used methacrylate monomers to selectively bind Gd and, in a similar polymer matrix, diethylene triamine pentaacetic acid (DPTA) derivative chelating monomers for La^{3+} [135]. Prasada Rao *et al.* have reviewed this topic recently [128, 136].

Since adsorption of metal ions by ion-imprinted polymers is governed almost entirely by chemisorption, using more sophisticated ligands yields higher specificity and more efficient adsorption of target analytes. The synthetic challenges have been to incorporate these sophisticated chelate functionalities into polymerizable molecules. It is sometimes possible to circumvent this challenge by using both polymerizable and non-polymerizable ligands to form the template–ligand complex. The non-polymerizable ligands becomes trapped in the polymer matrix even after the template species is removed [136].

Advantages of imprinted polymers are simple preparation, durability, and predetermined and predictable selectivity. They may be used as bulk polymer, but are more effective when incorporated into nanostructured materials. Ion-imprinted polymers may be cast as thin films, incorporated into membranes, or applied as coatings to solid supports [127, 129–132]. Applying the polymer to a high surface area material eliminates the mass transport limitation involved in using the material in bulk. Drawbacks to ion-imprinted polymers involve poor solubility of the template (metal ion) in the imprinting mixture.

This section has outlined just two particular classes of material that are being explored for trace metal adsorption from aqueous matrices. There is tremendous room for development of zeolites, ion-imprinted polymer-based materials, and other nanoporous materials like aluminophosphate materials and metal organic frameworks as sorbents for trace-level metal contaminants. Practical utilization of these nanomaterials in trace-level assay will depend upon the capability to tailor the materials into form factors that allow integration with devices and functionalize the surfaces to have high selectivity and affinity.

8.7
Concluding Thoughts

We have reviewed selected nanostructured materials, many with thiol functionalization, for the capture of softer heavy metals from aqueous systems for environmental and sensing and separation applications. It has been clearly shown that correctly constructed nanomaterials can be superior sorbents over conventional materials. These nanomaterials can be used for analytical and remediation applications. It should be reiterated that we chose to emphasize work employing thiol surface chemistry, since it provides a highly effective means for the capture of many toxic heavy metals from aqueous systems and it offers a useful baseline for comparing the performance of the various nanomaterials. Many other elegant surface chemistries exist and enable the use of these nanomaterials for applications to other analyte sets. Beyond simple performance, one remaining factor that merits discussion is the economics of using nanomaterials for collection and detection.

Nanostructured materials can be expensive. While industrial scale-up can bring the cost down, it does not address two fundamental problems. First, scaling the production volumes up from laboratory beakers to industrial tonnage is a non-trivial effort, particularly for advanced materials with complex structures. Second, the materials and methods needed to assemble sorbents with the useful nanostructure and surface chemistry will almost inevitably make them more expensive than bulk or traditional materials. For large-scale applications of nanomaterials, such as environmental remediation efforts, the performance of the materials must be sufficient to merit the additional cost. Viability will have to be assessed on a case by case basis, but will likely be more successful on high value applications such as those associated with nuclear materials or localized applications such as batch treatment. The much larger K_d and capacity values for functionalized nanoporous silica and nanoparticulate iron oxide shown in Tables 8.1, 8.2 and 8.4 suggest that despite higher costs than tradition bulk sorbents they may provide better value for select applications. Furthermore, the possibility of non-covalent modification of SAMMS and related nanomaterials (as in the chemisorbed phenyl-SAMMS materials described in Section 8.3.1) provides a route to mitigate the cost of the nanomaterial by providing a greater product lifetime and sorption capacity.

In contrast to large-scale separations, analytical applications require much smaller volumes of sorbent materials. In most cases analytical applications will require only milligrams (or conceivably even micrograms) of sorbent material and the material might be used for many repeated measurements. Information coming from analytical assays forms the basis for many high value decisions such as those made as a result of medical diagnostics, legal forensics, determination of clean water and food, or operation of an industrial process. Consequently, with high value results and small quantities of material needed, the cost of the sorbent material for analytical applications is generally irrelevant. For analytical applications, obtaining a stable supply of materials with dependable performance and effectively integrating them into the analytical method/device is more important than the nanomaterial cost. For some nanomaterials, such as quantum dot emitters and PANAM dendrimers, reliable industrial production and their wide spread analytical utilization has already been demonstrated. As commercial sources of nanomaterials become increasing available it is inevitable they will be integrated in the products as appropriate, with high value niche applications such as improved analytical devices/methods leading the way.

In conclusion, we have briefly discussed the material science and application of functionalized nanoporous silica, functionalized superparamagnetic particles, nanostructured carbon-based materials, and other structured materials such as zeolites. These materials only scratch the surface of possible nanomaterials that can be employed in sensing and remediation applications. Additional materials that are presently under development, as well as others not yet imagined, will provide new and relevant capabilities, enabling a range of analytical applications for trace-level measurements.

Acknowledgments

Funding for this work was provided in part by the Safer Nanomaterials Nanomanufacturing Initiative (SNNI) of Oregon Nanoscience and Microtechnologies Institute (ONAMI) and Pacific Northwest National Laboratory (PNNL). A portion of this research was performed using Environmental Molecular Sciences Laboratory (EMSL), a national scientific user facility sponsored by the Department of Energy's Office of Biological and Environmental Research located at PNNL. PNNL is operated for the U.S. Department of Energy by Battelle under contract DE-AC06-67RLO 1830. This work was also supported by grants from NIAID (R01-AI080502), NIOSH (1R21 OH008900), and NIEHS (R21-ES015620) and PNNL's LDRD program. Its contents are solely the responsibility of the authors and do not necessarily represent the official views of NIH.

D. W. J, S. A. F., and T. G. C. gratefully acknowledge support from the University of Oregon and an NSF-CAREER award (CHE-0545206). D. W. J is a Cottrell Scholar of Research Corporation. T. G. C. and S. A. F. acknowledge the National Science Foundation (NSF) for Integrative Graduate Education and Research Traineeships (DGE-0549503).

References

1 Poole, C.F. (2003) *Trends Anal. Chem.*, **22**, 362–373.
2 Fryxell, G.E., Lin, Y., Fiskum, S., Birnbaum, J.C., Wu, H., Kemner, K., and Kelly, S. (2005) *Environ. Sci. Technol.*, **39**, 1324–1331.
3 Fryxell, G.E., Liu, J., Mattigod, S.V., Wang, L.Q., Gong, M., Hauser, T.A., Lin, Y., Ferris, K.F., and Feng, X. (2000) in, *Environmental Issues and Waste Management Technologies in the Ceramic and Nuclear Industries V*, (eds G.T. Chandler and X. Feng), Ceramics Transactions, Vol. **107**, The American Ceramic Society, Westerville, OH, pp. 29–37.
4 Fryxell, G.E., Lin, Y., Wu, H., and Kemner, K.M. (2002), in *Studies in Surface Science and Catalysis*, Vol. **141** (eds A. Sayari and M. Jaroniec), Elsevier Science, pp. 583–590.
5 Fryxell, G.E., Addleman, R.S., Mattigod, S.V., Lin, Y., Zemanian, T.S., Wu, H., Birnbaum, J.C., Liu, J., and Feng, X. (2004) in *Dekker Encyclopedia of Nanoscience and Nanotechnology* (ed. J.A. Schwarz), Marcel Dekker, pp. 1135–1145.
6 Mellor, D.P. (1979) *Chemistry of Chelation and Chelating Agents*, Pergamon Press, New York.
7 Richens, D.T. (1997) *The Chemistry of Aqua Ions*, John Wiley & Sons, Ltd, Chichester.
8 Pearson, R.G. (1963) *J. Am. Chem. Soc.*, **85**, 3533–3539.
9 Herrin, R.T., Andren, A.W., Shafer, M.M., and Armstrong, D.E. (2001) *Environ. Sci. Technol.*, **35**, 1959–1966.
10 Powell, K.J., Brown, P.L., Byrne, R.H., Gajda, T., Hefter, G., Sjöberg, S., and Wanner, H. (2004) *Aust. J. Chem.*, **57**, 993–1000.
11 Byrne, R.H. (2002) *Geochem. Trans.*, **3**, 11–16.
12 Yantasee, W., Charnhattakorn, B., Fryxell, G.E., Lin, Y., Timchalk, C., and Addleman, R.S. (2008) *Anal. Chim. Acta*, **620**, 55–63.
13 Yantasee, W., Warner, C.L., Sangvanich, T., Addleman, R.S., Carter, T.G., Wiacek, R.J., Fryxell, G.E., Timchalk, C., and Warner, M.G. (2007) *Environ. Sci. Technol.*, **41**, 5114–5119.

14 Lloyd-Jones, P.J., Rangel-Mendez, J.R., and Streat, M. (2004) *Process Saf. Environ. Prot.*, **82**, 301–311.

15 Carter, T.G., Yantasee, W., Sangvanich, T., Fryxell, G.E., Johnson, D.W., and Addleman, R.S. (2008) *Chem. Commun.*, 5583–5585.

16 United States Pharmacopeia (1990) *USP United States Pharmacopeial Convention Inc.*, 22nd edn, United States Pharmacopeia, Rockville, MD.

17 Shargel, L., and Yu, A.B.C. (2002) *Encyclopedia of Pharmaceutical Technology*, Vol. **1**, 2nd edn (eds J. Swarbrick and J.C. Boylan), Informa Health Care, p. 164.

18 Corr, S.A., Rakovich, Y.P., and Gun'ko, Y.K. (2008) *Nanoscale Res. Lett.*, **3**, 87–104.

19 Hsing, I.M., Xu, Y., and Zhao, W. (2007) *Electroanalysis*, **19**, 755–768.

20 Huber, D.L. (2005) *Small*, **1**, 482–501.

21 Katz, E. and Willner, I. (2004) *Angew. Chem. Int. Ed.*, **43**, 6042–6108.

22 Katz, E., Willner, I., and Wang, J. (2004) *Electroanalysis*, **16**, 19–44.

23 Latham, A.H. and Williams, M.E. (2008) *Acc. Chem. Res.*, **41**, 411–420.

24 Laurent, S., Forge, D., Port, M., Roch, A., Robic, C., Vander Elst, L., and Muller, R.N. (2008) *Chem. Rev.*, **108**, 2064–2110.

25 Lu, A.H., Salabas, E.L., and Schueth, F. (2007) *Angew. Chem. Int. Ed.*, **46**, 1222–1244.

26 Magnani, M., Galluzzi, L., and Bruce, I.J. (2006) *J. Nanosci. Nanotechnol.*, **6**, 2302–2311.

27 Pankhurst, Q.A. (2006) *BT Technol. J.*, **24**, 33–38.

28 Pankhurst, Q.A., Connolly, J., Jones, S.K., and Dobson, J. (2003) *J. Phys. D Appl. Phys.*, **36**, R167–R181.

29 Rodriguez-Mozaz, S., de Alda, M.J.L., and Barcelo, D. (2006) *Anal. Bioanal. Chem.*, **386**, 1025–1041.

30 Salgueirino-Maceira, V. and Correa-Duarte, M.A. (2007) *Adv. Mater.*, **19**, 4131–4144.

31 Seydack, M. (2005) *Biosens. Bioelectron.*, **20**, 2454–2469.

32 Tartaj, P., Morales, M.D., Veintemillas-Verdaguer, S., Gonzalez-Carreno, T., and Serna, C.J. (2003) *J. Phys. D Appl. Phys.*, **36**, R182–R197.

33 Yantasee, W., Lin, Y., Hongsirikarn, K., Fryxell, G.E., Addleman, R.S., and Timchalk, C. (2007) *Environ. Health Perspect.*, **115**, 1683–1690.

34 Cushing, B.L., Kolesnichenko, V.L., and O'Connor, C.J. (2004) *Chem. Rev.*, **104**, 3893–3946.

35 Jun, Y., Choi, J., and Cheon, J. (2006) *Angew. Chem. Int. Ed.*, **45**, 3414–3439.

36 Neouze, M.A. and Schubert, U. (2008) *Monatsh. Chem.*, **139**, 183–195.

37 Park, J., Joo, J., Kwon, S., Jang, Y., and Hyeon, T. (2007) *Angew. Chem. Int. Ed.*, **46**, 4630–4660.

38 Deng, Y.-H., Wang, C.-C., Hu, J.-H., Yang, W.-L., and Fu, S.-K. (2005) *Colloids Surf. A*, **262**, 87–93.

39 Ennas, G., Musinu, A., Piccaluga, G., Zedda, D., Gatteschi, D., Sangregorio, C., Stanger, J.L., Concas, G., and Spano, G. (1998) *Chem. Mater.*, **10**, 495–502.

40 Lu, C.W., Hung, Y., Hsiao, J.K., Yao, M., Chung, T.H., Lin, Y.S., Wu, S.H., Hsu, S.C., Liu, H.M., Mou, C.Y., Yang, C.S., Huang, D.M., and Chen, Y.C. (2007) *Nano Lett.*, **7**, 149–154.

41 Santra, S., Tapec, R., Theodoropoulou, N., Dobson, J., Hebard, A., and Tan, W. (2001) *Langmuir*, **17**, 2900–2906.

42 Stober, W., Fink, A. and Bohn, E. (1968) *J. Colloid Interface Sci.*, **26**, 62–69.

43 Carpenter, E.E. (2001) *J. Magn. Magn. Mater.*, **225**, 17–20.

44 Mikhaylova, M., Kim, D.K., Bobry-sheva, N., Osmolowsky, M., Semenov, V., Tsakalakos, T., and Muhammed, M. (2004) *Langmuir*, **20**, 2472–2477.

45 Lo, C.K., Xiao, D., and Choi, M.M.F. (2007) *J. Mater. Chem.*, **17**, 2418–2427.

46 Lu, Q.H., Yao, K.L., Xi, D., Liu, Z.L., Luo, X.P., and Ning, Q. (2006) *J. Magn. Magn. Mater.*, **301**, 44–49.

47 Mandal, M., Kundu, S., Ghosh, S.K., Panigrahi, S., Sau, T.K., Yusuf, S.M., and Pal, T. (2005) *J. Colloid Interface Sci.*, **286**, 187–194.

48 Park, H.Y., Schadt, M.J., Wang, L.Y., Lim, I.I.S., Njoki, P.N., Kim, S.H.,

Jang, M.Y., Luo, J., and Zhong, C.J. (2007) *Langmuir*, **23**, 9050–9056.

49 Lim, J., Eggeman, A., Lanni, F., Tilton, R.D., and Majetich, S.A. (2008) *Adv. Mater.*, **20**, 1721–1726.

50 Xu, Z., Hou, Y., and Sun, S. (2007) *J. Am. Chem. Soc.*, **129**, 8698–8699.

51 Wang, L., Luo, J., Maye, M.M., Fan, Q., Rendeng, Q., Engelhard, M.H., Wang, C., Lin, Y., and Zhong, C.-J. (2005) *J. Mater. Chem.*, **15**, 1821–1832.

52 Gong, P., Li, H., He, X., Wang, K., Hu, J., Tan, W., Zhang, S., and Yang, X. (2007) *Nanotechnology*, **18**, 285604.

53 Kimishima, Y., Yamada, W., Uehara, M., Asaka, T., Kimoto, K., and Matsui, Y. (2007) *Mater. Sci. Eng. B*, **138**, 69–73.

54 Tang, D.P., Yuan, R., and Chai, Y.Q. (2006) *J. Phys. Chem. B*, **110**, 11640–11646.

55 Liu, C., Zhou, Z., Yu, X., Lv, B., Mao, J., and Xiao, D. (2008) *Inorg. Mater.*, **44**, 291–295.

56 Yantasee, W., Hongsirikarn, K., Warner, C.L., Choi, D., Sangvanich, T., Toloczko, M.B., Warner, M.G., Fryxell, G.E., Addleman, R.S., and Timchalk, C. (2008) *Analyst*, **133**, 348–355.

57 Santandreu, M., Sole, S., Fabregas, E., and Alegret, S. (1998) *Biosens. Bioelectron.*, **13**, 7–17.

58 Riskin, M., Basnar, B., Huang, Y., and Willner, I. (2007) *Adv. Mater.*, **19**, 2691–2693.

59 Liu, H.B., Guo, J., Jin, L., Yang, W.L., and Wang, C.C. (2008) *J. Phys. Chem. B*, **112**, 3315–3321.

60 Cai, J., Guo, J., Ji, M.L., Yang, W.L., Wang, C.C., and Fu, S.K. (2007) *Colloid Polym. Sci.*, **285**, 1607–1615.

61 Shin, S. and Jang, J. (2007) *Chem. Commun.*, 4230–4232.

62 Ngomsik, A., Bee, A., Draye, M., Cote, G., and Cabuil, V. (2005) *C. R. Chim.*, **8**, 963–970.

63 Yavuz, C.T., Mayo, J.T., Yu, W.W., Prakash, A., Falkner, J.C., Yean, S., Cong, L.L., Shipley, H.J., Kan, A., Tomson, M., Natelson, D., and Colvin, V.L. (2006) *Science*, **314**, 964–967.

64 Yean, S., Cong, L., Yavuz, C.T., Mayo, J.T., Yu, W.W., Kan, A.T., Colvin, V.L., and Tomson, M.B. (2005) *J. Mater. Res.*, **20**, 3255–3264.

65 Palecek, E. and Fojta, M. (2007) *Talanta*, **74**, 276–290.

66 de la Escosura-Muniz, A., Ambrosi, A., and Merkoci, A. (2008) *Trends Anal. Chem.*, **27**, 568–584.

67 EPA (2010) Drinking water contaminants, http://www.epa.gov/ogwdw/hfacts.html (accessed 20 January 2010).

68 Liu, G.D., Timchalk, C., and Lin, Y.H. (2006) *Electroanalysis*, **18**, 1605–1613.

69 Liu, G. and Lin, Y. (2005) *J. Nanosci. Nanotechnol.*, **5**, 1060–1065.

70 Bansal, R.C. and Goyal, M. (2005) *Activated Carbon Adsorption*, Taylor & Francis, New York.

71 Pelech, R., Milchert, E., and Bartkowiak, M. (2006) *J. Colloid Interface Sci.*, **296**, 458–464.

72 Jiang, Z.X., Liu, Y., Sun, X.P., Tian, F.P., Sun, F.X., Liang, C.H., You, W.S., Han, C.R., and Li, C. (2003) *Langmuir*, **19**, 731–736.

73 Satapathy, D. and Natarajan, G.S. (2006) *Adsorption - J. Int. Adsorption Soc.*, **12**, 147–154.

74 Babel, S. and Kurniawan, T.A. (2004) *Chemosphere*, **54**, 951–967.

75 Kannan, N. and Rengasamy, G. (2005) *Fresenius Environ. Bull.*, **14**, 435–443.

76 Kasaini, H., Everson, R.C., and Bruinsma, O.S.L. (2005), *Sep. Sci. Technol.*, **40** (1–3), 507–523.

77 Abdulkarim, M. and Abu Al-Rub, F. (2004) *Adsorption Sci. Technol.*, **22**, 119–134.

78 Rangel-Mendez, J.R. and Streat, M. (2002) *Water Res.*, **36**, 1244–1252.

79 Biniak, S., Pakula, M., Szymanski, G.S., and Swiatkowski, A. (1999) *Langmuir*, **15**, 6117–6122.

80 Yin, C.Y., Aroua, M.K., and Daud, W. (2007) *Colloids Surf. A*, **307**, 128–136.

81 Yusof, A.M., Rahman, M.M., and Wood, A.K.H. (2004) *J. Radioanal. Nucl. Chem.*, **259**, 479–484.

82 Yusof, A.M., Rahman, M.M., and Wood, A.K.H. (2007) *J. Radioanal. Nucl. Chem.*, **271**, 191–197.

83 Yin, C.Y., Aroua, M.K., and Daud, W. (2007) *Sep. Purif. Technol.*, **52**, 403–415.

84 Yantasee, W., Lin, Y.H., Alford, K.L., Busche, B.J., Fryxell, G.E., and Engelhard, M.H. (2004) *Sep. Sci. Technol.*, **39**, 3263–3279.

85 Yantasee, W., Lin, Y.H., Fryxell, G.E., Alford, K.L., Busche, B.J., and Johnson, C.D. (2004) *Ind. Eng. Chem. Res.*, **43**, 2759–2764.

86 Merrifield, R.B. (1963) *J. Am. Chem. Soc*, **85**, 2149.

87 Samuels, W.D., LaFemina, N.H., Sukwarotwat, V., Yantasee, W., Li, X.S., and Fryxell, G.E. (2010) *Sep. Sci. Technol.*, **45**, 228–235.

88 Gholivand, M.B., Ahmadi, F., and Rafiee, E. (2007) *Sep. Sci. Technol.*, **42**, 897–910.

89 Ryoo, R., Joo, S.H., and Jun, S. (1999) *J. Phys. Chem. B*, **103**, 7743–7746.

90 Joo, S.H., Choi, S.J., Oh, I., Kwak, J., Liu, Z., Terasaki, O., and Ryoo, R. (2001) *Nature*, **412**, 169–172.

91 Gierszal, K.P. and Jaroniec, M. (2006) *J. Am. Chem. Soc*, **128**, 10026–10027.

92 Lee, J., Sohn, K., and Hyeon, T. (2001) *J. Am. Chem. Soc*, **123**, 5146–5147.

93 Kim, C.H., Lee, D.K., and Pinnavaia, T.J. (2004) *Langmuir*, **20**, 5157–5159.

94 Lu, A.H., Kiefer, A., Schmidt, W., and Schuth, F. (2004) *Chem. Mater.*, **16**, 100–103.

95 Liang, C.D. and Dai, S. (2006) *J. Am. Chem. Soc*, **128**, 5316–5317.

96 Wang, X.Q., Liang, C.D., and Dai, S. (2008) *Langmuir*, **24**, 7500–7505.

97 Xia, Y.D. and Mokaya, R. (2005) *Chem. Mater.*, **17**, 1553–1560.

98 Xia, Y.D., Yang, Z.X., and Mokaya, R. (2004) *J. Phys. Chem. B*, **108**, 19293–19298.

99 Yang, Z.X., Xia, Y.D., Sun, X.Z., and Mokaya, R. (2006) *J. Phys. Chem. B*, **110**, 18424–18431.

100 Fulvio, P.F., Jaroniec, M., Liang, C.D., and Dai, S. (2008) *J. Phys. Chem. C*, **112**, 13126–13133.

101 Cotton, F.A. and Wilkinson, G. (1980) *Advanced Inorganic Chemistry*, 4th edn, John Wiley & Sons, Inc., New York.

102 Shin, Y., Fryxell, G.E., Engelhard, M.H., and Exarhos, G.J. (2007) *Inorg. Chem. Commun.*, **10**, 1541–1544.

103 Bahl, O.P., Shen, Z., Lavin, J.G., and Ross, R.A. (1998) *Carbon Fibers*, 3rd edn (eds S. Rebouillat, J.B. Donnet, T.K. Wang, and J.C.M. Peng), Marcel Dekker, New York, pp. 1–83.

104 Shin, Y., Fryxell, G.E., Johnson, C.A., and Haley, M.M. (2008) *Chem. Mater.*, **20**, 981–986.

105 Shin, Y., Wang, C., Engelhard, M.H., and Fryxell, G.E. (2009) *Microporous Mesoporous Mater.*, **123**, 345–348.

106 Shin, Y.S., Fryxell, G., Um, W.Y., Parker, K., Mattigod, S.V., and Skaggs, R. (2007) *Adv. Funct. Mater.*, **17**, 2897–2901.

107 Baerlocher, C., Meier, W.M., and Olson, D.H. (2001) *Atlas of Zeolite Framework Types*, 5th edn, Elsevier, Amsterdam.

108 Erdem, E., Karapinar, N., and Donat, R. (2004) *J. Colloid Interface Sci.*, **280**, 309–314.

109 Hartmann, M. and Kevan, L. (1999) *Chem. Rev.*, **99**, 635–664.

110 San Miguel, G., Lambert, S.D., and Graham, N.J.D. (2006) *J. Chem. Technol. Biotechnol.*, **81**, 1685–1696.

111 Zamzow, M.J., Eichbaum, B.R., Sandgren, K.R., and Shanks, D.E. (1990) *Sep. Sci. Technol.*, **25** (13-15), 1555–1569.

112 Um, W. and Papelis, C. (2004) *Environ. Sci. Technol.*, **38**, 496–502.

113 Um, W. and Papelis, C. (2003) *Am. Mineral.*, **88**, 2028–2039.

114 Dal Bosco, S.M., Jimenez, R.S., and Carvalho, W.A. (2005) *J. Colloid Interface Sci.*, **281**, 424–431.

115 Weitkamp, J. (2000) *Solid State Ionics*, **131**, 175–188.

116 Wingenfelder, U., Hansen, C., Furrer, G., and Schulin, R. (2005) *Environ. Sci. Technol.*, **39**, 4606–4613.

117 Shahwan, T., Zünbül, B., Tunusoglu, Ö., and Eroglu, A.E. (2005) *J. Colloid Interface Sci.*, **286**, 471–478.

118 Ören, A.H. and Kaya, A. (2006) *J. Hazard. Mater.*, **131**, 59–65.

119 Logar, N.Z. and Kaucic, V. (2006) *Acta Chim. Slov.*, **53**, 117–135.

120 Vaca Mier, M., López Callejas, R., Gehr, R., Jiménez Cisneros, B.E., and Alvarez, P.J.J. (2001) *Water Res.*, **35**, 373–378.

121 Payne, K.B. and Abdel-Fattah, T.M. (2005) *J. Environ. Sci. Health, Part A*, **40**, 723–749.

122 Dousová, B., Grygar, T., Martaus, A., Fuitová, L., Kolousek, D., and Machovic, V. (2006) *J. Colloid Interface Sci.*, **302**, 424–431.

123 Xu, Y.-H., Nakajima, T., and Ohki, A. (2002) *J. Hazard. Mater.*, **92**, 275–287.

124 Tao, Y.S., Kanoh, H., Abrams, L., and Kaneko, K. (2006) *Chem. Rev.*, **106**, 896–910.

125 Davis, M.E. (2002) *Nature*, **417**, 813–821.

126 Drews, T.O. and Tsapatsis, M. (2005) *Curr. Opin. Colloid Interface Sci.*, **10**, 233–238.

127 Valdés, M.G., Pérez-Cordoves, A.I., and Díaz-García, M.E. (2006) *Trends Anal. Chem.*, **25**, 24–30.

128 Prasada Rao, T., Daniel, S., and Mary Gladis, J. (2004) *Trends Anal. Chem.*, **23**, 28–35.

129 Nishide, H. and Tsuchida, E. (1976) *Makromol. Chem.*, **177**, 2295–2310.

130 Nishide, H., Deguchi, J., and Tsuchida, E. (1976) *Chem. Lett.*, 169–174.

131 Bae, S.Y., Southard, G.L., and Murray, G.M. (1999) *Anal. Chim. Acta*, **397**, 173–181.

132 Shirvani-Arani, S., Ahmadi, S.J., Bahrami-Samani, A., and Ghannadi-Maragheh, M. (2008) *Anal. Chim. Acta*, **623**, 82–88.

133 Biju, V.M., Gladis, J.M., and Rao, T.P. (2003) *Anal. Chim. Acta*, **478**, 43–51.

134 Rosatzin, T., Andersson, L.I., Simon, W., and Mosbach, K. (1991) *J. Chem. Soc. Perkin Trans. 2*, 1261–1265.

135 Garcia, R., Pinel, C., Madic, C., and Lemaire, M. (1998) *Tetrahedron Lett.*, **39**, 8651–8654.

136 Rao, T.P., Kala, R., and Daniel, S. (2006) *Anal. Chim. Acta*, **578**, 105–116.

9
Synthesis and Analysis Applications of TiO$_2$-Based Nanomaterials

Aize Li, Benjamen C. Sun, Nenny Fahruddin, Julia X. Zhao, and David T. Pierce

9.1
Introduction

Nanoscale titania (TiO$_2$) has a wide range of properties – from optical to electronic, chemical, and even structural – that make it an especially versatile material for performing chemical analysis. For example, TiO$_2$ has a finite resistance as a semiconductor that can change with adsorption of other species. This property has been used to build effective chemical-electrical transducers that form the basis of many chemical sensors [1–5]. TiO$_2$ (formally Ti^{IV}) is also susceptible to oxidation and its conversion into Ti^{III} is accompanied by a significantly increased conductance and redshift in adsorption wavelength. This behavior has been used for the analysis of certain oxidizing or reducing agents by monitoring changes in electrical [4, 6–8] and optical properties [9–13]. Other properties of TiO$_2$ nanomaterials that are favorable for trace analysis include high thermal and chemical stability, optical transparency in the visible and near-IR domain, photovoltaic properties, photo-cleaning capability, strong adsorption characteristics, and pH-dependent surface charges.

Under UV illumination, the electrons of TiO$_2$ nanomaterials can be pumped into a conduction band, leaving holes in a valence band. This photovoltaic effect has been used for voltammetric sensing when the nanomaterials are placed in an electrical circuit under light irradiation [13–16]. Alternatively, photo-generated electrons and holes can also act as strong reducing and oxidizing agents in the absence of an electric circuit and can participate in surface reactions. A useful application of this property, from the standpoint of analysis, is the photo-cleaning and recovery of titania-based sensors that have been poisoned by adsorption of organic contaminants [17–19].

The strong adsorption characteristics of nanostructured TiO$_2$ are attributed to the high chemical activities of these surfaces and result primarily from their amphiprotic nature [20–22]. While strong adsorption can sometimes lead to matrix interferences and poisoning issues, this property has been widely exploited to preconcentrate analytes by solid-phase extraction [20, 23–25] and to design sensitive platforms for trace analysis. For similar reasons, the surface of nanostructured TiO$_2$ can be readily functionalized and charges on the surface can be altered by

Trace Analysis with Nanomaterials. Edited by David T. Pierce and Julia Xiaojun Zhao
Copyright © 2010 WILEY-VCH Verlag GmbH & Co. KGaA, Weinheim
ISBN: 978-3-527-32350-0

simply controlling pH conditions [26]. For instance, the hydroxyl groups on the TiO_2 surfaces can react with many functional groups such as phosphate [27–29], sulfate [30], or carboxylic acid groups [31]. This functionality of TiO_2 nanomaterials allows specific targets to be enriched on the surfaces, effectively improving both the sensitivity and selectivity [32–38]. The ability to alter the surface charge of titania is useful for applications in which electrostatic interaction plays a role.

During the last decade, the nanostructure has been shown to have a considerable influence over the physical and chemical behavior of TiO_2. It is expected that findings of this nature will play an important role in directing how TiO_2 nanomaterials will be used for analysis in coming years. For example, changes in the size and shape of semiconductors like TiO_2 at the nanoscale can result in quantum confinement and thus affect the movement of electrons and holes [39–41]. Recent studies have shown that conductance among the polymorphs of TiO_2 follows the order of amorphous TiO_2 > anatase > rutile TiO_2 crystals, whereas the order with respect to the particle size is opposite [42–44]. On the other hand, nanotubular TiO_2 has a higher conductivity (ca. $7.9 \times 10^{-7}\,S\,cm^{-1}$) [45] than that of anatase or rutile nanoparticles (ca $10^{-9}\,S\,cm^{-1}$) [46]. In addition, as a consequence of small material size, the surface area and the chemical activities of TiO_2 nanomaterials can be dramatically increased [47, 48]. Since most TiO_2-based sensing methods involve surface reactions, this property is especially favorable for engineering particular interactions between target analytes and nanomaterials.

TiO_2 nanomaterials can be synthesized by several methods, such as sol–gel [22, 49, 50], hydrothermal [14, 15, 20, 23–26, 51–55], solvothermal [56], microwave [57], microemulsion/reverse microemulsion [58–62], direct oxidation [63], and electrochemical methods [64–66]. Several excellent papers have recently reviewed these syntheses [67–70]. Diverse nanostructures can be prepared to meet different sensing requirements, including nanoparticles, nanosheets, nanorods (nanofibers or nanowires), nanotubes, and mesoporous/nanoporous nanomaterials. Among them, nanostructures that present large surface areas such as mesoporous TiO_2 nanomaterials [71–74] and TiO_2 nanotubes [5, 69] gather much attention, due to the improved accessibility to targets and efficiency of separation of electron–hole pairs. In addition, TiO_2 nanomaterials can act as hosts to hybridize with various inorganic materials (e.g., metal ions or nanoparticles, or non-metal elements) and organic materials (e.g., dye molecules [15, 75] or proteins [27]) and thereby contribute their own properties and extend functionality. Furthermore, TiO_2 nanomaterials can also be easily transferred onto various solid substrates to accommodate detection in different situations (e.g., direct detection versus extraction) or with different platforms (e.g., optical versus electrical).

In this chapter the general synthetic methods to prepare some commonly used TiO_2 nanostructures in the sensing field are briefly reviewed, including TiO_2 nanoparticles, nanotubes, mesoporous TiO_2, hybrid TiO_2 nanomaterials, and TiO_2 nanofilms. The uses of these materials for trace analysis are then described in detail with a focus on applications for gaseous, aqueous, organic, and biological systems. In each of these cases, the sensing principles and performance characteristics of the TiO_2 nanomaterials employed will be discussed.

9.2
Synthesis of TiO$_2$ Nanostructures

9.2.1
TiO$_2$ Nanoparticles

The sol–gel method is probably the most common approach used to prepare TiO$_2$ nanoparticles. These processes initially involve the formation of an amorphous TiO$_2$ network by the hydrolysis and polymerization of titania precursors in solution, followed by evaporation to form a colloidal gel [76]. Additional treatment such as annealing, which is often carried out in an oven or an autoclave chamber, is usually required to convert the amorphous TiO$_2$ gel into a crystalline form. Isley *et al.* have found that certain conditions during sol–gel synthesis, especially pH, have a significant impact on the phase composition of the resulting titania nanoparticles and their size [50]. For example, lowering the pH resulted in more anatase content and increasing the pH resulted in an increase in particle size.

9.2.2
Mesoporous TiO$_2$

Mesoporous TiO$_2$ nanostructures are typically engineered through templating methods that use phosphates [20], amines [77] and other ions [72–78], block copolymers [79–81], non-ionic surfactants [82], and non-surfactants [83] as scaffolding materials. Template species are mixed with titania precursors and become embedded within the solid titania framework during hydrolysis and condensation. The template species are then removed by liquid extraction or annealing to yield an open, mesoporous structure [84]. However, some mesoporous nanostructures have been prepared without any apparent templating. For instance, hydrolysis of titania precursors at high ionic strength [74] and acidic pH [85] can produce mesoporous TiO$_2$ by other interactions. In the first case, high ionic strength of the hydrolysis solution reduces the surface potential of the growing TiO$_2$ colloid and favors uniform packing to form a mesoporous superstructure. In the second case, acidic conditions retard hydrolysis of titania precursors, leading to the formation of small monodispersed TiO$_2$ nanoparticles. Aggregation of the TiO$_2$ nanoparticles results in mesoporous frameworks.

9.2.3
TiO$_2$ Nanotubes

There are three general methods used to prepare TiO$_2$ nanotubes – anodization of titanium, template synthesis [86–89], and the alkaline hydrothermal method. A relatively simple method used to form nanotube arrays involves potentiostatic anodization of purified titanium foil [90–94]. The foil is typically oxidized at a constant potential in a two-electrode electrochemical cell and subsequently annealed at 500 °C. Morphology of the TiO$_2$ nanotubes prepared by this method can be controlled by several factors, such as the applied potential and presence of phosphate electrolytes [90]. Low applied voltage results in nanotubes with shorter lengths and smaller diameters. However, an increase in voltage is believed to increase ionic transport through the passivating oxide layer at the bottom of each pore, thus speeding the growth and increasing the total length of the nanotubes. A potential of 20–60 V is recommended to produce well-aligned nanotube arrays. The total growth will reach a maximum after the nanotube begins to degrade and lowering the voltage will cause this maximum to occur sooner. Phosphate electrolytes also affect the properties of the anodized titanium. Chemisorption of phosphate on TiO_2 apparently represses surface ionic mobility and thereby inhibits anatase-to-rutile transformation [95]. The net result is that TiO_2 nanotubes grown anodically in phosphoric acid display ordered structure and lengths up to 500 nm. This phenomenon suggests that nanotube length may be further tailored by modifying the electrochemical conditions of the anodization reaction.

Engineering TiO_2 nanotubes using template synthesis is similar to the method used to prepare mesoporous TiO_2 nanostructures. First, TiO_2 nanomaterials are attached to a template surface by either hydrolyzing a titanium precursor in the presence of templating agents or coating formed TiO_2 nanoparticles onto templates through dip-coating or atomic layer deposition (ALD) [87]. The ALD technique allows the final diameter and length of the nanotubes to become highly defined. The number of deposited layers is used to control the wall thickness. The templating agents, which are harder than the ones used for mesoporous materials, are then removed via annealing. In this case, the templates used to prepare TiO_2 nanotubes include aluminum oxide [96, 97], ZnO [98], electrospun fibers [99], and carbon nanotubes. Recently, to further increase the surface area of TiO_2, attempts have been made to combine hard and soft templates together to create mesoporous TiO_2 nanotubes. For instance, Qiu et al. [100] have incorporated poly(ethylene glycol) (a soft template) into a TiO_2 gel followed by dip-coating a ZnO nanorod (a hard template). Removal of the templates eventually left nanoporous structures on the wall of TiO_2 nanotubes, which increased their surface area to about twice that of pure TiO_2 nanotubes.

The third method, which actually was the first reported, for preparing TiO_2 nanotubes is based on alkaline hydrothermal treatment. Typically, TiO_2 nanotubes can be converted from the raw material of any polymorph (anatase, rutile, brookite, or amorphous forms) in concentrated NaOH at temperatures ranging from 110 to 150 °C. However, temperatures higher than 170 °C caused formation of TiO_2 nanofibers [101, 102]. The products are usually washed with HCl to substitute Na^+ with H^+ and thereby impart the materials with a strong ion-exchange capacity.

9.2.4
TiO$_2$-Based Nanohybrids

TiO_2-based nanohybrids are polycomponent nanomaterials and are useful for sensing applications in which TiO_2 alone cannot achieve the needed level of sensitivity, specificity, or reproducibility. For instance, TiO_2–metal hybrids are used in situations, such as electrochemical devices, where the metal nanoparticles offset the poor conductance of pure TiO_2 and retard electron–hole recombination on the TiO_2 surface. At the same time, combination with TiO_2 prevents metal nanoparticles from aggregating at elevated temperatures [103] and adsorption of TiO_2 enriches the local concentration of analyte. For TiO_2–SiO_2 hybrids, the combination with SiO_2 not only inhibits the anatase-to-rutile phase transition and improves thermal stability but also gives rise to some new optical properties that may be used for improved sensing [13].

9.2.4.1 TiO$_2$-Metal Nanoparticle Hybrids

Several techniques have been used to modify TiO_2 nanomaterials with metal NPs, such as impregnation [104, 105] and electrodeposition [106]. In these processes, because of weak acidity on the TiO_2 surface, metal-ion precursors can be adsorbed onto TiO_2 by ion exchange and then reduced by chemical or electrochemical means to form metallic nanoparticles. However, a shortcoming of these methods is their limited control over the morphology of the metal nanoparticle ad-layer and resulting aggregation of metal deposits. To obtain monodispersed metal nanoparticles on the TiO_2 surface, either metallic nanoparticles are pre-synthesized and transferred to the TiO_2 surface [107] or a soft templating method is used [84]. In the latter case, surfactant molecules are used to direct the formation and deposition of metal nanoparticles on the TiO_2 surface. Another newly developed strategy allows metal nanoparticles to be embedded in a TiO_2 network [108, 109]. In this one-pot synthesis, metal and titania precursors are mixed in a solution that contains surfactants and metal nanoparticles are formed while applying heat. As shown in Figure 9.1, the metal nanoparticles withstand the severe aggregation usually experienced at higher temperatures and become evenly distributed within

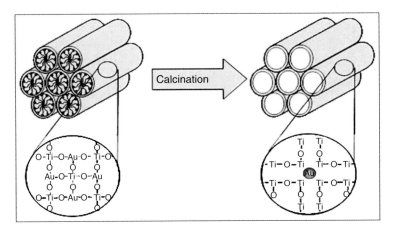

Figure 9.1 Mesoporous Au/TiO_2 nanocomposites with Au nanoparticles homogeneously embedded within crystalline TiO_2 framework. Reprinted with permission from Reference [108]. Copyright 2007 The American Chemistry Society.

the TiO_2 network. Removal of the surfactants leaves a uniform, porous structure of metal NPs evenly distributed within the TiO_2 nanomatrix.

9.2.4.2 TiO_2–SiO_2 Hybrids

Because of the relatively fast hydrolysis of titania precursors, the synthesis of silica-supported TiO_2 usually starts with dispersion of SiO_2 nanoparticles in a solvent such as ethyl alcohol [110] or toluene [111]. A small amount of water is then introduced before the addition of the titanium precursor(s) to form an adsorbed layer of water on the SiO_2 surface. The layer serves to localize the hydrolysis and condensation of titania precursors around the SiO_2 nanoparticles and thereby form a uniform shell [110]. Other methods used to prepare silica-supported titania nanocomposites include microemulsion [112–115], impregnation, precipitation, chemical vapor deposition [116, 117], and sol–gel formation [118, 119]. However, all of these methods are only suitable for preparing SiO_2 nanoparticles having only a thin layer of titania. If the TiO_2 loading needs to be more than 30 wt%, a multistep coating is recommended [120].

9.2.5
Fabrication of TiO_2 Nanofilms

TiO_2 films with nanoscale structures are recommended for applications that require or can benefit from a large sensing area [121–125]. Such films – when deposited on optically transparent or electrically conductive substrates – are notable for their broad applications in chemical sensing and determination [126–128]. Several approaches have been developed to construct TiO_2 nanofilms, including spin coating [11], chemical vapor deposition (CVD) [129–131], self-assembly [132], and matrix-assisted pulsed laser evaporation (MAPLE). Among these methods, spin coating appears to be the most simple and reproducible process. Film thickness can be controlled to within micrometer tolerances by varying the concentration of a precursor solution and the rotation rate of the substrate. CVD is less convenient because it requires a vacuum apparatus in which molecular precursors are condensed at low vapor pressure through selective chemical reactions. Self-assembly, like spin coating, can form highly reproducible films. In this case, nanostructured materials are usually deposited in multiple layers with a self-assembly reagent, such as the alternating layers of polycations and semiconductor nanoparticles used by Kotov *et al.* [133]. MAPLE is a recently developed technique that shows some advantages because of gentler transfer of precursor molecules into the vapor phase [134]. While MAPLE offers uniform substrate coverage and preservation of the composition and crystalline phase of the precursor material, the method needs better reproducibility before it can be used in any practical fashion for sensor fabrication.

An example that shows the level of control currently available for TiO_2 nanofilm preparation was recently reported by Yin *et al.* [71]. These workers developed a flexible synthesis based on an amphiphilic macromolecular template and a process similar to that used for mesoporous TiO_2 nanoparticles. The template was first

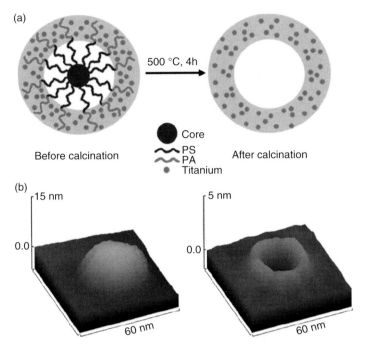

Figure 9.2 Dendritic amphiphilic macromolecules serve as templates to assist the formation of nanoporous TiO$_2$ films (a); surface of AFM image of TiO$_2$ templated by the macromolecules (b) before and (c) after calcination. Reprinted with permission from Reference [71]. Copyright 2008 Wiley-VCH Verlag GmbH & Co. KGaA.

dissolved in DMF with HCl and titanium tetraisopropoxide as the TiO$_2$ precursor. Spin coating and calcination produced a TiO$_2$ thin film with nanobubbles that collapsed to form more advantageous nanopores (Figure 9.2). By this method, the concentration of pores could be controlled with the concentration of the macromolecular template and their size could be adjusted with choice of macromolecule.

9.3
Applications of TiO$_2$-Based Nanomaterials for Chemical Analysis

9.3.1
Analysis of Gas-Phase Samples

Sensitive, real-time sensors are in high demand for several potentially hazardous or toxic species that may be present at low levels in the gas phase. Such species include H$_2$ and CO. In many situations the intended gas matrix is ambient air, which requires some consideration of possible interferences from common airborne components such as oxygen and water vapor. However, in certain cases even these interfering species need to be determined at low concentrations for systems closed to the air.

Only a robust sensor material could be adapted to all of these applications and TiO_2 is especially suited in this regard. The response of titania-based gas sensors can be geared to either optical [135] or electrical measurements, thus permitting a great deal of flexibility to make improvements. Recent trends in the development of titania gas-sensors are toward greater analyte sensitivity (and thereby lower detection limits), shorter response and recovery times, greater stability, higher selectivity, and lower operating temperatures [136, 137]. These goals have generally been achieved by the hybridization of TiO_2 with augmenting components, such as surface reagents or catalysts [129, 137–144]. It is also apparent that the nano- or mesoscopic structure of the TiO_2 or its hybridizing components usually plays a key role.

9.3.1.1 Hydrogen

Molecular hydrogen is a potentially explosive reagent or product in many industrial processes. Development of a direct, reliable H_2 sensor is of great importance for real-time monitoring of leaks into ambient airspace. As hydrogen is a redox-active species, both electrical and electrochemical detection are relatively straightforward and common approaches for sensing. The adsorption process of H_2 gas on the TiO_2 surface starts from the dissociation of H_2 molecules and the formation of OH groups with surface oxygen. Meanwhile, electrons are transferred to the titania conduction band, thus forming an electron-rich region along the TiO_2 surface and resulting in a decrease in electrical resistance.

A quite different approach for H_2 sensing has been to exploit certain unique optical properties of TiO_2. For instance, Maciak et al. have fabricated a chemochromic optical sensor in which the reaction with hydrogen species caused a change in the adsorption spectrum of TiO_2 [141]. These workers used a nanostructured dual-layer architecture composed of $Pd-TiO_2/NiO_x$ thin films. The palladium layer readily absorbed hydrogen from the atmosphere and also catalyzed its oxidation. The reversible double injection of H^+ and electrons onto the TiO_2 substrate Equation (9.1) caused a reduction of Ti^{IV} to Ti^{III} and a visible color change that was monitored by spectrophotometry. The sensor detection limit was approximately 2% (v/v) H_2 in air and response/recovery time was 30–40 s:

$$TiO_2 \text{ (bleached)} + H^+ + e^- \rightleftharpoons HTiO_2 \text{ (colored)} \tag{9.1}$$

Since interaction of a gas with TiO_2 is primarily a surface phenomenon, the specific surface area of the sensing material is an important factor for imparting sensitivity. Nanostructures with large surface areas, such as TiO_2 nanotubes or nanoporous TiO_2, have demonstrated remarkable performance compared to bulk TiO_2. However, the preference for using TiO_2 nanoparticles or nanotubes in sensing applications is not only because of their higher effective surface area but also because of enhanced conductance [145]. As mentioned earlier, the conductance of TiO_2 nanomaterials follows the order TiO_2 nanotubes > amorphous TiO_2 > anatase > rutile TiO_2 crystals [42, 43] Since most TiO_2-based gas sensors are based on the measurement of electrical resistance change with adsorption of analytes, improved conductance of the sensing material is also beneficial to achieving higher sensitivity and shorter response time. For instance, Carney et al. have found that sensors based on TiO_2 nanofibers had larger surface areas and were more sensitive to H_2 in the presence of O_2 than sintered SnO_2–TiO_2. The best sensitivity occurred at an operating temperature of 500 °C, along with short response/recovery times of 1 and 5 min [146]. Devi et al. have found that ordered mesoporous TiO_2 exhibited higher H_2 and CO sensitivities than sensors made from common TiO_2 powders due to increased surface area, and that loading of 0.5 mol % Nb_2O_5 further increased the sensitivity by maintaining the high surface area of TiO_2 at an operating temperature of 600 °C [147].

Other factors that may affect sensor sensitivity and selectivity include the morphology of TiO_2 nanostructures, humidity in the environment, and hybridization with other nanomaterials. For example, an increase in length of TiO_2 nanotubes from 380 nm to 1 μm was found to increase sensitivity nearly tenfold because of enlarged contact surface area. However, a decreased sensitivity and longer response/recovery times were observed with a 6 μm long sample, probably because of the prolonged time for the H_2 gas diffusion along the nanotube wall [69]. In a humid environment, adsorbed water molecules can block active sites where hydrogen adsorption occurs and thereby reduce the sensitivity (Figure 9.3) [148]. In

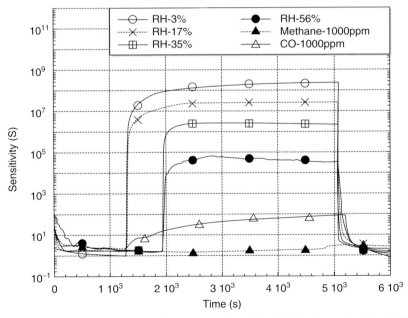

Figure 9.3 Hydrogen sensitivity of a 1.0 μm long nanotube array sample as a function of relative humidity. The sensitivity of the sample to carbon monoxide and methane in dry ambient is also shown. Reprinted with permission from Reference [69]. Copyright 2006 Elsevier.

general, hybrid TiO_2 nanomaterials show advantages over TiO_2 alone. Higher selectivity can be achieved by incorporating a metal component that can specifically facilitate H_2 dissociation on the TiO_2 surface at relatively low temperatures. Du et al. have developed a H_2 sensor that was composed of a nanostructured dual-layer film of titania and PtO-Pt [142]. The sensor showed preferential H_2 uptake against NH_3, CO, and CH_4 interference. Furthermore, due to the partial reduction of Ti^{IV} to Ti^{III} by the PtO-Pt nano-catalysts, more oxygen vacancies were formed along the film surface, which lowered resistance and improved sensitivity. In this case, the hybrid sensor could accurately measure H_2 in the range 1–10%.

9.3.1.2 Carbon Monoxide

The response of most TiO_2-based sensors to CO is dependent upon two competing factors: adsorption of CO and removal of adsorbed CO because of its reaction at the TiO_2 surface. Adsorption of CO molecules is accompanied by a decrease in resistance of the TiO_2. This effect is monitored most sensitively using thin TiO_2 nanofilms. Once adsorbed, CO is converted into CO_2 at elevated temperatures by a water-gas shift reaction mediated through adsorbed oxygen species, thereby recovering adsorption capacity and making the response reversible.

The performance of these CO sensors can be improved by adding catalysts to hybridize the TiO_2 surface. For instance, addition of both CuO and La_2O_3 has been observed to retard anatase-to-rutile phase transformation. Anatase crystals, because of their small particle size, present more oxygen defect sites on the surface, which can then trap more oxygen species and increase the reactivity of CO. Moreover, the presence of CuO improves the adsorption of CO while the increase in oxygen active sites created by La_2O_3 enhances CO reactivity [144]. Teleki et al. have shown recently that improved detection of CO can also be obtained by doping TiO_2 films with Nb or Cu via flame spray pyrolysis [138]. The Nb-doped TiO_2 films responded to CO levels in the range 50–750 g m^{-3} at 400 °C with an optimum Nb/TiO_2 composition of 4%. While Nb-doped sensors showed good stability and response times between 15 and 30 s, recovery times were much longer, approximately 100 s. By comparison, Cu-doped TiO_2 films performed much better, with response and recovery times of less than 15 s and higher overall sensitivity CO. In another case, introduction of Ga_2O_3 into TiO_2 by a sol–gel process gave a linear response to levels of CO between 25 and 400 g m^{-3} at 200 °C at 30% relative humidity [139].

Several design factors, such as film thickness, substrate, temperature, and humidity have also been evaluated for sensors based on unhybridized TiO_2 [136]. Surprisingly, resistance was found to decrease as film thickness increased and the times of both response and recovery were observed to decrease as CO concentration increased. Films deposited on a sapphire {001} crystal face gave the highest sensitivity compared to films deposited on glass or Si{100}. At elevated temperatures, the structure has fewer oxygen vacancies, resulting in a decrease in resistivity. Interestingly, the resistance was independent of humidity, a parameter that typically affects other TiO_2-based gas sensors. This independence was attributed to a lack of direct reaction between CO and H_2O because the shift reaction is actually mediated through adsorbed oxygen intermediates.

9.3.1.3 Oxygen

The adsorption of oxygen on the TiO_2 surface follows a similar pathway to H_2 adsorption. However, as an oxidizing agent, O_2 molecules are chemisorbed in the form of O_2^- on the TiO_2 surface, which effectively traps electrons from the titania conduction band and leads to an increased resistance [149, 150].

Hybridization has been used successfully to improve titania-based sensors for molecular oxygen. For example, Llobet et al. have recently found that integration of multiwalled carbon nanotubes (MWCNT) into titania yielded an oxygen sensor capable of trace-level detection (≤ 10 g m^{-3}) and having more than 50 times the sensitivity than the non-hybrid counterparts alone [151]. The authors considered that there was an n–p junction existing at the interface between TiO_2 layer and MWCNT, namely, n-TiO_2/p-MWCNT. The adsorption of oxygen on TiO_2 modified the width of the TiO_2 depletion layer, and in turn significantly changed another depletion layer at the n-TiO_2/p-MWCNT interface. Apparently, there was an amplification effect involved that caused a dramatic change in the resistance of the resulting hybrid sensor. Moreover, the increased sensitivity of the hybrid sensor might be attributed to the partial adsorption of oxygen on MWCNTs. Sharma and Bhatnagar have observed the beneficial effect of Cr-doping on a TiO_2-based gas sensor [137]. Their 0.4% Cr doped TiO_2 sensor showed a 13-fold improvement in sensitivity over undoped platforms, from which they proposed a direct chromium–oxygen interaction. Eventually, sensitivity and response time (down to 5 s) were optimized at 700 °C and 0.40 wt% Cr. Temperatures higher than the optimum caused an increased desorption of oxygen beyond the chemisorption capabilities of the sensor, resulting in a longer response time.

Besides conductance sensors, some hybrid oxygen sensors have been developed around changes in photoluminescence (PL) [15]. When TiO_2 nanoparticles are suspended in a sensing film that contains on oxygen-sensitive dye, the embedded TiO_2 nanoparticles can scatter incident light used to excite the dye. As a result, the PL intensity of the dye is increased up to tenfold and the time for the PL decay is significantly prolonged (Figure 9.4).

It has been suggested that certain nanostructures of TiO_2 may play a fundamental role in the performance of oxygen sensors. A revealing study by Lu et al. has shown that amorphous TiO_2 demonstrates better sensing capabilities than its anatase form [42]. It was proposed that the relative abundance of structural defects in amorphous TiO_2 compared to crystalline phases provides additional interaction sites and therefore greater sensitivity to oxygen. In this case, a detection limit of 200 g m^{-3} O_2 was obtained at an optimum temperature of 100 °C.

9.3.1.4 Water Vapor

When films of TiO_2 nanomaterials are exposed to a humid atmosphere, water molecules tend to adsorb on the TiO_2 surfaces causing a decrease in film resistance [7]. However, a complication arises because the mode of water adsorption on a TiO_2 film varies with humidity. At low relative humidity (RH), water molecules

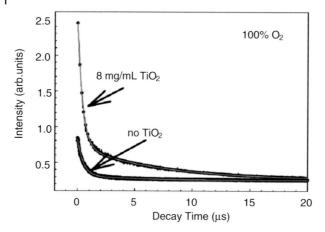

Figure 9.4 Gas-phase PL (photoluminescence) decay curves in 100% O_2 environments for two films—one without titania and the other with 8 mg mL^{-1} titania. Reprinted with permission from Reference [15]. Copyright 2007 Wiley-VCH Verlag GmbH & Co. KGaA.

are chemisorbed on the TiO_2 surface and conductance occurs by proton hopping between sites. At high RH, water molecules are only physisorbed and proton hopping between these sites is much more rapid. The result is a disproportionately high conductance when TiO_2 surfaces are exposed to a high humidity environment.

Other factors that affect the performance of water vapor sensors include surface area, morphology and crystal phase of TiO_2, density of oxygen vacancies on the surface, and hybridizing components. For example, TiO_2 in the anatase phase is more sensitive to humidity and desorbs H_2O more quickly than TiO_2 in the rutile phase. This is primarily due to the larger surface area and correspondingly higher water adsorption capacity of anatase TiO_2 [7]. With respect to effects of the nano- and mesoscopic structure of TiO_2, one-dimensional TiO_2 fibers appear to be better materials over two-dimensional films because the fiber structure not only offers a higher surface area but it also propagates the movement of charge at a faster rate [152]. The density of oxygen vacancies on the TiO_2 surface is another factor that influences the sensitivity of water vapor sensors because it limits the number of proton charge-carriers produced on the surface as well as their rate of transport. Such effects were recently suggested by Wang *et al.*, who found that TiO_2-B nanowires, the least dense polymorph of TiO_2 with a monoclinic form, possessed many oxygen vacancies on their surface and performed better over P25 TiO_2 [3]. Finally, unsurprisingly, the performance of water vapor sensors can also be improved by hybridization of the TiO_2 substrate. For example, Li *et al.* have found that doping and calcination with 30% lithium chloride altered the structure of electrospun TiO_2 nanofibers and produced a sensor that gave resistance response over three orders of magnitude between relative humidity levels of 2.5 and 65% [152].

9.3.2
Analysis of Aqueous Samples

TiO$_2$ nanomaterials, because of their amphoteric nature, are versatile components for augmenting the trace analysis of both inorganic and organic species in aqueous solution. Their powerful ion-exchange capabilities with metal ions ensure high adsorption capacities. Meanwhile, their OH functional groups promote adsorption of organic species and are easily functionalized with various recognizing agents to create hybrid nanomaterials. This section considers both native and hybridized TiO$_2$ nanomaterials in various sensing and analysis applications.

9.3.2.1 Ion Detection and Sensing

Hybridization of TiO$_2$ with coordinating dyes has led to the development of several effective colorimetric detection and sensing schemes for ions, particularly for mercury(II). For example, Palomares et al. have anchored the dye complex [bis(2,2'-bipyridyl-4,4'-dicarboxylato)ruthenium(II) bis-tetrabutylammonium bis-isothiocyanate] (known as N719) to nanocrystalline TiO$_2$ that was cast as a thin, transparent film. Within 10 min of immersing the hybridized film in a solution with mercuric ions, a color change from red-purple to yellow was evident (Figure 9.5) [153, 154]. This response was selective for Hg^{2+}. The authors attributed both the blue-shift of the dye's visible absorption as well its selectivity for Hg^{2+} to direct coordination with sulfur atoms of the dye complex. Detection limits obtained with the hybridized film were ca 20 µM Hg^{2+} by visual discrimination and as low as 3 µM (ca. 0.5 ppm) by spectrophotometric detection. However, a notable problem with this detection system was desorption of the dye from the film in aqueous solution, making it unstable as a sensor. More recently, Grätzel and coworkers have used a similar amphiphilic dye complex (N621) as well as nanocrystalline TiO$_2$ films to make a faster and fully reversible optical sensor that demonstrated detection limits for Hg^{2+} in aqueous solution down to 20 ppb [155]. Interestingly, Palomares and coworkers have also used this same colorimetric approach with alizarin complexone and a nanoporous TiO$_2$ film to achieve selective detection of the anions fluoride and cyanide. While detection limits reported for these anions were relatively high (ca 50 µM), the colorimetric response was rendered insensitive to pH by the intrinsic buffering effect of the TiO$_2$ substrate [36].

9.3.2.2 Metal Ion Extraction

The adsorption of metal ions from aqueous solution by TiO$_2$ is highly dependent on the solution pH. Figure 9.6 illustrates a mechanism for mesoporous TiO$_2$. In this case, two types of ion exchange groups are present on the TiO$_2$ surface – bridged and terminal OH [20]. Bridged OH groups are relatively acidic whereas terminal OH groups are less so. Under acidic conditions only bridged OH groups of TiO$_2$ are available to exchange protons with metal ions. However, under basic conditions both bridged and terminal OH groups can take part in the exchange and the amount of metal ions adsorbed is increased. This behavior has important

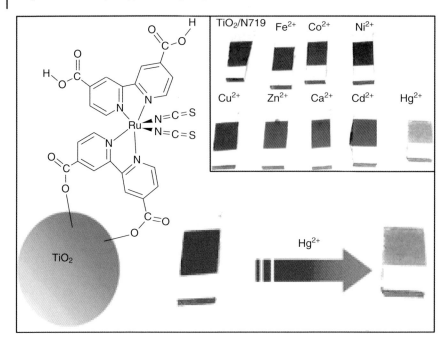

Figure 9.5 Structure of the $TiO_2/N719$ film, and the color change observed after the TiO_2/N719 film is dipped in Hg^{2+} aqueous solution. Inset: color change induced by either Hg^{2+} or other interference metal ions, indicative of the high selectivity of the sensor. Reprinted with permission from Reference [153]. Copyright 2004 The Royal Society of Chemistry.

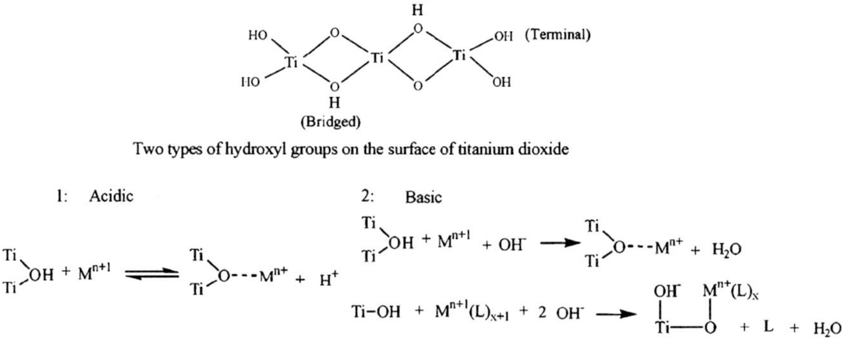

Figure 9.6 Adsorptive mechanism of mesoporous TiO_2 to metal ions. Reprinted with permission from Reference [20]. Copyright 2007 Elsevier.

implications for the separation, recovery or preconcentration of different metal ions by adjusting solution pH. Table 9.1 compares different TiO_2 materials and other kinds of sorbents used for preconcentration and determination of various waterborne metal ions.

Table 9.1 Comparison of sorption capacities (mg g^{-1}) for TiO$_2$ nanomaterials and other sorbents.

Element	Mesoporous TiO$_2$ [20]	Nano-sized TiO$_2$ [156]	Multiwalled carbon nanotubes [157]	Silica gel [158]
Cu^{2+}	8.1	6.89	–	0.96
Mn^{2+}	22.3	2.12	4.86	–
Cd^{2+}	8.1	–	6.89	3.92
Co^{2+}	13.9	–	–	0.59
Ni^{2+}	8.6	1.98	7.42	0.94
Cr^{3+}	14.8	7.58	–	–

Because of favorable adsorption behavior, both nano- and mesoscopic TiO$_2$ materials are often used as sorbents and supports for solid-phase extraction (SPE) of various metal ions. SPE, whether for preconcentration or removal of potential interferences, is a very useful technique for improving the sensitivity or selectivity of any subsequent analysis method. Typically, extraction is performed either off- or on-line with respect to the subsequent analysis. For on-line extractions, TiO$_2$ materials are packed into a micro-column [20–23] and aqueous samples containing metal ions are allowed to pass through. The adsorbed metal ions are then eluted with an acidic solution and introduced directly to the analysis device, such as an atomic emission spectrometer [20]. However, a significant problem with columns of this type is that fine TiO$_2$ nanomaterials can be lost during elution, making it difficult sustain the column. One solution is to hybridize the TiO$_2$ nanomaterials with a larger supporting matrix, such as silica gel [23, 24]. Another is to eliminate the column altogether. Here, adsorption and desorption are performed off-line by batch processes and the TiO$_2$ nanomaterials are filtered, leaving the preconcentrated sample available for subsequent analysis [159].

9.3.2.3 Organic Compounds

Because the TiO$_2$ surface is amphoteric in nature, organic compounds are also prone to adsorption. For this reason, TiO$_2$ nanomaterials have been combined with various electrochemical, mass-sensitive, and chromatographic techniques to determine trace organic species in aqueous samples.

As mentioned for other adsorbates, chemi- and physisorption of organic compounds can change the electrical properties of a TiO$_2$ substrate and thereby form the basis of a sensor. One electrochemical tool that has been applied in making sensors for ionizable organic compounds is the ion-sensitive field-effect transistor (ISFET). ISFETs have been used more than 30 years for selective ion sensing – the first application being pH measurements. This low-power, all solid-state platform can monitor the electrical properties of TiO$_2$ materials with high sensitivity upon their exposure to an aqueous sample. For example, Lahav et al. have used a sol–gel process to deposit a molecularly imprinted TiO$_2$ thin film on the SiO$_2$ gate of an

ISFET [160]. Such imprinting relies on self-assembly to impart nano- and even sub-nanostructure on a polymeric substrate. It is intended to selectively enhance the adsorption of a target molecule (Figure 9.7). In this case, the organic targets were 4-chlorophenoxyacetic acid and 2,4-dichlorophenoxyactic acid, two commonly used herbicides. The ISFET exhibited high selectivity for each of these compounds

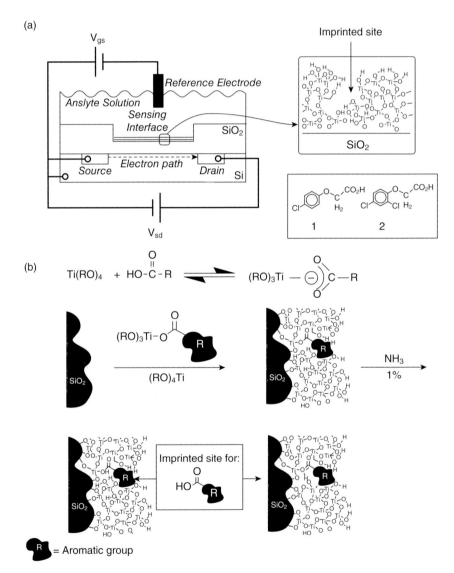

Figure 9.7 (a) Schematic configuration of the molecular-imprinted FET (field effect transistor) device; (b) preparation of molecular-imprinted sites for chloroaromatic acids in a TiO_2 thin film acting as the sensing interface on the FET gate. Reprinted with permission from Reference [160]. Copyright 2001 The American Chemistry Society.

as well as useful detection limits of 0.5 and 0.01 mM, respectively. In a more recent example, molecular imprinting was used to construct a TiO_2-based ISFET for various benzylphosphonic acid derivatives and thiophenol substrates [35].

Another sensitive technique that has been used to detect adsorption of organic compounds by TiO_2 nanomaterials is the quartz-crystal microbalance (QCM). The QCM uses a precise quartz resonator to detect frequency changes caused by adsorption of organic compounds. Adsorption results in a decrease in frequency, which is proportional to the mass of analyte molecules (Δm) deposited on the oscillating crystal surface. The relationship can be described for many situations by the well-known Sauerbrey equation, Equation (9.2) [32]:

$$\Delta f = \frac{2 f_0^2}{A \rho_q v_q} \Delta m \tag{9.2}$$

where

Δf is the observed frequency change (Hz),
f_0 is the fundamental resonant frequency of crystal,
A is the active area of the crystal,
ρ_q is the density of quartz,
v_q is the shear wave velocity in the quartz.

Nanostructured TiO_2, because of its large surface area and strong adsorption characteristics, has often been deposited on the crystal surface as a sensing layer. It was recently shown that a QCM sensor based on TiO_2 nanoparticles had over twice the sensitivity of bulk TiO_2 sol–gel due to better analyte accessibility to the nanoparticle surface [161]. As with other sensing systems, better selectivity can be achieved by the hybridization of the TiO_2. For instance, Zheng et al. have found that polyaniline-TiO_2 nanocomposites not only had higher selectivity for trimethylamine over ethanol, formaldehyde, and acetaldehyde, but also exhibit higher thermal stability [32]. Likewise, Lieberzeit et al. have found that TiO_2 nanoparticles modified with molybdenum disulfide (MoS_2) showed a high selectivity to organic compounds containing thiol groups [161]. Selectivity has also been controlled for native TiO_2 nanomaterials by adjusting pH and salt condition. For instance, the order of adsorption of analytes on the TiO_2 surface can completely switch for the selective detection of organic compounds such as salicylic acid (SA), phenol, and benzoic acid [160]. In another case, Nezu et al. found that lysozymes could adsorb strongly onto TiO_2 at pH 3 and desorbed at pH 9. Although this adsorption process was strongly pH dependent, the presence of 0.05 M NaCl was sufficient to disrupt the adsorption at pH 3 [163].

As with metal ions, nano- and meso-structured TiO_2 have been used for SPE of organic compounds, with a marked focus on the preconcentration of hazardous or toxic pollutants. In most cases the SPE was performed on-line with analysis by high-performance liquid chromatography (HPLC). An illustrative example is that of Zhou and coworkers [21]. In this case TiO_2 nanotubes were used for the extraction and enrichment of the environmentally banned pesticide o,p'-DDT [1,1,1-trichloro-2-(o-chlorophenyl)-2-(p-chlorophenyl)ethane] and related compounds from water samples. HPLC analysis of the extracts showed recoveries of over 80%, detection limits below ca 5 ng L^{-1} (parts per trillion), and limited interferences from natural substances such as humic acids. However, this example is somewhat unique because native, unmodified TiO$_2$ nanomaterials are not commonly used for the extraction of organic compounds. Better selectivity and adsorption capacity is usually achieved by hybridization of the TiO$_2$ matrix with complementary organic functionalities. In this regard, the approach of Kim et al. is more typical of the solid phases used for extraction of trace organic compounds, where a sol–gel titania-poly(dimethylsiloxane) (TiO$_2$-PDMS) coating was developed [22]. In this case the hybrid solid phase showed exceptional stability (both thermal and pH) and demonstrated sub-ppb detection limits for ketones and alkylbenzenes when the preconcentration was performed on-line with analysis by HPLC.

9.3.3
Biosensors

Applications of TiO$_2$ nanomaterials are covered in this section with an emphasis on voltammetric and optical methods. In general, voltammetry and its related electrochemical techniques offer several advantages for the construction of effective biosensors. These include good stability (electrodes can be easily cleaned or protected), high sensitivity, fast response, high spatial resolution (when using micro- and nanoelectrodes), and relatively low-cost instrumentation. However, a key challenge to integrating TiO$_2$ nanomaterials with any type of voltammetric analysis is their relatively high resistance (as a semiconductor) and potential to block effective electron transfer between analytes and electrodes. The examples provided below illustrate various adaptations to this difficulty and demonstrate present directions in the development of TiO$_2$-based electrochemical biosensors.

Biosensors that rely on optical measurements do not suffer from the conductivity issues that arise with electrochemical measurements. Accordingly, these devices may provide a more convenient and robust avenue for the analytical determination. So far, the development of TiO$_2$-based optical biosensors is still at an early stage. Sensors based on differences in light intensity, wavelength, phase, or polarization have yet to mature as they have with other nanomaterials. Nonetheless, titania-based optical biosensors are receiving greater interest because TiO$_2$ nanomaterials present some unique optical and photovoltaic properties.

9.3.3.1 Voltammetric Biosensors
Nanostructured TiO$_2$, because of its good biocompatibility, has been widely used as a matrix for the immobilization of various proteins. TiO$_2$ nanomaterials not only minimize the deactivation and structural change that usually arises from direct attachment of proteins on a metal electrode [164, 165] but they also greatly increase protein loading because of their large surface areas. For example, Topoglidis et al. have reported that nanoporous TiO$_2$ films prepared by screen-printing significantly increase the active surface area by a factor of 150 for adsorption of cytochrome c (cyc c) and hemoglobin (Hb) [166–168]. Similarly, Liu et al. have

found that the coverage of Hb on TiO$_2$ nanotube electrodes is two orders higher than its adsorption onto a TiO$_2$ monolayer [169]. Nevertheless, the poor conductivity of TiO$_2$ hinders efficient electron transfer between adsorbed proteins and an electrode. To tackle this problem, several approaches have been developed, including the use of a TiO$_2$ nanomaterial with defined morphologies, the incorporation of conductive components into the TiO$_2$ matrix, and the electroreduction of Ti^{IV} to its more conductive Ti^{III} form.

Biosensors with Defined TiO$_2$ Morphology It has been proposed that nanostructured TiO$_2$ alone can mediate electron transfer between redox-active sites buried in a protein and an electrode as long as the nanostructure has a controllable morphology [14]. As mentioned previously, TiO$_2$ nanotubes are more conductive than their nanoparticles counterparts. TiO$_2$ nanostructures with more ordered or defined morphology seem to facilitate direct electron transfer between biomolecules and electrodes due to either their inherent electric properties or spatial construction. Using a glucose biosensor as an example, a relatively high detection limit for glucose was reached (1 μM) when a normal nanostructured TiO$_2$ film was modified with glucose oxidase (GO$_x$), an enzyme for glucose oxidation [170]. In contrast, when an ordered, three-dimensional, and macroporous TiO$_2$ (prepared with a colloidal polystyrene sphere template, Figure 9.8) was used as the matrix for the GO$_x$ loading, the glucose sensor showed a much lower detection limit (0.02 μM) and an improved sensitivity. This was attributed to the large surface area and interconnected microenvironments of the three-dimensional TiO$_2$ network

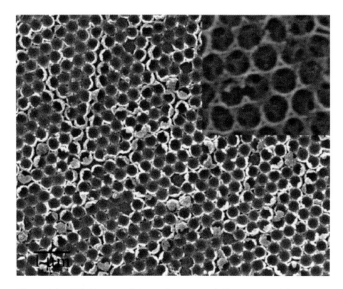

Figure 9.8 SEM image of titania inverse opal film. Inset: FCC structure at higher magnification. Reprinted with permission from Reference [171]. Copyright 2008 Wiley-VCH Verlag GmbH & Co. KGaA.

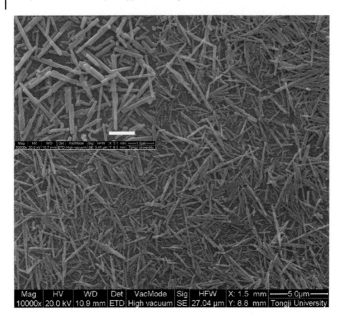

Figure 9.9 SEM images of TiO$_2$ nanoneedles film. The bar in the inset corresponds to 1.0 μm. Reprinted with permission from Reference [172]. Copyright 2009 The American Chemistry Society.

[171]. In another example, Luo et al. have used TiO$_2$ nanoneedles (Figure 9.9) as the matrix for the adsorption of cyc c and subsequent mediated electrochemical detection of H$_2$O$_2$ released from human liver cancer cells [172]. Other examples where nanostructured TiO$_2$ was used to mediate protein electron transfer with an electrode include Hb [173], myoglobin (Mb) [174], horseradish peroxidase (HRP) [175], and superoxide dismutase (SOD) [176].

Biosensors with Hybridized TiO$_2$ An alternative method to augment electron transfer between TiO$_2$ films and biomolecules is to hybridize the TiO$_2$ with conductive materials. These materials have included metal nanoclusters of various sizes and types (Au [27, 103, 177, 178] or Pt [179]), graphite powder [28], and conductive polymers [30, 180]. In terms of glucose biosensing, researchers have inserted carbon nanotubes (CNT) within TiO$_2$ nanotubes, and used Pt nanoparticles instead of GO$_x$ to catalyze the oxidation of glucose. Owing to the large surface area of TiO$_2$ nanotubes and good conductivity of CNTs, this sensor has shown a detection limit of 5.7 μM and a linear range of 0.006–1.5 mM [106]. Similarly, Benvenuto et al. have demonstrated the viability of multiple nanohybrid combinations that included GO$_x$, chitosan, Prussian Blue, Au nanoparticles, and TiO$_2$ nanotubes. The sensor gave a comparable detection limit (5 μM) [181].

With respect to H$_2$O$_2$ sensing, Cui et al. have electrodeposited Pt nanoparticles on a nanotubular TiO$_2$ film; the final biosensor presented a detection limit of

4.0 μM [179]. To further reduce the detection limit, attempts were made to incorporate both metal components and enzymes into a TiO$_2$ matrix. For instance, by coating a Au layer on a TiO$_2$ film via an argon plasma technique, followed by immobilizing HPR on the surface, the detection limit was reduced to 2.0 μM [177]. Li et al. have used a layer-by-layer technique to assemble negatively charged chitosan-stabilized gold nanoparticles on TiO$_2$ surfaces, with Hb covalently bound to the Au surfaces. Because of improved dispersion of the Au nanoparticles, the detection limit was further decreased to 0.37 μM [182]. In a similar manner, Guo et al. have used carbon-TiO$_2$ nanotube hybrids loaded with Hb to achieve a low detection limit of 0.92 μM [183].

Biosensors with Partially Reduced TiO$_2$ Considering that the Ti^{III} state is more conductive than Ti^{IV}, electro-reduction of TiO$_2$ to its Ti^{III} state without destroying its morphology provides another avenue to facilitate electron transfer between biomolecules and an electrode. The reduction process is the same as that shown in Equation (9.1). To construct a H$_2$O$_2$ sensor, Liu et al. have immobilized Hb onto TiO$_2$ nanotubes and partially reduced TiO$_2$ electrochemically to its Ti^{III} state. The sensor demonstrated a low detection limit of 1.5 μA – lower than sensors incorporating a conductive metal component – and a wide linear range of 0.0049–1.1 mM [169]. However, a study by Milsom et al. revealed that the improved conductance of TiO$_2$ resulting from this approach lacked long-term stability. They found that when a TiO$_2$ film was loaded with methemoglobin and subjected to electro-reduction treatment the increased conductance of TiO$_2$ decayed over the time. To maintain the same level of the conductance, the authors found that much milder negative potentials should be applied in the presence of oxygen [174].

The three approaches mentioned above have been shown to offset the limited conductance of TiO$_2$, and thus improve the sensitivity of biosensors based on these nanomaterials. However, selectivity is another important performance criterion that has been taken into account in biosensor design. One method used to improve the selectivity of voltammetric biosensors has been to select an appropriate hybridizing mediator that can shift the potential window for an analyte to a region where interferences are minimized [184, 185]. For instance, direct oxidation of H$_2$O$_2$ requires working potentials exceeding +0.6 V (vs. Ag/AgCl) to attain adequate sensitivity. However, at this extreme working potential other electrochemically active interfering species such as ascorbic acid (AA), uric acid (UA), and acetaminophen (AP) can be easily oxidized, thereby compromising selectivity. Introduction of an appropriate mediator protein can overcome the problem by lowering the oxidation potential of H$_2$O$_2$. For example, the H$_2$O$_2$ sensor previously described from Luo et al. (cyc c/TiO$_2$ nanoneedles) provided a high selectivity by shifting the oxidation potential of H$_2$O$_2$ to 0.0 V (vs. Ag/AgCl) [172]. Table 9.2 lists other proteins and enzymes that have been used to reduce the oxidation potential for H$_2$O$_2$ and thus suppress the interfering currents.

Coating nanostructured TiO$_2$ films with an ion-selective polymer layer has been another approach to increase sensor selectivity. For example, Yuan et al. have coated a porous nano-TiO$_2$ film with a layer of Nafion® and used the sensing

Table 9.2 Analytical performance of the H_2O_2 biosensors based on TiO_2 nanocomposites.

Nanocomposites[a]	Potential (mV vs SCE)	Linear range (µM)	Detection limit (µM)	Reference
Cytochrome c/TiO_2 nanoneedles	−45	0.85–24 000	0.26	[172]
Hb/CMC-TiO_2 nanotubes	−300	4–64	4.637	[186]
Mb/titanate nanotubes	(CV peak) ca −290	2–160	0.6	[187]
Mb/titanate nanosheets	(CV peak) ca −310	2–160	0.6	[188]
HRP/TiO_2 nanoparticles	0	7.5–123	2.5	[175]
HRP/Th-TiO_2 nanotubes	−645	10–3 000	–	[189]
HRP-PMS-TiO_2 sol–gel	−250	4–1 000	0.8	[180]

a) CMC, Th, Mb, and PMS represent carboxymethyl cellulose, thionine chloride, and 3-mercapto-1-propanesulfonate, respectively.

material to selectively detect dopamine and catechol, both electroactive neurotransmitters. In this case, the cationic form of dopamine exchanged with Na^+ in the Nafion film and was detected selectively while the anionic form of AA was excluded [191].

Still another approach used to increase selectivity of TiO_2-based biosensors has been to electrochemically activate specific surface chemistry. For analytes such as catechol that contain –OH functional groups, electrochemical oxidation allows their selective immobilization and detection (Scheme 9.1). The catechol sensor developed by Lunsford and coworkers using this chemistry showed high selectivity

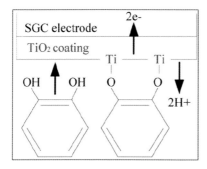

Scheme 9.1 Bonding and electron transfer in sonogel-carbon electrode modified with nanostructured titanium dioxide (SGC/TiO_2). Reprinted with permission from Reference [192]. Copyright 2007 Elsevier.

in the presence of AA and good reversibility because the immobilization reaction could be reversed electrochemically [192].

9.3.3.2 Optical Biosensors

One recent advent in the design of optical biosensors has been the use of silica-titania nanohybrids for detection of an analyte-induced fluorescence change. Shu et al. have demonstrated this effect at room temperature with TiO_2/SiO_2 binary oxides that normally emit phosphorescence from 450 to 650 nm at an excitation wavelength of 403 nm [13]. The authors noted that this phosphorescence could be quenched by H_2O_2 due to cleavage of the Si–O–Ti bonds (Scheme 9.2) and showed that a sensitive "turn-off" H_2O_2 sensor could be constructed based on this phenomenon. A detection limit of 0.16 µM was achieved, which was superior to the TiO_2-based electrochemical sensors mentioned above. In addition, the quenched phosphorescence could be recovered by simply dipping the device in a solution containing hydroxylamine hydrochloride, a reducing agent. This photochemical quenching was found to be quite selective, yielding essentially the same response for H_2O_2 in the presence of various inorganic ions and organic molecules.

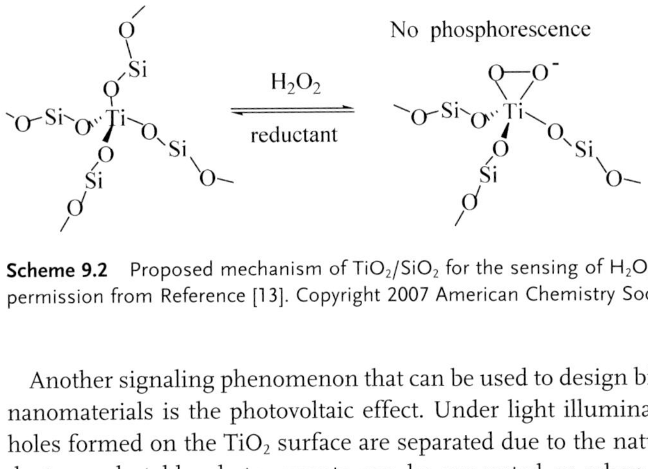

Scheme 9.2 Proposed mechanism of TiO_2/SiO_2 for the sensing of H_2O_2. Reprinted with permission from Reference [13]. Copyright 2007 American Chemistry Society.

Another signaling phenomenon that can be used to design biosensors with TiO_2 nanomaterials is the photovoltaic effect. Under light illumination, electrons and holes formed on the TiO_2 surface are separated due to the nature of the semiconductor and stable photocurrents can be generated as when external voltage is applied. Because the generated electrons and holes can mediate electron transfer between biomolecules and an electrode [193], this effect has been used for biosensor design. For example, the sensitivity of a TiO_2 nanotube/Pt NP electrode used to detect H_2O_2 showed a nearly twofold increase in sensitivity under UV illumination as well as improved amperometric response. Improvement was also observed when mediator proteins such as hemoglobin were used to hybridize the TiO_2 nanotubes. In this case, sensitivity was increased 3-fold with UV irradiation [194].

The photovoltaic effect of TiO_2 has also been used for the detection of DNA. Lu et al. have developed a TiO_2-based photoelectrochemical DNA sensor by attaching Au nanoparticles labeled with DNA probe strands on the TiO_2 [16]. Under UV

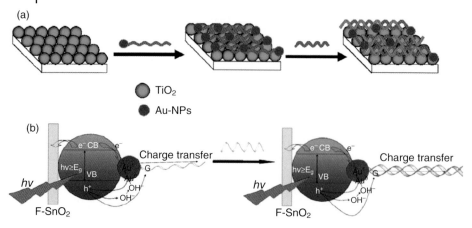

Figure 9.10 (a) Schematic of the fabrication of an Au-DNA probe-modified TiO$_2$ electrode and the detection of target DNA; (b) photo-induced process of electron–hole generation and charge transfer process. Reprinted with permission from Reference [16]. Copyright 2008 Elsevier.

irradiation the Au nanoparticles were able to trap the photo-induced holes and thereby promote charge separation on the TiO$_2$. As a result, the background photocurrent was significantly enhanced. When target DNA was introduced into the system, it hybridized with the complementary probe DNA on the Au NPs and decreased the photocurrent because of the hindered photoelectron transfer (Figure 9.10). By this method, a rather low detection limit of the target DNA (1 nM) was achieved.

9.4
Conclusions

TiO$_2$ nanomaterials offer useful optical and electronic properties, highly variable surface reactivity, good biocompatibility, and outstanding structural diversity. By integrating these characteristics with advanced instrumentation and analytical methods, an impressive range of sensors and techniques has been developed for the trace analysis of gaseous, aqueous, and biological systems.

TiO$_2$ nanomaterials provide many avenues for improving sensor and analysis performance, especially in terms of sensitivity, selectivity, and reversibility. Target molecules can be enriched on TiO$_2$ surfaces by either spontaneous adsorption or ion exchange, thereby increasing sensitivity. Because the surface areas of nanomaterials are much greater than their bulk counterparts, this enrichment is even more profound with nanostructured TiO$_2$ and its many polymorphs. TiO$_2$ nanomaterials can also be readily hybridized by a wide range of inorganic and organic components (some even having their own nanostructures) that complement the functionality of native TiO$_2$ and, more importantly, improve the selectivity of designed sensors and extraction systems. Finally, reversibility can be readily achieved with TiO$_2$ nanomaterials by varying solution pH or ionic strength and sensor response can be restored by irradiation.

Current trends in titania-based trace analysis are largely focused on engineering new nanostructured materials (both natural and hybrid) that provide desirable analytical features, such as enlarged surface area (e.g., mesoporous TiO$_2$), enhanced conductance (e.g., TiO$_2$ nanotubes), and improved mediation of electron transfer between biomolecules and electrodes (e.g., nanoneedles). One area that seems to offer favorable research prospects is in the development of optical biosensors. Relatively little work has taken place with TiO$_2$ nanomaterials in this area, especially towards sensors based on differences in light intensity, wavelength, phase, or polarization. However, as fundamental studies continue to provide new insights into the unique optical properties of TiO$_2$ nanostructures, it can be expected that new avenues will be developed for their application in biological detection and sensing.

Acknowledgments

This work was supported by the U.S. Department of Energy under grant DE-FG02-6ER46292, the National Science Foundation under grant CHE-0616878, and a Doctoral Dissertation Assistantship from University of North Dakota for A. L. Neither the United States Government or agency thereof, nor any of its employees, makes any warranty, express or implied, or assumes any legal liability or responsibility for the accuracy, completeness, or usefulness of any information disclosed. References herein do not necessarily constitute or imply endorsement, recommendation, or favoring by the United States Government or the agency thereof or by the State of North Dakota.

References

1 Qi, Q., Zhang, T., and Wang, L. (2008) *Appl. Phys. Lett.*, **93**, 023105.
2 Guirado-Lopez, R.A., Sanchez, M., and Rincon, M.E. (2007) *J. Phys. Chem. C*, **111**, 57–65.
3 Wang, G., Wang, Q., Lu, W., and Li, J. (2006) *J. Phys. Chem. B*, **110**, 22029–22034.
4 Kim, I.-D., Rothschild, A., Lee, B.H., Kim, D.Y., Jo, S.M., and Tuller, H.L. (2006) *Nano Lett.*, **6**, 2009–2013.
5 Varghese, O.K., Gong, D., Paulose, M., Ong, K.G., Dickey, E.C., and Grimers, C.A. (2003) *Adv. Mater.*, **15**, 624–627.
6 Shimizu, K.I., Kashiwagi, K., Nishiyama, H., Kakimoto, S., Sugaya, S., Yokoi, H., and Satsuma, A. (2008) *Sens. Actuators, B*, **130**, 707–712.
7 Biju, K.P. and Jain, M.K. (2008) *Thin Solid Films*, **516**, 2175–2180.
8 Beranek, R. and Kisch, H. (2007) *Photochem. Photobiol. Sci.*, **7**, 40–48.
9 Manera, M.G., Spadavecchia, J., Busoc, D., Fernández, C.J., Mattei, G., Martucci, A., Mulvaney, P., Pérez-Juste, J., Rella, R., Vasanelli, L., and Mazzoldi, P. (2008) *Sens. Actuators, B*, **132**, 107–115.
10 Manera, M.G., Davide Cozzoli, P., Leo, G., Lucia Curri, M., Agostiano, A., Vasanelli, L., and Rella, R. (2007) *Sens. Actuators, B*, **126**, 562–572.

11 Choi, S.Y., Mamak, M., Von Freymann, G., Chopra, N., and Ozin, G.A. (2006) *Nano Lett.*, **6**, 2456–2461.
12 Qi, Z.-M., Honma, I., and Zhou, H. (2006) *Anal. Chem.*, **78**, 1034–1041.
13 Shu, X., Chen, Y., Yuan, H., Gao, S., and Xiao, D. (2007) *Anal. Chem.*, **79**, 3695–3702.
14 Bao, S.J., Li, C.M., Zang, J.F., Cui, X.Q., Qiao, Y., and Guo, J. (2008) *Adv. Funct. Mater.*, **18**, 591–599.
15 Zhou, Z., Shinar, R., Allison, A.J., and Shinar, J. (2007) *Adv. Funct. Mater.*, **17**, 3530–3537.
16 Lu, W., Jin, Y., Wang, G., Chen, D., and Li, J. (2008) *Biosens. Bioelectron.*, **23**, 1534–1539.
17 Zhang, X., Fujishima, A., Jin, M., Emeline, A.V., and Murakami, T. (2006) *J. Phys. Chem. B*, **110**, 25142–25148.
18 Guan, K. (2005) *Surf. Coat. Technol.*, **191**, 155–160.
19 Mor, G.K., Carvalho, M.A., Varghese, O.K., Pishko, M.V., and Grimes, C.A. (2004) *J. Mater. Res.*, **19**, 628–634.
20 Huang, C., Jiang, Z., and Hu, B. (2007) *Talanta*, **73**, 274–281.
21 Zhou, Q., Ding, Y., Xiao, J., Liu, G., and Guo, X. (2007) *J. Chromatogr. A*, **1147**, 10–16.
22 Kim, T.-Y., Alhooshani, K., Kabir, A., Fries, D.P., and Malik, A. (2004) *J. Chromatogr. A*, **1047**, 165–174.
23 Liu, R. and Liang, P. (2007) *Anal. Chim. Acta*, **604**, 114–118.
24 Liu, Y., Liang, P., and Guo, L. (2005) *Talanta*, **68**, 25–30.
25 Liang, P., Shi, T., and Jing, L. (2004) *Int. J. Environ. Anal. Chem.*, **84**, 315–321.
26 Lee, D., Omolade, D., Cohen, R.E., and Rubner, M.F. (2007) *Chem. Mater.*, **19**, 1427–1433.
27 Qu, Y., Min, H., Wei, Y., Xiao, F., Shi, G., Li, X., and Jin, L. (2008) *Talanta*, **76**, 758–762.
28 Maroneze, C.M., Arenas, L.T., Luz, R.C.S., Benvenutti, E.V., Landers, R., and Gushikem, Y. (2008) *Electrochim. Acta*, **53**, 4167–4175.
29 McKenzie, K.J., King, P.M., Marken, F., Gardner, C.E., and Macpherson, J.V. (2005) *J. Electroanal. Chem.*, **579**, 267–275.
30 Kumar, S.A., Tang, C.F., and Chen, S.M. (2008) *Talanta*, **76**, 997–1005.
31 O'Regan, B. and Gratzel, M. (1991) *Nature*, **353**, 737–740.
32 Zheng, J., Li, G., Ma, X., Wang, Y., Wu, G., and Cheng, Y. (2008) *Sens. Actuators, B*, **133**, 374–380.
33 Wu, H.P., Cheng, T.L., and Tseng, W.L. (2007) *Langmuir*, **23**, 7880–7885.
34 Tai, H., Jiang, Y., Xie, G., Yu, J., and Zhao, M. (2007) *Int. J. Environ. Anal. Chem.*, **87**, 539–551.
35 Pogorelova, S.P., Kharitonov, A.B., Willner, I., Sukenik, C.N., Pizem, H., and Bayer, T. (2004) *Anal. Chim. Acta*, **504**, 113–122.
36 Palomares, E., Vilar, R., Green, A., and Durrant, J.R. (2004) *Adv. Funct. Mater.*, **14**, 111–115.
37 Li, C., Wang, C., Ma, Y., and Hu, S. (2004) *Microchim. Acta*, **148**, 27–33.
38 Baraton, M.I. and Merhari, L. (2004) *J. Eur. Ceram. Soc.*, **24**, 1399–1404.
39 Alivisatos, A.P. (1996) *Science*, **271**, 933–937.
40 Burda, C., Chen, X., Narayanan, R., and El-Sayed, M.A. (2005) *Chem. Rev.*, **105**, 1025–1102.
41 Murray, C.B., Kagan, C.R., and Bawendi, M.G. (2000) *Annu. Rev. Mater. Res.*, **30**, 545–610.
42 Lu, H.F., Li, F., Liu, G., Chen, Z.G., Wang, D.W., Fang, H.T., Lu, G.Q., Jiang, Z.H., and Cheng, H.M. (2008) *Nanotechnology*, **19**, 405504.
43 Tang, H., Prasad, K., Sanjinés, R., and Lévy, F. (1995) *Sens. Actuators, B*, **26**, 71–75.
44 Takaoka, G.H., Hamano, T., Fukushima, K., Matsuo, J., and Yamada, I. (1997) *Nucl. Instrum. Methods Phys. Res. B*, **121**, 503–506.
45 Thorne, A., Kruth, A., Tunstall, D., Irvine, J.T.S., and Zhou, W. (2005) *J. Phys. Chem. B*, **109**, 5439–5444.
46 Dittrich, T., Weidmann, J., Timoshenko, V.Y., Petrov, A.A., Koch, F., Lisachenko, M.G., and Lebedev, E. (2000) *Mater. Sci. Eng. B*, **69–70**, 489–493.
47 Alivisatos, A.P. (1996) *J. Phys. Chem.*, **100**, 13226–13239.

48 Chen, X., Lou, Y., Dayal, S., Qiu, X., Krolicki, R., Burda, C., Zhao, C., and Becker, J. (2005) *Nanosci. Nanotechnol.*, **5**, 1408–1420.

49 Telipan, G., Ignat, M., Tablet, C., and Parvulescu, V. (2008) *J. Optoelectron. Adv. Mater.*, **10**, 2138–2141.

50 Isley, S.L. and Penn, R.L. (2008) *J. Phys. Chem.*, **112**, 4469–4474.

51 Wang, Y., Du, G., Liu, H., Liu, D., Qin, S., Wang, N., Hu, C., Tao, X., Jiao, J., Wang, J., and Wang, Z.L. (2008) *Adv. Funct. Mater.*, **18**, 1131–1137.

52 Tian, G., Fu, H., Jing, L., Xin, B., and Pan, K. (2008) *J. Phys. Chem. C*, **112**, 3083–3089.

53 Wen, P., Itoh, H., Tang, W., and Feng, Q. (2007) *Langmuir*, **23**, 11782–11790.

54 Kobayashi, M., Petrykin, V.V., Kakihana, M., Tomita, K., and Yoshimura, M. (2007) *Chem. Mater.*, **19**, 5373–5376.

55 Yang, S. and Gao, L. (2005) *J. Am. Ceram. Soc.*, **88**, 968–970.

56 Wang, X., Zhuang, J., Peng, Q., and Li, Y. (2005) *Nature*, **437**, 121–124.

57 Peiro, A.M., Peral, J., Domingo, C., Domenech, X., and Ayllon, J.A. (2001) *Chem. Mater.*, **13**, 2567–2573.

58 Fernandez-Garcia, M., Wang, X., Belver, C., Hanson, J.C., and Rodriguez, J.A. (2007) *J. Phys. Chem. C*, **111**, 674–682.

59 Anukunprasert, T., Saiwan, C., Bartolomeo, E., and Traversa, E. (2007) *J. Electroceram.*, **18**, 295–303.

60 Andersson, M., Kiselev, A., Oesterlund, L., and Palmqvist, A.E.C. (2007) *J. Phys. Chem. C*, **111**, 6789–6797.

61 Andersson, M., Osterlund, L., Ljungstrom, S., and Palmqvist, A. (2002) *J. Phys. Chem. B*, **106**, 10674–10679.

62 Chhabra, V., Pillai, V., Mishra, B.K., Morrone, A., and Shah, D.O. (1995) *Langmuir*, **11**, 3307–3311.

63 Wu, J.-M., Zhang, T.-W., Zeng, Y.-W., Hayakawa, S., Tsuru, K., and Osaka, A. (2005) *Langmuir*, **21**, 6995–7002.

64 Mor, G.K., Shankar, K., Paulose, M., Varghese, O.K., and Grimes, C.A. (2006) *Nano Lett.*, **6**, 215–218.

65 Macak, J.M., Tsuchiya, H., Taveira, L., Aldabergerova, S., and Schmuki, P. (2005) *Angew. Chem. Int. Ed.*, **44**, 7463–7465.

66 Lei, Y., Zhang, L.D., and Fan, J.C. (2001) *Chem. Phys. Lett.*, **338**, 231–236.

67 Bavykin, D.V., Friedrich, J.M., and Walsh, F.C. (2006) *Adv. Mater.*, **18**, 2807–2824.

68 Pénard, A.-L., Gacoin, T., and Boilot, J.-P. (2007) *Acc. Chem. Res.*, **40**, 895–902.

69 Mor, G.K., Varghese, O.K., Paulose, M., Shankar, K., and Grimes, C.A. (2006) *Sol. Energy Mater. Sol. Cells*, **90**, 2011–2075.

70 Chen, X. and Mao, S.S. (2007) *Chem. Rev.*, **107**, 2891–2959.

71 Yin, M., Cheng, Y., Liu, M., Gutmann, J.S., and Müllen, K. (2008) *Angew. Chem. Int. Ed.*, **47**, 8400–8403.

72 Wang, D., Ma, Z., Dai, S., Liu, J., Nie, Z., Engelhard, M.H., Huo, Q., Wang, C., and Kou, R. (2008) *J. Phys. Chem. C*, **112**, 13499–13509.

73 Shimizu, Y., Hyodo, T., and Egashira, M. (2004) *J. Eur. Ceram. Soc.*, **24**, 1389–1398.

74 Lakshminarasimhan, N., Bae, E., and Choi, W. (2007) *J. Phys. Chem. C*, **111**, 15244–15250.

75 Yusoff, N.H., Salleh, M.M., and Yahaya, M. (2008) *Sains Malaysiana*, **37**, 249–253.

76 Testino, A., Bellobono, I.R., Buscaglia, V., Canevali, C., D'Arienzo, M., Polizzi, S., Scotti, R., and Morazzoni, F. (2007) *J. Am. Chem. Soc.*, **129**, 3564–3575.

77 Oki, A.R., Xu, Q., Shpeizer, B., Clearfield, A., Qiu, X., Kirumakki, S., and Tichy, S. (2007) *Catal. Commun.*, **8**, 950–956.

78 On, D.T. (1999) *Langmuir*, **15**, 8561–8564.

79 Luo, H., Wang, C., and Yan, Y. (2003) *Chem. Mater.*, **15**, 3841–3846.

80 Yu, J.C., Zhang, L., Zheng, Z., and Zhao, J. (2003) *Chem. Mater.*, **15**, 2280–2286.

81 Yun, H.-S., Miyazawa, K., Zhou, H., Honma, I., and Kuwabara, M. (2001) *Adv. Mater.*, **13**, 1377–1380.

82 Kluson, P., Kacer, P., Cajthaml, T., and Kalaji, M. (2001) *J. Mater. Chem.*, **11**, 644–651.

83 Zheng, J.-Y., Pang, J.-B., Qiu, K.-Y., and Wei, Y. (2001) *J. Mater. Chem.*, **11**, 3367–3372.

84 Sarkar, J., John, V.T., He, J., Brooks, C., Gandhi, D., Nunes, A., Ramanath, G., and Bose, A. (2008) *Chem. Mater.*, **20**, 5301–5306.

85 Liu, C., Fu, L., and Economy, J. (2004) *J. Mater. Chem.*, **14**, 1187–1189.

86 Guo, Y.G., Hu, J.S., Liang, H.P., Wan, L.J., and Bai, C.L. (2005) *Adv. Funct. Mater.*, **15**, 196–202.

87 Sander, M.S., Côté, M.J., Gu, W., Kile, B.M., and Tripp, C.P. (2004) *Adv. Mater.*, **16**, 2052–2057.

88 Shin, H., Jeong, D.K., Lee, J., Sung, M.M., and Kim, J. (2004) *Adv. Mater.*, **16**, 1197–1200.

89 Cottam, B.F. and Shaffer, M.S.P. (2007) *Chem. Commun.*, 4378–4380.

90 Kuang, D., Brillet, J., Chen, P., Takata, M., Uchida, S., Miura, H., Sumioka, K., Zakeeruddin, S.M., and Graetzel, M. (2008) *ACS Nano*, **2**, 1113–1116.

91 Zheng, Q., Zhou, B., Bai, F., Li, L., Jin, Z., Zhang, J., Li, J., Liu, Y., Cai, W., and Zhu, X. (2008) *Adv. Mater.*, **20**, 1044–1049.

92 Mor, G.K., Varghese, O.K., Paulose, M., and Grimes, C.A. (2005) *Adv. Funct. Mater.*, **15**, 1291–1296.

93 Liu, Z., Zhang, X., Nishimoto, S., Jin, M., Tryk, D.A., Murakami, T., and Fujishima, A. (2008) *J. Phys. Chem. C*, **112**, 253–259.

94 Chanmanee, W., Watcharenwong, A., Chenthamarakshan, C.R., Kajitvichyanukul, P., deTacconi, N.R., and Rajeshwar, K. (2008) *J. Am. Chem. Soc.*, **130**, 965–974.

95 Hirano, M., Ota, K., and Iwata, H. (2004) *Chem. Mater.*, **16**, 3725–3732.

96 Tan, L.K., Gao, H., Zong, Y., and Knoll, W. (2008) *J. Phys. Chem. C*, **112**, 17576–17580.

97 Ye, Z., Liu, H., Schultz, I., Wu, W., Naugle, D.G., and Lyuksyutov, I. (2008) *Nanotechnology*, **19**, 325303.

98 Lee, J.H., Leu, I.C., Hsu, M.C., Chung, Y.W., and Hon, M.H. (2005) *J. Phys. Chem. B*, **109**, 13056–13059.

99 Qiu, Y. and Yu, J. (2008) *Solid State Commun.*, **148**, 556–558.

100 Qiu, J., Yu, W., Gao, X., and Li, X. (2007) *Nanotechnology*, **18**, 295604.

101 Yuan, Z.-Y. and Su, B.-L. (2004) *Colloids Surf. A*, **241**, 173–183.

102 Bavykin, D.V., Parmon, V.N., Lapkina, A.A., and Walsh, F.C. (2004) *J. Mater. Chem.*, **14**, 3370–3377.

103 Milsom, E.V., Novak, J., Oyama, M., and Marken, F. (2007) *Electrochem. Commun.*, **9**, 436–442.

104 Zhu, B., Guo, Q., Wang, S., Zheng, X., Zhang, S., Wu, S., and Huang, W. (2006) *React. Kinet. Catal. Lett.*, **88**, 301–308.

105 Zou, J.J., Liu, C.J., Yu, K.L., Cheng, D.G., Zhang, Y.P., He, F., Du, H.Y., and Cui, L. (2004) *Chem. Phys. Lett.*, **400**, 520–523.

106 Pang, X., He, D., Luo, S., and Cai, Q. (2009) *Sens. Actuators, B*, **137**, 134–138.

107 Li, J. and Zeng, H.C. (2006) *Chem. Mater.*, **18**, 4270–4277.

108 Li, H., Bian, Z., Zhu, J., Huo, Y., Li, H., and Lu, Y. (2007) *J. Am. Chem. Soc.*, **129**, 4538–4539.

109 Lee, D.-W., Park, S.-J., Ihm, S.-K., and Lee, K.-H. (2007) *J. Phys. Chem. C*, **111**, 7634–7638.

110 Jiang, X. and Wang, T. (2008) *J. Am. Ceram. Soc.*, **91**, 46–50.

111 Bonelli, B., Cozzolino, M., Tesser, R., Serio, M., Piumetti, M., Garrone, E., and Santacesaria, E. (2007) *J. Catal.*, **246**, 293–300.

112 Pu, Y.Y., Fang, J.Z., Peng, F., and Li, H. (2007) *Chin. J. Inorg. Chem.*, **23**, 1045–1050.

113 Lee, M.S., Lee, G.D., Park, S.S., Ju, C.S., Lim, K.T., and Hong, S.S. (2005) *Res. Chem. Intermed.*, **31**, 379–389.

114 Kim, S.H., Kim, K.D., Song, K.Y., and Kim, H.T. (2004) *J. Ind. Eng. Chem.*, **10**, 435–441.

115 Hong, S.S., Lee, M.S., Ju, C.S., Lee, G.D., Park, S.S., and Lim, K.T. (2004) *Catal. Today*, **93–95**, 871–876.

116 de Souza, J.L., Fabri, F., Buffon, R., and Schuchardt, U. (2007) *Appl. Catal., A*, **323**, 234–241.

117 Zhang, H., Luo, X., Xu, J., Xiang, B., and Yu, D. (2004) *J. Phys. Chem. B*, **108**, 14866–14869.

118 Ismail, A.A., Ibrahim, I.A., Ahmed, M.S., Mohamed, R.M., and El-Shall, H.

(2004) *J. Photochem. Photobiol. A*, **163**, 445–451.
119 Chen, L.C. and Huang, C.M. (2004) *Ind. Eng. Chem. Res.*, **43**, 6446–6452.
120 Lee, J.W., Kong, S., Kim, W.S., and Kim, J. (2007) *Mater. Chem. Phys.*, **106**, 39–44.
121 Choi, H., Kim, Y.J., Varma, R.S., and Dionysiou, D.D. (2006) *Chem. Mater.*, **18**, 5377–5384.
122 Kim, I.-D., Rothschild, A., Yang, D.-J., and Tuller, H.L. (2008) *Sens. Actuators, B*, **130**, 9–13.
123 Zhang, Y., Li, J., and Wang, J. (2006) *Chem. Mater.*, **18**, 2917–2923.
124 Vargas-Florencia, D., Edvinsson, T., Hagfeldt, A., and Furo, I. (2007) *J. Phys. Chem. C*, **111**, 7605–7611.
125 Zukalova, M., Zukal, A., Kavan, L., Nazeeruddin, M.K., Liska, P., and Gratzel, M. (2005) *Nano Lett.*, **5**, 1789–1792.
126 Wang, D., Jakobson, H.P., Kou, R., Tang, J., Fineman, R.Z., Yu, D., and Lu, Y. (2006) *Chem. Mater.*, **18**, 4231–4237.
127 Qi, Z.-M., Honma, I., and Zhou, H. (2006) *J. Phys. Chem. B*, **110**, 10590–10594.
128 Carotta, M.C., Gherardi, S., Malagu, C., Nagliati, M., Vendemiati, B., Martinelli, G., Sacerdoti, M., and Lesci, I.G. (2007) *Thin Solid Films*, **515**, 8339–8344.
129 Barreca, D., Comini, E., Ferrucci, A.P., Gasparotto, A., Maccato, C., Maragno, C., Sberveglieri, G., and Tondello, E. (2007) *Chem. Mater.*, **19**, 5642–5649.
130 Grubert, G., Stockenhuber, M., Tkachenko, O.P., and Wark, M. (2002) *Chem. Mater.*, **14**, 2458–2466.
131 Ding, Z., Hu, X., Lu, G.Q., Yue, P.-L., and Greenfield, P.F. (2000) *Langmuir*, **16**, 6216–6222.
132 Li, C., Wang, C., Wang, C., and Hu, S. (2006) *Sens. Actuators, B*, **117**, 166–171.
133 Kotov, N.A., Dekany, I., and Fendler, J.H. (1995) *J. Phys. Chem.*, **99**, 13065–13069.
134 Caricato, A.P., Capone, S., Ciccarella, G., Martino, M., Rella, R., Romano, F., Spadavecchia, J., Taurino, A., Tunno, T., and Valerini, D. (2007) *Appl. Surf. Sci.*, **253**, 7937–7941.
135 Kumazawa, N., Rafiqul Islam, M., and Takeuchi, M. (1999) *J. Electroanal. Chem.*, **472**, 137–141.
136 Al-Homoudi, I.A., Thakur, J.S., Naik, R., Auner, G.W., and Newaz, G. (2007) *Appl. Surf. Sci.*, **253**, 8607–8614.
137 Sharma, R.K., Bhatnagar, M.C., and Sharma, G.L. (2000) *Rev. Sci. Instrum.*, **71**, 1500–1504.
138 Teleki, A., Bjelobrk, N., and Pratsinis, S.E. (2008) *Sens. Actuators, B*, **130**, 449–457.
139 Mohammadi, M.R., Fray, D.J., and Ghorbani, M. (2008) *Solid State Sci.*, **10**, 884–893.
140 Tan, J., Wlodarski, W., and Kalantar-Zadeh, K. (2007) *Thin Solid Films*, **515**, 8738–8743.
141 Maciak, E. and Opilski, Z. (2007) *Thin Solid Films*, **515**, 8351–8355.
142 Du, X., Wang, Y., Mu, Y., Gui, L., Wang, P., and Tang, Y. (2002) *Chem. Mater.*, **14**, 3953–3957.
143 Zakrzewska, K. (2001) *Thin Solid Films*, **391**, 229–238.
144 Dutta, P.K., Ginwalla, A., Hogg, B., Patton, B.R., Chwieroth, B., Liang, Z., Gouma, P., Mills, M., and Akbar, S. (1999) *J. Phys. Chem. B*, **103**, 4412–4422.
145 Xu, Y. and Zhou, X. (2000) *Sens. Actuators, B*, **65**, 2–4.
146 Carney, C.M., Yoo, S., and Akbar, S.A. (2005) *Sens. Actuators, B*, **108**, 29–33.
147 Devi, G.S., Hyodo, T., Shimizu, Y., and Egashira, M. (2002) *Sens. Actuators, B*, **87**, 122–129.
148 Morimoto, T., Nagao, M., and Tokuda, F. (1969) *J. Phys. Chem.*, **73**, 243–248.
149 Madou, M.J. and Morrison, S.R. (1989) *Chemical Sensing with Solid State Devices*, Academic Press, p. 556.
150 Wang, C.C., Akbar, S.A., and Madou, M.J. (1998) *J. Electroceram.*, **2**, 273–282.
151 Llobet, E., Espinosa, E.H., Sotter, E., Ionescu, R., Vilanova, X., Torres, J., Felten, A., Pireaux, J.J., Ke, X., Van Tendeloo, G., Renaux, F., Paint, Y., Hecq, M., and Bittencourt, C. (2008) *Nanotechnology*, **19**, 375501.
152 Li, Z., Zhang, H., Zheng, W., Wang, W., Huang, H., Wang, C., MacDiarmid, A.G., and Wei, Y. (2008) *J. Am. Chem. Soc.*, **130**, 5036–5037.

153 Palomares, E., Vilar, R., and Durrant, J.R. (2004) *Chem. Commun.*, 362–363.

154 Coronado, E., Galn-Mascars, J.R., Mart-Gastaldo, C., Palomares, E., Durrant, J.R., Vilar, R., Gratzel, M., and Nazeeruddin, M.K. (2005) *J. Am. Chem. Soc.*, **127**, 12351–12356.

155 Nazeeruddin, M.K., Di Censo, D., Humphry-Baker, R., and Grätzel, M. (2006) *Adv. Funct. Mater.*, **16**, 189–194.

156 Liang, P., Qin, Y., Hu, B., Peng, T., and Jiang, Z. (2001) *Anal. Chim. Acta*, **440**, 207–213.

157 Liang, P., Liu, Y., Guo, L., Zeng, J., and Lu, H. (2004) *J. Anal. At. Spectrom.*, **19**, 1489–1492.

158 Dutra, R.L., Maltez, H.F., and Carasek, E. (2006) *Talanta*, **69** (2), 488–493.

159 Zhang, L., Morita, Y., Sakuragawa, A., and Isozaki, A. (2007) *Talanta*, **72**, 723–729.

160 Lahav, M., Kharitonov, A.B., Katz, O., Kunitake, T., and Willner, I. (2001) *Anal. Chem.*, **73**, 720–723.

161 Lieberzeit, P.A., Afzal, A., Rehman, A., and Dickert, F.L. (2007) *Sens. Actuators, B*, **127**, 132–136.

162 Hidaka, H., Honjo, H., Horikoshi, S., and Serpone, N. (2006) *Catal. Commun.*, **7**, 331–335.

163 Nezu, T., Masuyama, T., Sasaki, K., Saitoh, S., Taira, M., and Araki, Y. (2008) *Dent. Mater. J.*, **27**, 573–580.

164 Holt, R.E. and Cotton, T.M. (1989) *J. Am. Ceram. Soc.*, **111**, 2815–2821.

165 Li, Q., Luo, G., Feng, J., Zhou, Q., Zhang, L., and Zhu, Y. (2001) *Electroanalysis*, **13**, 413–416.

166 Topoglidis, E., Cass, A.E.G., Gilardi, G., Sadeghi, S., Beaumont, N., and Durrant, J.R. (1998) *Anal. Chem.*, **70**, 5111–5113.

167 Topoglidis, E., Campbell, C.J., Cass, A.E.G., and Durrant, J.R. (2001) *Langmuir*, **17**, 7899–7906.

168 Topoglidis, E., Cass, A.E.G., O'Regan, B., and Durrant, J.R. (2001) *J. Electroanal. Chem.*, **517**, 20–27.

169 Liu, M., Zhao, G., Zhao, K., Tong, X., and Tang, Y. (2009) *Electrochem. Commun.*, **11**, 1397–1400.

170 Viticoli, M., Curulli, A., Cusma, A., Kaciulis, S., Nunziante, S., Pandolfi, L., Valentini, F., and Padeletti, G. (2006) *Mater. Sci. Eng. C*, **26**, 947–951.

171 Cao, H., Zhu, Y., Tang, L., Yang, X., and Li, C. (2008) *Electroanalysis*, **20**, 2223–2228.

172 Luo, Y., Liu, H., Rui, Q., and Tian, Y. (2009) *Anal. Chem.*, **81**, 3035–3041.

173 Topoglidis, E., Astuti, Y., Duriaux, F., Gratzel, M., and Durrant, J.R. (2003) *Langmuir*, **19**, 6894–6900.

174 Milsom, E.V., Dash, H.A., Jenkins, T.A., Opallo, M., and Marken, F. (2007) *Bioelectrochemistry*, **70**, 221–227.

175 Zhang, Y., He, P., and Hu, N. (2004) *Electrochim. Acta*, **49**, 1981–1988.

176 Luo, Y., Tian, Y., Zhu, A., Rui, Q., and Liu, H. (2009) *Electrochem. Commun.*, **11**, 174–176.

177 Kafi, A.K.M., Wu, G., and Chen, A. (2008) *Biosens. Bioelectron.*, **24**, 566–571.

178 Shi, Y.T., Yuan, R., Chai, Y.Q., and He, X.L. (2007) *Electrochim. Acta*, **52**, 3518–3524.

179 Cui, X., Li, Z., Yang, Y., Zhang, W., and Wang, Q. (2008) *Electroanalysis*, **20**, 970–975.

180 Xu, X., Zhao, J., Jiang, D., Kong, J., Liu, B., and Deng, J. (2002) *Anal. Bioanal. Chem.*, **374**, 1261–1266.

181 Benvenuto, P., Kafi, A.K.M., and Chen, A. (2009) *J. Electroanal. Chem.*, **627**, 76–81.

182 Li, W., Yuan, R., Chai, Y., Hong, C., and Zhuo, Y. (2008) *J. Electrochem. Soc.*, **155**, F97–F103.

183 Guo, C., Hu, F., Li, C.M., and Shen, P.K. (2008) *Biosens. Bioelectron.*, **24**, 825–830.

184 Zhu, L., Zhai, J., Guo, Y., Tian, C., and Yang, R. (2006) *Electroanalysis*, **18**, 1842–1846.

185 Fiorito, P.A., Gonçales, V.R., Ponzio, E.A., and Torresi, S.I.C. (2005) *Chem. Commun.*, 366–368.

186 Zheng, W., Zheng, Y.F., Jin, K.W., and Wang, N. (2008) *Talanta*, **74**, 1414–1419.

187 Liu, A., Wei, M., Honma, I., and Zhou, H. (2005) *Anal. Chem.*, **77**, 8068–8074.

188 Zhang, L., Zhang, Q., and Li, J. (2007) *Adv. Funct. Mater.*, **17**, 1958–1965.

189 Xiao, P., Garcia, B.B., Guo, Q., Liu, D., and Cao, G. (2007) *Electrochem. Commun.*, **9**, 2441–2447.

190 Lu, H., Yang, J., Rusling, J.F., and Hu, N. (2006) *Electroanalysis*, **18**, 379–390.

191 Yuan, S. and Hu, S. (2004) *Electrochim. Acta*, **49**, 4287–4293.

192 Lunsford, S.K., Choi, H., Stinson, J., Yeary, A., and Dionysiou, D.D. (2007) *Talanta*, **73**, 172–177.

193 Zhou, H., Gan, X., Liu, T., Yang, Q., and Li, G. (2006) *Bioelectrochemistry*, **69**, 34–40.

194 Zhou, H., Gan, X., Wang, J., Zhu, X., and Li, G. (2005) *Anal. Chem.*, **77**, 6102–6104.

10
Nanomaterials in the Environment: the Good, the Bad, and the Ugly
Rhett J. Clark, Jonathan G.C. Veinot, and Charles S. Wong

10.1
Introduction

Detection of environmental pollutants is an important first step toward ensuring the safety and viability of our natural environment. Unfortunately, most contaminants occur at trace levels. This presents the need to detect low concentrations selectively. In addition, straightforward, rapid methods suitable for field testing are desirable. Engineered nanomaterials (defined as man-made particles have at least one dimension between 1 and 100 nm) have been the focus of intense study over the past couple of decades due to their unique properties that differ from their bulk properties. Nanomaterials have become desirable for use (or potential use) in various consumer items such as pigments (i.e., paints), food additives, health care agents, and clothing, as well as electronics and photovoltaic devices. They have also gained increasing attention for sensors, many of which are suitable for use in detection of environmental pollutants [1–3].

Inorganic zero-dimensional (spherical) nanoparticles and quantum dots (usually below 12 nm in diameter) exhibit increased reactivity compared to their bulk counterparts, partly due to the high surface to volume ratio. Quantum effects that arise for particles in this size regime also contribute to properties that differ from microscale equivalent [4]. Noble metal nanoparticles, such as gold, display intense optical absorption in the visible region while quantum dots, such as silicon or cadmium selenide, fluoresce with high quantum yields. Selectivity of nanomaterial devices is bestowed through facile surface functionalization/modification, allowing tunable reactivity and surface characteristics without compromising the visual output. As a result, numerous analytes can be selectively detected using a common nanoparticle, simply by altering the surface chemistry. The intense optical properties in the visible region, meanwhile, make these materials ripe for use in sensor devices suitable for colorimetric field-based monitoring, leading to potentially smaller devices with lower detection limits [5].

Despite growing interest in the use of nanomaterials, there is growing concern over their potential effects once they inevitably reach the environment through disposal, leaching, or some other mechanism. To date, there are still numerous questions regarding the transport and toxicity of nanomaterials [6]. While naturally occurring nanomaterials are ubiquitous and have been present throughout time, engineered nanoparticles are of concern because of their unique, tailored properties. Great strides have been made in our understanding of nanomaterial behavior in the natural environment but most studies are relegated to controlled laboratory investigations [7]. Direct detection of nanomaterials *in situ* is challenging because of limitations in current analytical instrumentation and further complicated by modifications that result from aggregation or interactions with chemical species, light, and/or microorganisms. Currently, a large number of techniques have found use in the analysis of nanomaterials [8], but continued improvements in instrumentation technology will assist in the study of nanomaterials moving forward.

This chapter reviews various aspects of inorganic nanomaterials as they relate to the environment. Section 10.2 discusses the state-of-the-art in optical environmental sensors. In particular, the focus is on enhancements made by the incorporation of nanomaterials. The importance of surface functionality in improving selectivity will also be highlighted. Section 10.3 outlines factors affecting the fate and toxicity of engineered nanomaterials in the environment. Finally, Section 10.4 describes challenges facing quantification of nanomaterials in environmental matrices. The advantages and limitations of current instrumentation will be discussed.

10.2
The Good: Nanomaterials for Environmental Sensing

10.2.1
Colorimetric Detection

10.2.1.1 Noble Metal Nanoparticles
Noble metal engineered nanoparticles (ENPs), like gold and silver, are conductors and so they do not have a band gap. For this reason they do not typically luminesce, although luminescence is observed in very small Au-ENPs (~1 nm) [9]. Instead, these materials are characterized by intense, size-dependent surface plasmon absorption in the visible or near-ultraviolet region when the particle size is decreased below the de Broglie wavelength of 20 nm [10]. When electrons are promoted into the conduction band, they become trapped and exhibit a characteristic oscillation known as the surface plasmon band. If the wavelength of incident light corresponds to that of the conduction band electrons, enhanced oscillation of the electron cloud results. This is referred to as the localized surface plasmon resonance (LSPR) [3]. The result is a strong absorbance at an energy that is unique for sufficiently small (~5–20 nm) nanoparticles. Au-NPs absorb light of decreasing energy with increased size. The LSPR is sensitive to changes in refractive index that can be incurred by changes in particle size [11] and/or surface functionality. As interparticle distance decreases, as is the case when aggregation occurs, absorbance maxima shift to longer wavelengths. Binding of different analytes to

the surface causes changes in the LSPR, giving rise to shifts in the absorption spectrum. Similar to the case of fluorescence monitoring, the surface of a noble metal particle may be modified to bind a desired analyte selectively, giving a characteristic color change. The absorbance intensity of the newly formed color is proportional to the concentration of analyte present.

Surface plasmon formation gives rise to intense colors such as those seen in ancient stained glass windows [3]. Noble metal nanomaterials are frequently used as colorimetric detectors due to their strong absorbance characteristics in the visible region. SPR (surface plasmon resonance) band extinction coefficients of 2.7×10^8 and $1.5 \times 10^{10}\,M^{-1}cm^{-1}$ (at 520 nm) have been reported for 13 and 50 nm diameter gold particles, respectively [12]. As mentioned previously, changes to the surface structure shifts the observed wavelength of the LSPR and hence gives rise to different colors. The magnitude of this shift depends the mass of the substituent and so the addition of an analyte alone will typically not induce a detectable SPR shift. Further SPR enhancement, however, may be induced by aggregation of NPs in the presence of a specific analyte. If bonding between surface groups and the analyte occurs, NPs move from being free-standing systems to large aggregates. This process also causes notable shifts in the LSPR and produces a distinct and easily identifiable color change [13].

To design an effective sensor, surface groups must be chosen such that they will react selectively with an analyte of choice and with a high binding constant. The functionalization of NPs is typically a simple, one-step process. A single noble metal NP may be surface modified in various ways to detect many different analytes. However, unlike single molecule detectors that each have different chemical and physical properties, NP sensors possess similar optical properties because their optical response arises from a common core material. With this knowledge, noble metal NPs may be used to detect a wide range of analytes, including trace metals, pesticides, and gases by changing only the surface group. The most common NP core employed for colorimetric detection is Au; however, Ag has proven useful in some cases.

Au-NPs are typically synthesized by citrate reduction of $HAuCl_4$, the size of which can be controlled by the citrate concentration [14]. This synthetic procedure results in citrate "capped" Au-NPs whose size may be tuned between 5 and 20 nm. Larger particles, up to 40 nm, can be realized by adjusting the pH and the citrate to Au ratio [15]. Following synthesis and purification, citrate remains physisorbed (i.e., adhered via electrostatic interactions as opposed to chemical bonds) to the surface, which stabilizes the nanoparticles in aqueous solution. The citrate capping is known to interact with various metal ion species, resulting in aggregation-induced colorimetric response at pH 6.7 [16]. Increasing the pH to 11.2 has allowed selective detection of Pb^{2+} at μM levels. As the pH increased, insoluble metal hydroxides form. Above pH 10, Pb^{2+} redissolved and the $[Pb(OH)_3]^-$ ion so-formed could then interact with the NPs. The detection limit was subsequently improved from a rather poor value of 200 mM without electrolyte to a much better level of 11.5 μM at an ionic strength of 0.04 μM.

Owing to the weak electrostatic interaction between citrate and the Au-NP surface, citrate can be displaced easily, especially by mercapto groups in the case of Au, leading to functionalized NPs. Such surface-exchange reactions may be exploited to create single-step detectors using as-synthesized Au-NPs. H_2O_2 is a potent oxidizer and can act as a source of highly reactive radicals such as the hydroxyl radical ($^{\cdot}$OH). Even at µM concentrations in the atmosphere and rainwater, H_2O_2 can have adverse effects toward flora and fauna [17]. H_2O_2 has been detected by Wu et al. using citrate capped Au-NPs in the presence of horseradish peroxidase (HRP) and o-phenylenediamine (OPD) [18]. In the absence of H_2O_2, OPD chelated to the NP surface at the amino groups and no further interactions occurred, even in the presence of the HRP catalyst. When H_2O_2 was added, oxidation of OPD resulted in its dimerization. This caused aggregation of the Au-NPs, resulting in a color change that was dependent on H_2O_2 concentration and detectable with the naked eye to a limit of 5.0 µM.

NP detectors are more commonly developed by an initial functionalization step followed by exposure to the analyte of choice. Xiong and Li have functionalized Ag-NPs with p-sufonatocalix [4] arene for the sub-µM detection of optunal, an organophosphorus pesticide used in insecticides and biocides [19]. Guan et al. have functionalized Au-NPs with histidine to create a pH-dependant sensor for Fe^{3+} [20]. When the pH was below the pK_a value of 9.6 for the α-NH_2 of histidine, the amino group was protonated and unable to bind to other Fe^{3+} ions. As a result, the imidazole was left as the only Fe^{3+} binding site. Similar binding to imidazole from other Au-NPs was responsible for aggregation resulting in a red to blue color change at µM levels. When the pH was above the pK_a, the neutral NH_2 form of histidine was present to bind Fe^{3+} in a bidentate fashion and inhibit aggregation. Yoosaf et al. have used a "one-pot" synthetic method to create gallic acid functionalized Au (or Ag) NPs that were used to detect as low as 30 µM Pb^{2+} [21]. Gallic acid spontaneously reduced Au^+ or Ag^+ ions, forming zero-valent metal NPs and concomitantly capped the NPs via electrostatic interaction at the carboxylic acid site. Scheme 10.1 shows the oxidation of gallic acid to the quinine form and Figure 10.1 illustrates the electrostatic interaction between the quinine and the ENP. The addition of Pb^{2+} was colorimetrically determined via a bathochromic shift caused by selectively induced aggregation.

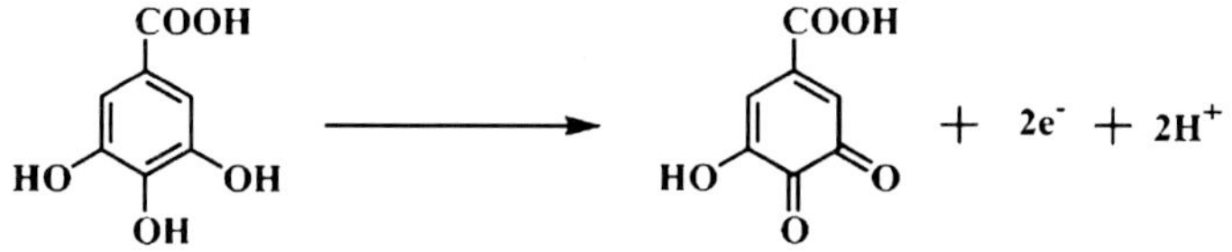

Scheme 10.1

Hybrid nanosystems have also been utilized as sensing motifs. For example, Raschke et al. showed that when an AuS core is surrounded by an Au nanoshell the SPR response was more sensitive to changes at the Au surface [22]. The addition of 16-mercaptohexaundecanoic acid to the surface was detected by surface plasmon resonance (SPR) spectroscopy without aggregation occurring.

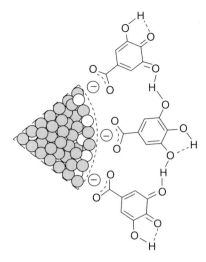

Figure 10.1 Electrostatic interaction between carboxylic acid and ENP (engineered nanoparticle) surface. Reprinted with permission from Reference [21]. Copyright 2007 American Chemical Society.

Non-aggregation induced SPR detection is remarkable, since aggregation is typically required to achieve a sufficiently detectable SPR shift. Huang *et al.* have used bimetallic Au–Ag nanorods functionalized with hexadecyltrimethylammonium bromide for Fe^{2+} detection [23]. The addition of Ag increased the length-to-diameter aspect ratio, giving rise to intense SPR absorption. In the presence of poly(sodium 4-styrenesulfonate) (PSS), an anionic polyelectrolyte, electrostatic interactions with the nanorods induced aggregation. Fe^{2+} was bound preferentially to PSS and the nanorod aggregates were destabilized. With this method, $1.0\,\mu M$ Fe^{2+} could be detected in 10 min.

NP functionality is not limited to the chemistry imposed by a single surface group. Bifunctional NPs expand the realm of possible applications. 3-Mercaptopropionic acid functionalized Au-NPs have been shown to interact with Hg^{2+} [24]. However, as the ionic strength was increased, the NPs precipitated out of solution, a property that could limit the practical application of these NPs to environmental analysis. Further functionalization with highly charged adenosine monophosphate allowed dispersion of NPs to be maintained in solutions of high ionic strength. This condition left the NPs available to aggregate only when exposed to reactive ions such as Hg^{2+} and produce a characteristic red to blue color change. By reducing aggregation that occurs in high ionic strength media when Hg^{2+} is not present, these NP systems are promising for use in environmental and biological detection.

Nanoparticles bearing multiple surface functionalities can also be incorporated to stack the deck in favor of performing chemistry that is induced by the analyte. "Click" chemistry [25] involves the Cu(I)-catalyzed cycloaddition between an azide and an alkyne (Scheme 10.2). Zhou et al. have created two sets of 13 nm Au-NPs, one bearing a terminal azide and the other a terminal alkyne [26]. Copper(II) was first reduced to Cu(I) by sodium ascorbate and subsequently catalyzed the click reaction between the two sets of NPs. The reaction led to an aggregation-induced color change. Since this reaction is specifically catalyzed by Cu(I), selectivity for Cu was achieved with a detection limit of 50 µM.

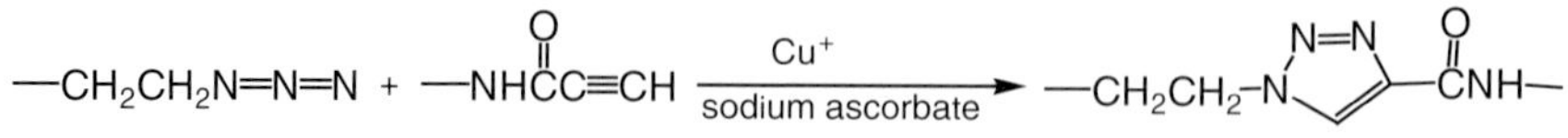

$$-CH_2CH_2N{=}N{=}N \;+\; -NHCC{\equiv}CH \xrightarrow[\text{sodium ascorbate}]{Cu^+} -CH_2CH_2-N{\underset{N=N}{\diagdown}}-CNH-$$

Scheme 10.2

The first colorimetric sensor using inorganic silica nanotubes has been developed by Lee et al. [27]. Nanotubes were functionalized with azo-coupled macrocyclic dyes that bind selectively with Hg^{2+} in a multidentate fashion. Although the mM detection limit is significantly higher than many fluorescence-based Hg^{2+} detectors (Section 10.2.2), the bioinert silica nanotube system is potentially advantageous for environmental use. Conceivably, these nanotubes could be incorporated into filtration structures that afford sequestration of Hg^{2+}. The nanotube morphology (i.e., µm scale length) also affords straightforward removal via filtration, which would facilitate Hg^{2+} removal. Furthermore, nanotube sensors have the potential to act as size-exclusion chromatographic columns, while responding to the separated analyte of choice.

10.2.1.2 DNAzymes

DNAzymes form a distinct subset of colorimetric nanomaterial-based detectors that offer a wealth of tailorability and selectivity. Typically, these nanostructures take the form of gold NPs functionalized with oligonucleotide chains of choice. Two primary mechanisms of detection exist for DNAzyme detectors. The first employs Au-NPs functionalized with complementary DNA containing a thymidine-thymidine mismatch (T-T mismatch) (Figure 10.2). Upon exposure to Hg^{2+}, thymidine sites bind to the metal ion selectively causing Au-NPs to aggregate, thereby producing a distinctive color change. The remaining surface DNA moieties bind to complementary strands on other particles within the aggregate, increasing the overall strength of the particle–particle interaction.

The second DNAzyme mechanism employs substrate–enzyme interaction (Figure 10.3). Au-NPs are functionalized with one of two oligonucleotides that are each complementary to different ends of a single substrate that is added to the solution. The substrate contains a single RNA moiety that acts as a cleavage site. An enzyme is also included that favorably interacts with the substrate and contains a reactive site that is selective for the target analyte. With no analyte present, the NPs will be aggregated. In the presence of the analyte, the substrate is cleaved and NPs redisperse, giving rise to a color change. This mechanism is more versatile than the T-T mismatch in light of the clear tunability of the enzyme, making analyte selectivity possible. Clearly, DNAzyme-induced aggregation shows promise

10.2 The Good: Nanomaterials for Environmental Sensing

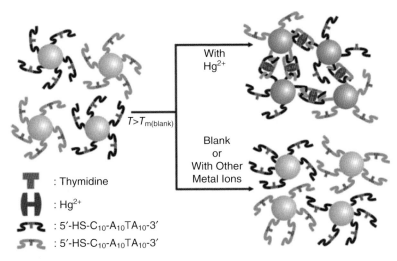

Figure 10.2 Colorimetric Hg^{2+} detection using DNA functionalized Au-ENPs containing a T-T mismatch. Reprinted with permission from Reference [28]. Copyright 2007 Wiley-VCH.

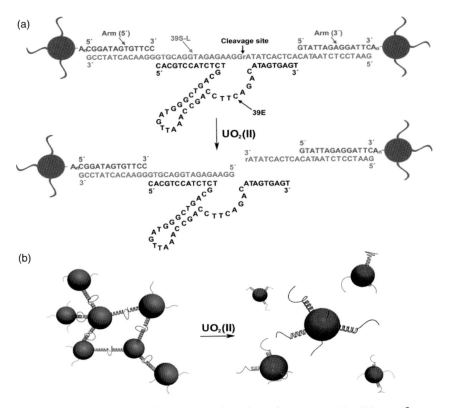

Figure 10.3 (a) DNA functionalized Au-ENPs bound to substrate containing UO_2-specific enzyme; (b) in the presence of UO_2 substrate cleavage causes dispersion of the Au-ENPs. Reprinted with permission from Reference [29]. Copyright 2008 American Chemical Society.

for sensors because of the selective ability to control interparticle distance and hence SPR absorption.

Lee et al. have used the T-T mismatch approach to sense Hg^{2+} [28]. It was found that aggregation occurred regardless of the presence or absence of the T-T mismatch. Upon melting of the DNA assemblies, NP aggregates dissociated, causing a purple to red color change. Selective complexation of Hg^{2+} with a T-T mismatch pair also caused aggregation of Au-NPs and the melting temperature increased proportionally to the Hg^{2+} concentration. The presence of Hg^{2+} (detection limit 100 nM) can be determined by heating the NP solution and observing the absence of a color change. The Hg^{2+} concentration was quantified by determining the temperature at which the color change occurred.

Various strategies have been employed to enhance the response of T-T mismatch sensors. One example employs Au-NPs bonded to a thymine-rich, single-stranded DNA sequence labeled with fluorescein dye. These NPs remain fully dispersed in NaCl solutions and appear pink, and no fluorescence was observed from the pendant fluorescein dye [30]. Upon exposure to Hg^{2+}, double-stranded DNA formed preferentially, leaving naked Au-NPs that aggregated in the high ionic strength medium. This resulted in a concomitant color change and appearance of fluorescein emission. Fluorescence measurements afforded a detection limit of 4.0×10^{-8} M with a linear range of 6.4×10^{-6} to 9.6×10^{-8} M. The fluorescence adaptation to DNAzyme sensing provides a sensitive, selective, and quantitative colorimetric means to detect Hg^{2+}.

A chip-based sensor has been developed by Lee and Mirkin in which the target DNA was tethered to a chip [31]. Au-NPs were functionalized with DNA containing a T-T mismatch. Binding of Hg^{2+} occurred as discussed previously. Washing of the chip with Ag^+/hydroquinone solution caused reduction of the Ag^+ to silver metal to yield silver spots. The intensity of scattered light from the silver spots, measured scanometrically, was proportional to the concentration of Hg^{2+} to a detection limit of 10 nM.

Liu and Lu have developed another DNAzyme sensor motif in which aggregated DNA-AuNPs were redispersed in the presence of Pb^{2+} to cleave the DNA, giving rise to a blue to red color change [32]. The initial design of this sensor incorporated a DNA substrate strand with an RNA linker. Au-NPs were then functionalized with single-stranded DNA such that binding to the substrate gave aggregation in a tail-to-tail formation. The enzyme, in the presence of Pb^{2+}, caused the substrate to cleave at the RNA site, giving free-standing particles and causing a color change [33]. Sensitivity was enhanced due to the catalytic nature of the metal toward cleavage. This design, however, was limited by a requisite heating and 2-hour cooling to facilitate aggregation. Two fundamental changes resulted in a much improved sensor. First, a change to a head-to-tail formation allowed aggregation to occur spontaneously at room temperature. Second, the use of larger particles (35 instead of 13 nm diameter) meant less aggregation was required to produce an equivalent signal intensity. The linear range for Pb^{2+} was 0.4–2 µM in about 10 min at 25 °C, making it suitable for use in the field. Using similar principles, Wei et al. have detected 500 nM Pb^{2+} in 20 min in 0.5 M NaCl without heating [34].

Detection of Pb^{2+} in the relatively high 0.5 M ionic strength solution indicates the robustness of the technique and demonstrates the potential for use in systems such as seawater (ionic strength of 0.7 M). By way of comparison, a Au electrode-bound DNAzyme electrochemical sensor achieved a detection limit of 0.3 µM [35]. Both sensors outperformed the 10 µM detection limit of a non-ENP based diaminoanthraquinone-linked polyazamacrocycle Pb^{2+} sensor [36]. Both DNAzyme sensors can detect well below the USEPA standard of 2.5 µM for products intended for use by infants or children [35].

Detection of Cu^{2+} using Au-NPs is also possible using Cu^{2+}-specific enzyme strands [37]. The ligation enzyme binds with the substrate in the presence of Cu^{2+} and initiates a red to purple color change. This method of detection has the advantage over cleavage based DNAzymes because there is minimal interference from non-specific cleavage. DNAzyme detection is also more effective than fluorescence-based methods for paramagnetic metals (e.g., Cu^{2+}) that quench fluorophore emission. Uranium detection can be achieved using selective DNAzyme detection of the most stable uranium species in water, uranyl (UO_2^{2+}). Colorimetric detection occurs as the Au-NPs disperse upon cleavage at the RNA linker (Figure 10.3) [29]. The 50 nM detection limit of this sensor is higher than the 45 pM limit of a similar DNAzyme fluorescence based sensor that does not incorporate NPs [38]. The NP sensor is, however, more selective against Th^{4+}, a common interfering species in uranyl detection. This selectivity is an improvement over inductively coupled plasma atomic emission spectroscopy in which Th^{4+} and UO_2^{2+} can scarcely be distinguished [38].

10.2.1.3 Monolithic Nanoporous Sensors

Nanoscale monolithic cages have shown promise in heavy metal colorimetric sensing. These structures are large nanoporous free-standing systems that typically act as a support. However, the underlying pore structure plays a key role in the detection limits and kinetics of metal detection. Application of nanoscale monoliths to sensing applications is a logical extension of mesoporous sensing materials that have been exploited for these purposes [39–42]. Substantial work has been performed by El-Safty *et al.* in which silica-based cage structures act as nanoporous supports for colorimetric detection of metal ions (see below). These systems offer the advantages of improved absorption capacity and increased chemical and thermal stability, while also providing efficient analyte transport to the probe sites. This particular class of sensors is also very versatile in the diversity of target species possible. Straightforward surface modification to include different chromophores has led to four distinct sensors composed of a common support. Pb^{2+}, Cd^{2+}, Sb^{2+}, and Hg^{2+} were all observed colorimetrically with limits of detection of 2.38, 13.5, 33.7, and 6.34 µg l^{-1}, respectively [43]. A similar multi-cation sensor was developed using a "building-block" approach upon the same type of silica surface [44]. The "building-block" approach refers to a bottom up assembly of monolayers, allowing the surface properties to be tailored with each layer. In this case, physisorption of dilauryl(dimethyl)ammonium bromide (DDAB) rendered the surface positively charged, allowing for the attachment of the desired chromophore probe. Hydrophobic probes were selective for Cr^{4+} and Pb^{2+}, while hydrophilic probes were used for the detection of Co^{2+} and Pb^{2+} [44]. The overall sensitivity, selectivity, and stability of the hydrophilic sensors were enhanced by the presence of a sodium dodecyl sulfate surfactant. This design allowed pM detection with 1–2 min of reaction time.

Unlike the multi-cation detectors discussed above, the following body of work focuses on probe designs for common analytes. This methodology not only optimizes how a particular analyte is measured but also may help to identify different properties that will be suitable for specific applications. Cd^{2+} was reversibly detected using two different approaches [45]. First, a direct grafting procedure was used to attach hydrophobic probes such as 4-n-dodecyl-6-(2-thiazolylazo)resorcinol onto the silica surface. The second approach was similar to the "building block" approach described earlier in which DDAB, for example, was layered onto the surface and the chromophore attached hence. In total, four sensors, two of each type, were all capable of $\mu g l^{-1}$ detection. Each detector produced a different, but distinct, color change at nM concentrations. The Sb^{3+} ion was also targeted via a building block approach [46]. The chromophore pyrogallol red was grafted to the N-trimethoxysilylpropyl-N,N,N-trimethylammonium chloride modified surface. A red to blue color change was observed with the addition of 1.36 nM Sb^{3+}. The Sb^{3+} could subsequently be removed via complexation with ethylenediaminetetraacetic acid in order to reuse the sensor.

Nanoporous silica monolithic structures have been used to develop a Hg^{2+} sensor [47]. Membrane strips were created and functionalized with azo chromophores such as 4-n-dodecyl-6-(2-pyridylazo)phenol. Colorimetric detection of Hg^{2+} of as little as $7.9\,\mu g l^{-1}$ was obtained in 10 min. This level of detection is approaching that of some fluorescence detectors (Section 10.2.2) but is still lagging behind. This method, though, provides a cheap, reusable, simple option for semi-quantitative field measurements.

10.2.2
Fluorescence-Based Detection

It is well established that optical response is a size-dependent characteristic of metallic and semiconducting (both indirect and direct bandgap) nanomaterials that is not manifested in the corresponding bulk systems. For indirect bandgap semiconductors, the lowest point of the conduction band and the highest point of the valence band occur at different wavevectors in k-space (i.e., different momentum coordinates) making the vertical optical transition forbidden [48]. For direct bandgap semiconductors, the highest point in the valence band and the lowest point in the conduction band occur at the same momentum coordinate and radiative recombination occurs readily. As particle size decreases below 100 nm, the continuous band structure changes to more discret energy levels and the bandgap increases. As the particle size decreases below the Bohr exciton radius (~5 nm for Si) the discret nature of the energy levels gives rise to quantum confinement effects (Figure 10.4) that prompt characteristic optical and electronic properties.

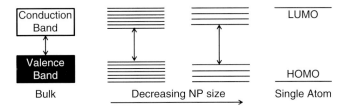

Figure 10.4 Quantum confinement: decreasing particle size leads to more discrete energy levels and increasing bandgap.

Zero-dimensional ENPs of sizes that exhibit quantum effects are termed quantum dots (QDs). An exciton is a charge–hole pair that results from the promotion of an electron from the valence band to the conduction band upon irradiation with a photon of greater energy than that of the band gap. For indirect bandgap materials this is manifest in phonon induced radiative recombination whereby the phonon is of sufficient energy to account for the misalignment of the bands. Both direct and indirect bandgap semiconducting nanomaterials with diameters below the Bohr radius are of interest because the bandgap is in the visible energy range and so these materials fluoresce with colors that can be modified, not only by changing the size but by functionalizing the surface.

Semiconductor nanocrystals or (QDs) offer distinct advantages over conventional molecular chromophores. They provide greater photoluminescent quantum yield, reduced photobleaching, and narrower spectral line widths [1], which reduce potential spectral overlap, resulting in reduced interference. Fluorescent QDs can be employed for fluorescence detection because their bandgap energy lies in the visible spectral region. The energy and intensity of photoluminescence (PL) can be altered by various factors, including particle size, morphology, crystal and surface defects, surface functionality, aggregation, and the solvent in which they are dispersed. Controlled adjustment of these factors allows exploitation of QDs as selective, sensitive tools for detection of a large variety of environmentally relevant pollutants.

CdSe nanocrystals have been used to create a single detector for three gaseous organic amines by Nazzal *et al.* [49]. Hexadecylamine functionalized CdSe NCs were spin cast into a thin film. This process quenched the photoluminescence (PL), but after photo-irradiation at 514.5 nm for 5–10 min the PL recovered fully to the levels seen in solution. Exposing the new photoluminescent film to triethylamine and benzylamine yielded an exposure-time dependent PL response. Triethylamine caused an intensity increase while benzylamine caused an intensity decrease. Both responses were reversible upon exposure to vacuum. Butylamine, which is structurally similar to hexadecylamine, did not produce a change in PL intensity. Unlike virtually all other devices, the analytes induced changes in the PL intensity and not to the wavelength of emission.

Nanoparticles may be used to enhance the signal obtained from another fluorescent label. Au-NP DNAzymes are effective colorimetric sensors (Section

10.2.1.2). This system can also be used to obtain a fluorescence-enhanced response to an analyte. Ye and Yin have functionalized Au-NPs with an oligonucleotide while the complementary strand was labeled with the fluorescent dye FAM (6-fluorescein-CE phosphoramidite). The oligonucleotides also contained a T-T mismatch for selective binding of the Hg^{2+} ion [50]. The binding of Hg^{2+} induced a fluorescent enhancement at concentrations as low as 1.0 nM (0.2 µg l^{-1}). This method is similar to other T-T mismatch DNAzyme sensors except that instead of inducing a color change from NP aggregation, the proximity of the Au-NP caused changes in the fluorescence characteristics of the dye. This detection limit was an order of magnitude lower than other Au-NP based fluorescence detectors [51, 52] but was similar to non-NP T-T mismatch DNA sensors [53, 54]. It is likely in this case that the oligonucleotide was sufficiently long to quell the fluorescence (or Förster) resonance energy transfer (FRET) that may otherwise cause quenching. FRET will be discussed further in Section 10.2.3. Instead, the increased molecular volume of the Au-NP decreases the rotational relaxation time of the fluorophore, which proportionally increases the fluorescence polarization (p) as per the Perrin equation [Equation (10.1)]:

$$\frac{1}{p} = \frac{1}{p_o} + \left(\frac{1}{p_o} - \frac{1}{3}\right)\frac{RT}{V}\frac{\tau}{\eta} \tag{10.1}$$

where

p_o represents the limiting polarization,
τ is the fluorescence lifetime,
η is the viscosity,
T is the temperature in kelvin,
R is the gas constant,
V is the volume of the rotating unit.

Detection of 0.4 µM Hg^{2+} was quantitatively detected from a spiked river sample by standard addition.

For obvious reasons, toxic trace metals such as Hg and Pb tend to draw most attention in terms of environmental detection. As illustrated throughout this chapter, NPs have shown substantial promise toward this end. Organic contaminants, however, are also candidates for NP detection given the ease with which NP surface chemistry may be diversified. Highly luminescent CdTe QDs show a dramatic loss in emission intensity when embedded in SiO_2 beads. SiO_2 stabilizes the QDs chemically and renders them inert. Li and Qu have attached calix[4]arene to the SiO_2 bead and reestablished the emission with only a slight shift in wavelength and then utilized this material as a probe for the pesticide methomyl [55]. In the presence of methomyl, a carbamate pesticide (detection limit 0.08 µM), the QD fluorescence intensity increased. The fluorescence intensity was not significantly enhanced by acetamiprid, another carbamate, or the organophosphate pesticides parathion-methyl, optunal, and fenamithion. Unlike methomyl, the other pesticides studied contain bulky aromatic groups that hinder effective binding to calix[4]arene and did not result in a significant increase in intensity. Similarly, the unfunctionalized CdSe/SiO$_2$ did not exhibit an increase in fluorescence intensity in the presence of any of the pesticides studied.

10.2.3
Fluorescence Quenching

Analyte detection based upon fluorescence quenching is qualitatively similar to fluorescence detection except the target analyte is identified by the loss of fluorescence signal. This sensor design utilizes energy transfer mechanisms such as FRET (Figure 10.5). Upon excitation of a fluorescent species, such as a QD, relaxation to the ground state typically occurs via radiative decay with a concomitant emission of a photon within the visible spectral region. If an acceptor chromophore is in close proximity (typically <10 nm), the emitted energy is transferred non-radiatively to the chromophore, causing quenching.

A fluorophore may also be quenched by the NP via the FRET mechanism. For example, the emission from Rhodamine B can be quenched by Au-NPs. This FRET process provides the basis of a Hg^{2+} sensor developed by Huang and Chang [51]. In the presence of Hg^{2+}, Rhodamine B was liberated from the Au-NP and its fluorescence was restored. Selectivity toward Hg^{2+} of 50-fold over other metals was

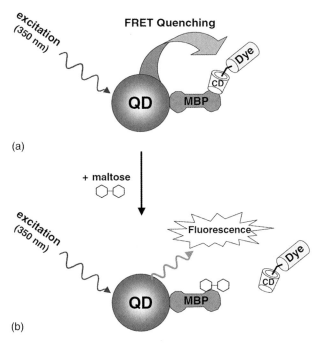

Figure 10.5 (a) Quantum dot fluorescence emission is quenched by FRET energy transfer to bound non-fluorescent dye; (b) fluorescence emission returns when dye is removed. Reprinted with permission from Reference [1]. Copyright 2006 Elsevier.

realized by functionalizing the Au-NP surface with thiol ligands such as mercaptopropionic acid and the chelating agent 2,6-pyridinecarboxylic acid. A detection limit of 2.0 µg l^{-1} was obtained in a pond water sample. Exploiting the same mechanism, Zheng et al. have modified Au-NP surface groups to thioglycolic acid without using a chelating agent, and reduced the detection limit of Hg^{2+} to 0.6 µg l^{-1} while maintaining excellent selectivity [52]. These examples clearly illustrate the importance of surface functionality as well as the versatility gained by the use of nanomaterials.

The versatility of NP sensors was further illustrated by Ruedas-Rama and Hall, who selectively detected Zn^{2+} and Mn^{2+} using NP with the same functionality and varying the mechanism of fluorescence quenching [56]. CdSe–ZnS core–shell quantum dots were functionalized with mercaptopropionic acid, poly(allylamine hydrochloride) and the chelating agent 2-carboxy-2-hydroxy-5-sulfoformazylbenzene (zincon), respectively, in successive layers. The addition of the zincon reduced the QD fluorescence, but complete quenching was not observed. Chelation of Mn^{2+} to the surface-bonded zincon disrupted zincon–QD interactions, causing increased fluorescence intensity. Conversely, the Zn^{2+}–zincon complex experienced increased FRET and fluorescence was more effectively quenched. Detection of one metal in the presence of another was effectively accomplished by adjusting the pH or by using a masking agent.

Shang et al. have capped CdSe QDs with triethanolamine to render them soluble in water [57]. NP fluorescence was effectively quenched by the Hg^{2+}/I$^-$ ion pair. While I$^-$ caused quenching unilaterally, high concentrations were required and quenching was not complete. Hg^{2+} alone also produced a limited quenching response. In the presence of I$^-$, Hg^{2+} caused significantly enhanced fluorescence quenching. This effect was proposed to arise from I$^-$ bonding with the amino N followed by complexation of I$^-$ with Hg^{2+}. It was also determined that quenching was not a result of the analyte ion displacing Cd^{2+} from the core, as has been previously reported [58].

Phenols, found in disinfectants and antiseptics, are toxic to mammals, fish, and other aquatic organisms, and therefore are another class of organic molecule of interest when monitoring environmental water samples. Typically, concentrations of phenols are evaluated using chromatographic analysis such as gas chromatography (GC) or capillary electrophoresis but sample derivatization, in the case of GC, and complex instrumentation are required. Li and Han have developed a sensor in which trioctylphosphine oxide capped CdSe–ZnS core–shell QDs were functionalized with α-, β-, or γ-cyclodextrin, cyclic oligosaccharides with six, seven, or eight glucopyranose units, respectively [59]. Upon binding of a phenol to the cyclodextrin, the complex was released from the QD surface, causing fluorescence quenching. Each cyclodextrin elicited a characteristic response. The γ-cyclodextrin was found to be essentially non-responsive to any of the phenols tested. Conversely, p-nitrophenol and 1-naphthol were detected by α- and β-cyclodextrin, respectively, via fluorescence quenching. The 7.92 and 4.83 nM detection limits approached those found using GC/MS for these compounds and far exceeded the μM detection of an electrochemical sensor [60]. While the detection limit may not be superior to that obtained using traditional instrumental analysis, the use of ENPs offers the potential for improvements in sensitivity in simple and portable devices.

Au-NPs are commonly used for colorimetric detection but, due to low quantum yields, they have not been explored for use as fluorescence sensors until very recently. Functionalization of ca. 25 nm diameter Au quantum dots with poly(amidoamine) dendrimers has been shown to increase the fluorescence quantum yield from as low as 10^{-5} [61] to as high as 0.7 [9, 10] with size dependent emission. Chen et al. have created fluorescent Au-NPs by capping with glutathione [62]. Glutathione is selective for Cu^{2+} and stabilizes Au-NPs in aqueous solution, making it potentially suitable for environmental sensing. The quantum yield of this system was determined to be 0.01, which sufficient to quantify Cu^{2+} by fluorescence quenching at concentrations as low as 3.6 nM. The initial fluorescence was completely recovered in a concentration dependent fashion by EDTA. This offers the potential for this NP system to be used as a sensor for species that complex copper.

10.3
The Bad: Environmental Fate of Nanomaterials

Engineered nanoparticles (ENPs) are already finding uses in consumer products, and the number of applications is continuing to increase. As ENPs become ubiquitous in everyday life, the likelihood of transport into environmental media becomes increasingly probable. For example, Ag-NPs, used as an antibacterial agent in socks, have been detected in wash water [63] and TiO_2 ENP pigment particles have been found in run-off from exterior paints [64]. With increased ENP usage, there is growing concern about the long-term effects of these materials in natural environments. At present, one might speculate that concentrations of ENPs in the environment are generally very low and that the risk of acute effects to the organisms that ingest these particles would be equally low. This postulate does not account for the risk of long-term effects at current levels or hazards that may be incurred as ENP concentrations increase. Complicating matters, there is little consensus whether toxicity of ENPs results due to particle size, composition, surface chemistry, or a combination of these parameters [6]. It is also reasonable that ENP toxicity will depend on the mechanism by which particles are transported through environmental media. Surface charge, pH, ionic strength as well as other factors affecting ENP aggregation and their interactions with naturally occurring materials (such as humic acids) must also be considered. It is also likely that the form in which the ENPs are present in the ambient environment is not the design created for the intended purpose. Changes in particle properties and solution dynamics may occur in the interim. The following discussion focuses on recent efforts toward elucidating the environmental fate and toxicity of ENPs.

10.3.1
Environmental Fate

10.3.1.1 Factors Affecting Aggregation

Particle aggregation/agglomeration will dramatically impact the mobility of ENPs in aqueous media. Larger particle assemblies may precipitate and become static in the solid phase of the environment. Conversely, well-dispersed ENPs can be more easily transported across large distances by remaining in the mobile aqueous phase. Consequently, concentrations of dissolved materials will inevitably increase. It is also logical for dispersed ENPs to be taken up by aquatic organisms. However, bottom-feeding organisms may still ingest ENPs associated with particulate matter [65]. In the context of all these concerns, the crucial question arises: What governs the size of a NP once it is released into the aquatic environment? Well-dispersed ENPs in controlled aqueous media may have nominal diameters as low as a few nanometers in the case of quantum dots and nanoparticles. Once distributed into uncontrolled systems, the dynamics of these mixtures may change drastically because individual ENPs may no longer exist. Much ongoing research focuses on the study of conditions that influence aggregation of various ENPs and how such particle assembly affects mobility and toxicity (Section 10.3.1.2).

The stability of colloid suspensions determines whether aggregation will occur. Stability can be determined both theoretically and experimentally. Mackay *et al.* have modeled particle stability using the stochastic methods of Gillespie [66, 67]. Experimentally, ENP stability may be understood in the context of Derjaguin–Landau–Verwey–Overbeek (DLVO) theory [68–70]. The stability, defined as resistance to aggregation, is measured by the sum of van der Waals attractive forces and electric double layer repulsive forces between two like-charged particles. At greater distances, repulsive forces will dominate. However, as the particles approach each other attractive forces will become increasingly stronger, resulting eventually in aggregation. Clearly, increasing ENP concentration will promote aggregation since particles are forced closer together [71]. In practice, the stronger the surface charge, the stronger the double layer repulsion and the less likely aggregation will occur. Zeta potential provides a direct measure of the strength of the electric double layer, specifically the potential at the slipping plane. Values greater than ±30 mV indicate a stable dispersion, while values closer to zero indicate increased probability of aggregate formation.

Ionic strength and pH are also key factors that contribute to zeta potential and hence aggregation [72]. Increasing the ionic strength affects interactions of charged particles by shielding the coulombic repulsion that would otherwise exist between particles. The zeta potential would approach zero with increasing ionic strength, because the increased charge concentration in the medium surrounding the ENP would neutralize the electric double layer. Aggregation therefore occurs faster and more completely in higher ionic strength solutions [73, 74]. However, the nature of the electrolyte (i.e., NaCl versus $MgCl_2$) has also been shown to effect aggregation [75]. As one might expect, neutral particles are not be affected by changes in the ionic strength since charge-based repulsive forces are not present.

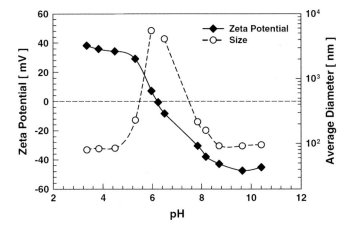

Figure 10.6 Graph of zeta potential and hydrodynamic diameter of TiO_2 ENPs in aqueous solution at an ionic strength of 0.001 M as a function of pH. Reprinted with permission from Reference [76]. Copyright 2009 Springer-Verlag.

Consequently, in these cases, the zeta potential will already likely be very low and aggregation arises as a result of dominant van der Waals interactions. Protonation or deprotonation of ENP surfaces can result from changing the pH of the medium. Figure 10.6 shows a plot of zeta potential versus pH. The point of zero charge, analogous to the endpoint of a titration, is the point at which the zeta potential is zero.

Generally, the zeta potential becomes more positive at lower pH and more negative at higher pH. The actual point of zero charge, however, is dependent on the composition of the ENP surface in question. Unfunctionalized Al_2O_3 NPs reached the point of zero charge at pH 7.9 [77] while Fe_2O_3 NPs reached maximum aggregation at pH 8.5 [78]. These pH values are significant because they are at or near pH values that may be encountered in various environmental settings, such as natural fresh waters in which the pH is typically between 6 and 8 [79]. By comparison, the point of zero charge is 2.1 for SiO_2 and 5.5 for TiO_2 [80]. Since surface functionality will change the surface properties, the nature of the functional group will play a large role in the aggregation properties. It is also important to consider the pK_a of the surface group that will dominate the surface charge, but this is often unpredictable. For example, citrate and acrylate capped Au-ENPs were largely aggregated at a pH of 2.0, but were far more dispersed at the higher pH of 12.5 [73]. Citrate and acrylate both contain carboxylate groups that will protonate at acidic pH. At neutral and basic pH, the surface will remain negatively charged, leading to an increased (negative) zeta potential and decreased aggregation.

The nature of the counter ion to the surface functional group may influence zeta potential and, consequently, aggregation properties. Zhang *et al.* have determined that thioglycolic acid functionalized CdTe QDs aggregated at lower concentrations of divalent versus monovalent cation [81]. The same relationship between zeta potential and cation concentration was seen for all cations at both pH 5 and 8. Closer examination of the presence of Ca^{2+} showed that bidentate binding with surface groups not only neutralized the particle charge but also induced aggregation by forming bridges between ENPs. Zeta potential approached zero at even lower concentration with the addition of the trivalent ion Al^{3+} at pH 5. At pH 8, the effect of cation concentration on zeta potential was reduced. The formation of the negatively charged $[Al(OH)_4(H_2O)_2]^-$ species did not neutralize the already negative surface charge.

In natural aqueous media, organic matter plays a vital role in NP stability. Humic and fulvic acids are naturally occurring, carbonaceous compounds that form high molecular weight aggregates due to extensive interparticle hydrogen bonding. Humic and fulvic acids can physisorb onto the surfaces of colloids [74, 82] and can drastically change their surface characteristics, leading to variations in stability. The propensity of citrate capped Au-ENPs to aggregate was found to increase with decreasing pH, a phenomenon attributed to protonation of the capping citrate ligand [73]. Humic acids physisorb to the surface of ENPs in two ways, either by coating or replacing the citrate, and in doing so reduce ENP aggregation. Presumably, the size and corresponding steric bulk of surface bonded humic acid prevented close approach of other ENPs. A similar effect was observed for Al_2O_3 ENPs at their point of zero charge. When humic acids were present aggregation was greatly reduced; however, aggregation still proceeded at acidic pH [77]. Domingos et al. have found the combination of TiO_2 NPs and fulvic acid led to stable dispersions using environmentally relevant pH, ionic strength, and fulvic acid concentrations [83]. At lower fulvic acid concentrations, though, increased aggregation was observed and attributed to bridging of the ENPs by fulvic acid. As one might expect, the point of zero charge of Fe_2O_3 ENPs shifted to lower values (i.e., from pH of 8.5 to pH ca. 4–5) in media containing high concentrations of humic acids [78]. High concentrations of humic acids were also found to disperse Fe_2O_3 ENP aggregates near the point of zero charge [71]. Incorporation of humic acids at neutral pH served to increase the surface charge and hence the repulsive forces of the NPs. Humic acid has also been shown to stabilize CdSe ENPs in natural water samples [84]. In the case of multiwalled carbon nanotubes, humic acid led to a two order of magnitude increase in stability due to steric repulsion [85].

10.3.1.2 Nanoparticles in Porous Media

Reports of the ENP fate and transport in porous media such as soils are few compared to the number outlining behavior in aqueous media. It is reasonable to presume that ENPs found in soil could be consumed by terrestrial organisms (including mammals), taken up by plants, or transported into aqueous systems [86]. Metal oxide ENPs will certainly physisorb to organic particles in soil and strong particle associations make characterization and detection of NPs difficult. Despite this apparent capture of ENPs by organic media in soils, their reactivity toward other environmental species may remain high [87]. As is the case for ENPs in aquatic systems, surface potential plays a crucial role in the degree of aggregation. As discussed previously, aggregation is greatly increased at the point of zero charge. At increased zeta potentials, when aggregation is not significant, particles may remain suspended and can more easily migrate through the porous soil medium [88]. A key parameter when studying ENPs in porous media is the attachment efficiency, which is a measure of how favorable the attachment of an ENP is to a particle in the medium (e.g., a sand grain). Larger latex particles have been found to exhibit greater attachment efficiency than smaller particles, contrary to the predictions of DLVO theory [74]. The presence of humic acids decreased the attachment efficiency. Mobility may be evaluated by passing ENPs through columns containing the porous media in question. Breakthrough curves (i.e., C/C_o vs pore volume) can be plotted, where C_o is the initial NP concentration and C is the concentration of NPs that exit the column. Particle composition, size, and velocity all affect the mobility [89, 90].

10.3.2
Toxicity

Human exposure to naturally occurring NPs and other small particles is not new; people have long inhaled soot from fires and volcanic ash that contain nanoparticulate matter [91] while aquatic NPs are ubiquitous [92]. Inhalation of naturally occurring particles can enter the body and various toxic effects may be induced [91]. Recently, the study and application of ENPs has grown, but probing their potentially harmful effects in biological systems is only recently gaining prominence, dating back little more than a decade [93].

Although ENPs are classified as such due to their size, toxicity will certainly be heavily dependent upon surface functionality [94]. As discussed in Section 10.3.1, numerous factors affect the mobility of ENPs in an environmental setting. Similarly, variables such as pH and ionic strength affect the stability of ENPs in a biological system, which could in turn affect the toxicity [95]. As such, ENPs cannot readily be grouped together based upon toxicity. Every ENP must be evaluated as a distinct entity based upon composition, size, surface functionality, etc. To date, the large majority of the NP toxicity literature concerns "as-prepared" ENPs, while surface functionality has been largely ignored.

Ag and TiO_2 ENPs are among the most common forms of ENPs used in commercial products today. Ag, with its well-documented antibacterial properties, is found in socks [63] and certain medical supplies while TiO_2 is used as a pigment in paints and cosmetics [96]. Aside from their intended uses, what effects might these materials produce? Starch-coated Ag ENPs were found to increase oxidative stress in human cells [97]. Griffit *et al.* have performed a series of studies involving the toxicity of various ENPs on differing organisms [98]. Ag, as well as Cu, ENPs have toxic effects on zebrafish, daphnids, and algae. TiO_2, however, has been found to be non-toxic by comparison. Further study indicated that Ag and Cu produce differing biological effects in zebrafish and toxicity is not strictly due to the release of soluble metal ions [99].

TiO_2 systems are the most extensively studied ENPs with regard to their effects on biological systems. This is due at least in part because of their ease of synthesis and generally low to negligible toxicity when ingested. Notably, however, TiO_2 can photo-catalytically produce reactive oxygen species that are damaging to DNA [100]. Nonetheless, TiO_2 ENPs serve as a suitable standard for evaluating other ENPs. Rainbow trout responded to treatments of 0.1–1.0 mg l^{-1} TiO_2 NPs in a sublethal manner, although some oxidative stress was observed [101]. TiO_2 produced decreased antioxidant enzyme activity in a dose-dependent fashion in terrestrial isopods, but no significant effects were observed at major high-level biomarkers [102]. TiO_2 exhibited greater antibacterial properties towards *Escherichia coli* and *Bacillus subtilis* than SiO_2, but less than ZnO. The toxic effects of TiO_2 were enhanced by light; however, bacterial growth inhibition was also noted for samples maintained in subdued light [103]. Conversely, Velzeboer *et al.* found no ecotoxicity toward bacteria, algae, crustaceans, or soil enzymes from either TiO_2, ZrO_2, Al_2O_3, or CeO_2 due to aggregation in the test media [104]. Clearly, numerous effects influence the interaction of ENPs with biological species.

Studying the toxicity of ENPs is challenging relative to equivalent evaluation of small molecule toxins. It is not enough to determine that introduction of an ENP into a biological system causes a toxic response. Future studies must establish if the reactivity of the ENP is responsible for the observed effect. Recall, it is also possible that dissolution of the ENP core or the release of a surface group is producing the response. For example, toxic Cd^{2+} and Se^{2-} ions could be released from a CdSe NP core. In practical terms, the influence of dissolution can be accounted for by running appropriate control experiments. Another key piece of the puzzle is the mode of toxicity: Where do the NPs distribute and how are they transported? Transmission electron microscopy (TEM) imaging can be used to locate the target organ of the ENPs while also establishing if the ENPs remained intact. Ravenzwaay *et al.* have demonstrated the deposition of inhaled TiO_2 into the lungs of rats using TEM [105]. The limitation of TEM analysis in this case was the need for extensive sample preparation. Fluorescence imaging is a more convenient means of monitoring ENP distribution, especially in small biological systems where the organism can be measured whole. Fluorescence studies are also useful for monitoring food chain transfer. Fluorescent latex ENPs (39 nm) were injected into medaka eggs. Upon hatching, ENPs were found in the gallbladder and liver of the medaka [106]. Fluorescence imaging was also used to monitor the transfer of carboxyl coated CdSe ENPs from the freshwater algae *Pseudokinchneriella subcapitata* to the invertebrate cladoceran *Ceriodaphnia dubia* using fluorescence imaging [107]. Fluorescence studies, however, are limited to use with emissive ENPs, or at least fluorescently labeled ENPs. Nonetheless, background emission from the organism itself is a complicating factor. Fluorescence imaging therefore does not directly measure the emission of the ENP moiety itself.

Work has begun, in earnest, to quantify the toxicological effects of nanomaterials, and it is clear most research is aimed at unfunctionalized or as-prepared NPs. In addition, ENP size is routinely reported in terms of TEM for dry samples or DLS for aqueous samples. However, these typical sizing techniques provide little information about the three-dimensional shape or porosity of the particle. It is also difficult to distinguish between large particles and aggregates containing few smaller particles. Overall, few reports exist that describe the effect of ENP surfaces on toxicity, but studies are now beginning to appear that incorporate changing surface characteristics. Ag ENP antibacterial properties were altered by the presence of various surface ligands. Compared with SO_4^{2-}, Cl^-, PO_4^{3-}, and EDTA, S^{2-} was the only surface-capping agent to significantly reduce toxicity [108]. Increasing ligand strength (i.e., surface binding strength) increased antibacterial effects. There is precedent that surface area could be more important in evaluating ENP toxicity than particle size. Warheit et al. have performed pulmonary bioassays that showed quartz particles ranging from 50 to 500 nm were cytotoxic to rats. The toxicity did not correlate with particle size, but did scale with surface area [109]. Stable suspensions of 12.5 and 27.0 nm SiO_2 ENPs were tested for effects on a freshwater green algae species. Although larger particles imparted greater influence on growth rate, there was an insignificant toxicity difference when surface area was evaluated [110]. Surface charge and functionality must be studied in detail as these factors will affect the transport of ENPs throughout an organism. Adsorption of an ENP to a protein changes the point of zero charge, while the surface charge will influence the ease with which the blood–brain barrier can be crossed [111].

It is not always the ENPs that must be evaluated for their effects in the environment. ENP-induced binding of other materials may lead to unprecedented or, at least, unexpected results. Limbach et al. have studied SiO_2 ENPs that were hybridized with varying amounts of Fe, Co, Mn, or TiO_2 [112]. Heavy metal uptake was enhanced using these hybrid systems and interference with bioassays was observed, suggesting a "Trojan-horse" type mechanism for oxidative stress is possible. ENPs, particularly Fe-based systems, have been targeted as remediation tools for pollutants, such as chlorinated organics, from aqueous environmental systems [113]. In this case, the Fe/Fe_xO_y ENPs can be removed after use via magnetic separation. SiO_2-TiO_2 NPs can be used to remove Hg^{2+} from coal combustion effluent. A study by Gao et al. shows Hg may remain bioactive while adsorbed onto the surface of the ENPs [114]. Methylation of Hg^{2+} was carried out while still bound to the Fe ENPs. Liberation of toxic materials from ENPs over time also cannot be discounted. It is therefore not enough to assume that ENP-bound toxins are safe. Instead ENP–toxin complexes may need to be further separated from the environmental medium to ensure effective environmental cleanup.

10.4
The Ugly: Detection of Nanomaterials in the Environment

The fate and effects of nanomaterials in the environment are still poorly understood despite recent, concerted efforts. This situation is mainly due to limitations in methods of characterization [8], the reasons for which are numerous. ENP concentrations are likely very low, perhaps ng/l or less, and interference from the complex matrix, including an abundance of naturally occurring NPs, must also be avoided [7]. Except for specific situations where ENPs are unusually concentrated, such as TiO_2 from paint in water runoff [64] or Ag NPs from socks after washing [63], unambiguous detection of ENPs in environmental samples is extremely difficult. One must also consider changes to surface coating, level of oxidation, or degree of aggregation once ENPs enter the environment [7]. During analysis, particle alteration, such as aggregation induced during evaporation as a part of sample preparation, may also occur, giving artefactual results [8]. Detection and analysis of ENPs in environmental samples is therefore troublesome and so laboratory studies looking at how ENPs aggregate and interact with other materials have become common [4].

Unlike other contaminants, there are at least two primary aspects to ENP characterization: elemental composition, which includes the core and surface groups, and size, including shape and morphology. Most common diagnostic analytical techniques, such as IR or mass spectrometry, provide molecular level information but cannot distinguish ENP sizes or morphologies. Energy dispersive X-ray analysis (EDX) provides compositional information and is often coupled with TEM instruments, which establish particle size and morphology.

However, common techniques capable of measuring particle size, such as TEM, SEM, or atomic force microscopy (AFM) are labor intensive and generally not compatible with high-throughput environmental analyses. In addition, solid-state techniques like TEM, SEM, and AFM require solvent evaporation that can lead to aggregation, possibly making sample analysis non-representative. One of the key limiting factors hindering effective ENP detection is identifying particles of varying nanoscale sizes simultaneously. There are numerous techniques, such as elemental analysis dispersive X-ray (EELS) and selected area electron diffraction (SAED), that have found some use in ENP analysis [115]. Here, we identify some new technical developments aimed at studying environmental ENPs.

Atmospheric ENPs can be counted by a condensation nucleus counter (CNC) [116]. This technique operates by growing particles to a larger size, making counting more straightforward. CNC does not characterize the ENP size. A micro-orifice uniform deposition impactor (MOUDI) was used effectively by Lin et al. for collecting and sorting particles of sizes ranging from 10 nm to 20 μm in a natural air sample [117]. Once collected, these samples were analyzed for composition by ICP-MS. Aqueous samples can be particularly difficult to analyze since solid-state techniques operate under high vacuum – hence results are potentially irrelevant to the original sample. Tiede et al. have developed environmental SEM that allows samples to be imaged at pressures of about 10–50 Torr [118]. Although imaging at atmospheric pressure cannot yet be performed, the realization of ESEM is an example of how techniques for monitoring ENP size in solution are advancing. Photothermal microscopy provides single particle detection in solution [119]. A major advantage of this method is that samples can be flowed through at a rate of up to $60\,pl\,s^{-1}$. This is an improvement over standard electron microscopy techniques in which the sample is static.

Dynamic light scattering (DLS) is an increasingly popular method for measuring ENP sizes in solution [76, 120]. DLS relies on the fact that particles in solution are in constant Brownian (random) motion. When the sample is irradiated with a laser, scattering of the incident light occurs. The dynamic motion of the particles causes the scattering intensity to fluctuate with time. Fluctuations are correlated to determine the average diffusion coefficient from which the hydrodynamic radius of solution-borne particles may be calculated. The hydrodynamic radius may be viewed as the radius of a hard sphere that moves through the liquid medium. This radius includes the particle as well as solvent layers surrounding the ENP. DLS is distinctly advantageous over solid-state particle sizing methods because little or no sample pretreatment is required. Representative samples can provide useful information about nanoparticulate matter present in solution. Unfortunately, no technique is without its disadvantages. DLS measurements are biased toward larger particles because of their increased scattering ability, while smaller particles may remain undetected. In addition, no morphological or compositional data is available, making further analysis necessary. Finally, DLS does not provide a direct measure of ENP size, rather the hydrodynamic radius is evaluated. Complementary information to DLS can be gained from zeta potential measurements. The zeta potential is a measure of the charge on the particle surface. Large surface charge (absolute value greater than 30 mV) indicates high electrostatic repulsion and hence a decreased propensity for the particles to aggregate. However, a greater surface charge when in a polar medium like water will lead to a greater level of solvation around the particle, resulting in a greater hydrodynamic radius. These measurements will also be affected by the pH and ionic strength of the solution since this will alter the charge and electrostatic interaction, respectively.

Field flow fractionation (FFF) is useful in separating particles over a size range of 1 nm up to 100 μm [121]. Sample is flowed through a narrow column (50–300 μm thick) while a perpendicular flow is applied, forcing particles to accumulate against the wall of the tube. Particles of similar size will accumulate and pass through the tube together where detection can be carried out via ultraviolet spectroscopy, light scattering techniques, or any other suitable detection method. More recently, inductively coupled plasma mass spectrometry has been coupled to FFF [122].

Despite the variety of techniques that are being established for size determination, Domingos *et al.* have shown that there are discrepancies among them [123]. Regardless of the technique employed, the user must always be vigilant of the true meaning and potential pitfalls of the results. The use of isotopic labeling, using both stable [124] and radioactive [125] isotopes, provides a means of monitoring NPs in complex environmental or biological media without painstaking pretreatment methods. Stable isotopes can be measured using techniques such as ICP-MS while radioactive decay can be detected by scintillation counting. Again, these methods provide no size, surface area, or morphological information. Tracking of ENPs to monitor their fate is more convenient using radioactive isotopes, but special handling procedures and laboratory regulations may restrict these types of experiments. QDs may also provide some hope for tracking experiments as they can be detected by their characteristic fluorescence [84]. Interactions that result in fluorescence quenching, such as oxidation of Si QDs, will limit the effectiveness.

Modeling of ENP transport will most certainly play a major role in deciphering the role of ENPs in the environment. Predictions based upon a combination of raw data, laboratory experiments, and chemical intuition can assist in directing future study. At this point, the limited information available and the uncertainties described herein hinder the effectiveness of current mathematical models to describe the fate of environmental ENPs. The more limited the raw data the more assumptions that must be made and the greater the uncertainty that arises in the results [126]. Consequently, great care and a critical eye are required when interpreting modeling results. As techniques for monitoring NPs improve and the quality and quantity of the raw data increases, the more powerful the models will become.

Clearly, no single technique will be the magic bullet for the detection of NPs in an environmental setting. As instrumentation improves and new methods are developed, our ability to analyze samples with lower concentrations and less sample preparation will increase. Perhaps it will even be NP-based sensors that hold the key to this problem. It may be possible to reverse engineer current sensors to selectively detect for the NP itself based on fluorescence changes or by colorimetric means whereby a NP analyte aggregates with a NP probe. Either way, environmental monitoring of NPs has only just begun.

10.5
Conclusions

The unique properties of various nanomaterials have been exploited in various "good" ways for improved environmental monitoring. This is an important initial step in being able to first understand the fate and then deal with the implications of contaminants in the natural environment. Noble metal NPs display intense color in the visible region due to surface plasmon effects. Induced aggregation gives rise to sharp color changes that can be exploited for sensitive analyte detection. DNAzymes have become an attractive functional group for noble metal NP sensors because of the fine control over selectivity and sensitive nature of the catalytic reactions that give rise to the colorimetric signal. Alternatively, quantum dots have been exploited for the intense fluorescence that they exhibit. Various methods have been described in this chapter in which specific analytes produce either a characteristic increase or quenching of fluorescence.

With the increasing popularity of engineered nanoparticles, the likelihood increases that they will appear in the environment at even higher concentrations. Although many nanomaterials are likely to be innocuous, many will undoubtedly exhibit "bad" influences when they are exposed to unintended targets. The fate and toxicity of many of these materials in a natural setting is still largely unknown and these effects will certainly vary based on the core composition, core size, surface functionality, and morphology.

The detection of NPs in the environment poses an "ugly" challenge for the future. Many analytical techniques are sensitive to chemical composition or size but not both. In addition, the complexity of environmental sample matrices means there will be a great deal of interference. The concentrations likely to be present, although potentially toxic to many organisms, will be exceptionally difficult to identify in such complex matrices. Our understanding of NPs and their behavior in an environmental setting will be enhanced as new methods of analysis are developed to monitor them. Perhaps these new techniques will utilize the beneficial properties of NPs for the analysis of NPs themselves.

Acknowledgments

This work was supported by the Natural Sciences and Engineering Research Council of Canada in the form of Discovery Grants and Strategic Grants to J .G. C. V. and C. S. W., and the Canada Research Chairs program to C. S. W.

References

1. Costa-Fernandez, J.M., Pereiro, R., and Sanz-Medel, A. (2006) *Trends Anal. Chem.*, **25**, 207–218.
2. Drake, C., Deshpande, S., Bera, D., and Seal, S. (2007) *Int. Mater. Rev.*, **52**, 289–317.
3. Riu, J., Maroto, A., and Rius, F.X. (2006) *Talanta*, **69** (2), 288–301.
4. Valcarcel, M., Simonet, B.M., and Cardenas, S. (2008) *Anal. Bioanal. Chem.*, **391**, 1881–1887.
5. Andreescu, S., Njagi, J., Ispas, C., and Ravalli, M.T. (2009) *J. Environ. Monit.*, **11**, 27–40.
6. Colvin, V.L. (2003) *Nat. Biotechnol.*, **21**, 1166–1170.
7. Simonet, B.M. and Valcarcel, M. (2009) *Anal. Bioanal. Chem.*, **393**, 17–21.
8. Lead, J.R. and Wilkinson, K.J. (2006) *Environ. Chem.*, **3**, 159–171.
9. Zheng, J., Petty, J.T., and Dickson, R.M. (2003) *J. Am. Chem. Soc.*, **125**, 7780–7781.
10. Zheng, J., Zhang, C.W., and Dickson, R.M. (2004) *Phys. Rev. Lett.*, **93**, 077402.
11. Fahlman, B.D. (2007) *Materials Chemistry*, Springer, Dordrecht.
12. Jin, R.C., Wu, G.S., Li, Z., Mirkin, C.A., and Schatz, G.C. (2003) *J. Am. Chem. Soc.*, **125**, 1643–1654.
13. Shipway, A.N., Katz, E., and Willner, I. (2000) *ChemPhysChem*, **1**, 18–52.
14. Frens, G. (1973) *Nat. Phys. Sci.*, **241**, 20–22.
15. Ji, X.H., Song, X.N., Li, J., Bai, Y.B., Yang, W.S., and Peng, X.G. (2007) *J. Am. Chem. Soc.*, **129**, 13939–13948.
16. Guan, J., Jiang, L., Zhao, L.L., Li, J., and Yang, W.S. (2008) *Colloids Surf. A*, **325**, 194–197.
17. Pena, R.M., Garcia, S., Herrero, C., and Lucas, T. (2001) *Atmos. Environ.*, **35**, 209–219.
18. Wu, Z.S., Zhang, S.B., Guo, M.M., Chen, C.R., Shen, G.L., and Yu, R.Q. (2007) *Anal. Chim. Acta*, **584**, 122–128.
19. Xiong, D.J. and Li, H.B. (2008) *Nanotechnology*, **19**, 465502.
20. Guan, J., Jiang, L., Li, J., and Yang, W.S. (2008) *J. Phys. Chem. C*, **112**, 3267–3271.
21. Yoosaf, K., Ipe, B.I., Suresh, C.H., and Thomas, K.G. (2007) *J. Phys. Chem. C*, **111**, 12839–12847.
22. Raschke, G., Brogl, S., Susha, A.S., Rogach, A.L., Klar, T.A., Feldmann, J., Fieres, B., Petkov, N., Bein, T., Nichtl, A., and Kurzinger, K. (2004) *Nano Lett.*, **4**, 1853–1857.
23. Huang, Y.F., Lin, Y.W., and Chang, H.T. (2007) *Langmuir*, **23**, 12777–12781.
24. Yu, C.J. and Tseng, W.L. (2008) *Langmuir*, **24**, 12717–12722.

25 Kolb, H.C., Finn, M.G., and Sharpless, K.B. (2001) *Angew. Chem. Int. Ed.*, **40**, 2004.

26 Zhou, Y., Wang, S.X., Zhang, K., and Jiang, X.Y. (2008) *Angew. Chem. Int. Ed.*, **47**, 7454–7456.

27 Lee, S.J., Lee, J.E., Seo, J., Jeong, I.Y., Lee, S.S., and Jung, J.H. (2007) *Adv. Funct. Mater.*, **17**, 3441–3446.

28 Lee, J.S., Han, M.S., and Mirkin, C.A. (2007) *Angew. Chem. Int. Ed.*, **46**, 4093–4096.

29 Lee, J.H., Wang, Z.D., Liu, J.W., and Lu, Y. (2008) *J. Am. Chem. Soc.*, **130**, 14217–14226.

30 Wang, H., Wang, Y.X., Jin, J.Y., and Yang, R.H. (2008) *Anal. Chem.*, **80**, 9021–9033.

31 Lee, J.S. and Mirkin, C.A. (2008) *Anal. Chem.*, **80**, 6805–6808.

32 Liu, J.W. and Lu, Y. (2004) *J. Am. Chem. Soc.*, **126**, 12298–12305.

33 Liu, J.W. and Lu, Y. (2003) *J. Am. Chem. Soc.*, **125**, 6642–6643.

34 Wei, H., Li, B.L., Li, J., Dong, S.J., and Wang, E.K. (2008) *Nanotechnology*, **19**, 095501.

35 Xiao, Y., Rowe, A.A., and Plaxco, K.W. (2007) *J. Am. Chem. Soc.*, **129**, 262–263.

36 Ranyuk, E., Douaihy, C.M., Bessmertnykh, A., Denat, F., Averin, A., Beletskaya, I., and Guilard, R. (2009) *Org. Lett.*, **11**, 987–990.

37 Liu, J. and Lu, Y. (2007) *Chem. Commun.*, **46**, 4872–4874.

38 Liu, J.W., Brown, A.K., Meng, X.L., Cropek, D.M., Istok, J.D., Watson, D.B., and Lu, Y. (2007) *Proc. Natl. Acad. Sci. U. S. A.*, **104**, 2056–2061.

39 Comes, M., Marcos, M.D., Martinez-Manez, R., Sancenon, F., Soto, J., Villaescusa, L.A., Amoros, P., and Beltran, D. (2004) *Adv. Mater.*, **16**, 1783.

40 Comes, M., Rodriguez-Lopez, G., Marcos, M.D., Martinez-Manez, R., Sancenon, F., Soto, J., Villaescusa, L.A., Amoros, P., and Beltran, D. (2005) *Angew. Chem. Int. Ed.*, **44**, 2918–2922.

41 Palomares, E., Vilar, R., and Durrant, J.R. (2004) *Chem. Commun.*, 362–363.

42 Palomares, E., Vilar, R., Green, A., and Durrant, J.R. (2004) *Adv. Funct. Mater.*, **14**, 111–115.

43 Balaji, T., El-Safty, S.A., Matsunaga, H., Hanaoka, T., and Mizukami, F. (2006) *Angew. Chem. Int. Ed.*, **45**, 7202–7208.

44 El-Safty, S.A., Ismail, A.A., Matsunaga, H., Hanaoka, T., and Mizukami, F. (2008) *Adv. Funct. Mater.*, **18**, 1485–1500.

45 El-Safty, S.A., Prabhakaran, D., Ismail, A.A., Matsunaga, H., and Mizukami, F. (2007) *Adv. Funct. Mater.*, **17**, 3731–3745.

46 El-Safty, S.A., Ismail, A.A., Matsunaga, H., and Mizukami, F. (2007) *Chem. Eur. J.*, **13**, 9245–9255.

47 El-Safty, S.A., Prabhakaran, D., Kiyozumi, Y., and Mizukami, F. (2008) *Adv. Funct. Mater.*, **18**, 1739–1750.

48 Veinot, J.G.C. (2006) *Chem. Commun.*, 4160–4168.

49 Nazzal, A.Y., Qu, L.H., Peng, X.G., and Xiao, M. (2003) *Nano Lett.*, **3**, 819–822.

50 Ye, B.C. and Yin, B.C. (2008) *Angew. Chem. Int. Ed.*, **47**, 8386–8389.

51 Huang, C.C. and Chang, H.T. (2006) *Anal. Chem.*, **78**, 8332–8338.

52 Zheng, A.F., Chen, J.L., Wu, G.N., Wei, H.P., He, C.Y., Kai, X.M., Wu, G.H., and Chen, Y.C. (2009) *Mikrochim. Acta*, **164**, 17–27.

53 Wang, J. and Liu, B. (2008) *Chem. Commun.*, 4759–4761.

54 Wang, Z.D., Lee, J.H., and Lu, Y. (2008) *Chem. Commun.*, 6005–6007.

55 Li, H.B. and Qu, F.G. (2007) *Chem. Mater.*, **19**, 4148–4154.

56 Ruedas-Rama, M.J. and Hall, E.A.H. (2009) *Analyst*, **134**, 159–169.

57 Shang, Z.B., Wang, Y., and Jin, W.J. (2009) *Talanta*, **78**, 364–369.

58 Xie, H.Y., Liang, H.G., Zhang, Z.L., Liu, Y., He, Z.K., and Pang, D.W. (2004) *Spectrochim. Acta, Part A*, **60**, 2527–2530.

59 Li, H.B. and Han, C.P. (2008) *Chem. Mater.*, **20**, 6053–6059.

60 Solna, R., Sapelnikova, S., Skladal, P., Winther-Nielsen, M., Carlsson, C., Emneus, J., and Ruzgas, T. (2005) *Talanta*, **65**, 349–357.

61 Bigioni, T.P., Whetten, R.L., and Dag, O. (2000) *J. Phys. Chem. B*, **104**, 6983–6986.

62 Chen, W.B., Tu, X.J., and Guo, X.Q. (2009) *Chem. Commun.*, 1736–1738.

63 Benn, T.M. and Westerhoff, P. (2008) *Environ. Sci. Technol.*, **42**, 4133–4139.
64 Kaegi, R., Ulrich, A., Sinnet, B., Vonbank, R., Wichser, A., Zuleeg, S., Simmler, H., Brunner, S., Vonmont, H., Burkhardt, M., and Boller, M. (2008) *Environ. Pollut.*, **156**, 233–239.
65 Nowack, B. and Bucheli, T.D. (2007) *Environ. Pollut.*, **150**, 5–22.
66 Mackay, C.E., Johns, M., Salatas, J.H., Bessinger, B., and Perri, M. (2006) *Integr. Environ. Assess. Manag.*, **2**, 293–298.
67 Gillespie, D.T. (1976) *J. Comput. Phys.*, **22**, 403–434.
68 Loux, N.T. and Savage, N. (2008) *Water Air Soil Pollut.*, **194**, 227–241.
69 Overbeek, J.T.G. (1952) *Stability of Hydrophobic Colloids and Emulsions*, Elsevier Publishing Co., Amsterdam.
70 Verwey, E.J.W. and Overbeek, J.T.G. (1948) *Theory of the Stability of Lyophobic Colloids*, Dover Publications, New York.
71 Baalousha, M. (2009) *Sci. Total Environ.*, **407**, 2093–2101.
72 Wnek, W.J. and Davies, R. (1977) *J. Colloid Interface Sci.*, **60**, 361–375.
73 Diegoli, S., Manciulea, A.L., Begum, S., Jones, I.P., Lead, J.R., and Preece, J.A. (2008) *Sci. Total Environ.*, **402**, 51–61.
74 Pelley, A.J. and Tufenkji, N. (2008) *J. Colloid Interface Sci.*, **321**, 74–83.
75 Smith, B., Wepasnick, K., Schrote, K.E., Bertele, A.H., Ball, W.P., O'Melia, C., and Fairbrother, D.H. (2009) *Environ. Sci. Technol.*, **43**, 819–825.
76 Jiang, J.K., Oberdorster, G., and Biswas, P. (2009) *J. Nanopart. Res.*, **11**, 77–89.
77 Ghosh, S., Mashayekhi, H., Pan, B., Bhowmik, P., and Xing, B.S. (2008) *Langmuir*, **24**, 12385–12391.
78 Baalousha, M., Manciulea, A., Cumberland, S., Kendall, K., and Lead, J.R. (2008) *Environ. Toxicol. Chem.*, **27**, 1875–1882.
79 Faust, S.D. and Aly, O.M. (1981) *Chemistry of Natural Waters*, Ann Arbor Science Publishers, Ann Arbor.
80 Zhang, X.T., Sato, O., Taguchi, M., Einaga, Y., Murakami, T., and Fujishima, A. (2005) *Chem. Mater.*, **17**, 696–700.
81 Zhang, Y., Chen, Y.S., Westerhoff, P., and Crittenden, J.C. (2008) *Environ. Sci. Technol.*, **42**, 321–325.
82 Franchi, A. and O'Melia, C.R. (2003) *Environ. Sci. Technol.*, **37**, 1122–1129.
83 Domingos, R.F., Tufenkji, N., and Wilkinson, K.J. (2009) *Environ. Sci. Technol.*, **43**, 1282–1286.
84 Navarro, D.A.G., Watson, D.F., Aga, D.S., and Banerjee, S. (2009) *Environ. Sci. Technol.*, **43**, 677–682.
85 Saleh, N.B., Pfefferle, L.D., and Elimelech, M. (2008) *Environ. Sci. Technol.*, **42**, 7963–7969.
86 Boxall, A.B., Tiede, K., and Chaudhry, Q. (2007) *Nanomedicine*, **2**, 919–927.
87 Theng, B.K.G. and Yuan, G.D. (2008) *Elements*, **4**, 395–399.
88 Guzman, K.A.D., Finnegan, M.P., and Banfield, J.F. (2006) *Environ. Sci. Technol.*, **40**, 7688–7693.
89 Lecoanet, H.F., Bottero, J.Y., and Wiesner, M.R. (2004) *Environ. Sci. Technol.*, **38**, 5164–5169.
90 Lecoanet, H.F. and Wiesner, M.R. (2004) *Environ. Sci. Technol.*, **38**, 4377–4382.
91 Buzea, C., Pacheco, I, and Robbie, K. (2007) *Biointerphases*, **2**, MR17–MR71.
92 Wigginton, N.S., Haus, K.L., and Hochella, M.F. (2007) *J. Environ. Monit.*, **9**, 1306–1316.
93 Oberdorster, G., Stone, V., and Donaldson, K. (2007) *Nanotoxicology*, **1**, 2–25.
94 Oberdorster, G., Oberdorster, E., and Oberdorster, J. (2005) *Environ. Health Perspect.*, **113**, 823–839.
95 Handy, R.D., von der Kammer, F., Lead, J.R., Hassellov, M., Owen, R., and Crane, M. (2008) *Ecotoxicology*, **17**, 287–314.
96 Chen, X. and Mao, S.S. (2007) *Chem. Rev.*, **107**, 2891–2959.
97 AshaRani, P.V., Mun, G.L.K., Hande, M.P., and Valiyaveettil, S. (2009) *ACS Nano*, **3**, 279–290.
98 Griffitt, R.J., Luo, J., Gao, J., Bonzongo, J.C., and Barber, D.S. (2008) *Environ. Toxicol. Chem.*, **27**, 1972–1978.
99 Griffitt, R.J., Hyndman, K., Denslow, N.D., and Barber, D.S. (2009) *Toxicol. Sci.*, **107**, 404–415.

100 Carlotti, M.E., Ugazio, E., Sapino, S., Fenoglio, I., Greco, G., and Fubini, B. (2009) *Free Radical Res.*, **43**, 312–322.

101 Federici, G., Shaw, B.J., and Handy, R.D. (2007) *Aquat. Toxicol.*, **84**, 415–430.

102 Jemec, A., Drobne, D., Remskar, M., Sepcic, K., and Tisler, T. (2008) *Environ. Toxicol. Chem.*, **27**, 1904–1914.

103 Adams, L.K., Lyon, D.Y., and Alvarez, P.J.J. (2006) *Water Res.*, **40**, 3527–3532.

104 Velzeboer, I., Hendriks, A.J., Ragas, A.M.J., and Van de Meent, D. (2008) *Environ. Toxicol. Chem.*, **27**, 1942–1947.

105 van Ravenzwaay, B., Landsiedel, R., Fabian, E., Burkhardt, S., Strauss, V., and Ma-Hock, L. (2009) *Toxicol. Lett.*, **186**, 152–159.

106 Kashiwada, S. (2006) *Environ. Health Perspect.*, **114**, 1697–1702.

107 Bouldin, J.L., Ingle, T.M., Sengupta, A., Alexander, R., Hannigan, R.E., and Buchanan, R.A. (2008) *Environ. Toxicol. Chem.*, **27**, 1958–1963.

108 Choi, O., Cleuenger, T.E., Deng, B.L., Surampalli, R.Y., Ross, L., and Hu, Z.Q. (2009) *Water Res.*, **43**, 1879–1886.

109 Warheit, D.B., Webb, T.R., Colvin, V.L., Reed, K.L., and Sayes, C.R. (2007) *Toxicol. Sci.*, **95**, 270–280.

110 Van Hoecke, K., De Schamphelaere, K.A.C., Van der Meeren, P., Lucas, S., and Janssen, C.R. (2008) *Environ. Toxicol. Chem.*, **27**, 1948–1957.

111 Karakoti, A.S., Hench, L.L., and Seal, S. (2006) *JOM*, **58**, 77–82.

112 Limbach, L.K., Wick, P., Manser, P., Grass, R.N., Bruinink, A., and Stark, W.J. (2007) *Environ. Sci. Technol.*, **41**, 4158–4163.

113 Zhang, W.X. (2003) *J. Nanopart. Res.*, **5**, 323–332.

114 Gao, J., Bonzongo, J.C.J., Bitton, G., Li, Y., and Wu, C.Y. (2008) *Environ. Toxicol. Chem.*, **27**, 808–810.

115 Farre, M., Gajda-Schrantz, K., Kantiani, L., and Barcelo, D. (2009) *Anal. Bioanal. Chem.*, **393**, 81–95.

116 Buseck, P.R. and Adachi, K. (2008) *Elements*, **4**, 389–394.

117 Lin, C.C., Chen, S.J., Huang, K.L., Hwang, W.I., Chang-Chien, G.P., and Lin, W.Y. (2005) *Environ. Sci. Technol.*, **39**, 8113–8122.

118 Tiede, K., Boxall, A.B.A., Tear, S.P., Lewis, J., David, H., and Hassellov, M. (2008) *Food Addit. Contam.*, **25**, 795–821.

119 Kulzer, F., Laurens, N., Besser, J., Schmidt, T., Orrit, M., and Spaink, H.P. (2008) *ChemPhysChem*, **9**, 1761–1766.

120 Murdock, R.C., Braydich-Stolle, L., Schrand, A.M., Schlager, J.J., and Hussain, S.M. (2008) *Toxicol. Sci.*, **101**, 239–253.

121 Giddings, J.C. (1993) *Science*, **260**, 1456–1465.

122 Stolpe, B., Hassellov, M., Andersson, K., and Turner, D.R. (2005) *Anal. Chim. Acta*, **535**, 109–121.

123 Domingos, R.F., Baalousha, M., Ju-Nam, Y., Reid, M.M., Tufenkji, N., Lead, J.R., Leppard, G.G., and Wilkinson, K.J. (2009) *Environ. Sci. Technol.* **43**, 7277–7284.

124 Gulson, B. and Wong, H. (2006) *Environ. Health Perspect.*, **114**, 1486–1488.

125 Oughton, D.H., Hertel-Aas, T., Pellicer, E., Mendoza, E., and Joner, E.J. (2008) *Environ. Toxicol. Chem.*, **27**, 1883–1887.

126 Mueller, N.C. and Nowack, B. (2008) *Environ. Sci. Technol.*, **42**, 4447–4453.

Part 3
Advanced Methods and Materials

Trace Analysis with Nanomaterials. Edited by David T. Pierce and Julia Xiaojun Zhao
Copyright © 2010 WILEY-VCH Verlag GmbH & Co. KGaA, Weinheim
ISBN: 978-3-527-32350-0

11
Electroanalytical Measurements at Electrodes Modified with Metal Nanoparticles
James A. Cox and Shouzhong Zou

11.1
Introduction

Modification of electrode surfaces with nanoparticles is a rapidly growing approach to electroanalytical measurements, as demonstrated by the scope of representative recent reviews [1–4]. Among the primary factors that motivate these investigations are enhancement of electron-transfer rates, favorable geometry (e.g., area-to-volume ratio) over bulk materials, and compatibility with design of nanoarrays. Regarding the last point, if a given electron-transfer reaction does not occur at the bulk electrode and is catalyzed by nanoparticles (NPs) immobilized thereon, the modified electrode can have characteristics of a nano-electrode array. The influence of mass transport on the limiting current at such an array will depend on the size and spacing of the NPs. An electrochemical reaction at the NPs will result in a layer, δ, around each NP that has a lower concentration of analyte than the bulk solution; δ is termed the "diffusion layer." Where D is the diffusion coefficient ($cm^2 \cdot s^{-1}$) and t (s) is the time during which the electrochemical reaction occurs, the thickness of the diffusion layer is given by $\delta \approx (Dt)^{1/2}$. If $\delta > r$, the radius of the NP, the resulting hemispherical diffusion layer will have a constant δ value, so a steady-state current, i_{ss}, will be developed. The ratio of the faradaic current to the background (capacitive) current in a potentiodynamic experiment such as cyclic voltammetry is increased under these conditions relative to that achieved under linear diffusion, where δ increases with time [5, 6]. Compton and coworkers [7, 8] have investigated these two-dimensional nanoarrays by digital simulation with the spacing and the geometry of the particles as variables. With low particle density and a high degree of order, electrochemical experiments can be performed so that the diffusion layers around each NP do not overlap. With non-overlapping δ, $i_{ss} - jnFDrC_A$, where n is the number of electrons transferred, F is the Faraday constant, C_A is the analyte concentration, and j is a proportionality constant. In addition to providing an improved ratio of faradaic to capacitive current, the small r relative to diffusion distances results in less tendency for these nanoscale electron-transfer moieties to become passivated by adsorption of oxidation or reduction products [9].

Trace Analysis with Nanomaterials. Edited by David T. Pierce and Julia Xiaojun Zhao
Copyright © 2010 WILEY-VCH Verlag GmbH & Co. KGaA, Weinheim
ISBN: 978-3-527-32350-0

This chapter is focused on the synthesis of metal NPs, their incorporation with electrodes, and their use in electrochemical trace analysis. Fundamental studies of electrocatalysis at metal nanoparticles benefit from using systems that are well defined in terms of spatial distribution and particle size. Control of these factors in the fabrication of NP-modified electrodes is discussed in Section 11.2. In addition, the shape and composition of metal NPs can influence rates of electron-transfer reactions at their surfaces, as discussed in Section 11.3. Representative analytical applications are summarized in Section 11.4. The NP-modified electrodes used for analytical purposes include both two- and three-dimensional dispersions. In many cases the metal NP is chemically modified with a monolayer sheath that can simply block aggregation of the NPs but also can act as a catalysis and assist electrode modification, for example, by serving as a tether that connects the NP to the electrode. Throughout this chapter, when a specific metal (e.g., gold) comprises the NP, the abbreviation will be compounded with the symbol of the element (i.e., AuNP).

11.2
Modification of Electrodes with Nanoparticles

Methods for the modification of electrodes with metal NPs depend to some extent on whether the NP is bare or monolayer-protected and on whether the targeted application benefits from two- or three-dimensional geometry. The subsections below describe the interaction among the method of synthesis, the approach to attachment, and the distribution of the NPs in the modified electrode.

11.2.1
Fabrication of Two-Dimensional Arrays of Nanoparticles

Two-dimensional arrays of NPs on electrodes have been fabricated by assembly of NPs from solutions containing colloidal metals or monolayer-protected NPs as well as by directly synthesizing the NPs on the surface by chemical or electrochemical reduction of an adsorbed metal complex. An important variation of the latter is seed-mediated synthesis, in which the nucleation and growth steps are performed sequentially and under different conditions. The electrode surface on which the fabrication occurs can be bare, modified by a monolayer of an agent that facilitates the assembly process, or patterned with a material to which neither NPs nor their precursor complexes adhere. These approaches yield very different NP-modified electrodes in terms of organization of the NPs, the size distribution of the particles, and nature of the NP surface. In some cases the patterning agent or the tether is removed prior to electrochemical application of the NP deposits. Examples of this wide variety of fabrication methods are described below.

11.2.1.1 Seed-Mediated Formation of a Two-Dimensional Array of Nanoparticles
Seed-mediated growth is a chemical method of preparing metal nanoparticles and nanorods on electrode surfaces. Initially, the method was developed primarily as

a means of growing nanorods in aqueous solution [10–12]. In the first step, silver [11] and gold [12] spherical NPs with a 4 ± 2 nm diameter were prepared in aqueous solution by reduction of $AgNO_3$ or $HAuCl_4$ with $NaBH_4$ in the presence of citrate, which served as a capping agent. These NPs were the seeds that were expanded in a growth solution that contained cetyltrimethylammonium bromide (CTAB) in addition to metal ion and ascorbic acid. Controlled quantities of seed solution were added, and the pH was changed by adding NaOH to initiate nanorod growth. The CTAB provided micelles that served as growth templates ([13] and citations therein). An adaptation of the method was used to synthesize AuNPs on an indium tin oxide (ITO) electrode [14, 15]. By utilizing defect sites for the adsorption of $HAuCl_4$, the seed NPs were formed directly on the electrode surface using $NaBH_4$ as the reducing agent. The platform was transferred to the growth medium described above. Field-emission scanning electron microscopy (FE-SEM) imaging showed that the seed AuNPs were 4 ± 1 nm in size; after 15 min and 24 h in the growth medium, the size was increased to 14 ± 2 nm and 22 ± 2 nm, respectively. This method did not yield an array with controlled interparticle spacing.

Seed-mediated methods that employ direct electrochemical deposition of nanoparticles on bare electrodes by double potential-pulse voltammetry have been developed. The initial pulse, which is well beyond the formal potential of the metal deposition half-reaction, is used to generate nuclei, and the second pulse, which is at a significantly lower overpotential, controls the growth of the NPs. Penner and coworkers have developed a model [16] and experimentally demonstrated [17] that the size dispersion of AgNPs and AuNPs deposited by the double-pulse method correlates to whether the diffusion fields around the particles during growth are coupled (overlap). Uncoupled diffusion-controlled growth is favored by a low density of nuclei; it results in less size dispersion. Moreover, for AgNPs with a low density of nuclei ($<10^7$ centers cm^{-2}) on pyrolytic graphite electrodes, the final particle size is larger than observed with a high density of nuclei ($>10^9$ centers cm^{-2}). This observation is consistent with the prediction that interparticle diffusion coupling lowers the growth rate [16, 17]. Typical experimental conditions with a solution consisting of 1 mM $AgClO_4$, 0.1 M $LiClO_4$ in acetonitrile were the following: nucleation, -500 mV versus $Ag^+|Ag$ for 5 ms; growth, -70 mV versus $Ag^+|Ag$ for 0.05–120 s. With the shorter deposition time, the AgNPs were in the range 42–100 nm [17].

Studies of the double-pulse method by Plieth and coworkers [18, 19] related the size dispersion to the value of the applied potential relative to the critical potential for generating nuclei. A linear relationship between the number of nuclei generated and the overpotential was observed with an ITO electrode. Because of aggregation, increasing the time at the growth overpotential decreased the number of particles when the nucleation step yielded a high density of centers. With a low density of nuclei, increasing the time at the growth overpotential did not change the number of particles, which demonstrated that nuclei were not formed under low overpotential (growth) conditions of the double-pulse method. During the time of application of the growth overpotential, nuclei below a critical radius dissolved. This phenomenon led to a narrower size distribution in the final array.

Potentiostatic deposition compromises both the order of the two-dimensional array and size control of NPs relative to the double-pulse method; however, it has been used to fabricate arrays suited for electroanalytical determinations. AuNPs were deposited on ITO from $AuCl_4^-$ in 0.5 M H_2SO_4 at −0.045 V versus SCE [20]. For a fixed deposition time, the size of the AuNPs increases with $AuCl_4^-$ concentration over the investigated range 0.01–1.0 mM, and at a fixed concentration, 0.1 mM, the size increases with deposition time over the range, 50–300 s. With 0.1 mM $AuCl_4^-$, a deposition time of 300 s yielded 70–309 nm (average diameter, 154 nm) AuNPs. The size distribution was narrower when several deposition pulses of shorter duration were employed. Under comparable conditions but with a glassy carbon electrode, smaller AuNPs were obtained [21]; here, with 1 mM $AuCl_4^-$ in 0.5 M H_2SO_4 a 300-s deposition at −0.27 V versus Ag|AgCl yielded AuNPs with an average size of 50 nm.

As suggested in Section 11.1, control of spatial distribution of two-dimensional arrays of NPs is important for certain applications. To address this objective, a method has been developed that involves formation of mixed monolayers of generation-four polyamidoamine (PAMAM) dendrimers on ITO electrodes; specifically, monolayers were assembled from mixtures of metal-free and NP-containing PAMAM [22]. NP-PAMAM was synthesized in accord with methods reported by Crooks and coworkers [23, 24]. First, $PtCl_4^{2-}$ or $AuCl_4^-$ were complexed with hydroxyl-terminated PAMAM via the protonated amine sites in its interior. Upon reduction with $NaBH_4$, the metal ions were converted into NPs in the size domain <4 nm. The NP-PAMAM was separated from the reaction vessel and subsequently mixed with metal-free PAMAM to yield a solution where the mole ratio, x, of NP-PAMAM relative to total PAMAM was controlled. An ITO electrode was modified with a monolayer of aminopropyltrimethoxysilane (APTES). Upon immersion in the mixture, the protonated amine sites of APTES electrostatically assembled the hydroxy-terminated dendrimers to the electrode, yielding ITO|APTES| [(NP-PAMAM)$_x$ + (PAMAM)$_{1-x}$]. The organic components were ashed at 350 °C, a temperature determined by thermogravimetric analysis. The resulting array of AuNP had a characteristic separation between centers of 200 nm when $x = 0.06$. Atomic force microscopy showed that the interparticle distance varied with x.

When the objective is the fabrication of a highly ordered array of NPs, the method based on the mole fraction of NP-PAMAM provides less control than the use of a template with nanoscale voids on the surface of the electrode. An example of such a template is a micellar block copolymer film based on poly(styrene)-*block*-poly(2-vinylpyridine), PS-*b*-P2VP [25, 26]. The PS block is preferentially dissolved in toluene. As a result, in that solvent the PS-*b*-P2VP organizes into micelles [25]. This system has a glass transition temperature above 100 °C, so a film on an electrode can be formed and dried without changing the structure of the micelles. Moreover, attaching $AuCl_4^-$ to the pyridine units and chemically reducing the complex to AuNPs affords highly ordered two-dimensional arrays, which are stable even at elevated temperature [26]. For example, 10-nm (height) AuNPs on SiO_2 were not altered even by a 12-h treatment at 800 °C. The size of the micelles and, hence, of the resulting NPs can be varied via the lengths of the polymer chains.

11.2 Modification of Electrodes with Nanoparticles

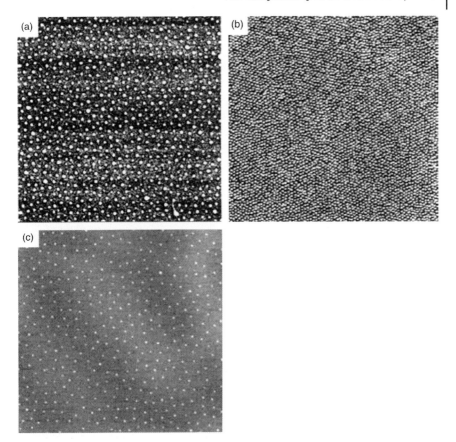

Figure 11.1 Scanning force microscopy images of arrays of gold nanoparticles formed in micellar templates assembled with a diblock polymer, poly(styrene)-*block*-poly(2-vinylpyridine), PS-*b*-P2VP, in toluene. The PS-*b*-P2VP was removed with an oxygen plasma prior to imaging. Images (a)–(c) result from changing polymer chain lengths using PS(800)-*b*-P2VP(860), PS(325)-*b*-P2VP(75), and PS(1700)-*b*-P2VP(450), respectively. Each was loaded with $HAuCl_4$ that was reduced to Au prior to forming the film on mica. Reprinted with permission from Reference [25]. Copyright 2000 The American Chemical Society.

Controlled variation of the interparticle distance and the size of the AuNPs are exemplified by results in Figure 11.1 [25] and Table 11.1 [26], respectively.

11.2.1.2 Direct Deposition of Nanoparticles on Bare Electrodes

A common method of modifying electrodes employs an initial conversion of the bare surface with an alkyl organic compound terminated on one end with a species that reacts with the electrode material and, on the other, by a charged functionality. For example, gold reacts with thiol groups, so in the presence of hexanethiol, R-SH, formation of AuSR results in monolayer coverage of gold [27]. When the

Table 11.1 Metal/polymer ratio used in the synthesis, and particle size and interparticle distance data from TEM images. Reprinted with permission from Reference [26]. Copyright 2009 The American Chemical Society.

Polymer template	HAuCl$_4$/P2VP mole ratio	Particle size[a] (nm)	Interparticle distance[a] (nm)
PS(23.6)-b-P2VP(10.4)	0.1	2.4 ± 1.0	26.8 ± 3.0
PS(23.6)-b-P2VP(10.4)	0.2	3.1 ± 0.5	28.3 ± 3.7
PS(25)-b-P2VP(15)	0.2	4.2 ± 0.4	28.5 ± 3.8
PS(25)-b-P2VP(15)	0.4	6.1 ± 0.4	33.1 ± 4.0
PS(25.5)-b-P2VP(23.5)	0.2	7.2 ± 0.5	38.3 ± 5.7
PS(25.5)-b-P2VP(23.5)	0.4	9.0 ± 0.9	36.3 ± 6.6
PS(81)-b-P2VP(14.2)	0.2	4.3 ± 0.4	60.0 ± 5.4
PS(188)-b-P2VP(16)	0.2	4.3 ± 0.6	80.0 ± 10.0

a) Average value of at least 200 measurements for the particle size and 100 measurements for the interparticle distance of each particle array, reported with the corresponding standard deviation.

terminal group carries a charge, such as when the monolayer is mercaptoundecanoic acid, the surface contains sites that can electrostatically assemble a second layer, including NPs with charged protecting groups.

Willner and coworkers [28] were among the first to systematically investigate and characterize electrodes prepared by tethering metal NPs to the surface. ITO electrodes were modified with monolayers of (aminopropyl)siloxane and of (mercaptopropyl)siloxane. Gold colloids of various sizes were bound to the modified ITO via the amine or the thiol functionality. The surface coverage was determined by further reaction of the AuNPs with a reversibly redox-active bipyridinium monolayer. The electrode modified with the amino functionality yielded a surface coverage of 0.5×10^{-10} mol cm^{-2}, whereas that modified with the thiol functionality had a surface coverage of 0.02×10^{-10} mol cm^{-2}. When 25-nm AuNPs were used in conjunction with (aminopropyl)siloxane, atomic force microscopy revealed a virtually continuous coverage of the electrode by AuNPs. In contrast, AuNPs at sizes up to 120 nm showed discontinuities in the coverage, perhaps a reflection of imperfections in the ITO that were smoothed by the former system. Overall, the coverage by gold was 27% higher with the aminopropyl than with the mercaptopropyl tether.

With ITO as the base electrode, a common modification is by a monolayer of APTES or the methoxy analog, APTMS. Cheng et al. [29] have used this approach to put positive ion-exchange sites on ITO for the subsequent assembly of negatively charged, citrate-protected particles (AuNP-Cit) [30]. The residual negative charge of the assembly, ITO|APTMS|AuNP-Cit, served as a base for growth of a three-dimensional layer-by-layer electrostatic assembly [29]. This method produced a film of a rather uniform dispersion of AuNPs (Figure 11.2). When ITO|APTMS|AuNP-Cit was thermally treated at 600 °C for 5 h, the particles increased in size and were

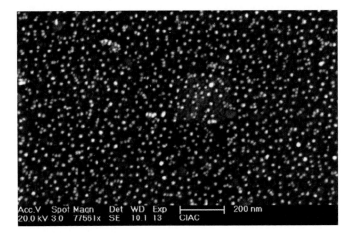

Figure 11.2 Field-emission scanning electron microscopy image of citrate-protected AuNPs assembled on an indium tin oxide (ITO) electrode modified with a monolayer of (3-aminopropyl)trimethoxysilane (APTMS). Reprinted with permission from Reference [29]. Copyright 2004 The American Chemical Society.

more dispersed than the surface shown in Figure 11.2 [29]. The data suggest surface diffusion of the NPs. The process is apparently dependent on the NP size and on the substrate. For example, single, 4-nm nanocrystals of CdSe on mica and silicon oxide do not show evidence of motion when heated on the order of hours at 120 °C; however, after 4 min at 90 °C, aggregates containing 80–100 of the 4-nm particles on graphite can move about 500 nm [31]. A study on the diffusion of gold nanorods that were coated with a monolayer of 2-mercaptomethanesulfonic acid on a surface modified with 2-mercaptoethylamine (MEA) showed a pH dependence on the motion [32]. When the MEA was protonated, no diffusion was observed over a time scale of hundreds of seconds.

Cysteine (cys) has been used to assemble AuNPs from a colloidal suspension to glassy carbon electrodes (GCEs) [33]. The hypothesis was that the amine site interacted with the electrode surface, and the thiols were free to react with AuNPs. X-Ray photoelectron spectroscopy and SEM were used to characterize the modified electrode. The results verified the formation of the Au–S bond and demonstrated that AuNPs were not immobilized on GCEs in the absence of cys. GCE|cys-AuNP was stable during 30 min of sonication and promoted the reproducible oxidation of uric acid over an 8-week period when stored in 0.1 M phosphate buffer at pH 7. With cys as the tether for AuNPs it is important to limit the positive potential excursion because at about 1.0 V, depending on the pH, AuNPs catalyze the oxidation of cysteine.

Related to tethering NPs to the surface is direct adsorption of monolayer-protected NPs. The process is illustrated with the polyoxometalate $PMo_{12}O_{40}^{3-}$ (PMo_{12}). This polyanion interacts with Au so strongly that it undergoes a place-exchange reaction with the hexanethiol (RSH) protecting group on AuNPs [34].

With AuNP-RSH in organic solvent (prepared by the well-characterized Brust–Schiffrin method [35, 36]) and the PMo_{12} in aqueous solution, liquid–liquid extraction in conjunction with the place-exchange reaction converts AuNP-SR$_{(org)}$ into AuNP-PMo$_{12(aq)}$ without a change in the core size [34]. This approach to obtaining AuNPs in aqueous solution permits the preparation of NPs of controlled size in the range below 10 nm. In this case, the AuNP–PMo$_{12}$ size was 4.4 ± 1.8 nm. Strong interaction between PMo_{12} and various electrode materials, including glassy carbon (GC), results in adsorption of AuNP–PMo$_{12}$ to a GCE. An attractive feature of GCE|AuNP–PMo$_{12}$ is that both the NP and the linking group are catalytically active. Evidence of synergism between catalysis by AuNP centers and PMo_{12} in the reduction of bromate was obtained [34]. The voltammetric peak current for the reduction of bromate at GCE|AuNP–PMo$_{12}$ was 2.2-times greater than that at GCE|PMo$_{12}$, and the overpotential for the reduction of bromate was lower at the former electrode.

A second variation of the tethering approach is to co-deposit either a polymer and NP or a polymer and precursor to a NP (such as $AuCl_4^-$) from a mixture. A typical example is the co-deposition of poly(4-aminothiophenol), PATP, and AuNPs on a GCE [37]. Initially, an inclusion complex of 4-aminothiophenyl (ATP) and β-cyclodextrin (CD) in dimethylformamide (DMF) was prepared. A film was deposited on the GCE by placing μl-quantities on the surface and evaporating the DMF. The GCE|ATP, CD was put into 0.5 M H_2SO_4 that contained 0.2 mM $HAuCl_4$. Cyclic voltammetry between 1.0 and −0.1 V versus SCE (100 scans, 50 mV s^{-1}) resulted in the simultaneous formation of PATP and AuNPs that had an average size in the range 8–10 nm. Cyclic voltammetry (Figure 11.3) and field-emission transmission electron microscopy (FETEM) supported the prediction that a PATP-AuNP film forms and grows. The FETEM image showed that the AuNPs are well dispersed in the polymer. The data in Figure 11.3 were interpreted as the adsorption with subsequent reduction of $AuCl_4^-$ on the first negative-going scan, followed by oxidative polymerization of ATP on the initial positive-going scan. The increased current with subsequent scans was attributed to increased formation of PATP. The role of the CD in the performance of the GC|PATP, CD, AuNP was not stated, but its structure can impart conductivity, even to thin films of an insulation material on an electrode [38].

Finally, two-dimensional deposits have been obtained by cooperative binding of a mixture of monolayer-protected NPs of opposite charge [39]. Using Si that was oxidized with an air plasma as the base and pH 7 solution as the medium, AuNPs with a protecting layer of mercaptoundecanoic acid (negatively charged) and AgNPs with a protecting layer of N,N,N-trimethyl(11-mercaptoundecyl)ammonium chloride (positively charged) were adsorbed in various combinations. Immersion of the oxidized Si in a solution containing only the negatively charged AuNPs did not give evidence of NPs on the surface. The analogous experiment with positively charged AgNPs showed only widely scattered NPs on the surface. However, when a mixture was used, the surface showed a dense surface coverage of NPs. With the mixture of NPs, changing the pH to either 4 or 10 resulted in a much lower surface coverage, which is consist with the need to have opposite charges

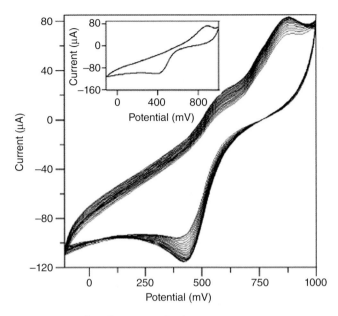

Figure 11.3 Cyclic voltammetry of a glassy carbon electrode modified with a film of an inclusion complex of 4-aminothiophenol and β-cyclodextrin in 0.2 mM HAuCl$_4$, 0.5 M H$_2$SO$_4$. Scan rate, 50 mV s^{-1}; scan range, 1.0 to −0.1 V versus SCE. Inset: first scan. Reprinted with permission from Reference [37]. Copyright 2007 Elsevier.

for the cooperative deposition. The method was extended to deposition of multilayers through repeated immersions; here, notably, multilayers were not formed unless the prior assembly (including the case with only a single layer) was rinsed and dried before next immersion. The assemblies achieved were not applied to electrochemistry, but the method has promise for achieving multifunctional nanoparticle-modified electrodes.

11.2.2
Deposition of Three-Dimensional Films Containing Metal Nanoparticles

Three-dimensional arrays of nanoparticles are of interest because they add to the number of catalytic centers relative to a monolayer (or sub-monolayer) coverage film on an electrode. The use of multilayer assemblies also can yield multifunctional films on electrode surfaces. The challenge in designing three-dimensional composites is to obtain facile mass and charge transport to the electrode. Several methods have been used to fabricate such composites. Among the most important are layer-by-layer (LbL) electrostatic assembly and electrochemical deposition of a conducting polymer containing dispersed nanoparticles. Other methods such as deposition of networks of conducting components generally are variations of these basic approaches.

11.2.2.1 Layer-by-Layer Electrostatic Assemblies Containing Metal Nanoparticles

The deposition of multilayers of oppositely charged polyanions was initially described in detail by Decher [40]. Electrostatic interaction was favored as the driving force because of the relative independence of steric factors in bonding. The assembly process involves immersion of the substrates in alternating solutions of polyelectrolytes of opposite charge with careful rinsing (and often drying) between steps. This process not only prevents contamination but also stabilizes weakly adsorbed layers. The repulsion effect between adjacent layers tends to result in single monolayers deposited in a given cycle. The outer layer after each step carries extra charges, so the deposition sequence is planned to give the system net anionic or cationic characteristics. The structure of LbL assemblies is typically depicted as a spatially well defined set of monolayers but is actually highly intercalated or "fuzzy" [40]. The intercalation can be a factor in obtaining high conductivity.

Various methods for incorporating metal nanoparticles into LbL assemblies have been described. Gold nanoparticles functionalized with mixed protecting layers consisting of 2-(12-mercaptododecyloxy)methyl-15-crown-5 ether and citrate (AuNP-CE/Cit) have been synthesized and assembled on an ITO electrode modified with APTES [41]. Subsequently, alternating layers of PAMAM and AuNP-CE/Cit were deposited. Quartz crystal microbalance measurements were used to show that single monolayers were assembled in each step. The ability to electrostatically fabricate PAMAM-containing films in combination with the synthesis of PAMAM-encapsulated metal nanoparticles [42] provides a convenient route for incorporation of NPs in LbL assemblies. For example, the LbL assembly of AuNPs encapsulated in PAMAM and myoglobin (Mb) on a pyrolytic graphite electrode (PGE) resulted in a surface at which the direct electrochemistry of Mb was observed (Figure 11.4) [43]. Linear assembly of 1–8 bilayers, n, of (PAMAM-Au|Mb)$_n$ was established by UV/visible spectrophotometry of the heme group. An important result was that compared to an analogous assembly, but with citrate-protected

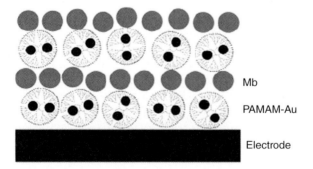

Figure 11.4 Layer-by-layer electrostatic assembly of gold nanoparticles encapsulated in generation-six polyamidoamine (PAMAM-Au) and of myoglobin (Mb). Reprinted with permission from Reference [43]. Copyright 2007 The American Chemical Society.

AuNPs [44], the dendrimer-encapsulated system showed significantly greater stability and resisted agglomeration (no agglomeration after 12 adsorption cycles versus a limit of 2–3 cycles with the citrate-protected system). The PGE|PAMAM-Au|Mb)$_n$ assembly was used to monitor H_2O_2 at the 10-µM level by amperometry at −0.1 V versus SCE at pH 7.0.

11.2.2.2 Fabrication of Conducting Polymer Films Doped with Metal Nanoparticles

Modifying electrodes by electrochemical deposition of films of conducting polymers is a well-established approach to fabricating platforms for such diverse uses as fundamental studies of electron transfer and biosensing. A logical extension of this methodology is to incorporate metal nanoparticles, particularly to improve conductivity and to serve as electrochemical catalysts. Regarding the latter, polyaniline (PANI) has been employed extensively as a film on electrodes; however, its utility for electroanalytical studies is limited by the need for an acidic environment (pH <3) to display voltammetric redox activity [45]. One method to impart conductivity at higher pH is to use PANI in a composite with poly(styrenesulfonate) (PSS) [46]. A direct comparison was made between charge transport across PANI/PSS and PANI/AuNP composites on gold electrodes in pH 7.5 phosphate (0.1 M) buffers [47]; the respective film thicknesses were 90 ± 5 and 95 ± 5 nm. In PANI/AuNP the nanoparticles occupied 25% of the film volume. The 4-nm AuNPs were functionalized with 2-mercaptoethane sulfonic acid via attachment of the thiol to the gold, thereby resulting in negatively charged particles. The PANI/PSS and PANI/AuNP films were deposited by voltammetry (100 cycles between −0.1 and 0.9 V versus SCE at 100 mV s^{-1}). The redox of PANI was monitored by potential-step chronoamperometry. Analysis of the current–time curves demonstrated that linear diffusion limited the current. The rate of charge propagation in the PANI/AuNP film was about 250-times greater than that with PANI/PSS.

A variation of the general method of occlusion of NPs during the electrochemical deposition of a conducting polymer was demonstrated for the poly(3,4-ethylenedioxythiophene) (PEDOT) system. PEDOT has gained attention as a conducting polymer for electrode modification because its conductivity is maintained at neutral pH. As in the above case with PANI, the deposition of PEDOT films on electrodes by continuous cyclic voltammetry of 3,4-ethylenedioxythiophene (EDOT) is an established method [48]. Inclusion of EDOT-protected AuNPs in the precursor solution resulted in electrochemical deposition of a composite of PEDOT and AuNPs (10 ± 0.1 nm based on TEM with a sampled area containing 30 particles) [49], but the nanoparticles were bound to the PEDOT structure rather than occluded therein. This deposition was from acetonitrile with the potential scanned at 30 mV s^{-1} between −0.5 and 1.9 V versus a Ag electrode.

Analogous to the deposition of metal NPs on bare electrodes (described in Section 11.2.2.1), electrodes modified with conducting polymers can serve as the base for electrochemical nucleation and growth of metal deposits, including nanoparticles. An example is to first electrochemically modify a carbon substrate with PEDOT and then use potentiostatic conditions to reduce a metal complex to the corresponding NPs [50]. With 2.0 mM H_2PtCl_6 in 0.1 M H_2SO_4 as the precursor solution, electrolysis at 0.1 V versus SCE deposited 2–3 nm PtNPs on a PEDOT-modified electrode. Although the NPs tended to aggregate in 30–50 nm domains on PEDOT, deposition on bare carbon significantly increased both the size of the aggregates and of the individual PtNPs. Similar results were reported for the electrochemical deposition of mixed Pt–Ru nanoparticles [51] and of PtNPs [52] onto PANI.

An extension of the systems described in Section 11.2.2.1 is to fabricate networks of multiple components such as mixed polymers, metal NPs, and one or more other catalysts as films on electrodes. Kulesza et al. [53] have described the fabrication of a composite of PANI, PtNPs, and PMo_{12} on glassy carbon and Pt electrodes. The strong adsorption of PMo_{12} onto Pt was employed to convert commercially available Pt microclusters into PtNPs by preparing a suspension of the former in an acidic PMo_{12} solution. The product was 5–10 nm PMo_{12}-protected PtNPs. The network was formed by methodology analogous to LbL electrostatic assembly. The electrode was sequentially immersed in PtNP–PMo_{12} and 0.07 M aniline, 0.5 M H_2SO_4. After a 10-min immersion in the latter, cyclic voltammetry (50 mV s^{-1}) between −0.1 and 0.85 V versus Ag|AgCl formed PANI by oxidation. Up to ten bilayers of PtNP–PMo_{12}|PANI were formed in this manner. The system formed a network because the PtNP–PMo_{12} was linked to the PANI.

11.3
Geometric Factors in Electrocatalysis by Nanoparticles

In most applications of nanoparticles to electroanalytical measurements, the identity of the core metal, the presence/absence of the protecting group (if any), the size, and in some cases the interparticle spacing are the only structural considerations regarding the NP. This section discusses the roles of NP size and of the identity of the core metal(s). The influence of NP shape on the electrocatalytic behavior is also considered. Although these variables are of concern mainly in the design of electrodes for fuel cells at present, they are predicted to emerge as important factors in future electroanalytical studies.

11.3.1
Particle Size Effects on Electrocatalysis

Electrochemical studies on the influence of size of NPs on catalysis have focused primarily on the oxygen reduction reaction (ORR) and CO oxidation on PtNPs. Early reports on Pt particle and grain size effects with the ORR as the test system gave contradictory results. For example, Bregoli [54], Sattler and Ross [55], and Peuckert et al. [56] have concluded that the activity decreased with decreasing particle size, but Bett et al. [57] reported that there is no correlation between particle size and O_2 reduction activity. In a study of the ORR on Pt thin films with nanoscale grains, Poirier and Stoner reported that both the specific activity (current density per cm² Pt at 0.9 V versus the reversible hydrogen electrode, RHE) and the mass activity (current density per mg Pt at 0.9 V versus RHE) decreased with increasing grain size [58]. That the apparent specific activity does not depend on the particle size but rather on the interparticle distance was suggested by Watanabe et al. [59]. When the interparticle distance was greater than ca 20 nm, the specific activity of Pt particles is maximized due to the absence of overlap of diffusion fields defined in Section 11.1. Kinoshita attributed the particle size dependence of O_2 reduction activity to the change of relative fraction of Pt surface atoms on (111) and (100) facets as the particle size decreases [60, 61]. However, this explanation is based upon the assumption that the particles have a cubo-octahedral shape [62]; as discussed in Section 11.3.2, this is not true in all cases.

More recent results have confirmed that the specific activity of PtNPs as catalysts of the ORR decreases with decreasing particle size [63, 64] The size-dependent activity was explained in terms of the stronger oxygen–Pt interaction of the smaller particles [63, 64]. Of particular interest is the measurement of the potential of zero total charge (PZTC). By using the CO replacement method, Mayrhofer et al. [63] have examined the PZTC of PtNPs of four different sizes, 1, 2, 5, and 30 nm. The PZTC was found to decrease with particle size, from 0.285 V (versus RHE) for 30-nm PtNPs to 0.245 V for 1-nm PtNPs. The decrease of PZTC increases the OH adsorption. The adsorbed OH occupies the surface Pt sites and, therefore, decreases the active sites for the ORR.

The negative shift of PZTC agrees with a model proposed by Plieth [65]. From thermodynamic considerations, Plieth showed that the redox potential and the potential of zero charge (PZC) for metal particles in the 1–10 nm range shifts in the negative direction inversely proportional to the particle radius:

$$\Delta E = -\frac{2 f V_M}{zF} \frac{1}{r} \tag{11.1}$$

where

ΔE is the equilibrium redox potential difference between the particle and bulk metal electrode,
f is the surface energy of the particles,
V_M is the molar volume of the metal,
z is the charge of the metal ion,
r is the radius of the particle.

Figure 11.5 shows an example based on Plieth's model for PtNPs. A decrease in potential greater than 100 mV is shown for PtNPs <5 nm. The prediction is larger than the value measured by Mayrhofer et al. [63], mainly due to the uncertainty in the surface energy. Corroborating the general prediction of Equation (11.1) are studies of AgNPs supported on a surface. Electrochemical Ostwald ripening resulted in oxidation and dissolution of small particles in conjunction with re-deposition of Ag^+ onto larger particles [66].

The dependence of CO oxidation on particle size has also been reported. Stimming and coworkers have studied the oxidation of irreversibly adsorbed CO (CO_{ads})

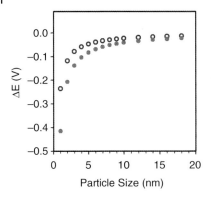

Figure 11.5 Pt nanoparticle redox potential shift as a function of particle size for the reaction Pt → Pt^{2+} + $2e^-$. ΔE is the potential difference between the particle and the bulk. Two particle surface energies were used: empty circle, 2.5 N m^{-1}; filled circle, 4.4 N m^{-1}.

on 1–4 nm PtNPs [67–69]. The studies showed that CO oxidation occurs at more positive potentials as the particles become smaller; that is, smaller particles are less active for the reaction. This was attributed to the stronger binding and hence less mobility of CO on smaller particle surfaces [67, 68]. Markovic and coworkers have examined CO oxidation on 1–30 nm PtNPs that were pretreated by two different methods [70–72]. The first was "oxide annealing," where the particles were subjected to potential cycling between 0.05 and 1.2 V (versus RHE) in Ar-saturated 0.1 M $HClO_4$ until a stable voltammogram was obtained. This procedure created defect sites on the particle surface. The second was "CO annealing," where the potential cycling was conducted in CO-saturated 0.1 M $HClO_4$ for 5 min. CO annealing removed the surface defects [72]. On the particles treated by oxide annealing, a negative shift in the potential for the oxidation of CO_{ads} was observed as the particle size increased, whereas the CO-annealed PtNPs did not show size dependence for particles smaller than 5 nm. Consistent with these results, the oxidation product, CO_2, probed by infrared spectroscopy, appeared at lower potentials as the particle size was increased with oxide-annealed particles, but the appearance potential was not size dependent with CO-annealed particles [70, 71]. On the basis of these observations, it was concluded that a size-dependent potential for the oxidation of CO_{ads} was related to the number of defect sites rather than the CO–Pt bond strength. In terms of reaction pathway, the CO oxidant appears to be adsorbed OH, (OH_{ads}), which is the species formed at the defect sites. Notably, the oxidation of dissolved CO showed an opposite particle size dependence [70, 71]. This was demonstrated by voltammetry under conditions where the initial potential was greater than 0.9 V, the surface was populated by OH_{ads}, and the solution was CO-saturated 0.1 M $HClO_4$. As the potential is scanned toward negative values, the amount of OH_{ads} decreases. In this case, the reaction rate is determined by the ability to sustain OH_{ads}. The 1-nm PtNPs promote an appreciable CO oxidation current at 0.6 V, whereas the 30-nm PtNPs are inactive in terms of CO oxidation at this potential. This observation corroborates the interpretation of the PZTC shift discussed above [63].

Electrocatalytic oxidation of methanol, formic acid, and formaldehyde on PtNPs of different sizes has been examined [73]. The rate of methanol oxidation decreases with decreasing particle size, but the formic acid oxidation rate increases with decreasing particle size. These results were explained by the ensemble effect [73]. For methanol oxidation, the cleavage of C–H bonds requires an ensemble of active sites. As the PtNP size decreases, the availability of such sites decreases. Consequently, methanol oxidation, whether or not CO was formed, was impeded. In the case of formic acid oxidation, the diminishing number of ensembles of small PtNPs results in less CO formation, which in turn decreases the extent of surface poisoning by adsorbed CO. More active sites are available for formic acid oxidation through the pathway not involving formation of CO. Unlike methanol oxidation, the non-CO formation pathway does not require an ensemble of Pt active sites for formic acid oxidation. The size-dependent methanol oxidation rates have also been explained by stronger CO and OH adsorption [74].

11.3.2
Particle Shape Dependence

In both electrocatalysis and heterogeneous chemical catalysis, monocrystalline surfaces have been extensively used to elucidate the structure–reactivity relationship [73, 75]. Many reactions relevant to the fuel cell technology were found to be structure sensitive [72, 75, 76]. Of importance is whether this knowledge is relevant for practical nm-scale catalysts. To address this issue, it is necessary to have shape-control of NPs. Particles with different shapes are enclosed with different facets. For example, Pt cubes are enclosed by six {100} facets; but Pt octahedra are terminated with eight {111} facets.

Investigations of electrocatalysis on shape-controlled NPs have been initiated recently. Feliu and coworkers [77] have studied formic acid and methanol oxidation on 8.2-nm and 8.6-nm PtNPs with preferential (100) and (111) faces, respectively. The catalytic activity for formic acid oxidation in terms of current density per Pt site follows the order (111) > (100) > polyoriented PtNPs. Although the intrinsic activity of (100)-oriented PtNPs is higher than that of (111)-oriented PtNPs, the former surface is poisoned much more rapidly by CO, the intermediate of the reaction. For methanol oxidation, the catalytic activity at low overpotentials follow the order (111) > polyoriented > (100) PtNPs; the (111)-oriented PtNPs are nearly three-times as active as the (100)-oriented PtNPs. Tong and coworkers [78] have studied methanol oxidation on 10-nm Pt nano-octahedra/tetrahedra and nanocubes. The octahedral/tetrahedral particles are enclosed with {111} facets. Similar to the results reported by Feliu et al. [77], the activity for methanol oxidation is higher on Pt nano-octahedra/tetrahedra than on nanocubes.

These shape-dependent activities are in general agreement with the observations made on extended single-crystal surfaces [77, 78]; however, some differences are apparent when the results obtained on the NP are compared to those on the extended single-crystal surfaces with the same crystal orientation. The peak current density of formic acid oxidation on the Pt(100)-extended electrode is nearly ten-times that of the (100)-oriented PtNPs [77]. For methanol oxidation, in contrast, the maximum activity on (111)-oriented PtNPs is more than five-times that of Pt(111)-extended surface [77]. In addition, PtNPs with similar shapes but prepared by different methods also show distinctly different methanol oxidation activity [77, 78]. Several factors may contribute to these observations. Residual surfactant, such as PVP, used in the synthesis of NPs may influence the catalytic efficiency. The quantity is significant; after treatment with NaOH, the PtNPs prepared in PVP micelles had more than 10 wt% PVP [78]. The effect of this residue on catalytic activity has not been explored in detail. Second, there is uncertainty in the surface atomic arrangement of NPs. The shapes are determined typically by TEM; ideal (111) or (100) surface atomic arrangements are assumed for octahedra/tetrahedra and cubes, respectively. This assumption does not always hold as defects on particles may form. In addition, TEM is a local probe, only sampling a small fraction of the total NPs.

It is well known that the adsorption and desorption of underpotential-deposited hydrogen atoms (H_{UPD}) on single crystal surfaces of Pt-group metals exhibit sharp current peaks in voltammograms [79]. The peak potentials and shapes are characteristic of different surface orientations and can be used to characterize the Pt surface. The fractions of different surface sites are readily estimated from approximating the areas of different peaks [79, 80], but deconvolution of overlapping peaks to separate contributions from different crystal faces is required for quantitative assessment. In addition, the contribution of the (111) terrace spreads over a large potential range, which makes accurate evaluation difficult. Taking advantage of the fact that bismuth and tellurium adsorb exclusively on (111) terraces and germanium adsorbs on (100) terraces of Pt, Feliu and coworkers have recently developed a method to probe the fraction of a particle surface with (111) and (100) orientation [80]. Using Bi adsorption on the (111) terrace as an example, a full coverage of irreversibly adsorbed Bi (Bi_{ads}) was formed on a Pt surface, and the charge related to Bi_{ads} oxidation, q_{Bi}, in a voltammogram was obtained (Figure 11.6) [80]. This charge was converted into the equivalent charge of H_{UPD} desorption from (111) sites, denoted as $q^{t-1}_{(111)}$. The total H_{UPD} desorption charge is obtained by integration of current under the H_{UPD} peaks. The ratio of $q^{t-1}_{(111)}$ to the total charge is the fraction of Pt(111) sites. The relationship between q_{Bi} and $q^{t-1}_{(111)}$ is [80]:

$$q_{Bi} = 0.64 q^{t-1}_{(111)} \tag{11.2}$$

This calibration equation was obtained by correlating q_{Bi} on Pt stepped surfaces with (111) terraces of various widths and plotting these charges versus $q^{t-1}_{(111)}$ (Figure 11.7). The experimental slope, 0.64, agrees with the theoretical value, 0.66, which comes from the fact that one Bi can block H adsorption on three Pt sites and the Bi is oxidized to $Bi(OH)_2$. Through a similar process using the oxidation of irreversibly adsorbed Ge, the fraction of (100) sites is obtained [80]. This semiempirical means to *in situ* probing of the Pt surface structure is very useful for understanding electrocatalysis with NPs.

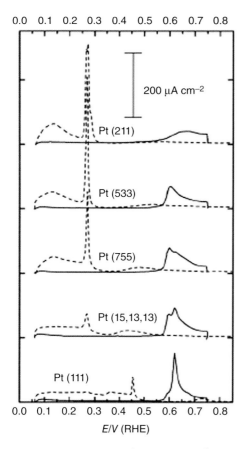

Figure 11.6 Linear scan voltammograms of Pt step single-crystal electrodes measured in 0.5 M H_2SO_4 with (solid curve) and without (dashed curve) irreversibly adsorbed Bi. Scan rate, 50 mV s^{-1}. Reprinted with permission from Reference [80]. Copyright 2008 The Royal Society of Chemistry.

Because of a higher density of step and kink sites, a surface plane with a high Miller index generally has a greater catalytic activity than one with a low Miller index [81, 82]. A challenge in optimizing the geometry of NPs for catalysis is that particles with high-index facets are difficult to synthesize. The surface energy of high-index facets is higher than that of low-index facets. Consequently, the growth rates perpendicular to the high-index facets are faster than those of the low-index facets, resulting in the disappearance of high-index facets. By using a variation of the double-pulse electrochemical method described in Section 11.2.1.1, Sun and coworkers [83] were able to produce Pt tetrahexahedra, Pt(THH), particles with a controlled size over the range 20–220 nm. Geometrically, THH can be viewed as a cube with a square-based pyramid on each face, so each particle contains 24 faces. Each of these faces is a {730} facet. By using a similar electrochemical pulse

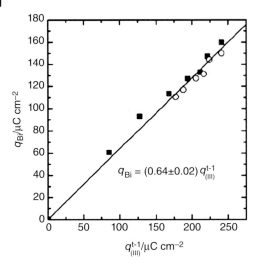

Figure 11.7 Linear relationship between Bi_{ads} redox charge and the equivalent H adsorption charge from (111) sites on the terrace of Pt step surfaces. ■ Pt (n, n, n–2) electrodes; ○ Pt (n+1, n–1, n–1) electrodes. Reprinted with permission from Reference [80]. Copyright 2008 The Royal Society of Chemistry.

deposition approach, Pt and Pd particles with other shapes have been made [84, 85]. The particles enclosed by higher Miller index planes may have higher catalytic activity and selectivity than those with basal planes. Compared with Pt/C or Pt spheres, the Pt(THH) particles show a 2–4 times enhancement in catalytic activity towards methanol and formic acid oxidation [83].

11.3.3
Particle Composition Dependence

It has been long recognized in surface science that the catalytic activity and selectivity of a given metal can be significantly modified by alloying with a second metal. This modification is explained by the d-band theory, which was developed mainly by Nørskov and coworkers [86–89]. The key concept is that the adsorption energy of adsorbates depends on the metal d-band center, and, when a metal is alloyed, the d-band center (average) energy is changed [86]. Whether the d-band energy is increased or decreased depends on two cumulative factors, the strain effect and the ligand effect [88, 89]. The strain effect arises from the difference in the lattice constant of the alloying metals. This concept can be explained by considering the example of a metal monolayer epitaxially grown on top of another metal. If the top layer metal has a smaller lattice constant than the substrate metal, the interatomic distance of the neighboring atoms of the top layer will be longer than that of the bulk metal. The monolayer is then under tensile strain. In contrast, if the metal in the top layer has a lattice constant that is larger than that of the substrate, the metal film will be under compressive strain. When the surface atoms are under tensile strain, the d-orbital overlap is decreased; consequently, the d-band becomes narrower. To conserve the d-band filling and the number of d states, the d-band center energy increases [86]. In terms of electrochemical potential, the d-band center becomes more negative. The increase of the d-band center facilitates the donation of metal d-electrons to the lowest unoccupied orbital of adsorbates, thus increasing the adsorption energy. Conversely, the compressed strain causes the broadening of the d-band and therefore a downshift of the d-band center energy [87]. The adsorption energy is then lower compared to the bulk metal.

The ligand effect mainly comes from the electronic interaction of the two metals [88, 89]. In metal alloys or layered-metal thin films, the metal atoms have hetero-nuclei metal–metal bonds. Their electronic structures will be different from those of the corresponding pure metals where only homo-nuclei metal–metal bonds exist. The change of the electronic structure modifies the chemical properties of the metal as well. The strain effect and the ligand effect can change the d-band energy in either the same or the opposite direction.

The d-band theory is used increasingly to explain the observed electrocatalytic activity changes caused by either alloying or deposition of layered metals [90–92]. The theory is used to explain the activity decrease toward methanol oxidation of Pt sub-monolayer thin films deposited on Au electrodes or Au nanoparticles [93, 94]. It also is employed to understand O_2 reduction on Pt-based alloy particles and Pt monolayers deposited on non-Pt surfaces [90–92]. Correlation between the d-band center position and the O_2 reduction activity has been demonstrated on Pt monolayer-covered single crystal surfaces and on $Pt_3Ni(111)$ alloy surfaces [90–92]. The reduction of O_2 on Pt monolayers deposited on Au(111), Rh(111), Pd(111), Ru(0001), and Ir(111) surfaces provides an example application of this theory [95]. Pt films deposited on Pd(111) show the highest catalytic activity toward O_2 reduction, and the Pt films on Ru(0001) have the lowest activity. The observation of the volcano-shape of the plot of activity versus d-band center energy was explained by invoking two competing factors [95]. The reduction of O_2 to H_2O involves the breaking of O–O bond and the formation of the O–H bond. Increasing the d-band center energy facilitates the breaking of O–O bond due to stronger adsorption energy of the product, atomic oxygen (O). However, the stronger adsorption energy slows the hydrogenation of O and OH, leading to the accumulation of adsorbed O and OH on the surface, thereby hindering further O_2 adsorption. This demonstration of how the activity of a catalyst is modified by balancing two competing factors suggests a general route for the rational design of high-performance catalysts.

Enhancement of the ORR was observed on Pt-based alloys and Pt-monolayer covered non-noble metal NPs [90–92]. The presence of a second metal changes the Pt–Pt inter-atomic distance [96] and, therefore, the d-band position. PtNi and PtCo alloy nanoparticles were among the first examples of enhancement of the ORR by alloying [96, 97]. Recently, Strasser and coworkers showed that de-alloying of PtCu nanoparticles by electrochemically dissolving Cu from the particles significantly increases the rate of the ORR [98–100]. The enhanced activity was believed to arise from the presence of a Pt-rich surface on top of a PtCu core.

Another effect of inclusion of a second metal in a NP is to introduce active sites that facilitate the removal of strongly adsorbed reaction intermediates that can otherwise poison the surface. In this regard, addition of an oxyphilic metal, such as Ru, to a Pt catalyst was reported to enhance methanol oxidation [101]. The increase in activity was attributed mainly to the "bifunctional mechanism," in which the role of Ru is to provide adsorbed oxygen-containing species (generally believed to be OH_{ads}). OH_{ads} is an oxidant toward CO_{ads}, which is an intermediate of methanol oxidation. The presence of CO_{ads} on Pt poisons the surface by blocking active sites for methanol oxidation. The formation of OH_{ads} mitigates the poisoning of the surface. Because OH_{ads} formation occurs at a much more positive potential on Pt than on Ru, the addition of Ru to a PtNP decreases the potential required for electrochemical formation of OH_{ads} and consequently facilitates removal of CO_{ads}. In addition, the presence of Ru modifies the electronic structure of Pt in a manner that weakens CO adsorption on Pt [102–105]. The significance of this second effect is still under debate. Surface-enhanced Raman spectroscopic studies reveal that the Pt–CO stretch vibration frequency is not changed by the presence of Ru [106]. However, a temperature-programmed desorption experiment [104] and density function calculations [107, 108] both show that the CO adsorption energy is smaller on a Pt monolayer deposited on a Ru(0001) surface than on pure Pt. The contradiction may result from the difference in these surface structures. In the former case, Ru is on top of Pt, while in the latter Pt is on top of Ru. In fact, the activity toward methanol oxidation of Pt-covered Ru is higher than that of PtRu alloy catalysts, underscoring the structural sensitivity [92].

In summary, the foregoing discusses the effects of nanoparticle size, shape, and composition on electrocatalysis. It is demonstrated that all these factors strongly influence the catalytic activity of nanoparticles. Basic understanding of how these factors affect electrocatalysis has been obtained. Although the discussions are mostly in the context of fuel cell reactions, it is foreseeable that these factors can play important roles in improving the sensitivity and selectivity of nanoparticle-based electrochemical measurements. Rational design of nanoparticles with size, shape, and composition control to target specific analytical problems arguably is next step in improving the performance of nanoparticle-based analytical methods.

11.4
Analytical Applications of Electrodes Modified with Metal Nanoparticles

Metal nanoparticles are applied to electroanalytical methodology primarily in three ways. First, they can serve as electron-transfer catalysts, thereby permitting the oxidation or reduction of analytes that otherwise undergo a slow electrode reaction or are not electrochemically active in the available potential window. Second, they can enhance the electrochemical utility of certain modified electrodes by increasing the conductivity of the modifier or extending the pH range of its conductivity. Third, they can improve the stability of modified electrodes by anchoring electron-transfer mediators to the system. In the last case, synergism in catalysis by the nanoparticle and the anchored mediator has been hypothesized. The applications are categorized according to the type of analyte, except that enzyme-based sensors that include NPs are treated as a special case. The examples cited below do not comprise a comprehensive review of the literature. Among the reviews available on this topic are those by Guo et al. [2], Katz et al. [4], and Hernandez-Santos et al. [109].

11.4.1
Determination of Inorganic Analytes

Nanoparticle-assisted electrochemical determinations have been reported for various inorganic analytes, including As^{III}, Sb^{III}, H_2O_2, nitrite, nitric oxide, hydrazine, Cr^{VI}, transition metal cations, and oxyhalogenates. Reports on the determination of As^{III} illustrate the scope of NP-assisted electroanalytical methods. Voltammetric determinations based on the catalyzed oxidation of As^{III} are the most common of these applications. Electrocatalytic oxidation of As^{III} by PtNPs [110, 111] and mediation of this oxidation by cobalt oxide NPs on glassy carbon are representative examples [112].Figures 11.8 and 11.9 show the voltammetric signatures of simple catalysis and catalysis via mediation of electron transfer.

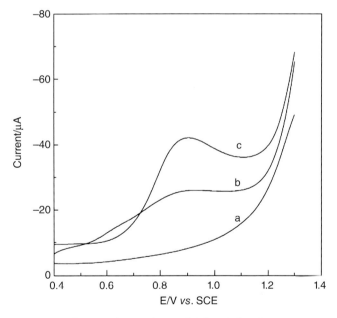

Figure 11.8 Electrocatalytic oxidation of As^{III} at a PtNP-coated glassy carbon electrode (GCE) containing carbon nanotubes (CNTs) on the surface: (a) CNT/GCE, (b) PtNP|GCE, and (c) PtNP|CNT/GCE. Solution, 0.04 mM As^{III} in 0.1 M H_2SO_4; initial potential, 0.4 V; scan rate, 100 mV s^{-1}. Reprinted with permission from Reference [110]. Copyright 2008 Elsevier.

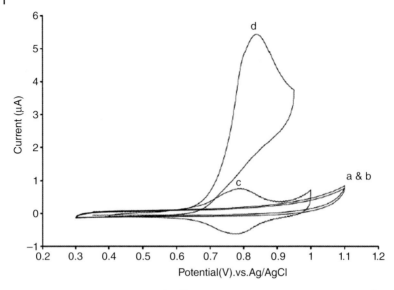

Figure 11.9 Mediated oxidation of As^{III} at a glassy carbon electrode (GCE) modified with cobalt oxide nanoparticles. (a) bare GCE in pH 7 buffer, (b) bare GCE in 0.07 mM As^{III} in pH 7 buffer, (c) CoO–NP|GCE in pH 7 buffer, (d) CoO–NP|GCE in 0.07 mM As^{III} in pH 7 buffer. Scan rate, 20 mV s^{-1}. Reprinted with permission from Reference [112]. Copyright 2008 Elsevier.

Regarding the former, As^{III} is not oxidized at bare GCE, but at GCE|PtNP an anodic peak corresponding to As^{III} concentration is developed (Figure 11.8) [110]. The potential of the anodic peak labeled b in Figure 11.8 is not related to a process involving oxidation of Pt, which precludes catalysis of As^{III} oxidation by a mediated pathway involving Pt species. Also illustrated is that inclusion of carbon nanotubes (CNTs) with the PtNPs (Figure 11.8, curve c) increases the current for the oxidation of As^{III}. The anodic peak potential was the same with GCE|PtNP and GCE|PtNP-CNT, so the likely role of the CNTs is to amplify the anodic current because of an increase in surface area. In support of this conjecture, the peak current was proportional to the square root of scan rate with both GCE|PtNP and GCE|PtNP-CNT electrodes, which is indicative of a diffusion-limited process; therefore, the CNTs do not amplify the current by increasing the rate of electron transfer. In contrast, the data in Figure 11.9 [112] clearly indicate a mediated pathway such as that in (11.3) and (11.4):

$$Co^{II} \rightarrow Co^{III} + e^{-} \text{ (electrochemical)} \tag{11.3}$$

$$2Co^{III} + As^{III} \rightarrow 2Co^{II} + As^{V} \text{ (chemical)} \tag{11.4}$$

The regeneration of Co^{II} increases the peak current related to Equation (11.3) in proportion to the As^{III} concentration.

The general behavior of As^{III} at NP-modified electrodes permitted extensions of the voltammetric determination of that analyte. An anodic stripping procedure has been developed for As^{III} determinations using a GCE|AuNP electrode [113]. The potentiostatic reduction of As^{III} to As^0 in 1 M HNO_3 served as the accumulation step; the anodic stripping was performed by square wave voltammetry. A detection limit of $0.25\,\mu g\,l^{-1}$ was reported. The voltammetry of As^{III} at an electrode modified with AuNPs capped with poly(L-lactide) provided a route for the indirect determination of hydrogen sulfide [114]. Sulfide inhibits As^{III} oxidation at AuNPs via the formation of As_2S_3. Perhaps as important as the analytical result was the demonstration that the expected inhibition from formation of Au–S bonds did not occur, which suggests that poly(L-lactide) is so strongly bonded to AuNPs that a place exchange reaction with sulfide was blocked.

The voltammetry of the $Sb^{0,III}$ couple at a AgNP-modified electrode was analogous to that of $As^{0,III}$ at a AuNP-modified electrode [115]. This permitted the determination of Sb^{III} by anodic stripping voltammetry. Here, carbon screen-printed electrodes (CSPEs) were modified with 50-nm AgNPs that were electrochemically deposited, a process that yielded nm-scale aggregates of AgNPs. The Sb^0 was deposited from a pH 2 solution of Sb^{III} at $-0.6\,V$ versus Ag|AgCl. The oxidation of Sb^0 was observed at $-0.18\,V$. When either bare CSPEs or solid Ag electrodes were used, a stripping current was not observed. A detection limit of 0.7 nM was reported when a 200-s preconcentration time was used. In this regard, with a $10^{-8}\,M$ Sb^{III} sample (presumably with a 200-s preconcentration time), the precision was 3.5% (ten trials). Bi^{III} generally interferes with the voltammetry of Sb^{III} because of similar electrochemical behavior, but at CSPE|AgNP separate stripping peaks were observed when both ions were $0.1\,\mu M$.

The electrochemistry of hydrogen peroxide is of importance to diverse areas of science. In fuel cell technology it is an intermediate in the four-electron reduction of O_2, so optimizing this electrode reaction requires consideration of catalysis for the reduction of hydrogen peroxide to water. Designs of amperometric biosensors that employ oxidase enzymes in conjunction with O_2 as the oxidizing agent likewise need to consider this reduction process. Several investigations have employed metal NPs as catalysts for the reduction of hydrogen peroxide, including Au [116], Pt [117], Ag [118], and Cu [119]. Films cast from methyltrimethoxysilane-derived sol–gels on GCEs were doped with AuNPs by including them in the precursor sol [116]. The quantity of the AuNPs was sufficient to make the sol–gel film highly conductive. This dopant not only catalyzed the reduction of hydrogen peroxide but also lowered the resistance of the sol–gel film. In pH 7.0 phosphate buffer (0.1 M), the onset of the reduction was observed near 0.0 V versus SCE, and the limiting current was developed near $-0.5\,V$ when point-by-point chronoamperometry was used to obtain the current–voltage curve. The difference between the onset potential and the point at which a limiting current was observed was indicative of a low electron-transfer rate constant even when AuNPs were used as a catalyst. Consistent with this result, cyclic voltammetry did not show a peak current prior to the onset of reduction of the supporting electrolyte when the electrode was modified with a AuNP-doped sol–gel film [116]. Nevertheless, a linear calibration curve over the range $2.5-45\,\mu M$ was obtained by chronoamperometry at $-0.5\,V$. The stability of the electrode was at least one week.

A composite film for the determination of hydrogen peroxide that consisted of PtNPs imbedded in a graphite-like matrix was fabricated by co-sputtering Pt and carbon on a Si substrate [117]. Based on its oxidation at 0.57 V versus Ag|AgCl a linear calibration plot was reported over the range 2 μM–5 mM H_2O_2. A composite film that gave possible synergism between two species as catalysts was prepared by depositing AgNPs (20–40 nm) by a chemical reduction method on multiwalled carbon nanotubes (MWCNTs) [118]. These AgNP-decorated MWCNTs were used to modify GCE for a study of the voltammetric reduction of H_2O_2. The H_2O_2 reduction potential at GCE|AgNP-MWCNT in pH 7.0 phosphate buffer, −0.64 V versus SCE, was less negative than that at GCE|AgNP, which suggests synergism between AgNPs and MWCNTs as catalysts.

Synergism between NPs and a second catalyst on a modified electrode can result in a lower overpotential, such as in the above example, and/or an increase in sensitivity. The latter occurs if the NP increases the rate of electron transfer from a value that gives a kinetic-limited current to a value where the current is mass-transport limited. Clearly, synergistic catalysis has analytical merit in this case. If the only role of the NP is to lower the overpotential, the primary analytical significance is to potentially improve the selectivity; this presumes that electrocatalysis of processes involving concomitants in a sample are not also catalyzed. Cases can be envisioned where the NPs increase the effective area of the electrode; for example, if the second catalyst is bound to the NP and the electrochemical reaction does not occur except at the NP centers, the effective area will be determined by the size and population of the NPs. Here, the sensitivity will be increased by the presence of NPs even if the electrochemical reaction is mass-transport limited when only the second catalyst is present.

Modification of a Nafion-coated Pt electrode by imbedding CuO-NPs (1–3 nm) resulted in a surface at which the mediated reduction of H_2O_2 occurred at −0.3 V versus SCE in 0.1 M NaOH [119]. That is, the electrode process was analogous to that shown in (11.3) and (11.4). A chronoamperometric calibration curve showed a linear range of 15–80 μM. The detection limit (3σ criterion) was reported as 60 nM. The modified electrodes retained 90% of their original activity after storage at 4 °C for nearly 4 weeks. For the purpose of determining H_2O_2, electrodeposition in conjunction with a templated electrode surface has been used to obtain a homogeneous distribution of AgNPs on GCE [120]. Type I collagen in the range 40–400 ng l^{-1} forms three-dimensional networks on GCE. The concentration influences the interparticle distance, NP size, and the degree of aggregation. For applications to the reduction of H_2O_2, a network was formed with 100 ng l^{-1} collagen, and the AgNPs were deposited for 80 s at −0.1 V versus SCE from a 5 mM $AgNO_3$, 0.1 M KNO_3 solution. The current–time curve at −0.3 V reached steady-state in 2 s when the sample solution was 1 mM H_2O_2 in pH 6.5 phosphate buffer; the electrode system retained its original sensitivity after 30-day storage in a N_2 atmosphere at 4 °C. The chronoamperometric detection limit (3σ criterion) was 0.7 μM. Of these characteristics, the stability is the more noteworthy feature of this modified electrode.

The determination of oxides of nitrogen is an important application of NPs to electroanalytical chemistry. Seed-mediated growth of AuNPs on ITO [12] has been used to fabricate a platform for the sensing of nitrite on the basis of its electrocatalytic oxidation. A chronoamperometric calibration curve obtained at 0.93 V versus Ag|AgCl was linear over the range 1.0 μM–0.5 mM ($r = 0.9988$), and the detection limit (3σ criterion) was 0.65 μM. The electrode reaction was the AuNP-catalyzed oxidation of NO produced from disproportionation of HNO_2 at pH <3. Similar results were reported with seed-mediated growth of AuNPs on a GCE [121]. The analytical figures-of-merit with GCE|AuNP [121] were higher by a factor of 10 than with ITO|AuNP [12], but cyclic voltammetry, which gives a greater capacitance current than chronoamperometry, was used to obtain the data in the study with ITO.

Metal ions have also been determined at NP-modified electrodes. The determination of Cr^{VI} in the presence of Cr^{III} has been achieved at an AuNP|ITO electrode [122]. A series of cyclic voltammetry experiments verified that AuNPs catalyze the reduction of Cr^{VI} (Figure 11.10). The reduction at bare ITO and a continuous film of Au on ITO occurred at about −0.05 V versus Ag|AgCl. Subsequently, ITO|AuNP was fabricated by staircase cyclic voltammetry over the range 1.5 V (initial potential) to −0.5 V (switching potential) at 100 mV s^{-1} in 1 mM $HAuCl_4$, 0.01 M H_2SO_4, 0.01 M Na_2SO_4 solution. On the first scan nucleation and initial deposition of AuNPs was observed in the range 0.2–0.0 V. Subsequent scans increased the size of the AuNPs without initiating additional nucleation. With five deposition cycles, the AuNPs were in the range 20–30 nm. Potentiostatic deposition results in larger

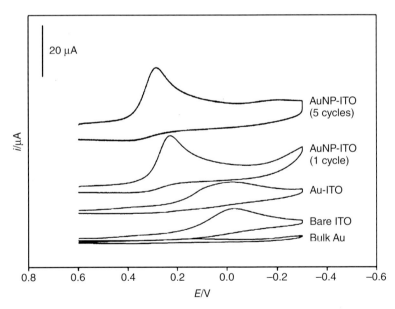

Figure 11.10 Staircase cyclic voltammetry of Cr^{VI} at bulk gold and at indium tin oxide in various stages of modification by gold nanoparticles. Solution, 50 μM Cr^{VI} in 0.01 M NaCl, 0.01 M HCl; scan rate, 100 mV s^{-1}. Reprinted with permission from Reference [122]. Copyright 2008 Elsevier.

AuNPs (100–150 nm) and a greater size distribution. Chronoamperometry in stirred solution with 0.2 V as the applied potential gave a linear dynamic range of 0.5–50 μM and a detection limit (3σ criterion) of 0.1 μM.

The scope of NP-based electroanalytical methods has been extended by Yantsee et al. [123], who used superparamagnetic Fe_3O_4 NPs (5.8 ± 0.9 nm) that were functionalized with dimercaptosuccinic acid (DMSA), which is a general complexing agent. Exchange of metal ions such as Cd^{II}, Pb^{II}, Cu^{II}, and Ag^{I} onto Fe_3O_4-DMSA nanoparticles was followed by collection of these NPs on a magnetic electrode (GCE with an electromagnetic core). The preconcentrated metal ions were reduced at −0.85 V versus Ag|AgCl for 60 s in 0.5 M HCl. The analytical signal was the re-oxidation current obtained by square wave voltammetry during a positive-going potential scan. The ability to use this method to measure Pb^{II} in rat urine is illustrated by the results in Figure 11.11. The method detected the background level of lead in rat urine (0.5 ± 0.2 μg Pb L^{-1}, measured by ICP-MS).

11.4.2
Determination of Organic and Biologically Important Analytes

Electroanalytical methods for organic compounds often require a catalyst and some means of protecting the electrode from passivation by adsorption of the analyte, its electrolysis product, and/or concomitant species in the sample. Metal NPs are being used for the former purpose and can mitigate passivation in some

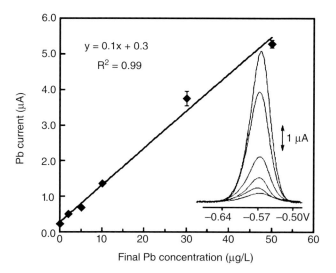

Figure 11.11 Staircase voltammetry for the oxidation of lead preconcentrated as a Pb^{II} complex on superparamagnetic Fe_3O_4 nanoparticles modified with dimercaptosuccinic acid and reduced at −0.85 V prior to stripping; matrix, 25% rat urine. After attracting the nanoparticles that were distributed in the matrix, the magnetic electrode was transferred into 0.5 M HNO_3 prior to voltammetry. Reprinted with permission from Reference [123]. Copyright 2008 The Royal Society of Chemistry.

11.4 Analytical Applications of Electrodes Modified with Metal Nanoparticles

cases by altering the electrode reaction to yield a product that is not surface active. This section describes some examples of how NPs are being used to provide improved electroanalytical methodology.

Modification of an electrode with a combination of a conducting polymer and metal NPs is an approach to addressing the limitations of electroanalytical methods for organic compounds. An example is to use AuNPs with polyaniline (PANI). However, PANI can present a problem in that it is not conductive in contact with solutions at pH's greater than 3–4. Doping PANI with AuNPs alleviated the problem of pH-dependent conductivity [124]. Here, the AuNPs were capped with mercaptosuccinate (MSA) and used in conjunction with positively charged PANI to fabricate a film on a gold electrode by LbL electrostatic assembly [40]. Initially, a 3-mercapto-1-propanesulfonate (MPS) monolayer was formed on the Au electrode. Alternating immersions in 1 mM, pH 2.6 PANI and 30 µg l^{-1} AuNP-MSA solutions (15 min each) resulted in the assembly of a bilayer. Using a Au|MPS| (PANI/AuNP-MSA)$_5$ assembly, the oxidation of NADH in pH 7.1 phosphate buffer was observed at 0.05 V versus Ag|AgCl. The peak current for the oxidation of NADH increased with bilayer number, n, over the range 5–12. The carboxylate group of the MSA on AuNP-MSA permitted attachment of various biomolecules to the modified electrode. By attachment of amino-terminated DNA "catcher probes," the system was used to electrochemically detect DNA hybridization and was able to discriminate complementary from non-complementary DNA sequences [124].

Because it maintains its conductivity at neutral pH, PEDOT [poly(3,4-ethylenedioxythiophene)] is an attractive alternative to PANI as a conducting polymer for electrode modification. Electrochemical formation of PEDOT that is dopted with AuNPs capped with a layer of 3,4-ethylenedioxythiophene (EDOT) has been reported. This procedure was analogous to the Brust–Schiffrin method [35, 36] except EDOT rather than an alkanethiol was the protecting group [125]. The AuNPs were in the 50–100 nm range. The composite was used for the electrochemical determination of dopamine (DA) in the presence of excess ascorbic acid (AA) [125]. Modification of a GCE with AuNP-PEDOT was by cyclic voltammetry of AuNP-EDOT in acetonitrile (potential range −0.5 to 1.9 V versus a Ag electrode). A characteristic of PEDOT is that it has both hydrophobic and hydrophilic regions. Oxidation of DA to dopamine-o-quinone occurred in the hydrophobic region, and the oxidation of AA to dehydroascorbic acid occurred in the hydrophilic region. The DA oxidation at pH 7.4 was at 0.12 V at GCE, GCE|PEDOT, and GCE|PEDOT-AuNP electrodes. Relative to GCE, the peak current was increased by 33% and 48% at GCE|PEDOT and GCE|PEDOT-AuNP. The role of the AuNPs was attributed to weak adsorption of DA thercon; however, alternative explanations are that the effective surface area of the electrode was increased and/or the rate of electron transfer was increased by the AuNPs. In support of the latter explanation, the AuNPs decreased the separation of the anodic and cathodic peaks in the cyclic voltammetry of DA, and with GCE|PEDOT-AuNP the oxidation approached a diffusion-limited process (Figure 11.12). In mixtures, the oxidations of DA and AA occurred at about the same potential at GCE but were

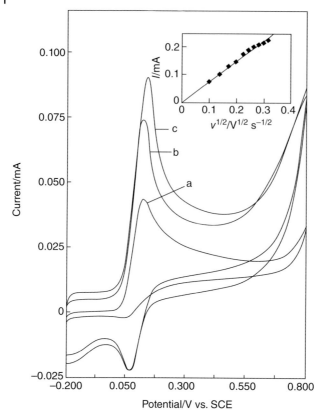

Figure 11.12 Cyclic voltammetry of 0.5 mM dopamine in pH 7.4 phosphate buffer at (a) a bare glassy carbon electrode (GCE), (b) GCE|PEDOT, and (c) GCE|PEDOT-AuNP. Scan rate, v, 50 mV s^{-1}; inset: anodic peak current versus $v^{1/2}$. Reprinted with permission from Reference [125]. Copyright 2005 Elsevier.

separated by 230 mV on GCE|PEDOT and GCE|PEDOT-AuNP. Pulse voltammetry for the determination of DA gave a linear calibration curve in the nM range with a detection limit of 2 nM.

The simultaneous determination of DA and AA also was achieved at a carbon paste electrode (CPE) modified with a film of carbon nanofibers that were decorated with PdNPs [126]. Electrospinning a polyacrylonitrile (PAN), Pd(acetate)$_2$ solution followed by heating in air at 230 °C, in H$_2$–Ar at 300 °C, and finally at 1100 °C in H$_2$–Ar carbonized the PAN into nanofibers and reduced the Pd^{II} to PdNP. A film of PdNP-decorated nanofibers was cast onto a CPE. At pH 4.5, a ternary mixture showed separate anodic peaks at 195, 372, and 565 mV for AA, DA, and uric acid (UA), respectively. By differential pulse voltammetry, DA was determined in the µM range in the presence of a 2000-fold excess of AA and a 1000-fold excess of UA. Ternary mixtures of these compounds were also determined at a GCE modified with choline (Ch) and AuNPs [127]. The GCE was initially modified by covalent linkage of Ch via hydroxyl groups using cyclic voltammetry (six cycles between −1.7 and 1.8 V versus SCE in 1 mM Ch, 10 mM KCl at pH 7.0). The AuNPs were electrochemically deposited by a 5-min electrolysis at −0.2 V in HAuCl$_4$ (0.2 mg ml^{-1}). The difference in cyclic voltammetric peak potentials for pairs of components were the following: 140 (DA and AA), 140 (DA and UA), and 280 mV (AA and UA). Differential pulse voltammetry was used to determine simultaneously DA and UA at the μM level in the presence of a 1000-fold excess of AA.

Applications that employ mediated electron-transfer can be problematic in terms of selectivity as the potential at which an electrode reaction occurs is determined primarily by the formal potential of the mediator; hence, coupling electrochemistry to a separation method such as HPLC is an important analytical method. An example is provided by the determination of homocysteine (hcys) and penicillamine at a CPE modified with cystamine (Cyst) to which a layer of AuNPs is attached [128]. Cyclic voltammetry at bare CPE and CPE|Cyst did not yield a signal for the oxidation of 0.1 mM hcys. With CPE|Cyst-AuNP, anodic peaks were observed at 600 and 950 mV versus Ag|AgCl. The former was due to oxidation of hcys, and the latter to the oxidation of AuNP. A hydrodynamic voltammogram was obtained with point-by-point variation of the applied potential and amperometric detection in a flow-injection system. A potential of 600 mV was determined to yield the highest signal-to-background ratio. But at this potential other compounds related to hcys gave an amperometric response. Using CPE|Cyst-AuNP in an amperometric detector for HPLC with a pH 2 mobile phase provided baseline resolution of the following compounds (detection limits, 3σ criterion, in parentheses): cys (20 nM), hcys (30 nM), glutathione (90 nM), penicillamine (120 nM), and N-acetylcysteine (30 nM).

11.5 Conclusions

Metal nanoparticles have been used in electroanalytical methodology as catalysts, to enhance conductivity of films on electrodes, and to provide conductive tethers between electrodes and species that serve analytical functions such as capturing agents for preconcentration and electron-transfer mediators. The catalytic function generally duplicates, rather than extends, the performance of conventional electrode modifiers. However, nanoparticles are advantageous in terms of utilization of mass of expensive materials, particularly when they are used as dispersed, two-dimensional nanoarrays. The nanoarrays are also of interest in terms of improving the ratio of faradaic to capacitive current in voltammetry, a factor that can optimize detection limits in potentiodynamic methods. Nanoparticles as dopants in an electrocatalytic composite also have promise in terms of synergistic catalysis. Geometric factors are anticipated to receive greater attention in future electroanalytical applications as the size, shape, and composition of nanoparticles continue to be demonstrated to have a strong influence over catalytic behavior in fundamental investigations.

References

1. Welch, C.M. and Compton, R.G. (2006) *Anal. Bioanal. Chem.*, **384**, 601–619.
2. Guo, S. and Wang, E. (2007) *Anal. Chim. Acta*, **598**, 181–192.
3. Castañeda, M.T., Alegret, S., and Merkoçi, A. (2007) *Electroanalysis*, **19**, 743–753.
4. Katz, E., Willner, I., and Wang, J. (2004) *Electroanalysis*, **16**, 19–44.
5. Reller, H., Kirowa-Eisner, E., and Gileadi, E. (1984) *J. Electroanal. Chem.*, **161**, 247–268.
6. Penner, R.M. and Martin, C.R. (1987) *Anal. Chem.*, **59**, 2625–2630.
7. Streeter, I., Baron, R., and Compton, R.G. (2007) *J. Phys. Chem. C*, **111**, 17008–17014.
8. Streeter, I. and Compton, R.G. (2007) *J. Phys. Chem. C*, **111**, 18049–18054.
9. Raj, C.R., Okajima, T., and Ohsaka, T. (2003) *J. Electroanal. Chem.*, **543**, 127–133.
10. Busbee, B.D., Obare, S.O., and Murphy, C.J. (2003) *Adv. Mater.*, **15**, 414–416.
11. Jana, N.R., Gearheart, L., and Murphy, C.J. (2001) *Chem. Commun.*, 617–618.
12. Jana, N.R., Gearheart, L., and Murphy, C.J. (2001) *J. Phys. Chem. B*, **105**, 4065–4067.
13. Törnblom, M. and Henriksson, U. (1997) *J. Phys. Chem. B*, **101**, 6028–6035.
14. Zhang, J., Kambayashi, M., and Oyama, M. (2004) *Electrochem. Commun.*, **6**, 683–688.
15. Zhang, J. and Oyama, M. (2005) *Anal. Chim. Acta*, **540**, 299–306.
16. Fransaer, J.L. and Penner, R.M. (1999) *J. Phys. Chem. B*, **103**, 7643–7653.
17. Liu, H. and Penner, R.M. (2000) *J. Phys. Chem. B*, **104**, 9131–9139.
18. Sandmann, G., Dietz, H., and Plieth, W. (2000) *J. Electroanal. Chem.*, **491**, 78–86.
19. Ueda, M., Dietz, H., Anders, A., Kneppe, H., Meixner, A., and Plieth, W. (2002) *Electrochim. Acta*, **48**, 377–386.
20. Dai, X. and Compton, R.G. (2006) *Anal. Sci.*, **22**, 567–570.
21. El-Deab, M.S., Sotomura, T., and Ohsaka, T. (2006) *Electrochim. Acta*, **52**, 1792–1798.
22. Ca, D.V., Sun, L., and Cox, J.A. (2006) *Electrochim. Acta*, **51**, 2188–2194.
23. Niu, Y., Yeung, L.K., and Crooks, R.M. (2001) *J. Am. Chem. Soc.*, **123**, 6840–6846.
24. Kim, Y.-G., Oh, S.-K., and Crooks, R.M. (2004) *Chem. Mater.*, **16**, 167–172.
25. Spatz, J.P., Mössmer, S., Hartmann, C., Möller, M., Herzog, T., Krieger, M., Boyen, H.-G., Ziemann, P., and Kabius, B. (2000) *Langmuir*, **16**, 407–415.
26. Kumar, S. and Zou, S. (2009) *Langmuir*, **25**, 574–581.
27. Murray, R.W. (2008) *Chem. Rev.*, **108**, 2688–2720.
28. Doron, A., Katz, E., and Willner, I. (1995) *Langmuir*, **11**, 1313–1317.
29. Cheng, W., Dong, S., and Wang, E. (2004) *J. Phys. Chem. B*, **108**, 19146–19154.
30. Grabar, K.C., Freeman, R.G., Hommer, M., and Natan, M.J. (1995) *Anal. Chem.*, **67**, 735–743.
31. Ge, G. and Brus, L.E. (2001) *Nano Lett.*, **1**, 219–222.
32. Angelo, S.K.S., Waraksa, C.C., and Mallouk, T.E. (1995) *Adv. Mater.*, **15**, 400–402.
33. Hu, G., Ma, Y., Guo, Y., and Shao, S. (2008) *Electrochim. Acta*, **53**, 6610–6615.
34. Ernst, A.Z., Sun, L., Wiaderek, K., Kolary, A., Zoladek, S., Kulesza, P.J., and Cox, J.A. (2007) *Electroanalysis*, **19**, 2103–2109.
35. Brust, M., Walker, M., Bethell, D., Schiffrin, D.J., and Whyman, W. (1994) *J. Chem. Soc. Chem. Commun.*, 801–802.
36. Brust, M., Fink, J., Bethell, D., Schiffrin, D.J., and Kiely, C. (1995) *J. Chem. Soc. Chem. Commun.*, 1655–1656.
37. Gopalan, A.I., Lee, K.-P., Manesh, K.N., Santhosh, P., Kim, J.H., and Kang, J.S. (2007) *Talanta*, **71**, 1774–1781.
38. Wandstrat, M.M., Spendel, W.U., Pacey, G.E., and Cox, J.A. (2007) *Electroanalysis*, **19**, 139–143.
39. Smoukov, S.K., Bishop, J.M., Kowalczyk, B., Kaisin, A.M., and Grzybowski, B.A. (2007) *J. Am. Chem. Soc.*, **129**, 15623–15630.

40 Decher, G. (1997) *Science*, **277**, 1232–1237.
41 Cox, J.A., Kittredge, K.W., and Ca, D.V. (2004) *J. Solid State Electrochem.*, **8**, 722–726.
42 Zhao, M. and Crooks, R.M. (1999) *Adv. Mater.*, **11**, 217–220.
43 Zhang, H. and Hu, N. (2007) *J. Phys. Chem. B*, **111**, 10583–10590.
44 Zhang, H., Lu, H., and Hu, N. (2006) *J. Phys. Chem. B*, **110**, 2171–2179.
45 Diaz, A.F. and Logan, J.A. (1980) *J. Electroanal. Chem.*, **111**, 111–114.
46 Bartlett, P.N. and Wang, J.H. (1996) *J. Chem. Soc. Faraday Trans.*, **92**, 4137–4143.
47 Granot, E., Katz, E., Basnar, B., and Willner, I. (2005) *Chem. Mater.*, **17**, 4600–4609.
48 Martin, C.R. and van Dyke, L.S. (1992) *Molecular Design of Electrode Surfaces* (ed. R.W. Murray), John Wiley & Sons, Inc., New York, pp. 403–424.
49 Mathiyarasu, J., Senthilkumar, S., Phani, K.L.N., and Yegnaraman, V. (2008) *Mater. Lett.*, **62**, 571–573.
50 Patra, S. and Munichandraiah, N. (2009) *Langmuir*, **25**, 1732–1738.
51 Kim, S. and Park, S.J. (2008) *Solid State Ionics*, **178**, 1915–1921.
52 Domínguez-Domínguez, S., Arias-Pardilla, J., Berenguer-Murcia, A., Morallón, E., and Cazorla-Amorós, D. (2008) *J. Appl. Electrochem.*, **38**, 259–268.
53 Kulesza, P.J., Chojak, M., Karnicka, K., Miecznikowski, K., Palys, B., Lewera, A., and Wieckowski, A. (2004) *Chem. Mater.*, **16**, 4128–4134.
54 Bregoli, L.J. (1978) *Electrochim. Acta*, **23**, 489–492.
55 Sattler, M.L. and Ross, P.N. (1986) *Ultramicroscopy*, **20**, 21–28.
56 Peuckert, M., Yoneda, T., Betta, R.A.D., and Boudart, M. (1986) *J. Electrochem. Soc.*, **133**, 944–947.
57 Bett, J., Lundquist, J., Washington, E., and Stonehart, P. (1973) *Electrochim. Acta*, **18**, 343–348.
58 Poirier, J.A. and Stoner, G.E. (1994) *J. Electrochem. Soc.*, **141**, 425–430.
59 Watanabe, M., Sei, H., and Stonehart, P. (1989) *J. Electroanal. Chem.*, **261**, 375–387.
60 Kinoshita, K. (1990) *J. Electrochem. Soc.*, **137**, 845–848.
61 Giordano, N., Passalacqua, E., Pino, L., Arico, A.S., Antonucci, V., Vivaldi, M., and Kinoshita, K. (1991) *Electrochim. Acta*, **36**, 1979–1984.
62 Kinoshita, K. (1992) *Electrochemical Oxygen Technology*, John Wiley & Sons, Inc., New York.
63 Mayrhofer, K.J.J., Blizanac, B.B., Arenz, M., Stamenkovic, V.R., Ross, P.N., and Markovic, N.M. (2005) *J. Phys. Chem. B*, **109**, 14433–14440.
64 Yano, H., Inukai, J., Uchida, H., Watanabe, M., Babu, P.K., Kobayashi, T., Chung, J.H., Oldfield, J.H.E., and Wieckowski, A. (2006) *Phys. Chem. Chem. Phys.*, **8**, 4932–4939.
65 Plieth, W.J. (1982) *J. Phys. Chem.*, **86**, 3166–3170.
66 Redmond, P.L., Hallock, A.J., and Brus, L.E. (2005) *Nano Lett.*, **5**, 131–135.
67 Maillard, F., Eikerling, M., Cherstiouk, O.V., Schreier, S., Savinova, E., and Stimming, U. (2004) *Faraday Discuss.*, **125**, 357–377.
68 Maillard, F., Savinova, E.R., Simonov, P.A., Zaikovskii, V.I., and Stimming, U. (2004) *J. Phys. Chem. B*, **108**, 17893–17904.
69 Maillard, F., Schreier, S., Hanzlik, M., Savinova, E.R., Weinkauf, S., Stimming, U., and Matter, S. (2005) *Phys. Chem. Chem. Phys.*, **7**, 385–393.
70 Arenz, M., Mayrhofer, K.J.J., Stamenkovic, V., Blizanac, B.B., Tomoyuki, T., Ross, P.N., and Markovic, N.M. (2005) *J. Am. Chem. Soc.*, **127**, 6819–6829.
71 Mayrhofer, K.J.J., Arenz, M., Blizanac, B.B., Stamenkovic, V., Ross, P.N., and Markovic, N.M. (2005) *Electrochim. Acta*, **50**, 5144–5154.
72 Markovic, N.M. and Ross, P.N. (2002) *Surf. Sci. Rep.*, **45**, 117–229.
73 Park, S., Xie, Y., and Weaver, M.J. (2002) *Langmuir*, **18**, 5792–5798.
74 Takasu, Y., Iwazaki, T., Sugimoto, W., and Murakami, Y. (2000) *Electrochem. Commun.*, **2**, 671–674.
75 Wieckowski, A., Savinova, E.R., and Vayenas, C.G. (eds) (2003) *Catalysis and*

76 Lipkowski, J. and Ross, P.N. (eds) (1998) *Electrocatalysis*, Wiley-VCH, New York.

77 Solla-Gullon, J., Vidal-Iglesias, F.J., Lopez-Cudero, A., Garnier, E., Feliu, J.M., and Aldaza, A. (2008) *Phys. Chem. Chem. Phys.*, **10**, 3689–3698.

78 Susut, C., Chapman, G.B., Samjeske, G., Osawa, M., and Tong, Y. (2008) *Phys. Chem. Chem. Phys.*, **10**, 3712–3721.

79 Clavilier, J. (1999) *Interfacial Electrochemistry* (ed. A. Wieckowski), Marcel Dekker, New York, pp. 231–248.

80 Solla-Gullon, J., Rodriguez, P., Herrero, E., Aldaz, A., and Feliu, J.M. (2008) *Phys. Chem. Chem. Phys.*, **10**, 1359–1373.

81 Housmans, T.H.M. and Koper, M.T.M. (2003) *J. Phys. Chem. B*, **107**, 8557–8567.

82 Shin, J.W. and Korzeniewski, C. (1995) *J. Phys. Chem.*, **99**, 3419–3422.

83 Tian, N., Zhou, Z.Y., Sun, S.G., Ding, Y., and Wang, Z.L. (2007) *Science*, **316**, 732–735.

84 Tian, N., Zhou, Z.Y., and Sun, S.G. (2008) *J. Phys. Chem. C*, **112**, 19801–19817.

85 Zhou, Z.Y., Tian, N., Huang, Z.Z., Chen, D.J., and Sun, S.G. (2008) *Faraday Discuss.*, **140**, 81–92.

86 Hammer, B., Morikawa, Y., and Nørskov, J.K. (1996) *Phys. Rev. Lett.*, **76**, 2141–2144.

87 Hammer, B. and Nørskov, J.K. (2000) Theoretical surface science and catalysis – calculations and concepts, in *Impact of Surface Science on Catalysis* (eds B.C. Gates and H. Knoezinger), Advances in Catalysis, vol. **45**, Academic Press, San Diego, pp. 71–129.

88 Kitchin, J.R., Nørskov, J.K., Barteau, M.A., and Chen, J.G. (2004) *Phys. Rev. Lett.*, **93**, 156801–156804.

89 Kitchin, J.R., Nørskov, J.K., Barteau, M.A., and Chen, J.G. (2004) *J. Chem. Phys.*, **120**, 10240–10246.

90 Stamenkovic, V.R., Fowler, B., Mun, B.S., Wang, G.F., Ross, P.N., Lucas, C.A., and Markovic, N.M. (2007) *Science*, **315**, 493–497.

91 Stamenkovic, V.R., Mun, B.S., Arenz, M., Mayrhofer, K.J.J., Lucas, C.A., Wang, G.F., Ross, P.N., and Markovic, N.M. (2007) *Nat. Mater.*, **6**, 241–247.

92 Adzic, R.R., Zhang, J., Sasaki, K., Vukmirovic, M.B., Shao, M., Wang, J.X., Nilekar, A.U., Mavrikakis, M., Valerio, J.A., and Uribe, F. (2007) *Top. Catal.*, **46**, 249–262.

93 Du, B.C. and Tong, Y.Y. (2005) *J. Phys. Chem. B*, **109**, 17775–17780.

94 Kumar, S. and Zou, S. (2007) *Langmuir*, **23**, 7365–7371.

95 Zhang, J.L., Vukmirovic, M.B., Xu, Y., Mavrikakis, M., and Adzic, R.R. (2005) *Angew. Chem. Int. Ed.*, **44**, 2132–2135.

96 Mukerjee, S., Srinivasan, S., Soriaga, M.P., and McBreen, J. (1995) *J. Phys. Chem.*, **99**, 4577–4589.

97 Mukerjee, S., Srinivasan, S., Soriaga, M.P., and McBreen, J. (1995) *J. Electrochem. Soc.*, **142**, 1409–1422.

98 Koh, S., Hahn, N., Yu, C.F., and Strasser, P. (2008) *J. Electrochem. Soc.*, **155**, B1281–B1288.

99 Koh, S. and Strasser, P. (2007) *J. Am. Chem. Soc.*, **129**, 12624–12625.

100 Mani, P., Srivastava, R., and Strasser, P. (2008) *J. Phys. Chem. C*, **112**, 2770–2778.

101 Watanabe, M. and Motoo, S. (1975) *J. Electroanal. Chem.*, **60**, 267–273.

102 Krausa, M. and Vielstich, W. (1994) *J. Electroanal. Chem.*, **379**, 307–314.

103 Buatier de Mongeot, F., Scherer, M., Gleich, B., Kopatzki, E., and Behm, R. (1998) *Surf. Sci.*, **411**, 249–262.

104 Frelink, T., Visscher, W., and van Veen, J.A.R. (1996) *Langmuir*, **12**, 3702–3708.

105 Tong, Y., Kim, H.S., Babu, P.K., Waszczuk, P., Wieckowski, A., and Oldfield, E. (2002) *J. Am. Chem. Soc.*, **124**, 468–473.

106 Yang, H., Yang, Y., and Zou, S. (2007) *J. Phys. Chem. C*, **111**, 19058–19065.

107 Koper, M.T.M., Shubina, T.E., and van Santen, R.A. (2002) *J. Phys. Chem. B*, **106**, 686–692.

108 Liao, M.S., Cabrera, C.R., and Ishikawa, Y. (2000) *Surf. Sci.*, **445**, 267–282.

109 Hernandez-Santos, D., Gonzalez-Garcia, M.B., and Garcia, A.C. (2002) *Electroanalysis*, **14**, 1225–1235.

110 Xu, H., Zeng, L., Xing, S., Yian, Y., and Lin, L. (2008) *Electrochem. Commun.*, **10**, 551–554.

111 Hrapovic, S., Liu, Y., and Luong, J.T. (2007) *Anal. Chem.*, **79**, 500–507.

112 Salimi, A., Mamkhezri, H., Hallaj, R., and Soltanian, S. (2008) *Sens. Actuators, B*, **129**, 246–254.

113 Majid, E., Hrapovic, S., Liu, Y., Maie, K.M., and Luong, J.T. (2006) *Anal. Chem.*, **78**, 762–769.

114 Song, Y.-S., Muthuraman, G., and Zen, J.M. (2006) *Electrochem. Commun.*, **8**, 1369–1374.

115 Domínguez Renedo, O. and Arcos Martínez, M.J. (2007) *Electrochem. Commun.*, **9**, 820–826.

116 Maduraiveeran, G. and Ramaraj, R. (2007) *J. Electroanal. Chem.*, **608**, 52–58.

117 You, T., Niwa, O., Horiuchi, T., Tomita, M., Iwasaki, Y., Ueno, Y., and Hirono, S. (2002) *Chem. Mater.*, **14**, 4796–4799.

118 Yang, P., Wei, W., Tao, C., Xie, B., and Chen, X. (2008) *Mikrochim. Acta*, **162**, 51–56.

119 Miao, X.M., Yuan, R., Chai, Y.-Q., Shi, Y.-Y., and Yuan, Y.-Y. (2008) *J. Electroanal. Chem.*, **612**, 157–163.

120 Song, Y., Cui, K., Wang, L., and Chen, S. (2009) *Nanotechnology*, **20**, 105501.

121 Cui, Y., Yang, C., Zeng, W., Oyama, M., Pu, W., and Zhang, J. (2007) *Anal. Sci.*, **23**, 1421–1425.

122 Tsai, M.C. and Chen, P.-Y. (2008) *Talanta*, **76**, 533–539.

123 Yantasee, W., Hongsirikarn, K., Warner, C.L., Choi, D., Sangvanich, T., Toloczko, M.B., Warner, M.G., Fryxell, G.E., Addleman, R.S., and Timchalk, C. (2008) *Analyst*, **133**, 348–355.

124 Tian, S., Liu, J., Zhu, T., and Knoll, W. (2004) *Chem. Mater.*, **16**, 4103–4108.

125 Kumar, S.S., Mathiyarasu, J., and Phani, K.L. (2005) *J. Electroanal. Chem.*, **578**, 95–103.

126 Huang, J., Liu, Y., Hou, H., and You, T. (2008) *Biosens. Bioelectron.*, **24**, 632–637.

127 Wang, P., Li, Y., Huang, X., and Wang, L. (2007) *Talanta*, **73**, 431–437.

128 Agüí, L., Peña-Farfal, C., Yáñez-Sedeño, P., and Pingarrón, J.M. (2007) *Talanta*, **74**, 412–420.

12
Single Molecule and Single event Nanoelectrochemical Analysis

Shanlin Pan and Gangli Wang

12.1
Introduction

In the past decade, the concept of single-molecule detection has been extended to the identification and quantification of individual small molecules, biomacromolecules, nanoparticles, or single interaction events such as protein–protein interactions and conformation changes of individual proteins and ribonucleic acid (RNA) molecules [1–7]. Currently most single-molecule studies rely on radioactive labeling and optical methods such as photoluminescence and surface enhanced Raman spectroscopy [8, 9]. The challenges of single-molecule detection using electrochemical methods reside in the weak signals associated with the charge transfer or mass transport processes of individual redox molecules or ions and high background current [10, 11]. The electrochemical signal-to-noise ratio has to be enhanced, normally by a signal amplification mechanism or using other techniques such as photoluminescence for indirect measurements on single-molecule electrochemical activities. Meanwhile, nanoelectrodes have been studied intensely recently, including research into new fabrication methods and the understanding of their electrochemical performance [12, 13]. Nanoelectrodes have dimensions in the nanometer range, which is comparable to the thickness of the electrical double layer or the size of the analytes. Therefore, improved electrochemical detection sensitivity and the spatial resolution of electrochemical imaging can be achieved. Applications of individual nanoelectrodes in electrochemical trace analysis such as single redox molecule detection [10] and biomolecule sensing have been demonstrated [14].

In this chapter, we first briefly introduce the concept of nanoelectrode electrochemistry and then review recent work on the ultrasensitive electrochemical detection of single molecules and single nanoparticles using various the combined techniques of electrochemistry and single-molecule spectroscopy. We then review recent work on single nanoelectrode fabrications and applications for high-resolution electrochemical imaging and biological applications. Lastly, localized delivery and imaging by using single nanopipette-based conductance techniques will be discussed. It is not our intention to cover all significant aspects in this area

Trace Analysis with Nanomaterials. Edited by David T. Pierce and Julia Xiaojun Zhao
Copyright © 2010 WILEY-VCH Verlag GmbH & Co. KGaA, Weinheim
ISBN: 978-3-527-32350-0

exclusively. Rather we will focus on studies that have employed at least one of the concepts of nanoelectrode, nanodomain activity, or nanoscale signals that are relevant to the scope of this book. We refer the reader to the relevant references as cited for further information.

12.2 Basic Concepts

12.2.1 Electrochemistry

The working electrode in many electrochemistry applications is composed of a conductor shrouded by an insulating material. The insulating shroud defines the geometry of the electrode, and thus it has an impact on mass transfer behavior. The desired signal, often current associated with a certain potential waveform, is dependent on this mass transfer behavior as well as electron-transfer kinetics at the electrode–solution interface [15]. There are two categories of current generated in electrochemistry when a potential is applied at a working electrode. The faradaic current is generated when a molecule changes its redox state by electron gain (reduction) or loss (oxidation) at the working electrode surface. Non-faradaic or charging current arises due to simple charge movement and accumulation at the electrode, a process similar to charging a capacitor. In this chapter, electrochemical measurements based on redox-induced current are referred to as voltammetry while those focused on charging are referred to as conductivity.

12.2.2 Nanoelectrodes

The surface of a working electrode is obviously critical for signal detection. Based on its relative surface area, it can be defined as a macroelectrode, a microelectrode, or a nanoelectrode. A smaller electrode surface area has a smaller electrical double-layer capacitance (C_{EL}) and correspondingly lower RC time constant, which allows analysis at low or even no additional electrolyte. When the size of an electrode is decreased to tens of nanometers, the flux at the electrode surface is dominated by radial instead of linear diffusion and the overall mass transfer rate is enhanced correspondingly. When the mass transfer is fast enough and no longer limits the current signal, electron transfer (ET) at the electrode–solution interface becomes the rate-determining step. Therefore, fast ET kinetics can be resolved using nanoelectrodes.

The novel electrochemical properties of nanoelectrodes have had an enormous impact on the development of modern electrochemistry [13]. Great efforts have been made to fabricate and apply these small electrodes to characterize unstable redox intermediates, to measure single electrochemical turnover levels of chemical reactions, and to image surfaces with versatile techniques such as scanning electrochemical microscopy (SECM) [16]. Precise measurements of ET rate constants with techniques that have a high time resolution and sensitivity, such as fast scan voltammetry, are only possible with small electrodes due to their negligible C_{EL} and ohmic drop. Nanoelectrodes have been used as a SECM tip to obtain detailed electrochemical information with extraordinary spatial resolution. They are also capable of wiring the redox centers of biomacromolecules for the purpose of ultrasensitive biomacromolecule detection [17]. Recent research results show that nanoelectrodes can be used to measure single-molecule chemical reactions and catalytic activities of a single catalytic nanoparticle. Significant progress has also been made in studying single-molecule and single-nanoparticle electrochemistry [18].

12.3
Single-Molecule Electrochemistry

In this section we discuss several recent electrochemical approaches that target or have achieved single-molecule sensing, including individual nanoelectrode methods and single-molecule spectroelectrochemistry. In these techniques, the use of nanoelectrodes for single-molecule detection provides appreciable faradaic current due to the high turnover rate and repeated ET reactions of one or only a few molecules [19]. Single-molecule spectroelectrochemistry collects single-molecule photoluminescence (PL). Changes in this PL with redox reactions can be used to probe single-molecule electrochemical activities such as interfacial charge transfer, charge storage, and electroluminescence. Our discussion excludes scanning tunneling microscopy (STM)-based single-molecule imaging and conductance measurements that have been studied intensely, including both measurements and theories [20, 21].

12.3.1
Single-Molecule Electrochemistry Using Nanoelectrodes

Single-molecule electrochemical activity is very difficult to detect by using a conventional electrochemical method because of the low turnover rate of the molecule and high background current (noise). A high turnover rate, as high as 10^7 Hz, is required to achieve measurable current at a level of 2 pA. Appreciable signal can be achieved by minimizing the electrode surface area to nanoscopic dimensions and finding a way to amplify the faradaic current. Nanometer-scale electrodes with various shapes and geometries can be fabricated using approaches shown in Section 12.5. The first example of single redox molecule electrochemistry was demonstrated by Bard and coworkers in 1995 [10, 11]. In their work single-molecule activities were investigated by trapping a small volume of solution in a gap formed by a nanoelectrode and a conductive substrate. The target molecule

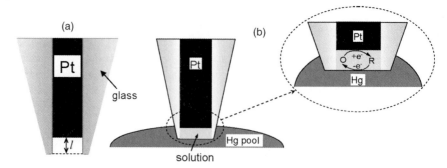

Figure 12.1 Recessed disk-type platinum nanoelectrode (a) used to form a nanometer-sized cylindrical thin layer cell by immersing the nanoelectrode into a pool of Hg for a single-molecule redox reaction (b). Reprinted with permission from Reference [22]. Copyright 2008 American Chemical Society.

dissolved at low concentration in this solution was allowed to cycle its redox state between the counterpoised electrodes. The distance between the nanoelectrode and the substrate was controlled by using the positioning drives of a SECM. Stochastic changes in tip current were observed and attributed to single redox molecule diffusion in and out of the small gap. Recently, Peng and Mirkin [22] have reported electrochemistry of individual molecules in zeptoliter volumes. This was achieved by performing the electrochemical reaction in a nanometer-sized thin layer cell formed by first etching back the surface of a disk-type platinum nanoelectrode (Figure 12.1).

A mercury pool was then used as the counter electrode and to seal the recessed nanoelectrode. High quality steady-state voltammograms of a few trapped molecules were obtained for different redox species. Small-gap experiments such as these allow the measurement of single-molecule activities with relatively small current fluctuation because the enclosed system prevents diffusion of redox molecules into or out of the small trapped volume. However, the use of such confined-electrode geometries in broader applications such as molecular sensing appears challenging since the signal measurement has to be conducted using a nanogap formed by two electrodes and the redox molecules must diffuse rapidly to deliver detectable charges. Using lithographically fabricated Au nanoelectrodes with dimensions a tens of nanometers, Lemay and coworkers [23] have studied a highly active [NiFe]-cluster isolated from *Allochromatium vinosum* and immobilized on a polymyxin-modified Au electrode (Figure 12.2). These workers successfully demonstrated a distinct catalytic response from less than 50 enzyme molecules and proved that studying redox enzymes at the macromolecule level is possible. Single-molecule electrochemistry at a nanoelectrode in such an open geometry might be achievable by optimizing the catalytic reaction and electrode geometry to increase the single molecule turnover rate while keeping background current to a minimum.

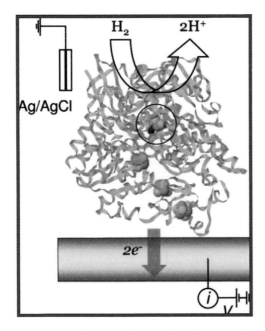

Figure 12.2 Schematic of a nanoelectrode approach toward single macromolecule electrochemistry: [NiFe]-hydrogenase protein film voltammetry at nanoelectrodes. Reprinted with permission from Reference [23]. Copyright 2008 American Chemical Society.

12.3.2
Single-Molecule Spectroelectrochemistry

Recent developments in optical microscopy and ultrasensitive photon detection allow the measurement of chemical and physical activities of individual molecules and particles, both in solids and under physiological conditions, as shown by recent review articles [3–7]. For instance, conformational changes of a single macromolecule such as RNA can be measured using single-molecule fluorescence resonance energy transfer (FRET) efficiency [24–26]. In addition, single protein expression in living cells can be tracked using single-molecule fluorescence [27, 28]. Single-molecule Raman spectroscopy has been achieved by using the enormous enhancements provided by localized surface plasmon resonance [29–32]; specifically, by controlling the geometry and aggregation states of the plasmonic antennas [33–35]. In contrast, it is very hard to collect enough faradaic current from single molecules for single-molecule detection due to their normally small redox turnover rates and the high background current. Recently, the groups of Barbara and Bard introduced single-molecule spectroelectrochemistry (SMS-EC), a powerful new technique for studying single-molecule electrochemical kinetics in highly heterogeneous systems. It is able to reveal dynamic and heterogeneous behaviors of individual molecules [36]. The instrumentation for SMS-EC consist of a standard potentiostat for potential control, an optical microscope, a quality laser as light source, and an ultrasensitive charge-coupled device (CCD) camera for single-molecule fluorescence imaging at a fast frame rate in a wide-field imaging geometry. A sensitive avalanche photodiode (APD) can be used to replace the CCD camera for single-molecule detection when a scanning stage such as a nanopositioner is used to provide high-resolution scanning over the sample in a confocal geometry. The confocal geometry for single-molecule detection has a slower frame rate but provides better spatial resolution and exposes the samples with less incident light than the wide-field collection mode. Single molecules under investigation are supported by an optically transparent electrode such as indium tin oxide (ITO) in order to have their electrochemical reactions under control when fluorescence signals are recorded simultaneously. Using SMS-EC one can measure the electrochemical kinetics of one molecule at a time through the changes in its fluorescence under applied potential. For example, single-molecule electrochemistry of cresyl violet has been investigated using SMS-EC by Ackerman and coworkers [37]. They used clay nanoparticles to form a transparent electrode, which provided a platform for the spontaneous adsorption of cresyl violet and subsequent observation of single-molecule fluorescence and spectro-electrochemical activities. Because cresyl violet does not fluoresce in its reduced state but has a strong fluorescence when oxidized, single-molecule fluorescence dynamics can be correlated to electrochemical potential using single-molecule spectroscopy (Figure 12.3). Fluorescence from a single conjugated polymer chain was totally quenched upon single-hole injection using single-molecule electrochemical measurements [36].

Interfacial charge transfer and catalytic activities of individual nanoparticles can be also studied using the single-molecule technique. For example, heterogeneous reaction pathways and catalytic dynamics of single gold nanoparticles have been

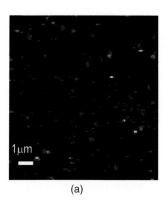

(a)

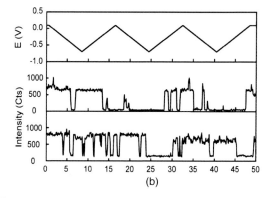

(b)

Figure 12.3 Photoluminescence image of individual cresyl violet⁺ molecules (a) formed by electrochemical modulation and single-molecule fluorescence trajectory under modulation of electrode potential (b). Reprinted with permission from Reference [37]. Copyright 2009 American Chemical Society.

studied recently by Chen and coworkers [38, 39]. This kind of measurement at the single molecule level is important to understand the interfacial charge transfer activities of fuel cell catalysts and photocatalysts since the catalytic activities of these electrochemical systems are highly dependent on particle size, shape, surface modification, and crystal structure. More recently, Chen et al. [40] have studied the single-molecule electrochemistry of resazurin using a single-molecule fluorescence technique. As shown in Figure 12.4, the non-fluorescence resazurin can be reduced at a SWNT surface selectively and its fluorescence can therefore be switched on. This typical electrocatalytic reaction occurs at discrete sites on SWNTs supported by a transparent ITO electrode, and therefore single-molecule fluorescence kinetic analysis can be applied to reveal the reaction mechanism. Heterogeneities in the catalytic reaction at individual reactive sites can be accessed using

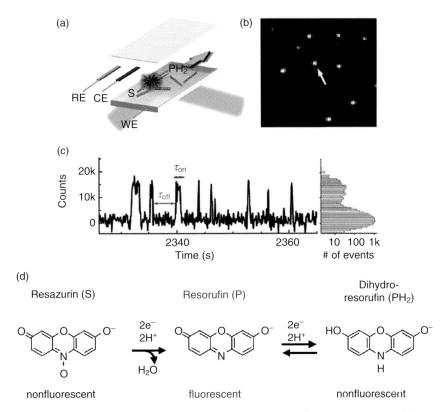

Figure 12.4 (a) Three-electrode flow cell using ITO-supported single-walled carbon nanotubes (SWNTs) as working electrode for (b) real-time detection of single-molecule fluorescence activated by SWNT electrocatalysis; (c) single-molecule fluorescence trajectory from the fluorescence spot marked by the arrow in (b) and histogram statistics of the trajectory; (d) redox reaction of resazurin whose fluorescence can be switched on through a two-electron reduction reaction catalyzed by SWNT. Reprinted with permission from Reference [40]. Copyright 2009 American Chemical Society.

the standard single-molecule fluorescence technique. Other workers have investigated singlet-oxygen radical generation at photocatalyst surface such as TiO_2 [41] and interfacial charge transfer dynamics of a porphyrin complex of zinc [42].

Combined single-molecule spectroscopy and SECM methods has been used to probe real-time pH microscopy at the single molecule level [43]. Single-molecule FRET activities of a fluorophore was used to sense the local pH changes around a SECM tip, which can produce local pH gradients by generation of H^+ or OH^- during redox reactions. Local pH profiles depend on tip diameter, and the buffer/mediator concentration ratio, and the tip–surface distance can therefore be studied by measuring the fluorescence energy transfer efficiency of single molecule. These single-molecule electrochemistry studies based on photoluminescence heavily rely on a large light absorption cross-section and rapid radiative decay of fluorophores. Special electrode treatments and well-defined geometry designs need to be applied for systems that have small light absorption cross-sections and low quantum efficiencies (e.g., Raman and electroluminescence).

12.4
Single-Nanoparticle Electrochemical Detection

12.4.1
Single-Nanoparticle Detection Using Nanoparticle Collision at a Microelectrode

Nanoparticles (NPs) represent a new class of materials that have unique optical, electrical, magnetic, chemical, and mechanical properties. NPs have been used as key components of heterogeneous catalysis [44], fuel storage [45], solar cells [46], nano-electronics [47], medicine and biology sensors [48], light-emitting devices [49], high-density information storage devices [50], nonvolatile memory devices [51], waveguide elements [52], batteries [53], and substrates enabling enormous spectral enhancements for studying the photophysics of molecules [54–56]. The unique characteristics of nanostructures in these widespread applications were always assessed by ensemble measurements. The performance of individual NPs could not be deconvoluted from the overall contributions made by the ensemble. Although several new optical approaches have been developed that can characterize individual NPs immobilized on a solid substrate [57, 58] or diffusing in solution [59], most of these techniques rely on electromagnetic properties of the NPs, such as light scattering, photoluminescence, and optical properties of small molecules attached to the surface of a NP. Electrochemical methods need to be used to reveal the heterogeneities and dynamics existing in single-nanoparticle electrochemistry. Electrochemical detection of a single NP can be studied using a similar detection scheme to that for single-molecule electrochemistry, as discussed above. For example, Xiao and Bard [60] have demonstrated single-nanoparticle electrochemistry by measuring the collision activities of individual platinum nanoparticles at a microelectrode (Figure 12.5). In this method, when a platinum nanoparticle

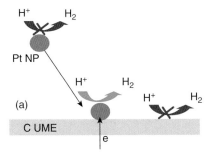
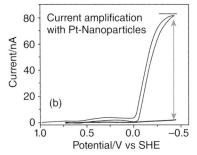

Figure 12.5 Schematic of a single platinum nanoparticle collision event to catalyze proton reduction since proton reduction reaction is favorable at a platinum surface in comparison with a carbon fiber electrode. Reprinted with permission from Reference [60]. Copyright 2007 American Chemical Society.

collides and sticks to the glassy carbon electrode, an electrocatalytic reaction occurs and produces a current that depends on the particle size and the reactant concentration in solution. Electrochemical reactions such as hydrazine oxidation, iodine reduction, and proton reduction were used as model systems to study the platinum nanoparticle collision activities since these reactions occur at a significantly higher rate at Pt than other electrodes (e.g., Au, C, or ITO electrodes). The current responses of particle collision were used to estimate the particle size, size distribution, surface activity, and the nanoparticle diffusion constant. Electroluminescence or photoluminescence of dye molecules caused by nanoparticle activities [61] have also been used to probe single-nanoparticle collision activities. For example, the unusual electrogenerated chemiluminescence (ECL) of graphene oxide nanoparticles was used to detect their collision activities [62]. The advantages of using ECL for single-nanoparticle detection in comparison with the methods described above include a small background signal and no required light source. The disadvantages of using ECL are poor quantum efficiency and electrode stability. These difficulties can be overcome to some extent by modifying metallic nanoparticles to prevent fluorescence surface quenching and by optimizing the ECL detection scheme for a stable ECL co-reactant reagent and high fluorescence quantum yield dye molecules. A single nanoparticle (e.g., Pt, Ni) can be generated and detected by using SECM-controlled electrodeposition on carbon fiber electrodes (CFEs) [63]. A catalytic reaction such as Fe^{3+} reduction at individual nanoparticles can be used to detect a single nanoparticle at the conductive carbon surface. In addition, single-nanoparticle electrochemistry can also be measured by probing individual nanoparticles attached to a conductive substrate using a STM tip or carbon nanoelectrode [64–67]. This method can measure single particle electrochemistry and geometry simultaneously and provide more information on the single-nanoparticle electrochemistry than other methods described previously.

12.4.2
Single-Nanoparticle Electrochemistry Using Single-Molecule Spectroscopy

Using techniques similar to SMS-EC, single-nanoparticle electrochemistry can be investigated indirectly. For instance, conjugated polymer nanoparticle fluorescence [68] and ECL can be used to understand charge storage and redox activities of single nanoparticles made of conjugated polymer [69]. Light scattering characteristics of single molecules or nanoparticles have also been used to collect single-molecule events. For example, the growth kinetics of individual gold nanorods by electroless deposition has been studied using single-particle surface plasmon spectroscopy [70]. In addition, it might be possible to study the electrochemical response of a single redox molecule using Raman spectroscopy if a well-defined nanostructure could be fabricated to supply a strong local field enhancement [29–32].

12.5
Nanoelectrodes for Ultrasensitive Electrochemical Detection and High-Resolution Imaging

Nanoelectrodes have displayed high sensitivities in trace analysis and can be used for single-molecule detection, as demonstrated in previous sections. The size and geometry of the nanoelectrodes play a vital role in their electrochemical performances. Very fast mass transfer at these nanometer-scale electrodes can be achieved [71], enabling one to obtain high resolution constant-height and constant-current imaging of both topography and local electrochemical activities with one electrode simultaneously [72]. The electrochemical performances and electron-transfer kinetic parameters are also highly dependent on the electrode material and redox species used [73]. In this section, we discuss recent work on novel nanoelectrode fabrication and nanoelectrode electrochemistry. Combined instrumental work performed with nanoelectrode and other techniques such as scanning probe microscopy (SPM) and optical microscopy technique for high-resolution imaging will be discussed. Our discussion will mainly focus on electrochemistry and applications of single nanoelectrodes with well-defined geometries and controlled electrode potential. These electrodes are primarily used for single-molecule detection, SECM and SPM probes that can be used to target phenomena or processes at nanoscale spatial resolution. Our discussions exclude electrochemistry of nanoelectrode arrays and nanoparticle ensembles as described explicitly in other chapters in this book [12].

12.5.1
Nanoelectrode Fabrication

Recent breakthroughs in nanotechnology have made electrodes of various materials and geometries at the nanoscale available. A nanometer-sized carbon fiber electrode has been used as SECM tip for surface micropatterning of active enzyme with a high resolution [74]. Such carbon fiber nanoelectrodes of nanometer size can be fabricated by electrochemical etching of carbon fibers followed by the deposition of electrophoretic paint [75]. Electrochemical behavior in a picoliter microenvironment has been probed with a single-nanotube based nano-needle electrode [76]. A membrane template deposition–dissolution procedure has been developed to create a porous Au nanowire electrode [77]. Recessed Pt nanoelectrodes can be fabricated by using direct-write local ion milling of a silicon nitride overlayer of Pt electrode [78]. Electrochemical characterizations show that the diffusion of redox species to the pore mouth of the recessed Pt nanoelectrode is the mass-transport limiting step. A gold nanoelectrode has been fabricated by electrochemical etching of a gold wire, insulating the wire with a varnish and removing the apex using ultrashort pulse etching in strong acid [79]. A carbon nanoelectrode can be made by exposing a single-carbon nanotube tip protected by an insulating material [80]. Fundamental voltammetric behavior at a platinum nanoelectrode embedded in polyacrylic acid has been studied [81]. In addition, a simple yet highly reproducible bench-top method for fabricating glass-sealed nanodisk electrodes, glass nanopore electrodes, and glass nanopore membranes of controlled size has been developed [82, 83]. Li and coworkers [84] have studied the electrochemical response of 1–3 nm Pt disk electrodes. The nanoelectrode was prepared by sealing Pt microwire into a bilayer quartz capillary and pulling using a laser-assisted pulling process to create an ultrasharp Pt nanowire sealed in a silica tip. The ultrasharp tip is then sealed into a piece of glass tubing, which is manually polished to expose the Pt nanodisk. Needle-type Pt-disk nanoelectrode can also be fabricated using this method [85].

12.5.2
Mass Transfer near a Nanoelectrode

A fundamental understanding of and an ability to control mass transfer behavior at nanoelectrodes are critical, especially for applications of trace analysis. Figure 12.6 illustrates a new concept of controlling mass transfer behavior at nanoscaled electrodes and membranes [86]. In this experiment, mass transfer is controlled by a glass nanopore electrode structure (Figure 12.6a) and chemically modified surface at the orifice region (Figure 12.6b). The mass transfer properties of nanoscaled devices are highly sensitive to surface features since their surface/volume ratio is high. Spiropyran molecules were coupled to the surface amine groups on the nanopore interior surface. The transport of charged molecules through the sub-100 nm nanopore electrodes can be controlled by electrostatic interaction by the surface electric field, which is imparted by the photochemical reaction of spiropyran molecules on the nanopore surface: UV light irradiation converts the neutral spiropyran into positive protonated merocyanine form while visible light reverses the process. The transport of positively charged species can be switched on and off by UV and visible irradiation. The transport of neutral molecules is not affected. A photochemical trap can be created by the same concept, in that the diffusion through the pore orifice of the charged molecules initially inside the

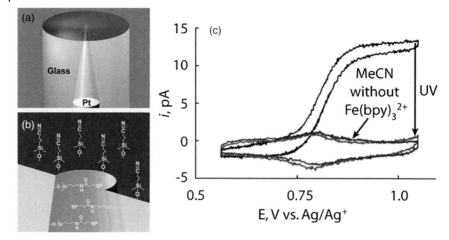

Figure 12.6 Schematic of a glass nanopore electrode (a) with interior and exterior glass surfaces modified, respectively, with spiropyran and cyanopropyl groups (b); (c) electrostatic trapping of $Fe(bpy)_3^{2+}$ inside a 30-nm-radius, 6.2-μm-deep nanopore electrode. Reprinted with permission from Reference [86]. Copyright 2006 American Chemical Society.

nanopore would be limited as well. Similar to the electrochemical response of thin layer cell or adsorption process, a symmetric current response was observed after UV irradiation (Figure 12.6c). The disappearance of the steady-state limiting current of the nanoelectrode signals the blockage of the redox transfer. Since slow scan voltammetry was used, the slow turnover rate limited the sensitivity, as a relatively large number of redox molecules (~10 million) were required to produce an appreciable current (picoamperes).

A combined sealing and pulling process has been employed to create a single nanotip channel [87]. Transport of nanoparticles and DNA were monitored by conductivity measurements based on the concepts of resistive pulse sensing. As a nanoparticle passes through the nanopore it blocks the pore and causes the mass transfer resistance, or current signal, to change. The pattern of the current change corresponds to single-particle translocation behavior and reveals thermodynamic and kinetic information.

12.5.3
Combined Optical and Electrochemical Imaging

To collect both electrochemical and optical signals at nanoscale domains, a nanosized optical fiber electrode has been fabricated for SECM and near-field optical microscopy imaging (Figure 12.7) [14]. The authors reproducibly made pencil-shaped and triple-tapered electrodes with nanometer radii by a selective chemical etching method. Electrochemical performance of the tip was tested by

12.5 Nanoelectrodes for Ultrasensitive Electrochemical Detection and High-Resolution Imaging | 331

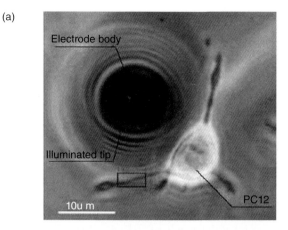

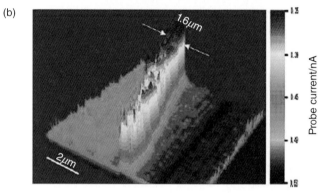

Figure 12.7 Pencil-shaped electrode fabricated using an optical fiber after selective chemical etching and coating for simultaneous electrochemical and optical measurements. Optical micrograph (a) of a differentiated PC12 cell and a nanometer-sized electrode used for SECM 3D imaging (b) of a differentiated PC12 cell using the nanoelectrode. Reprinted with permission from Reference [14]. Copyright 2006 American Chemical Society.

SECM. In a combined approach with optical and near-field scanning microscopy, single-cell SECM imaging and optical information can be measured simultaneously. Demaille and coworkers have modified a gold nanoelectrode with PEG polymer labeled with the redox molecule ferrocene [88]. They claimed that local electrochemical reactions at nanoscale can be probed in an atomic force microscopy (AFM)-SECM mode (Figure 12.8). The long flexible PEG chain reduces the size of the tip–substrate interaction area, which allows the AFM/SECM to efficiently probe local electrochemical reactivity of the sample substrate. In principle, such modified electrodes can be used to investigate the dynamics and heterogeneities of single biomolecule activities and for high-resolution electrochemical imaging.

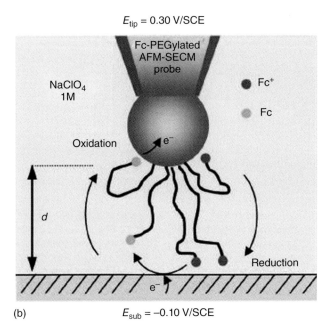

Figure 12.8 (a) Surface modification of an AFM probe with Fc-PEG disulfide molecule for AFM-SECM experiment; (b) principle of AFM-SECM microscopy, showing the reduced size of the tip–substrate interaction area because of the flexible Fc-PEG backbone and the Fc heads used to efficiently "sense" locally the electrochemical reactivity of the substrate. Reprinted with permission from Reference [88]. Copyright 2009 American Chemical Society.

12.6
Electrochemical Detection in Nanodomains of Biological Systems

Microelectrodes (e.g., carbon fibers) have been applied to directly measuring neurotransmitters such as dopamine in the brain (behaving mouse) by voltammetric methods [89, 90]. Single-wall carbon nanotubes have been used to modify the carbon-fiber microelectrode to enhance the detection sensitivity of serotonin, another neurotransmitter. Biofouling was found to be suppressed after the nanotube modification [91]. To probe nanoscale biological domains and achieve better electrochemical performance (e.g., sensitivity and spatial electrochemical imaging resolution) than the microelectrodes alone, the overall dimension of the electrode has to be miniaturized to nanometer scale. Kueng *et al.* [92, 93] have demonstrated the integration of nanoelectrodes with AFM for the investigation of laterally resolved information on biological activity. Both topographical and electrochemical information on soft surfaces (e.g., an enzyme-modified micropattern) and polymers can be obtained. Mirkin *et al.* [94] have used a nanoelectrode in combination with the SECM to image cultured human breast cells. The nanoelectrode can penetrate a cell and travel inside without causing damage to the cell membrane. Therefore, the transmembrane charge transport rate, membrane potential, and redox properties can be measured at not only a single cell but also at subcellular level. The nanoelectrode-based imaging method needs to be validated for future *in vivo* sensing by evaluating such criteria as time resolution, chemical resolution, selectivity for the analyte of interest, and sensitivity. Compared to other voltammetric methods, fast scan and potential reversal waveforms offer advantages of background subtraction and analyte identification. Owing to the complex chemical environment in tissues of behaving animal, unequivocal selectivity of a specific analyte is unlikely for any single method. Therefore, coordinated signal verification by multiple methods is essential for *in vivo* sensing at the nanometer scale.

Nanoelectrodes could also be used for the electrochemical sensing of biomolecule folding and specific binding [95, 96] with less sample and greater sensitivity than macro- and micro-electrodes or even nanoelectrode arrays. Many biological activities such as protein folding and DNA binding are associated with conformation changes. Various strategies are being developed to probe these features, both for fundamental understanding as well as biosensor development. Among these strategies, electrochemical methods will enjoy practical advantages such as low cost and field-friendly instrumentation.

12.7
Localized Delivery and Imaging by Using Single Nanopipette-Based Conductance Techniques

The use of nanoelectrodes for trace analysis in biological system is nontrivial. The signals or responses in biological system often require a series of trigging events. With nanoscopic electrodes and pipettes available, significant advances can be achieved as subcellular or other nanodomains can be selectively activated for mechanistic studies. By utilizing a pipette tip with a nanosize opening, Klenerman and coworkers have achieved sub-micron local delivery of reagents in a controllable fashion [97]. The experiment was performed with scanning ion conductance microscopy (SICM). SICM belongs to the scanning probe microscopy family, distinguishing itself from other methods, including SECM, in that its probe is a hollow nanopipette [98]. The ionic current flowing through the nanopipette tip is monitored continuously while the nanopipette is scanned over the sample surface (often biological cells). The current variation reflects changes in the local ionic flux and provides noncontact imaging of cell surfaces. In addition to monitoring cell responses at subcellular domains, the hollow nanopipette can be used to release specific reagents at desired location as well – a concept similar to dip-pen nanotechnology [99, 100]. Figure 12.9 shows the experimental apparatus. The SICM is coupled with total internal reflection fluorescence (TIRF) optical imaging for independent quantification of each delivery dose. After the pipette tip–surface distance, bias potential, and other factors are calibrated, the delivery dosage is driven by a pulsed voltage as short as 10 ms and can affect local areas within a radius of 1–30 micron.

Very impressive proof-of-concept experiments have been presented for the melting of individual DNA molecules by localized delivery of H^+ ions (Figure 12.10). In this demonstration, single-stranded DNA was complexed with complementary DNA bearing a FRET donor. Local delivery of 1 M HCl caused the dis-

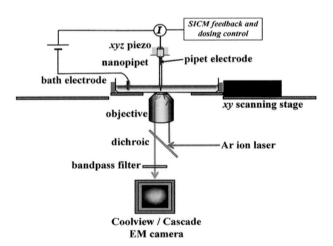

Figure 12.9 Schematic of the apparatus used for localized dosing in scanning ion conductance microscopy (SICM). The pipets with 100 and 320 nm diameter opening were prepared by pulling a glass capillary with a micropipette puller. A metal wire electrode was inserted at the back side of the glass pipet to control the bias potential applied. Reprinted with permission from Reference [97]. Copyright 2008 American Chemical Society.

(a)

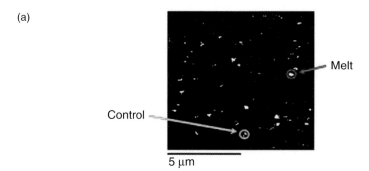

(b)
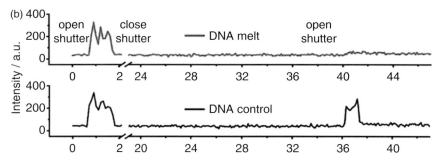

Figure 12.10 Confocal imaging of single DNA molecule dissociation by local unfolding. Biotinylated single-strand DNA was immobilized on the surface with FRET donor attached. The complementary 12 base-pair strand with FRET acceptor was then dosed to form DNA duplex. The shutter was briefly opened and then closed to confirm the presence of the DNA molecule with the acceptors. Reprinted with permission from Reference [97]. Copyright 2008 American Chemical Society.

sociation of DNA duplexes, thus separating the donor from the acceptor, which lowers the FRET efficiency. The confocal image (Figure 12.10a) and time-base signals (Figure 12.10b) illustrate that a local dose of 1 M HCl effectively reduced the acceptor's signal to background level while adjacent molecules were not affected. Photobleaching was ruled out as duplexes spaced microns apart remained detectable. While single-molecule resolution was achieved in this case by fluorescence detection instead of electrochemical means, the powerful combination of electrochemical nanodomain excitation and optical detection holds promise for other applications of nanoelectrochemistry in biology.

12.8
Final Remarks

Single-molecule, single-nanoparticle, and single-event electrochemical activities have been studied using nanoelectrodes with well-defined geometries and single-molecule spectroelectrochemistry techniques. Rich information can be obtained by the combined approaches of optical (e.g., photoluminescence, light scattering, ECL, and Raman) and electrochemical methods (e.g., voltammetry, SECM, SICM, and patch clamp). These powerful techniques can be applied to study the interactions of macromolecules and their activities in living cells and potentially in behaving animals. Optical methods provide spectroscopic information, allowing one to study single-molecule conformation changes, photophysics, and translocation in living cells. Electrochemical methods provide a unique platform to facilitate electrochemical reactions of single molecules and single nanoparticles containing redox molecules. More importantly, mass transfer at the nanoscale can be manipulated by designed nanoelectrode devices and applied potential waveforms. This allows one to actively control the analytical process. The combined approach has other merits in addition to high resolution and sensitivity. Conventional electrochemical measurements on concentrated samples only provide an ensemble average of the properties of interest and provide no means of assessing the heterogeneity and local information of a complex system. A combined optical and electrochemical approach, in contrast, can provide insight into the behavior of each redox molecule and individual nanoparticles in a complex electrochemical environment.

Recent advances in nanoscience and nanotechnology lead us to believe that electrodes with any imaginable geometry and composition could be fabricated to seek answers previously unachievable, especially when interfaced with versatile techniques such as SECM, SPM, and spectroscopic methods. As yet, single-molecule detection using single nanoelectrodes remains a challenging topic due to the slow turnover rate of redox molecules and high background signals. This problem is expected to be solved by selecting an ideal co-reactant to undergo homogeneous catalytic reaction to enhance the turnover rates of single redox molecule, and modifying the geometry and/or surface of the nanoelectrode with a special functional group to help facilitate efficient charge transfer. Some other technical challenges and scientific questions that need to be addressed in this area are (i) how to achieve single-molecule ECL using single-molecule electrochemical techniques; (ii) how to prepare nanoelectrodes with well-defined geometry and capable of studying single-molecule electrochemistry when interfaced with non-fluorescent techniques; (iii) how to develop new methodology for ultrasensitive detection of macromolecule interactions and conformation changes using nanoelectrodes; and (iv) how to apply these techniques to better understand broadly defined biological systems.

Acknowledgments

S. P. is grateful to The University of Alabama College of Arts and Sciences for start-up funds. G. W. would like to acknowledge support from the Department of Chemistry and Research Initiation Award of Georgia State University.

References

1. Kapanidis, N. and Strick, T. (2009) *Trends Biochem. Sci.*, **34** (5), 234–243.
2. Brand, L. and Johnson, M.L. (2008) *Methods in Enzymology, Acadamic Press*, **450** (1), 129–157.
3. Karunatilaka, K.S. and Rueda, D. (2009) *Chem. Phys. Lett.*, **476** (1–3), 1–10.
4. Raj, A. and van Oudenaarden, A. (2009) *Annu. Rev. Biophys. Biomol. Struct.*, **38**, 255–270.
5. Ipowsky, I.R., Beeg, J., Dimova, R., Klumpp, S., Liepelt, S., Mueller, M.J.I., and Valleriani, A. (2009) *Biophys. Rev. Lett.*, **4** (1/02), 77–137.
6. Eremchev, Y., Naumov, A.V., Vainer, Y.G., and Kador, L.J. (2009) *J. Chem. Phys.*, **130** (18), 184507/1–184507/5.
7. Budini, A.A. (2009) *Phys. Rev. A At. Mol. Opt. Phys.*, **79** (4 Pt. B), 043804/1–043804/17.
8. Etchegoin, P.G. and Le Ru, E.C. (2008) *Phys. Chem. Chem. Phys.*, **10** (40), 6079–6089.
9. Etchegoin, P.G., Le Ru, E.C., and Meyer, M. (2009) *J. Am. Chem. Soc.*, **131** (7), 2713–2716.
10. Fan, F.-R.F. and Bard, A.J. (1995) *Science*, **267**, 871–874.
11. Fan, F.-R.F., Kwak, J., and Bard, A.J. (1996) *J. Am. Chem. Soc.*, **118**, 9669–9675.
12. Murray, R.W. and Royce, W. (2008) *Chem. Rev.*, **108** (7), 2688–2720.
13. Arrigan, D.W.M. (2004) *Analyst*, **129**, 1157–1165.
14. Maruyama, K., Ohkawa, H., Ogawa, S., Ueda, A., Niwa, O., and Suzuki, K. (2006) *Anal. Chem.*, **78** (6), 1904–1912.
15. Bard, A.J. and Faulkner, L.R. (2001) *Electrochemical Methods: Fundamentals and Applications*, 2nd edn, John Wiley & Sons, Inc., New York, p. xxi, 833 pp.
16. Mirkin, M. and Bard, A.J. (2001) *Scanning Electrochemical Microscopy*, Marcel Dekker, New York.
17. Itamar, W. (2003) *Science*, **299**, 1877–1881.
18. Bard, A.J. (2008) *ACS Nano*, **2**, 2437–2440.
19. Amatore, C., Gruen, F., and Maisonhaute, E. (2003) *Angew. Chem. Int. Ed.*, **42** (40), 4944–4947.
20. Metzger, R.M. (2008) *J. Mater. Chem.*, **18**, 4364–4396.
21. Zhang, J.D., Kuznetsov, A.M., Medvedev, I.G., Chi, Q.J., Albrecht, T., Jensen, P.S., Ulstrup, J.K., and Jens., K.L. (2008) *Chem. Rev.*, **108** (7), 2737–2791.
22. Peng, S. and Mirkin, M.V. (2008) *J. Am. Chem. Soc.*, **130**, 8241–8250.
23. Hoeben, F.J.M., Meijer, F.S., Dekker, C., Albracht, S.P., Heering, H.A., and Lemay, S.G. (2008) *ACS Nano*, **2**, 2497–2504.
24. Landes, C.F., Zeng, Y.N., Liu, H.W., Musier-Forsyth, K., and Barbara, P.F. (2007) *J. Am. Chem. Soc.*, **129** (33), 10181–10188.
25. Bokinsky, G., Rueda, D., Misra, V.K., Rhodes, M.M., Gordus, A., Babcock, H.P., Walter, N.G., and Zhuang, X.W. (2003) *Proc. Nat. Acad. Sci. U. S. A.*, **100** (16), 9302–9307.
26. Bates, M., Blosser, T.R., and Zhuang, X.W. (2005) *Phys. Rev. Lett.*, **94** (10), 108101.
27. Selvin, P.R., and Ha, T. (2007) *Single Molecule Techniques, A Laboratory Manual*, CSHL Press, New York.
28. Gell, C., Brockwell, D., and Smith, A. (2006) *Handbook of Single Molecule Fluorescence Spectroscopy*, Oxford University Press, Oxford.
29. Liz-Marzan, L.M. (2006) *Langmuir*, **22**, 32–41.
30. Xia, Y. and Halas, N.J. (2005) *J. Mater. Res. Bull.*, 30.
31. Daniel, M.C. and Astruc, D. (2004) *Chem. Rev.*, **104**, 293–346.
32. Chang, S.S., Shih, C.W., Chen, C.D., Lai, W.C., and Wang, C.R.C. (1999) *Langmuir*, **15**, 701–709.
33. Nie, S.M. and Emory, S.R. (1997) *Science*, **275**, 1102–1106.
34. Yeo, B.S., Stadler, J., Schmid, T., Zenobi, R., and Zhang, W.H. (2009) *Chem. Phys. Lett.*, **472** (1–3), 1–13.

35 Etchegoin, P.G. and Le Ru, E.C. (2008) *Phys. Chem. Chem. Phys.*, **10** (40), 6079–6089.
36 Palacios, R.E., Fan, F.-R.F., Bard, A.J., and Barbara, P.F. (2006) *J. Am. Chem. Soc.*, **128**, 9028–9029.
37 Lei, C.H., Hu, D.H., and Ackerman, E. (2009) *Nano Lett.*, **9** (2), 655–658.
38 Xu, W.L., Kong, J.S., Yeh, Y.T.E., and Chen, P. (2008) *Nat. Mater.*, **7**, 992–996.
39 Chen, P., Xu, W., Zhou, X., and Panda, D. (2009) *Chem. Phys. Lett.*, **470**, 151–157.
40 Xu, W., Shen, H., Kim, Y., Yoon, J., Zhou, X., Liu, G., Park, J., and Chen, P. (2009) *Nano Lett.*, **9**, 3968–3973.
41 Naito, K., Tachikawa, T., Fujitsuka, M., and Majima, T. (2009) *J. Am. Chem. Soc.*, **131**, 934–936.
42 Wang, Y.M., Wang, X.F., Ghosh, S.K., and Lu, H.P. (2009) *J. Am. Chem. Soc.*, **131**, 1479–1487.
43 Boldt, F.M., Heinze, J., Diez, M., Petersen, J., Boersch, M., and Michael, F. (2004) *Anal. Chem.*, **76**, 3473–3481.
44 Park, K.W., Choi, J.H., Kwon, B.K., Lee, S.A., Sung, Y.E., Ha, H.Y., Hong, S.A., Kim, H.S., and Wieckowski, A. (2002) *J. Phys. Chem. B*, **106** (8), 1869–1877.
45 Langhammer, C., Zorić, I., Kasemo, B., and Clemens, B.M. (2007) *Nano Lett.*, **7** (10), 3122–3127.
46 Lee, Y.L., Huang, B.M., and Chien, H.T. (2008) *Chem. Mater.*, **20** (22), 6903–6905.
47 Becerril, H.A., Stoltenberg, R.M., Wheeler, D.R., Davis, R.C., Harb, J.N., and Woolley, A.T. (2005) *J. Am. Chem. Soc.*, **127** (9), 2828–2829.
48 Scodeller, P., Flexer, V., Szamocki, R., Calvo, E.J., Tognalli, N., Troiani, H., and Fainstein, A. (2008) *J. Am. Chem. Soc.*, **130** (38), 12690–12697.
49 Tan, M., Munusamy, P., Mahalingam, V., and van Veggel, F.C.J.M. (2007) *J. Am. Chem. Soc.*, **129** (46), 14122–14123.
50 Chen, M., Kim, J., Liu, J.P., Fan, H., and Sun, S. (2006) *J. Am. Chem. Soc.*, **128** (22), 7132–7133.
51 Begtrup, G.E., Gannett, W., Yuzvinsky, T.D., Crespi, V.H., and Zettl, A. (2009) *Nano Lett.*, **9** (5), 1835–1838.
52 Cai, W., Sainidou, R., Xu, J., Polman, A., and Javier Garcia de Abajo, F. (2009) *Nano Lett.*, **9** (3), 1176–1181.
53 Shaju, K.M. and Bruce, P.G. (2008) *Chem. Mater.*, **20** (17), 5557–5562.
54 Brechignac, C., Lahmani, M., and Houdy, P. (2008) *Nanomaterials and Nanochemistry*, Springer, Dordrecht.
55 Schmid, G. (ed.) (1994) *Clusters and Colloids*, VCH, Weinheim.
56 Hayat, M.A. (ed.) (1991) *Colloidal Gold: Principles, Methods, and Applications*, Academic Press, San Diego.
57 Palacios, R.E., Fan, F.-R.F., Grey, J.K., Suk, J., Bard, A.J., and Barbara, P.F. (2007) *Nat. Mater.*, **6**, 680–685.
58 Sönnichsen, C., Reinhard, B.M., Liphardt, J., and Alivisatos, A.P. (2005) *Nat. Biotechnol.*, **23**, 741–745.
59 Ignatovich, V., Topham, D., and Novotny, L. (2006) *J. Sel. Top. Quantum Electron.*, **12**, 1292–1300.
60 Xiao, X.Y. and Bard, A.J. (2007) *J. Am. Chem. Soc.*, **129**, 9610–9612.
61 Fan, F.R.F. and Bard, A.J. (2008) *Nano Lett.*, **8**, 1746–1749.
62 Fan, F.R.F., Park, S.J., Zhu, Y.W., Ruoff, R.S., and Bard, A.J. (2009) *J. Am. Chem. Soc.*, **131**, 937–939.
63 Tel-Vered, R. and Bard, A.J. (2006) *J. Phys. Chem. B*, **110** (50), 25279–25287.
64 Eikerling, M., Meier, J., and Stimming, U.Z. (2003) *Phys. Chem.*, **217**, 395–414.
65 Chen, S., Kucernak, C.S., and Kucernak, A. (2004) *J. Phys. Chem. B*, **108**, 13984–13994.
66 Chen, S., Kucernak, C.S., and Kucernak, A. (2003) *J. Phys. Chem. B*, **107**, 8392–8402.
67 Meier, J., Schiot, M.J., Schiot, J., Liu, P., Norskov, J.K., and Stimming, U. (2004) *Chem. Phys. Lett.*, **390**, 440.
68 Fan, F.R.F., Grey, J.K., Suk, J.D., Bard, A.J., and Barbara, P.F. (2007) *Nat. Mater.*, **6**, 680–685.
69 Chang, Y.L., Palacios, R.E., Fan, F.R.F., Bard, A.J., and Barbara, P.F. (2008) *J. Am. Chem. Soc.*, **130**, 8906–8907.
70 Novo, C., Funston, A.M., and Mulvaney, P. (2008) *Nat. Nanotechnol.*, **3**, 598–602.

71. Penner, R.M., Heben, M.J., Longin, T.L., and Lewis, N.S. (1990) *Science*, **250** (4984), 1118–1121.
72. Laforge, F., Velmurugan, J., Wang, Y.X., and Mirkin, M.V. (2009) *Anal. Chem.*, **81** (8), 3143–3150.
73. Velmurugan, J., Sun, P., and Mirkin, M.V. (2009) *J. Phys. Chem. C*, **113** (1), 459–464.
74. Li, X., Geng, Q.H., Wang, Y.Y., Si, Z.K., Jiang, W., Zhang, X.L., and Jin, W.R. (2007) *Electrochim. Acta*, **53** (4), 2016–2024.
75. Chen, S.L. and Kucernak, A. (2002) *J. Phys. Chem. B*, **106** (36), 9396–9404.
76. Yum, K., Cho, H.N., Hu, J., and Yu, M.F. (2007) *ACS Nano*, **1** (5), 440–448.
77. Laocharoensuk, R., Sattayasamitsathit, S., Burdick, J., Kanatharana, P., Thavarungkul, P., and Wang, J. (2007) *ACS Nano*, **1** (5), 403–408.
78. Lanyon, Y.H., De Marzi, G., Watson, Y.E., Quinn, A.J., Gleeson, J.P., Redmond, G., and Arrigan, D. (2007) *Anal. Chem.*, **79** (8), 3048–3055.
79. Woo, D.H., Kang, H., and Park, S.M. (2003) *Anal. Chem.*, **75**, 6732–6736.
80. Quinn, B., Bernadettes, M., and Lemay, G. (2006) *Adv. Mater.*, **18** (7), 855–859.
81. Watkins, J.J., Chen, J., White, H.S., Abruna, H.D., Maisonhaute, E., and Amatore, C. (2003) *Anal. Chem.*, **75** (16), 3962–3971.
82. Zhang, B., Galusha, J., Shiozawa, P.G., Wang, G., Bergren, A.J., Jones, R.M., White, R.J., Ervin, E., Cauley, C.C., White, H.S., and Henry, S. (2007) *Anal. Chem.*, **79** (13), 4778–4787.
83. Zhang, B., Zhang, Y.H., and White, H.S. (2004) *Anal. Chem.*, **76** (21), 6229–6238.
84. Li, Y.X., Bergman, D., and Zhang, B. (2009) *Anal. Chem.*, **81**, 5496–5502.
85. Katemann, B.B. and Wolfgang, W.S. (2002) *Electroanalysis*, **14** (1), 22–28.
86. Wang, G., Bohaty, A.K., Zharov, I., and White, H.S. (2006) *J. Am. Chem. Soc.*, **128**, 13553–13558.
87. Zhang, B., Wood, M., and Lee, H. (2009) *Anal. Chem.*, **81** (13), 5541–5548.
88. Anne, A., Demaille, C., and Goyer, C. (2009) *ACS Nano*, **3** (4), 819–827.
89. Robinson, D.L., Hermans, A., Seipel, A.T., and Wightman, R.M. (2008) Monitoring rapid chemical communication in the brain. *Chem. Rev.*, **108**, 2554–2584.
90. Phillips, P.E.M. and Wightman, R.M. (2003) *Trends Anal. Chem.*, **22**, 509–514.
91. Swamy, B.E. and Venton, B.J. (2007) *Analyst*, **132** (9), 876–884.
92. Kranz, C., Kueng, A., Lugstein, A., Bertagnolli, E., and Mizaikoff, B. (2004) *Ultramicroscopy*, **100** (3–4), 127–134.
93. Vo-Dinh, T. (2008) *Protein Nanotechnology*, Humana Press, **300**, 403–415.
94. Sun, P., Laforge, F.O., Abeyweera, T.P., Rotenberg, S.A., Carpino, J., and Mirkin, M.V. (2008) *Proc. Nat. Acad. Sci. U. S. A.*, **105** (2), 443–448.
95. Fan, C., Plaxco, K.W., and Heeger, A.J. (2003) *Proc. Natl. Acad. Sci. U. S. A.*, **100**, 9134.
96. Lubin, A.A., Vander Stoep Hunt, B., White, R.J., and Plaxco, K.W. (2009) *Anal. Chem.*, **81**, 2150–2158.
97. Piper, J.D., Li, C., Lo, C.-J., Berry, R., Korchev, Y., Ying, L., and Klenerman, D. (2008) *J. Am. Chem. Soc.*, **130**, 10386–10393.
98. Hansma, P.K., Drake, B., Marti, O., Gould, S.A., and Prater, C.B. (1989) *Science*, **243**, 641–643.
99. Piner, R.D., Zhu, J., Xu, F., Hong, S., and Mirkin, C.A. (1999) *Science*, **283**, 661–663.
100. Lee, B., Park, S.J., Mirkin, C.A., Smith, J.C., and Mrksich, M. (2002) *Science*, **295** (5560), 1702–1705.

13
Analytical Applications of Block Copolymer-Derived Nanoporous Membranes
Takashi Ito and D.M. Neluni, T. Perera

13.1
Introduction

Nanoporous media have been widely used for separation and detection methods. For example, entangled polymer gels and media have been employed as the stationary phase in size-exclusion chromatography that has been widely used to separate biological and polymer molecules based on size [1, 2]. Chemically-modified electrodes based on charged polymer films offer charge-based selectivity in detection of small redox-active molecules [3]. Although these commonly-used nanoporous media can be prepared by simple procedures, their polydisperse and flexible porous structures may limit their separation performance. Additionally, it is challenging to develop theoretical models of molecular motion within such heterogeneous media in order to understand quantitatively experimental mass transport behavior. In contrast, monolithic membranes containing arrays of uniform cylindrical nanopores have been employed as model systems for quantitative theoretical studies of molecular mass transport within nanoporous media because of their simpler, well-defined structures [4, 5]. In addition, the uniform diameters and well-defined pore structures in such membranes make it possible to tune their separation efficiency in a controlled manner [5, 6]. In this chapter, we will briefly overview previous achievements in investigations of chemical separation and detection based on molecular transport through membranes containing cylindrical nanopores. Subsequently, we briefly explain fabrication methods of block copolymer (BCP)-derived nanoporous monoliths, and describe their applicability for trace analysis that takes advantage of well-controlled molecular transport through the uniform cylindrical nanopores.

13.2
Monolithic Membranes Containing Arrays of Cylindrical Nanoscale Pores

As monolithic membranes containing uniform, cylindrical nanopores, two types of commercially available membranes have been widely used: track-etched

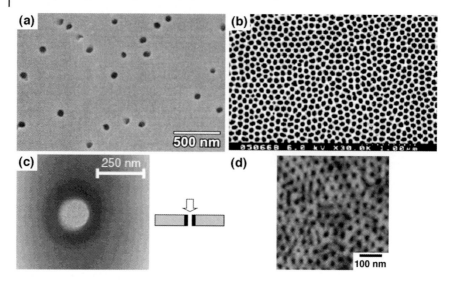

Figure 13.1 (a) SEM image of a track-etched polycarbonate membrane. Reprinted with permission from Reference [6]. Copyright 2005 Taylor & Francis. (b) SEM image of a nanoporous anodic alumina membrane. Reprinted with permission from Reference [6]. Copyright 2005 Taylor & Francis. (c) TEM image of a single carbon nanotube channel. Reprinted with permission from Reference [15]. Copyright 2004 American Chemical Society. (d) AFM image of a 71K PS-b-PMMA-derived nanoporous film on gold. Reprinted with permission from Reference [8]. Copyright 2008 American Chemical Society.

polymer membranes (TEPMs) and nanoporous anodic alumina membranes (NAAMs) [7].

TEPMs (Figure 13.1a) are fabricated by chemical etching of damage tracks created by bombarding a sheet of polycarbonate or polyester film (ca 10 μm thick) with nuclear fission fragments. The distribution of the pores is random and, generally, the pore density is relatively low (≤10 pores per μm^2). The pore diameter can be controlled by varying the etching time, and membranes with several different pore diameters (10 nm–10 μm) are commercially available. Additionally, the pore diameter can be decreased by electroless deposition of gold onto the pore surface [5, 6]. By controlling the bombarding and etching conditions, TEPMs containing cone-shaped pores at a controlled density can also be fabricated [6, 9]. The pore surface can be tailored via amidation for polyester-based TEPMs as well as via thiolate self-assembled monolayer (SAM) modification on gold-coated nanopores [5, 6].

NAAMs (Figure 13.1b) are fabricated by anodization of Al metal (usually, ca 60 μm thick Al foils). The hexagonal distribution of nanopores results in very high porosity (up to 10^3 pores μm^{-2}). The pore diameter, which is larger than 5 nm, can be controlled by changing the anodization conditions, and can be decreased by immobilizing sol–gel thin films on the nanopore surface [6]. The surface properties of nanopores can be tailored by immobilizing organosilanes on the pore surface [6].

These membranes with controlled molecular-sized diameter and surface functionality exhibited unique selectivity in molecular separation, including distinct size-based separation of small molecules and proteins, charge-based molecular separation based on permselective transport, hydrophobicity-based separation, and enhanced separation of chiral molecules and DNA based on chemical interactions [6]. These membranes when incorporated into microfluidic devices could be used as potential-controlled gates [10] and for the electrokinetic preconcentration of molecules [11]. For designing detection methods, these membranes were used as templates to fabricate nanodisk electrode arrays that can detect redox-active species at much higher sensitivity than a regular disk electrode due to the overlapped diffusion layers extended from individual electrodes [12]. Highly sensitive detection of molecules that adsorb onto the nanopore surface was possible by measuring the molecular permeability through these nanoporous membranes [6].

More recently, unique mass transport behavior through carbon nanotube (CNT) pores was demonstrated using monolithic membranes incorporating vertically aligned CNTs in solution-impermeable matrixes. In contrast to nanopores within TEPMs and NAAMs, CNT nanopores are perfectly cylindrical, chemically homogeneous, and atomically smooth, and thus will provide ideal model systems for understanding molecular mass transport through nanoporous media. In addition, these unique characteristics of a CNT nanopore lead to unique mass transport behavior that can be applied for trace analysis methods. For example, a monolithic polymer membrane containing a single CNT pore (Figure 13.1c) permitted us to simultaneously measure the size and penetration time of individual nanoparticles based on the principle of a Coulter counter [13–16]. Owing to the negligible surface charge of a CNT pore, nanoparticles pass through the pore via electrophoresis, making it possible to determine the surface charges of individual nanoparticles from their penetration times [14–16]. CNT-array membranes exhibited enhanced fluid flow due to the frictionless molecular transport through the atomically-smooth surface of CNT pores [17–19] in addition to distinct size-based molecular separation [17, 19]. However, these membranes have not been studied extensively, probably due to the technical difficulties in their fabrication.

The above overview indicates that molecular-sized cylindrical nanopores with uniform diameters offer enhanced size-based, charge-based, and interaction-based selectivity in chemical separation and detection. Single-pore membranes can be uniquely used to detect individual nanoscale objects based on the Coulter counting principle, but are not suitable for chemical separation. Multipore membranes with high porosity integrate high selectivity with sensitive detection and efficient separation, and thus are suitable for designing novel methods for trace analysis of molecular species.

13.3
BCP-Derived Monoliths Containing Arrays of Cylindrical Nanopores

Another type of a monolithic membrane containing a vertical array of cylindrical nanopores can be prepared using cylinder-forming BCPs (Figure 13.1d) [20–22]. As with the aforementioned nanoporous membranes, the pore diameter and surface properties in BCP-derived nanopores are controllable, allowing one to take advantage of the unique features based on mass transport through well-defined nanoscale pores, both for chemical separation and detection. These nanoporous monoliths with high porosity can be prepared by simple annealing processes and subsequent chemical etching without expensive and sophisticated instruments [21, 22]. In contrast to the above conventional nanoporous membranes with thin membrane structures, the geometry of BCP-derived nanoporous monoliths can be controlled flexibly to optimize the size, structure, and performance of the analytical devices. We think these characteristics will make BCP-derived nanoporous monoliths promising materials to develop novel trace analysis methods.

BCP consists of two or more chemically distinct polymer fragments [20–22]. If the constituent polymers are immiscible and the volume fraction of the minor fragments is around 0.3, the minor fragments form cylindrical domains via self-assembly [20–22]. Table 13.1 and Figure 13.2 summarize representative BCPs that have been used to fabricate monoliths containing arrays of cylindrical nanopores. First, monolithic materials are fabricated as thin films prepared by spin coating, or as blocks formed in molds. Subsequently, cylindrical domains are usually formed by heating a BCP monolith under an inert atmosphere at a temperature higher than the glass transition temperature of the BCP. Finally, the cylindrical domains are chemically etched to obtain a nanoporous monolith. For polystyrene

Table 13.1 Representative BCPs (block copolymers) used to fabricate monolithic membranes/films containing arrays of cylindrical nanopores.

Block copolymer	Etched domain	Etching conditions	Reference
PS-PMMA (1)	PMMA	1. UV irradiation (254 nm) 2. Acetic acid	[23]
PS-PEO (2)	PEO	57% HI, 60 °C	[24]
PS-PE (3)	PS	Fuming nitric acid	[25]
PS-PLA (4)	PLA	0.5 M NaOH in water–methanol (60 : 40 v/v), heated to 65 °C	[26]
PLA-PDMA-PS (5)	PLA	0.5 M NaOH in water–methanol (60 : 40 v/v), heated to 65 °C	[27]
PS–PI-PLA (6)	PLA	Aqueous solution of 0.1 wt% SDS (sodium dodecyl sulfate) in 0.5 M NaOH at 50 °C	[28]

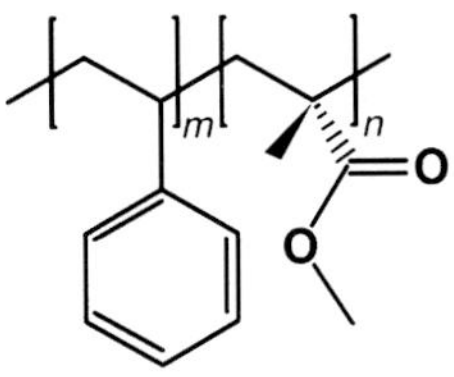

13.3 BCP-Derived Monoliths Containing Arrays of Cylindrical Nanopores

1 (PS-PMMA)

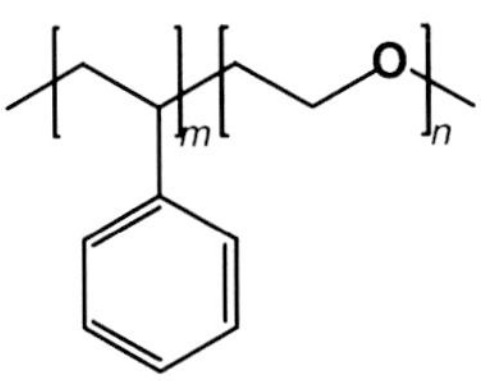

2 (PS-PEO)

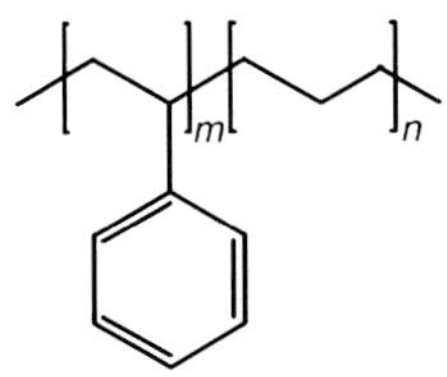

3 (PS-PE)

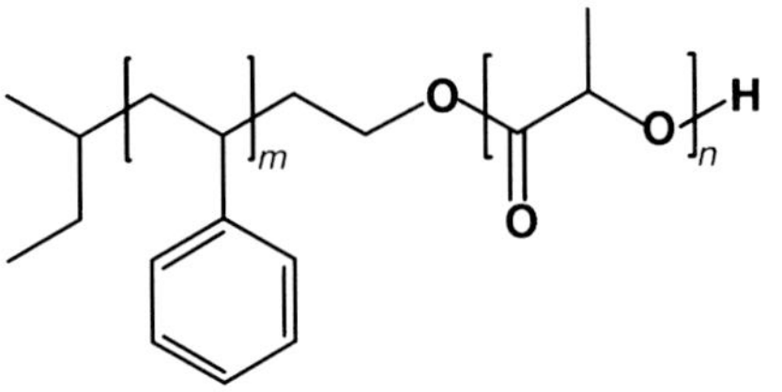

4 (PS-PLA)

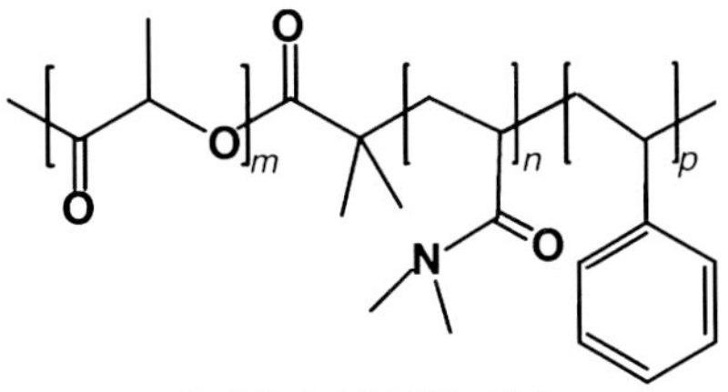

5 (PLA-PDMA-PS)

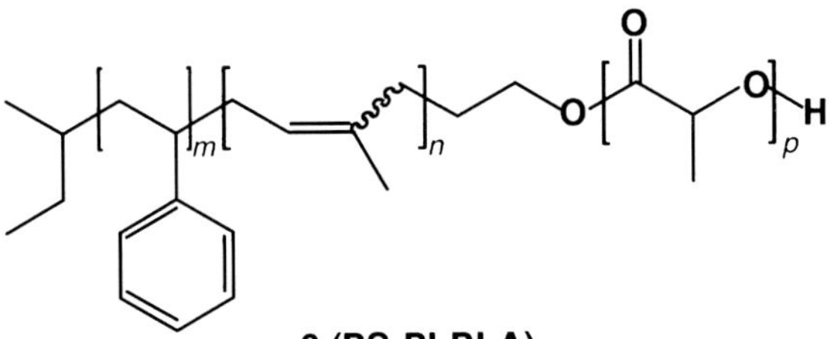

6 (PS-PI-PLA)

Figure 13.2 Chemical structures of representative BCPs (block copolymers) employed to prepare monolithic membranes/films containing arrays of cylindrical nanopores. PS: polystyrene; PMMA: poly(methyl methacrylate); PEO: poly(ethylene oxide); PE: polyethylene, PLA: polylactide; PDMA: polydimethylacrylamide, PI: polyisoprene.

(PS)-containing BCPs, UV irradiation leads to crosslinking of the PS matrix in addition to the decomposition of the minor domains, which improves the mechanical and chemical stability of the monolith [29, 30]. The diameters of the nanopores are uniform (relative standard deviation ≤20%), and can be controlled in the range between 10–100 nm by changing the molecular weight of the BCP [31]. The resulting nanoporous materials have been used mainly as templates for nanowire synthesis [32, 33] and as masks for lithography [34].

To employ BCP-derived nanoporous monoliths for the aforementioned applications, the orientation of the cylindrical nanopores in a monolith needs to be controlled [35]. For many analytical applications, cylindrical nanopores should vertically penetrate through a monolith from one surface to the other. However, it is often challenging to obtain vertically oriented domains in a BCP film supported on a substrate, because the orientation of cylindrical domains is basically determined by interactions at the BCP–substrate interface and at the free surface of the BCP film: preferential wetting of one fragment at an interface leads to horizontal domain orientation due to the segregation of the fragment at the interface. Thus, to vertically align cylindrical domains to an underlying substrate, the substrate surface has often been chemically tailored with a brush layer of a PS-poly(methyl methacrylate) (PMMA) random copolymer having an appropriate volume fraction to balance the affinities of the BCP fragments [36, 37]. In addition, the vertical alignment of cylindrical PMMA domains in a film of PS-b-PMMA (1 in Figure 13.2) can be improved by optimizing film thickness [38, 39], by adding PMMA homopolymers [40], by controlling solvent-evaporation conditions [41], and by applying an electric field during annealing [42]. The roughness of an underlying substrate induces the vertical orientation of cylindrical domains in a PS-b-PMMA film, although the orientation is metastable [43].

13.4
Surface Functionalization of BCP-Derived Cylindrical Nanopores

In addition to the domain orientation, the surface chemistry of the cylindrical nanopores needs to be controlled for their analytical applications. The surface functional groups will make it possible to control mass transport selectivity based on electrostatic and more specific interactions between nanopore wall and molecules being transported. In addition, nonspecific adsorption needs to be reduced for separation and detection of biomolecules.

One approach to control the nanopore surface chemistry is to design and synthesize new BCPs that offer known surface functional groups upon the formation of nanoporous structures. For example, Hillmyer's group have synthesized a series of BCPs containing polylactide (PLA) as the minor component (4–6 in Figure 13.2) [26–28]. Nanoporous structures were formed by the hydrolysis of cylindrical polylactide domains, which simultaneously offered surface –COOH groups on the resulting nanopores. The surface –COOH groups could be tailored via esterification [26] and amidation [27], which were shown using spectroscopic methods. In addition, they fabricated nanoporous materials having derivatizable alkene groups from PS-polyisoprene (PI)-PLA triblock copolymers [28]. More recently, Thayumanavan's group have synthesized a series of diblock copolymers whose two fragments were connected by a disulfide bond [44]. The disulfide moieties in these new types of BCPs can be cleaved using a reducing agent under mild conditions to obtain nanoporous structures. The sulfide groups formed on the resulting nanopores will allow immobilization of chemical moieties on the nanopore surface.

Another approach is to employ surface functional groups produced during chemical etching processes for surface functionalization. Recently, using cyclic voltammetry (CV), we showed the presence of surface –COOH groups on nanopores derived from PS-b-PMMA [45]. The –COOH groups were produced during the UV irradiation, probably due to oxidation of the surface of etched PMMA domains. The surface –COOH groups can be further tailored via amidation mediated by 1-ethyl-3-(3-dimethylaminopropyl)carbodiimide (EDC) (Figure 13.3) [8]. The surface modification of PS-b-PMMA-derived nanoporous monoliths permit-

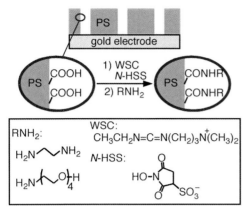

Figure 13.3 Schematic illustration of surface functionalization of PS-b-PMMA-derived nanopores via EDC-mediated amidation. Reprinted with permission from Reference [8]. Copyright 2008 American Chemical Society.

ted us to control the surface charge of the nanopores [8], to shrink the effective pore diameter [8], and to reduce the nonspecific adsorption of biomolecules [46], which were shown using CV. These studies showed the potential applicability of these materials, as described below.

13.5
Investigation of the Permeation of Molecules through BCP-Derived Nanoporous Monoliths and their Analytical Applications

The molecular permeability of BCP-derived monoliths containing arrays of cylindrical nanopores has been investigated to determine their potential applicability for chemical separation and detection. So far, most of the results that will be explained here have been obtained on nanoporous monoliths derived from PS-b-PMMA as BCPs, probably because these nanoporous monoliths can be fabricated using procedures well-established by Russell *et al.* [23]. In addition to small solvated molecules, larger nanoparticles and biomacromolecules have been employed to assess their permeability of the monolith, which demonstrates the analytical applicability of these materials for biosensing.

13.5.1
Permeation of Small Molecules through PS-b-PMMA-Derived Nanoporous Monoliths

The permeation of small solvated molecules through BCP-derived nanoporous monoliths, which is essential for chemical separation and detection, has been assessed using electrochemical methods and measurements of solution flow.

CV provides a simple means for assessing the permeability of nanoporous media, although it requires nanoporous monoliths to be immobilized on an electrode surface. Since a polymer film is supported on a solid electrode, the film should be mechanically stable within a very wide thickness range (e.g., ≥10 nm). Furthermore, analysis of electrochemical data based on theoretical models has permitted us to obtain various information [47, 48]. CV data of ferrocene in acetonitrile solutions showed the formation of a PS-*b*-PMMA-derived nanoporous film on a gold substrate [49]. CV measurements of 1,1′-ferroecedimethanol [$Fc(CH_2OH)_2$] in aqueous solution provided a simple means for monitoring the extent of the chemical etching of the PMMA domains [45]. Furthermore, the density of nanopores penetrating through a film could be estimated using similar CV measurements, which allowed us to discuss the orientation of cylindrical PMMA domains within the film [43]. The permeation of $Fc(CH_2OH)_2$ through PS-*b*-PMMA-derived nanopores decreased upon immobilization of molecules on the nanopore surface [8], suggesting that it is possible to detect molecules trapped within nanopores by electrochemically monitoring the membrane permeability.

Measurements of the flow of liquid through a stand-alone nanoporous membrane can be used to determine directly the density of open nanopores by analyzing the data using the Hagen–Poiseuille law. The flow of water measured for a stand-alone nanoporous monolith derived from a PLA-PDMA-PS triblock copolymer made it possible to estimate an average pore diameter [50]. It was shown that the flow of water through a stand-alone PS-*b*-PMMA-derived nanoporous membrane was two orders of magnitude higher than that through a TEPM, which could be explained by the much higher porosity of the BCP-derived nanoporous membrane in addition to the smaller membrane thickness [51].

All the results reported so far indicate that the liquid flow rate and solute diffusion rate through the BCP-derived nanopores are similar to those expected from the bulk parameters of the liquid and solute, suggesting that the nanoporous monoliths can be used as efficient filter membranes.

13.5.2
Regulation of Molecular Permeability Based on Electrostatic Interactions

Molecular permeability through nanoporous media can be controlled based on electrostatic interactions. In particular, permselective transport based on electrostatic repulsion between molecules and nanoporous media plays an essential role in various membrane technologies, including separation of charged molecules, desalination membranes, and fuel cell membranes. Permselective transport was observed through nanopores having radii close to the thickness of the diffuse electrical double layer (i.e., on the order of 10 nm or smaller) [5, 6], including conventional nanoporous membranes described in Section 13.2.

Similarly, PS-*b*-PMMA-derived nanoporous films have exhibited pH-dependent changes in permeation of charged redox molecules [8, 45]. Figure 13.4 shows CVs of $[Fe(CN)_6]^{3-}$ and $[Ru(NH_3)_6]^{3+}$ at different pH on a gold electrode coated with a native PS-*b*-PMMA-derived nanoporous film (pore diameter ~20 nm). The redox

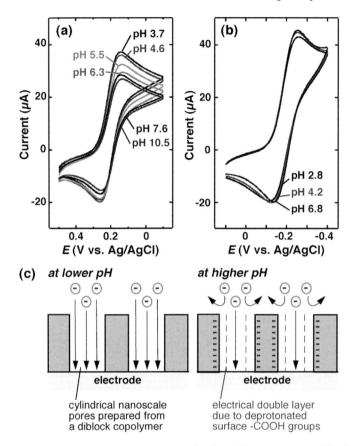

Figure 13.4 (a) CVs (scan rate: 0.02 V s^{-1}) of 3.0 mM K$_3$Fe(CN)$_6$ in 0.1 M KNO$_3$ at different pH on a gold substrate coated with a native 57K PS-b-PMMA-derived nanoporous film (~30 nm thick); (b) CVs (scan rate: 0.02 V s^{-1}) of 3.0 mM Ru(NH$_3$)$_6$Cl$_3$ in 0.1 M KNO$_3$ at different pH on a gold substrate coated with a native 57K PS-b-PMMA-derived nanoporous film (~30 nm thick); the geometrical electrode area was 0.34 cm^2; (c) schematic of the pH-dependent change in CV for K$_3$Fe(CN)$_6$.

currents of anionic [Fe(CN)$_6$]$^{3-}$ (Figure 13.4a) decreased at higher pH, whereas those of cationic [Ru(NH$_3$)$_6$]$^{3+}$ (Figure 13.4b) and uncharged Fc(CH$_2$OH)$_2$ [8, 45] were very similar regardless of the pH. The pH range that showed the decrease in redox current of [Fe(CN)$_6$]$^{3-}$ (pH 4.5–6.3) corresponded to the pK_a of -COOH groups, indicating that the current decreased due to electrostatic interactions with deprotonated -COOH groups on the nanopore surface (Figure 13.4c) [45]. In addition, the pH-dependent change in the redox reaction was affected by the supporting electrolyte concentration [45], as expected for the control of molecular permeation based on electrostatic interactions.

The pH-dependent permeability of BCP-derived nanopores can be controlled by chemical functionalization of the nanopore surface via amidation. Figure 13.5

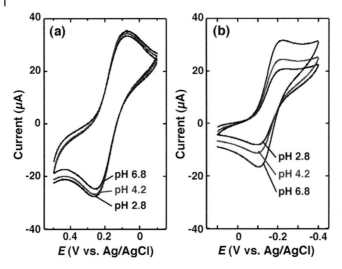

Figure 13.5 CVs (scan rate: 0.02 V s^{-1}) of (a) 3.0 mM $K_3Fe(CN)_6$ and (b) 3.0 mM $Ru(NH_3)_6Cl_3$ in 0.1 M KNO_3 at different pH on a gold substrate coated with a 57K PS-b-PMMA-derived nanoporous film (~30 nm thick) after EDC-mediated amidation with ethylenediamine. The geometrical electrode area was 0.34 cm^2.

shows CVs of $[Fe(CN)_6]^{3-}$ and $[Ru(NH_3)_6]^{3+}$ at three different pH on a gold electrode coated with a native PS-b-PMMA-derived nanoporous film after EDC-mediated amidation with ethylenediamine [8]. In contrast to CVs in Figure 13.4, the redox current of $[Ru(NH_3)_6]^{3+}$ decreased at lower pH (Figure 13.5b), whereas a pH-dependent change in the redox current of $[Fe(CN)_6]^{3-}$ was not observed (Figure 13.5a). Again, the pH-dependent change in the redox current of $[Ru(NH_3)_6]^{3+}$ reflects electrostatic repulsion with positively-charged nanopores at lower pH, indicating the successful immobilization of terminal -NH_2 groups on the nanopore surface. These results indicate the potential applicability of BCP-derived nanoporous monoliths for charge-based molecular separation.

13.5.3
Influence of Supporting Electrolyte Concentration to Effective Nanopore Diameter

We also found that the effective pore diameter of PS-b-PMMA-derived nanopores monoliths was affected by the supporting electrolyte concentration [45]. Figure 13.6 shows CVs of uncharged $Fc(CH_2OH)_2$ at three different supporting electrolyte concentrations (0.01, 0.1, and 1 M KNO_3) on a gold electrode coated with a native PS-b-PMMA-derived nanoporous film. At higher supporting electrolyte concentration, the redox current was larger, which indicates that the solvated molecules could pass through the nanopore more efficiently. This observation may suggest the presence of a thin flexible layer formed by incomplete degradation of PMMA that can stretch and shrink according to the salt concentration like a gel. The

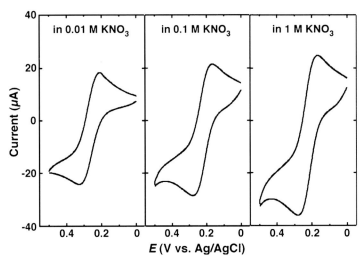

Figure 13.6 CVs (scan rate: 0.02 V s^{-1}) of 3.0 mM 1,1′-ferrocenedimethanol in 0.01 M, 0.1 M and 1 M KNO$_3$ (pH 6.3) on a gold substrate coated with a native 57K PS-*b*-PMMA-derived nanoporous film (~30 nm thick). The geometrical electrode area was 0.34 cm^2.

resulting control of the effective pore diameter may lead to development of chemical sensors and chemical valves, which were demonstrated on gel-forming BCPs [52].

13.5.4
Permeation of Nanoparticles, Polymers, and Biomacromolecules through BCP-Derived Nanopores

BCP-derived nanoporous monoliths contain arrays of cylindrical nanopores with relatively large, uniform diameters (5–100 nm), and thus allow for the penetration of biomolecules and nanoparticles. Indeed, metal and semiconductor nanoparticles were deposited within PS-*b*-PMMA-derived vertical nanopores immobilized on supporting material, both when aided by electrophoresis [53] and chemical interactions with SAMs immobilized at the bottom of the nanopores [54].

PS-*b*-PMMA-derived nanoporous monoliths have also exhibited size-based separation capability for biomolecules. For example, we have investigated the size-exclusion properties of such nanoporous monoliths (15 and 20 nm in diameter; 30–40 nm thick) immobilized on gold substrates by measuring CVs of the iron-storing protein ferritin (12 nm in diameter) [46]. As shown in Figure 13.7, the redox current of ferritin was observed for 20-nm nanopores, indicating that ferritin passed through the nanopores and then was immobilized on the underlying gold electrode. In contrast, no redox current of ferritin was observed for 15-nm nanopores prepared from PS-*b*-PMMA with a smaller molecular weight, indicating that ferritin could not reach the underlying electrode because the size of ferritin was

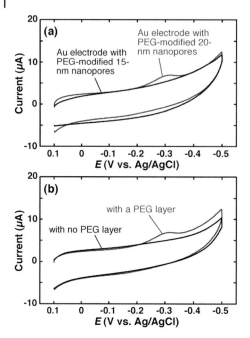

Figure 13.7 CVs (scan rate: 0.1 V s^{-1}) on gold substrates coated with PS-b-PMMA-derived nanoporous films in 0.1 M KH$_2$PO$_4$–K$_2$HPO$_4$ buffer (pH 7.0) after immersion of a ferritin solution (5 mg mL^{-1}, pH 7) for 2–12 h. The geometrical electrode area was 0.34 cm^2, and the gold surface was modified with cysteamine. (a) CVs on gold substrates coated with PEG-modified nanoporous films derived from 57K and 71K PS-b-PMMA (effective pore diameter: ~15 and ~20 nm, respectively); (b) CVs on gold electrodes coated with nanoporous films derived from 71K PS-b-PMMA with and without PEG layer (effective pore diameter: ~20 nm and ~24 nm, respectively). Reprinted with permission from Reference [46]. Copyright 2009 American Chemical Society.

too close to the pore diameter. In addition, we showed that nonspecific adsorption of the biomolecules was decreased upon covalent modification of the nanopores with poly(ethylene glycol) (PEG), suggesting that PEG modification is required to use the nanoporous monoliths for filtration of biomolecules [46].

The groups of Kim and Jang have reported the potential applicability of PS-b-PMMA-derived nanoporous membranes for filtration of human rhino virus (HRV) [51]. Figure 13.8 shows a HRV (30 nm in diameter) trapped by PS-b-PMMA-derived nanopores (~15 nm in diameter) in a stand-alone nanoporous membrane (80 nm thick) immobilized on a microporous support. In contrast to HRV, smaller protein molecules (bovine serum albumin) could pass through the nanopores. Interestingly, the PS-b-PMMA-derived nanoporous membrane could completely capture HRV in contrast to NAAMs containing pores of 20 nm in nominal diameter because of the absence of larger nanopores. In addition, the membrane has comparable filtration efficiency to TEPMs with higher flux. These results

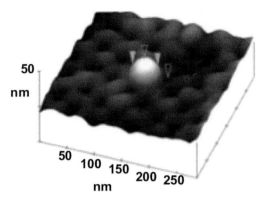

Figure 13.8 AFM image of a HRV (human rhino virus) deposited at the entrance of a PS-b-PMMA-derived nanopore. Reprinted with permission from Reference [51]. Copyright 2006 Wiley-VCH Verlag GmbH & Co. KGaA.

indicate that the PS-b-PMMA-derived nanoporous membranes incorporate the advantageous features of TEPMs (highly selective size-based separation due |to the narrow pore size distribution) and NAAMs (high flux due to the high porosity).

Kim et al. have also developed more mechanically and chemically robust nanoporous membranes derived from PS-b-PMMA for virus filtration [30]. Crosslinking of the PS matrix and mixed nanopore orientation inside the membrane improved the stability so that the membrane was not damaged by application of pressure (2 bar) or by exposure to acidic, basic, and organic solutions.

Uehara et al. have fabricated mechanically robust, stand-alone nanoporous membranes (15–30 μm thick) from PS-b-polyethylene (PE), and demonstrated size-selective permeability of the membranes [55]. Nanoporous structures could be formed by selective etching of the PS domains in a PS-b-PE membrane by fuming nitric acid. The average pore size could be controlled in the range of 5–30 nm by varying the etching time. The most unique feature of the nanoporous membrane was its very high mechanical robustness and deformability. Figure 13.9 shows pictures of a 15-μm thick membrane containing 30-nm diameter nanopores. Even after folding, the monolith was not broken. The high mechanical stability may originate from the crystalline nature of the PE moieties. Although the nanopores in the membrane were not cylindrical, membranes containing pores of 10 nm average diameter exhibited selective diffusion of glucose over bovine serum albumin (7.1 nm in diameter). The authors anticipated that the mechanically robust nanoporous membranes could be adapted for a millimeter-size implantable glucose sensor.

Nuxoll et al. have demonstrated that a microfabricated silicon substrate could be used as a support of an 80-nm thick stand-alone nanoporous membrane prepared from a BCP [56]. Using photolithography and reaction ion etching

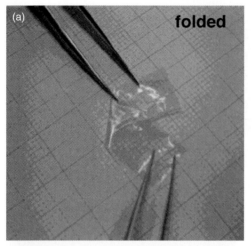
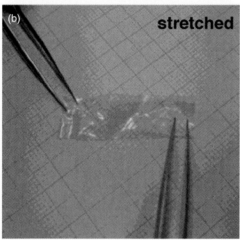

Figure 13.9 Photographs of a stand-alone nanoporous membrane prepared from PS-b-PE: (a) folded and (b) stretched. Reprinted with permission from Reference [55]. Copyright 2009 American Chemical Society.

techniques, a PS-PI-PLA-derived membrane containing cylindrical nanopores of 43 nm in diameter could be successfully fabricated on 20 μm square holes (Figure 13.10a). The resulting membrane was mechanically robust and exhibited efficient size-exclusion of dextran blue (molecular weight = 2 MDa; Stokes radius = 27 nm) without hindering diffusion of small molecules (methyl orange). The permeability of methyl orange across the membrane was 1500 times higher than that of dextran blue (Figure 13.10b), whereas the diffusion coefficient of methyl orange in solution is only 30 times larger than that of dextran blue. In addition to the high size-based selectivity, the diffusion of the small molecules across the nanoporous membrane was very fast because of the small thickness (80 nm) of the membrane.

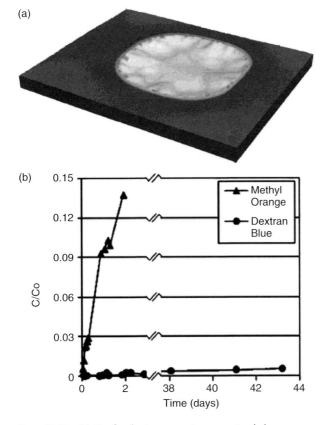

Figure 13.10 (a) Confocal microscope image of a PS-PI-PLA-derived nanoporous film supported by a microfabricated 20-μm diameter hole in a silicon support structure; (b) diffusion of methyl orange and dextran blue (ca 27 nm in diameter) across a stand-alone nanoporous membrane derived from a PS-PI-PLA (pore diameter ~43 nm; thickness ~80 nm). Reprinted with permission from Reference [56]. Copyright 2009 American Chemical Society.

Considering the tunability of nanopore diameter by varying the molecular weight of a BCP, these results indicate that BCP-derived nanoporous membranes can be used to fabricate filtration membranes for biomolecules with distinct size-based selectivity. The separation speed can be improved by using thinner membranes with higher porosity, as demonstrated by Nuxoll et al. [56].

13.6 Conclusions

This chapter has described fabrication of BCP-derived monoliths containing an array of cylindrical nanoscale pores and their analytical applicability. The nanoporous monoliths with controlled pore diameters and shapes can be obtained by simple and inexpensive processes once the domain orientation can be controlled. They have shown fast flow of solution due to their high porosity, charge-based control of molecular permeability due to their nanoscale pore diameter, and distinct size-based separation selectivity due to their narrow distribution of pore diameters in a monolith. Chemical functionalization of the nanopore surface with more sophisticated binding moieties [57, 58] will make it possible to separate and detect biomolecules based on their chemical properties in addition to their size, which will make trace analysis of target biomolecules possible. The flexibility of the monolith geometry will permit us to fabricate analytical microdevices that can be used for *in vivo* monitoring of bioactive species. Furthermore, the nanoporous monoliths may allow us to develop highly-sensitive detection methods, as demonstrated for conventional nanoporous membranes. We are currently working on employing these membranes for developing microfabricated bioanalytical devices.

Acknowledgments

The authors thank past and present group members who worked on the projects related to this chapter, including Shaida Ibrahim for critically reading the manuscript. The authors gratefully acknowledge the ACS-PRF (ACS PRF #46192-G5), Kansas NSF EPSCoR (NSF 43529), Terry C. Johnson Center for Basic Cancer Research, and Kansas State University for financial support of this work.

References

1 Irvine, G.B. (1997) *Anal. Chim. Acta*, **352**, 387–397.
2 Righetti, P.G. and Verzola, B. (2001) *Electrophoresis*, **22**, 2359–2374.
3 Murray, R.W., Ewing, A.G., and Durst, R.A. (1987) *Anal. Chem.*, **59**, 379A–390A.
4 Deen, W.M. (1987) *AIChE J.*, **33**, 1409–1425.
5 Martin, C.R., Nishizawa, M., Jirage, K., and Kang, M. (2001) *J. Phys. Chem. B*, **105**, 1925–1934.
6 Baker, L.A., Jin, P., and Martin, C.R. (2005) *Crit. Rev. Solid State Mater. Sci.*, **30**, 183–205.
7 Martin, C.R. (1994) *Science*, **266**, 1961–1966.
8 Li, Y. and Ito, T. (2008) *Langmuir*, **24**, 8959–8963.
9 Siwy, Z.S. (2006) *Adv. Funct. Mater.*, **16**, 735–746.
10 Gatimu, E.N., Sweedler, J.V., and Bohn, P.W. (2006) *Analyst*, **131**, 705–709.
11 Dai, J., Ito, T., Sun, L., and Crooks, R.M. (2003) *J. Am. Chem. Soc.*, **125**, 13026–13027.
12 Menon, V.P. and Martin, C.R. (1995) *Anal. Chem.*, **67**, 1920–1928.
13 Sun, L. and Crooks, R.M. (2000) *J. Am. Chem. Soc.*, **122**, 12340–12345.
14 Ito, T., Sun, L., and Crooks, R.M. (2003) *Anal. Chem.*, **75**, 2399–2406.
15 Ito, T., Sun, L., Henriquez, R.R., and Crooks, R.M. (2004) *Acc. Chem. Res.*, **37**, 937–945.
16 Ito, T., Sun, L., Bevan, M.A., and Crooks, R.M. (2004) *Langmuir*, **20**, 6940–6945.
17 Hinds, B.J., Chopra, N., Rantell, T., Andrews, R., Gavalas, V., and Bachas, L.G. (2004) *Science*, **303**, 62–65.
18 Majumder, M., Chopra, N., Andrews, R., and Hinds, B.J. (2005) *Nature*, **438**, 44.
19 Holt, J.K., Park, H.G., Wang, Y., Stadermann, M., Artyukhin, A.B.,

Grigoropoulos, C.P., Noy, A., and Bakajin, O. (2006) *Science*, **312**, 1034–1037.

20 Li, M., Coenjarts, C.A., and Ober, C.K. (2005) *Adv. Polym. Sci.*, **190**, 183–226.

21 Hillmyer, M.A. (2005) *Adv. Polym. Sci.*, **190**, 137–181.

22 Olson, D.A., Chen, L., and Hillmyer, M.A. (2008) *Chem. Mater.*, **20**, 869–890.

23 Thurn-Albrecht, T., Steiner, R., DeRouchey, J., Stafford, C.M., Huang, E., Bal, M., Tuominen, M., Hawker, C.J., and Russell, T.P. (2000) *Adv. Mater.*, **12**, 787–791.

24 Mao, H. and Hillmyer, M.A. (2005) *Macromolecules*, **38**, 4038–4039.

25 Uehara, H., Yoshida, T., Kakiage, M., Yamanobe, T., Komoto, T., Nomura, K., Nakajima, K., and Matsuda, M. (2006) *Macromolecules*, **39**, 3971–3974.

26 Zalusky, A.S., Olayo-Valles, R., Wolf, J.H., and Hillmyer, M.A. (2002) *J. Am. Chem. Soc.*, **124**, 12761–12773.

27 Rzayev, J. and Hillmyer, M.A. (2005) *J. Am. Chem. Soc.*, **127**, 13373–13379.

28 Bailey, T.S., Rzayev, J., and Hillmyer, M.A. (2006) *Macromolecules*, **39**, 8772–8781.

29 Jeong, U., Ryu, D.Y., Kim, J.K., Kim, D.H., Russell, T.P., and Hawker, C.J. (2003) *Adv. Mater.*, **15**, 1247–1250.

30 Yang, S.Y., Park, J., Yoon, J., Ree, M., Jang, S.K., and Kim, J.K. (2008) *Adv. Funct. Mater.*, **18**, 1371–1377.

31 Xu, T., Kim, H.-C., DeRouchey, J., Seney, C., Levesque, C., Martin, P., Stafford, C.M., and Russell, T.P. (2001) *Polymer*, **42**, 9091–9095.

32 Thurn-Albrecht, T., Schotter, J., Kastle, G.A., Emley, N., Shibauchi, T., Krusin-Elbaum, L., Guarini, K., Black, C.T., Tuominen, M.T., and Russell, T.P. (2000) *Science*, **290**, 2126–2129.

33 Kim, H.-C., Jia, X., Stafford, C.M., Kim, D.H., McCarthy, T.J., Tuominen, M., Hawker, C.J., and Russell, T.P. (2001) *Adv. Mater.*, **13**, 795–797.

34 Park, M., Harrison, C., Chaikin, P.M., Register, R.A., and Adamson, D.H. (1997) *Science*, **276**, 1401–1404.

35 Lazzari, M. and De Rosa, C. (2006) *Block Copolymers in Nanoscience* (eds M. Lazzari, G. Liu, and S. Lecommandoux), Wiley-VCH Verlag GmbH, Weinheim, pp. 191–231.

36 Mansky, P., Russell, T.P., Hawker, C.J., Pitsikalis, M., and Mays, J. (1997) *Macromolecules*, **30**, 6810–6813.

37 Huang, E., Russell, T.P., Harrison, C., Chaikin, P.M., Register, R.A., Hawker, C.J., and Mays, J. (1998) *Macromolecules*, **31**, 7641–7650.

38 Fasolka, M.J. and Mayes, A.M. (2001) *Annu. Rev. Mater. Res.*, **31**, 323–355.

39 Guarini, K.W., Black, C.T., Milkove, K.R., and Sandstrom, R.L. (2001) *J. Vac. Sci. Technol. B*, **19**, 2784–2788.

40 Jeong, U., Ryu, D.Y., Kho, D.H., Kim, J.K., Goldbach, J.T., Kim, D.H., and Russell, T.P. (2004) *Adv. Mater.*, **16**, 533–536.

41 Xuan, Y., Peng, J., Cui, L., Wang, H., Li, B., and Han, Y. (2004) *Macromolecules*, **37**, 7301–7307.

42 Thurn-Albrecht, T., DeRouchey, J., Russell, T.P., and Jaeger, H.M. (2000) *Macromolecules*, **33**, 3250–3253.

43 Maire, H.C., Ibrahim, S., Li, Y., and Ito, T. (2009) *Polymer*, **50**, 2273–2280.

44 Klaikherd, A., Ghosh, S., and Thayumanavan, S. (2007) *Macromolecules*, **40**, 8518–8520.

45 Li, Y., Maire, H.C., and Ito, T. (2007) *Langmuir*, **23**, 12771–12776.

46 Li, Y. and Ito, T. (2009) *Anal. Chem.*, **81**, 851–855.

47 Bard, A.J. and Faulkner, L.R. (2001) *Electrochemical Methods, Fundamentals and Applications*, 2nd edn, John Wiley & Sons, Inc., New York.

48 Ito, T., Audi, A.A., and Dible, G.P. (2006) *Anal. Chem.*, **78**, 7048–7053.

49 Jeoung, E., Galow, T.H., Schotter, J., Bal, M., Ursache, A., Tuominen, M.T., Stafford, C.M., Russell, T.P., and Rotello, V.M. (2001) *Langmuir*, **17**, 6396–6398.

50 Phillip, W.A., Rzayev, J., Hillmyer, M.A., and Cussler, E.L. (2006) *J. Membr. Sci.*, **286**, 144–152.

51 Yang, S.Y., Ryu, I., Kim, H.Y., Kim, J.K., Jang, S.K., and Russell, T.P. (2006) *Adv. Mater.*, **18**, 709–712.

52 Liu, G., Ding, J., Hashimoto, T., Kimishima, K., Winnik, F.M., and Nigam, S. (1999) *Chem. Mater.*, **11**, 2233–2240.

53 Zhang, Q., Xu, T., Butterfield, D., Misner, M.J., Ryu, D.Y., Emrick, T., and Russell, T.P. (2005) *Nano Lett.*, **5**, 357–361.
54 Bandyopadhyay, K., Tan, E., Ho, L., Bundick, S., Baker, S.M., and Niemz, A. (2006) *Langmuir*, **22**, 4978–4984.
55 Uehara, H., Kakiage, M., Sekiya, M., Sakuma, D., Yamanobe, T., Takano, N., Barraud, A., Meurville, E., and Ryser, P. (2009) *ACS Nano*, **3**, 924–932.
56 Nuxoll, E.E., Hillmyer, M.A., Wang, R., Leighton, C., and Siegel, R.A. (2009) *ACS Appl. Mater. Interf.*, **1**, 888–893.
57 Lee, S.B., Mitchell, D.T., Trofin, L., Nevanen, T.K., Soderlund, H., and Martin, C.R. (2002) *Science*, **296**, 2198–2200.
58 Kohli, P., Harrell, C.C., Cao, Z., Gasparac, R., Tan, W., and Martin, C.R. (2004) *Science*, **305**, 984–986.

14
Synthesis and Applications of Gold Nanorods
Carrie L. John, Shuping Xu, Yuhui Jin, Shaina L. Strating, and Julia Xiaojun Zhao

14.1
Introduction

The need for sensitive determinations of trace analytes has driven the rapid development of various novel photoactive nanomaterials [1]. Metallic gold nanomaterials are one of the most prominent signaling reagents for achieving highly sensitive detection. In fact, gold particles are traditional metallic photoactive materials that have been used for many years. The advances in nanotechnology have greatly increased applications of gold nanomaterials within the past few decades. This increase has resulted in the development of new types of gold nanomaterials, including gold nanorods (AuNRs).

The term nanorod first appeared in a publication on carbide nanorods in 1995 [2]. Two years later, AuNRs were developed [3]. Since then, the syntheses and applications of AuNRs have grown tremendously. Various synthesis methods have been developed for making AuNRs, including electrochemical, photochemical, and seed-mediated growth. These AuNRs have demonstrated great potential for the determination of trace analytes, especially biomedical analytes.

Biomedical applications of AuNRs are favorable because their absorption wavelength falls within the near-infrared (NIR) region of the electromagnetic spectrum. Unlike the visible region, the NIR portion (700–1000 nm) has shown two desirable advantages for biosamples: low auto-luminescence and absorption of radiation as well as low light scattering. These features result in high sensitivity and deep penetration of NIR radiation, which allows for retrieval of the innermost structural information of biosamples [4, 5].

Traditional Au nanomaterials, such as gold colloids and gold particles, are photoactive in the visible region. Their absorption wavelength is size dependent. The larger the size of the nanoparticle, the longer absorption wavelength appears. This optical property is the basis for many imaging studies of biological samples using gold nanoparticles as labels. However, the wavelength changes of gold nanoparticles are limited to the visible region.

The absorption wavelength of AuNRs is adjustable by changing their aspect ratio (length to width). In addition to absorption signals, luminescence signals of

Trace Analysis with Nanomaterials. Edited by David T. Pierce and Julia Xiaojun Zhao
Copyright © 2010 WILEY-VCH Verlag GmbH & Co. KGaA, Weinheim
ISBN: 978-3-527-32350-0

AuNRs have been recently realized for trace analysis. Employing these two types of optical signals, AuNRs have been used directly as labels for analysis of biological samples. However, metallic AuNRs are employed more often as a substrate to enhance detection signals based on surface plasmon resonance. In both cases, surface modification of AuNRs is needed for targeting specific biological targets, such as cells or proteins.

This chapter will first discuss three major AuNR synthesis methods. The principles of AuNR enhancement of optical signals and the luminescence of the AuNR itself will also be covered. Applications of AuNRs in trace analysis will include the fabrication of sensors for optical and electrochemical measurements of biological samples. Finally, the applications of AuNRs as electrocatalysts and in photothermal therapy will be briefly discussed.

14.2
Au Nanorod Synthesis

Three major AuNR synthesis methods are discussed in this section – electrochemical, photochemical, and seed-mediated growth. Seed-mediated growth is the most commonly used pathway. This method can produce AuNRs with desired aspect ratios by simply adjusting mild reaction conditions. Photochemical synthesis is simple but has not been frequently used. Electrochemical synthesis produces uniform sizes and shapes of nanorods, but a template and additional instruments are needed for this method. Nanorods can be made from various materials. The methods discussed in this section are suitable not only for developing AuNRs but also are adaptable to other types of nanorods, including silver [6], ZnO [7, 8], CdSe [9], CdS [10], and MnO_x [11] nanorods as well as ZnS:Mn/ZnS core–shell nanorods [12].

14.2.1
Electrochemical Synthesis

Electrochemical synthesis is a traditional method for making desired materials. Advantages of this method include low cost, simple protocols, and fast reactions. Employing these advantages, AuNRs have been synthesized through an electrochemical reduction of aurate ions on a cathode surface. The product of the reduction is elemental Au. The Au can be formed into a rod shape when a template is placed on the cathode. These templates can be categorized as soft and hard templates. The soft templates are micelles in contrast to the solid hard templates which are physically coated onto the electrode surface. Here, we introduce the electrochemical synthesis of AuNRs using both the hard and soft templates. The methods of forming the templates will also be discussed.

14.2.1.1 Electrochemical Synthesis Employing a Hard Template
AuNRs can be produced by reducing aurate ions and "casting" gold atoms on a nanosized hard template [13–16]. Prior to synthesis, the template is coated onto

the electrode to provide the porous cylindrical environment for the gold deposition. After the deposition, the gold continues growing inside the template pores. The template regulates the physical dimension of the AuNRs by confining this growth. The final size and morphology of AuNRs greatly depend on the quality of template. Thus, the preparation of hard templates is the critical step in AuNR synthesis.

The classic template is porous alumina, which has been developed by electrochemically etching the aluminum film onto the electrode [17, 18]. Figure 14.1 shows the formation of the nanoporous alumina. First, aluminum is deposited on the substrate of electrode to form a membrane. The thickness of this alumina membrane is manipulated by multistep etching. Second, pores are formed that penetrate throughout the aluminum membrane. This process is accomplished by a second round of electrochemical etching. The pore size is controlled by a pore-widening approach using 5% phosphoric acid. The final depth of nanopore is determined by the membrane thickness.

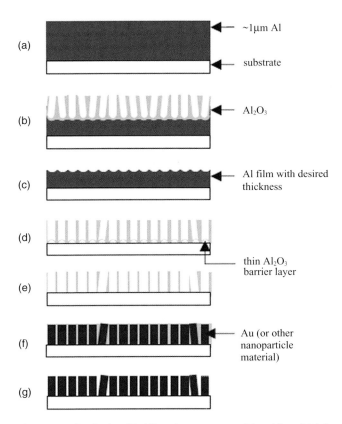

Figure 14.1 Synthesis of AuNRs using nanoporous alumina templates. (a) Deposition aluminum membrane onto the substrate; (b) and (c) adjusting the thickness of aluminum membrane by electrochemical etching; (d) and (e) formation of nanoporous alumina; (f) and (g) growth of AuNRs on the alumina template. Reprinted with permission from Reference [18]. Copyright 2003 Wiley-VCH Verlag GmbH & Co. KGaA.

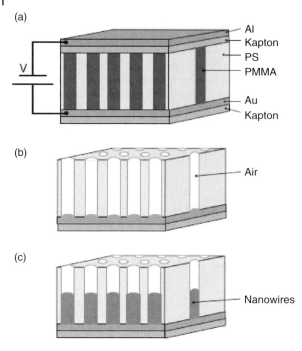

Figure 14.2 Growth of nanowires in a nanoporous polymer membrane template. (a) Formation of cylindrical array of PMMA surrounded by PS; (b) construction of nanopores on the membrane through a UV-irradiation process; (c) synthesis of nanowires in the polymer template. Reprinted with permission from Reference [19]. Copyright 2000 American Association for the Advancement of Science.

Hard, porous templates have also been prepared by assembling and selectively etching a diblock polymer, namely, polystyrene (PS) and poly(methyl methacrylate) (PMMA) (Figure 14.2) [19]. The two polymers were first arranged into a structure as shown in Figure 14.2a. Ordered hexagonal cylindrical PMMA domains were surrounded by a PS matrix. Then, UV irradiation was used to simultaneously degrade the PMMA domains and crosslink the PS matrix. This crosslinking strengthened the PS matrix. As the PMMA debris was washed away, nanopores were formed in the firm PS matrix. This hard template method is universal for the deposition of many metals and even conductive polymers. If the diameter of the pore is smaller and the length of the pore is longer, the template can also be used for making nanowires. Figure 14.3 shows a scanning electron microscopy (SEM) image of blocks of nanowires [20].

14.2.1.2 Electrochemical Synthesis Employing a Soft Template

The electrochemical synthesis of AuNRs using a soft template was first reported in 1997 by Wang's group [3, 21]. Instead of hard templates, a surfactant was used to produce micelles that served as soft templates for the shape development of

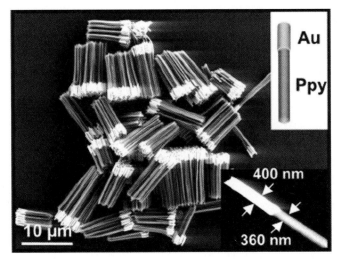

Figure 14.3 SEM image of diblock nanowires composed of gold and polypyrrole (Ppy). Reprinted with permission from Reference [20]. Copyright 2004 American Association for the Advancement of Science.

gold nanorods. Surfactant molecules, such as hexadecyltrimethylammonium bromide (CTAB) and tetradodecylammonium bromide ($TC_{12}AB$), usually assemble into micellar structures as soft templates for the synthesis and stabilization of AuNRs. In addition, a small amount of acetone is believed to facilitate the formation of cylindrical micelles.

The function of the soft template is the same as the hard template, which is to shape and hold the AuNRs [3]. However, the configuration of the electrochemical system is different than that of the hard template. With the soft templates, a gold anode is scarified to generate aurate ions and feed the platinum cathode for the reductive growth of AuNRs. Throughout the synthesis process, continuous sonication is required to loosen the newly formed gold nanorods from the cathode surface and refresh the surface for another cycle of nanorod growth.

Aspect ratio is one of the most important structural characteristics of AuNRs. Unlike the hard template synthesis, in which the shape and size of AuNRs are completely confined by the template, the AuNR aspect ratio in soft template synthesis can be controlled by the concentration of silver ions. The silver ions are produced from a silver plate in the electrochemical system. Both the concentration and the rate of silver ions introduced to the solution are significant for controlling the aspect ratio of the gold nanorods [3].

When CTAB is used as a soft-template surfactant, special attention must be given to the final condition of the AuNRs. CTAB is toxic to living samples. It has been shown that unpurified AuNRs containing CTAB caused a high rate of cell death. An appropriate protective layer of poly(ethylene glycol) (PEG) has been found to reduce this toxic effect [22–24]. Thus, to ensure low toxicity of AuNRs, complete washing of AuNRs is needed after the synthesis is complete [25]. In addition, embedding AuNRs within a polymer film or performing a ligand exchange can improve the safety of AuNRs in biological applications.

14.2.2
Photochemical Synthesis

Photochemical synthesis is an important synthetic approach in chemistry and material sciences. Although this synthesis is not widely used for obtaining AuNRs at present, it is more straightforward than the electrochemical method [26]. In addition, the photochemical approach can be used to investigate the mechanism of AuNR formation.

The photochemical reaction occurs in a solution containing surfactants, acetone, cyclohexane, aurate, and silver ions, which is quite similar to the soft-template electrochemical synthesis. However, differing from other methods, the photochemical approach employs a UV irradiation source to reduce the gold ions and generate the nanorods with the assistance of silver. The aspect ratio of the AuNRs can be manipulated as the amount of silver ions is adjusted. However, the silver is hardly observed in the final nanorod structures. In the absence of silver ions, a similar approach forms only spherical gold nanoparticles. This indicates that although silver is not incorporated in the nanorods formed, it is of great importance in regulating the shape of AuNRs.

14.2.3
Seed-Mediated Growth

The seed-mediated growth approach is the most popular synthetic method for AuNRs mainly because the aspect ratio is easily adjustable in the reaction process [27–31]. The basis of this method is to produce small gold seeds and expose them to a growth medium – usually containing CTAB or other soft-template surfactants [27] – for the anisotropic growth of the seeds into rods. The growth is based on the reduction of tetrachloroauric acid onto the surface of the gold seeds in a given direction. The type and concentration of the surfactant determines the growth direction.

In this synthesis pathway, two reactions are needed. The first is the reduction of gold ions for the formation of seeds. By adding a strong reducing agent (sodium borohydride), this reaction can be completed quickly. To allow the nanorods to grow to a designed size and to minimize further nucleation of new gold particles, the second reaction should be carried out by employing a weak reducing agent, usually ascorbic acid.

Seed-mediated growth can produce different aspect ratios of AuNRs as needed [32, 33]. The few factors that usually affect the aspect ratio include growth time [33], amount of $AgNO_3$ [28, 29], type of surfactant [34], temperature [35], and pH. As the aspect ratio increases the longitudinal plasmon wavelength displays a redshift into and within the NIR region [36].

To further manipulate the size of AuNRs, overgrowth [37] and shortening [38] of the AuNRs are approaches used to change the aspect ratio in a secondary step [39]. Shortening of AuNRs is commonly achieved by oxidation using H_2O_2. The extent of the oxidation can be controlled by the initial concentration of H_2O_2 and the exposure time (Figure 14.4). Longer exposure time and higher concentration of H_2O_2 result in smaller aspect ratios and the greater blue-shift of the absorption wavelengths of AuNRs. Another approach for the shortening of AuNRs is the use of O_2 along with heating. Overgrowth has been accomplished by blocking the ends of the AuNRs using cysteine while adding additional Au precursor to the growth medium. The extent of overgrowth is related to the amount of additional Au precursor (Figure 14.5).

The aspect ratio of AuNRs can also be controlled by the concentration and type of additives used. However, the additives usually lead to other anisotropic Au structures, including stars, tetrapods, blocks, and cubes [27, 40].

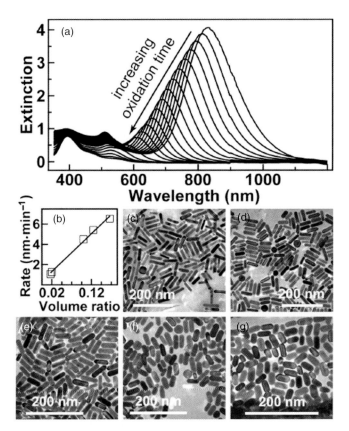

Figure 14.4 (a) The absorption spectra display both a blue-shift and a decrease in intensity for increased oxidation times; the blue-shift is consistent with the shortening of the aspect ratio; (b) the increased oxidation time leads to shorter AuNRs, as confirmed by TEM images (c–g). Reprinted with permission from Reference [39]. Copyright 2008 The American Chemical Society.

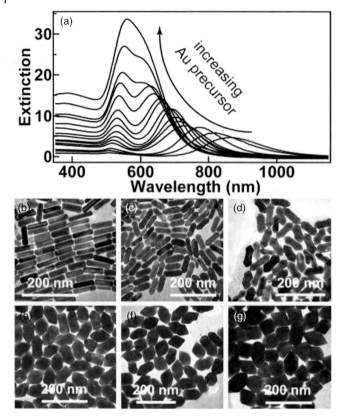

Figure 14.5 Overgrowth of AuNRs produces wider NRs when more Au precursor is present. (a) The blue-shift of absorption spectra shows a decreasing length to width ratio confirmed by TEM images (b–g) when additional precursor is added. Reprinted with permission from Reference [39]. Copyright 2008 The American Chemical Society.

Besides size and shape, surface modifications play an important role in AuNR applications. Several methods for the modification of AuNRs have been reported. Replacing the surfactant used in the synthesis process with more desirable and biologically favorable ligands has been the focus of several research groups [22, 23, 25, 41–44]. Electrostatic attraction is a simple and common approach for surface modification. For example, the positive charge from the surfactant can adsorb a negatively charged component to the surface for further applications [32].

Characterization of the AuNRs relies on typical instrumentations used for photoactive nanomaterials, including time-resolved spectroscopy [45], absorption spectroscopy [33, 36], and electron microscopy [32]. In some cases the first test is to directly observe the color changes of the synthesis solution, which indicates nanorod growth.

14.3
Signal Enhancement

Frequently, AuNRs are employed as signal enhancement substrates for obtaining high detection sensitivity in trace analysis. In these cases, the detection signal is from an additional signaling probe. The signaling probe is linked to the AuNR substrate prior to the determination. In this section, two typical enhancement techniques will be discussed. One is surface-enhanced Raman scattering (SERS) and the other is the enhancement of the luminescence signal from dyes. In both cases, the plasmon resonance of AuNRs plays a key role. Thus, characteristics of the plasmon resonance band will be discussed first.

AuNRs are also directly employed as signaling probes. Their spectral absorption in the visible and NIR regions is a commonly used optical signal. Recently, luminescence signals of AuNRs were found to be detectable. With appropriate conditions, the luminescence signal can be greatly enhanced for trace analysis [25]. The last part of this section will cover this new finding.

14.3.1
Plasmon Resonance

Localized surface plasmon resonance (LSPR) is a common phenomenon of metallic materials under light irradiation. Within the nanoscale, collective oscillations of metallic free electrons are limited by the nanostructure boundaries and thus form surface plasmon waves along the interface. When the nanomaterial interface is irradiated by an incident light beam, the surface plasmon wave resonates with the optical wave at an optimal wavelength, resulting in the greatest absorption of the incident light. As a result, an enlarged localized electromagnetic field is manifested around the nanostructures, providing extra energy for signaling reagents that may be present within this zone [1].

LSPR spectra contain abundant information about the metallic nanomaterials themselves and their surrounding environments [46, 47]. The size and shape of AuNRs, as well as the refractive index of the surrounding medium, affect the LSPR spectrum [48–52]. When targets bind to AuNRs the change in refractive index on the AuNR surface results in remarkable shifts in the LSPR signals. Therefore, numerous biosensing applications using AuNRs have been based on LSPR determinations [27].

Both simulations and experimental findings demonstrate that the plasmon resonance of AuNRs presents two distinctive bands in its absorption spectrum. One band with lower intensity appears at about 520 nm, which originates from transverse electronic oscillation perpendicular to the long axis of the AuNRs. The other, higher intensity band is in the NIR region. This NIR band is produced by longitudinal electronic oscillation along the long axis of the AuNRs. Frequently, plasmon resonance bands with different wavelengths are obtained by changing aspect ratios of AuNRs.

14.3.2
Surface-Enhanced Raman Scattering

One remarkable example of metallic signal enhancement is SERS. Using Au or Ag nanomaterials, SERS signals have been enlarged by a factor 10^{10} [53–57]. The energy level of the electromagnetic field strongly depends on the metallic plasmon property of the nanomaterial [48–52]. This property is determined by several factors, including characteristics of the metal noumenon (size, shape, structure, and dielectric constant), the surrounding medium (dielectric constant), and the incident light (direction and wavelength). Among these, the shape, size, and structure of the metal nanomaterials are critical to achieving controllable plasmonic materials and powerful surface-enhanced matrices [58, 59].

Raman scattering spectrometry has been used for many years in trace analysis. Several metallic nanomaterials can dramatically enhance Raman scatting signals. The pathways by which AuNRs enhance Raman scattering is usually divided into two aspects: chemical enhancement and physical (or electromagnetic) enhancement.

The chemical effect is conveyed by the surface energy of different facets of AuNRs [60]. This surface energy determines the adsorbing ability of an analyte. Further, the amount of analyte adsorbed on the AuNRs and the molecular orientation of the analyte determine the intensity of the SERS signals. Nikoobakht et al. [61, 62] have observed an enhancement factor of 10^5 for determination of 2-aminothiophenol by AuNRs while no enhancement was observed by gold nanospheres. They believed that four {110} facets along the longitudinal axes of AuNRs possessed much higher surface energy than other shaped particles. Thus, the chemical interaction between the {110} facets of AuNRs and the analyte was stronger.

Studies on physical enhancement of SERS by AuNRs have focused on the effective coupling between the plasmon resonance band and the incident wavelength of radiation. When the peak of the plasmon band overlaps with the incident wavelength, SERS enhancement occurs [63, 64]. As previously discussed, the plasmon resonance band can be adjusted from the visible to NIR range to match the incident wavelength of radiation. It has been proven that overlapping SPR and the excitation line maximized the SERS enhancement of 4-mercaptobenzoic acid from both experiments and theoretical electromagnetic simulation [63]. Murphy et al. [64] have studied SERS enhancement of AuNRs with different aspect ratios. They found the AuNRs with an aspect ratio of 1.7 have given $10–10^2$ greater SERS signal than other aspect ratios.

Similar to gold nanoparticles, aggregates of AuNRs generate higher signal enhancement for SERS than the contribution of individual AuNRs. For instance, Wang and Dong [65] have assembled multilayers of AuNRs into a nanostructure to evaluate their SERS activity. The results showed that the strong inter-AuNR coupling among the closely positioned multilayers contributed significantly to the SERS signals. Xu and Zhang [66] have proven that the physical enhancement is larger than the chemical enhancement by using a two-dimensional standing AuNR array as the SERS substrate. Overall, the AuNR-induced SERS is significant in both chemical and physical enhancement. This is the foundation for sensitive determination of trace analytes using AuNRs in SERS [67].

14.3.3
Luminescence Enhancement of Dye Molecules

Several AuNR-dye complexes have been developed as enhanced luminescence probes in recent years [68–71]. Unlike a SERS probe directly adsorbing on the AuNR surface, a spacer is needed between a dye molecule and AuNR to obtain enhanced luminescence of AuNR–dye complexes [70]. Meso-structured silica films [68] and biomolecules (e.g., human serum albumin [69]) have been employed to separate the fluorophores from AuNRs. Without spacers, the luminescence signal would be quenched by the surface plasmon-coupled emission [72]. Energy transfer occurs from the excited dyes to the metal, which causes the quenching of the luminescence signals of dye molecules.

Several pathways for metal enhancement of dye luminescence have been proposed. Similar to SERS, the occurrence of surface plasmon resonance leads to a strongly enhanced absorption of the incident light. Firstly, when the surface plasmon resonance band of a metal nanostructure overlaps with the excitation wavelength of the fluorophore, energy is transferred from the metal to the fluorophore molecules, thereby increasing the possibility of excitation of the dye molecule. Secondly, the metal nanostructure can change the radiative deactivation rate of the dye. Thus, the fluorescence lifetime and the quantum yield of dye molecules are changed [4, 73]. By shortening the excited-state lifetime, the photostability of the dye is improved. Thirdly, the scattering of the metallic nanostructures affects the coupling efficiency of the fluorescence emission in the far-field from the metal surface [74, 75]. In each of these paths, a common requirement is coincidence of the plasmon resonance band and the fluorophore excitation or emission band. When this condition is met, luminescence enhancement can be obtained [76].

In general, these three enhancement pathways should be accessible with AuNRs. However, due to the limited number of studies in this direction, it is too early to draw conclusions about the effect of AuNRs on the luminescence of dye molecules. Nevertheless, some promising results have been reported. Halas et al. [69] have evaluated the luminescence enhancement ability of AuNRs with short axes of 11 nm and long axes of 46 nm. The lifetime of a NIR fluorophore molecule, IR800, was greatly reduced from 563 to 121 ps when AuNRs were linked to the IR800. Meanwhile, the luminescence of IR800 was enhanced by ninefold and its quantum yield increased from 7% to 74%. Wang et al. [68] have prepared an AuNR-silica core–shell structure and doped dye molecules in the silica shell. They found that luminescence enhancement occurred only for selective dyes. However, in the work presented by Okamoto et al. [71], an individual AuNR showed a quenching effect under an aperture-type scanning near-field optical microscope, especially for AuNRs with high aspect ratios. Overall, more fundamental studies are needed in this area to better understand the effects of AuNRs on molecular luminescence.

14.3.4
Enhanced Luminescence of Au Nanorods

Absorption of visible radiation is the most apparent optical property of gold nanomaterials. When gold nanomaterials are used as an optical probe for trace analysis, the magnitude of their absorption peak is measured; this is usually proportional to the analyte concentration. The luminescence of gold nanomaterials is used much less for this purpose. However, AuNRs exhibit luminescent characteristics [77–79]. As the aspect ratio increases, the luminescence intensity is increased. The enhancement can be as high as 10^6-fold greater than regular gold metal [77]. The quantum efficiency of AuNR increases linearly with the square of the length of longitudinal axes [77]. Luong et al. [79] have observed intense luminescence from long AuNRs (aspect ratio larger than 13) at 743 nm (stronger) and 793 nm (weaker). Similar to dye molecules, the luminescence intensity of AuNRs was concentration-dependent (Figure 14.6).

The luminescence mechanism of gold is similar to that of quantum dots. The inter-band recombination between the electron (sp-conduction band) and hole (d-valence band) produces luminescence. El-Sayed et al. have further discussed the mechanism of luminescence enhancement of AuNRs [78]. The coupling of the surface plasmon resonance with the incoming and outgoing electric field leads to the enhanced emission. The luminescence power (P) can be calculated by Equation (14.1) [80]:

$$P = 2^4 \beta_1 |E_0|^2 V |L^2(\omega_{ex}) L^2(\omega_{em})| \tag{14.1}$$

where

β_1 is a proportionality constant,
E_0 is the incident electric field,

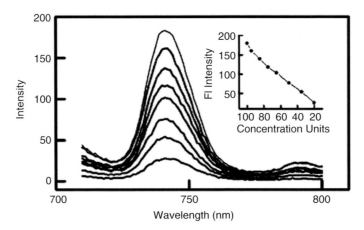

Figure 14.6 Relationship between luminescence intensity and the concentration of AuNRs. Reprinted with permission from Reference [79]. Copyright 2005 The Royal Society of Chemistry.

V is the volume of the AuNR [equal to $(4/3)\pi ab^2$, where a and b are the height and radius of the AuNR],

ω_{ex} and ω_{em} indicate the frequency of excitation and emission, respectively,

$L(\omega)$ is the ω-dependent local field correction factor.

According to this equation, longer AuNRs should possess stronger luminescence and a redshifted emission band.

AuNRs can enhance two-photon luminescence as well [81]. In two-photon luminescence, excitation wavelengths are much longer than emission wavelengths due to the two-photon mechanism. As a result, the excitations usually fall within the NIR region. Thus, advantages of the NIR region can be fully utilized, making AuNRs promising luminescent probes for bioanalysis and bioimaging. Figure 14.7 shows an example of two-photon luminescence measurement using AuNRs.

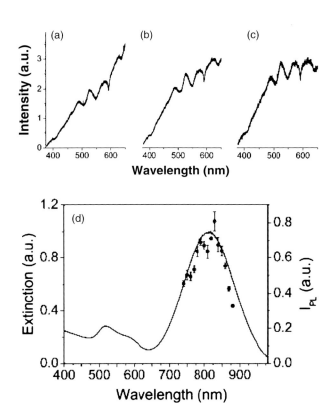

Figure 14.7 Two-photon luminescence spectra of AuNRs with excitation wavelengths of (a) 730, (b) 780, and (c) 830 nm; (d) excitation profile of the intensity of two-photon luminescence (right coordinate axes) compared with the extinction spectrum of AuNRs (left coordinate axes). Reprinted with permission from Reference [81]. Copyright 2005 National Academy of Sciences, USA.

14.4
Applications of Au Nanorods in Trace Analysis

AuNRs have a broad range of applications in trace analysis. The detection targets are particularly numerous within the biomedical and biological fields. Two major advantages make AuNRs promising photoactive reagents in bioanalysis. First is their high signal enhancement ability. The enhanced signals provide high sensitivity, making trace analysis possible. The second major advantage is their low toxicity. Gold is known for being biologically compatible. Detailed studies of gold nanomaterial toxicity were recently reviewed by Murphy et al. [24].

In this section we first discuss methods for fabricating different types of sensors employing AuNRs. Then, bioanalysis based on optical and electrochemical measurements will be discussed.

14.4.1
Fabrication of Au Nanorod-Based Sensors

14.4.1.1 Fabrication of SERS and LSPR Sensors
The process of fabricating SERS sensors varies for different applications. However, films of AuNRs are usually the key sensing component in most designs. The films can be assembled through layer-by-layer deposition on glass substrate [65], or aggregated at the interface of water and toluene [29]. This type of sensor is not affected by the length or the aspect ratio of AuNRs [66]. It has been determined that the AuNR tip is the active site and the diameter of the rod significantly affects sensing ability. The enhancement of the SERS signal using AuNR films was about 10^5-fold, while using the 2D AuNR array the enhancement was between 10^7- and 10^9-fold. Several analytes, including small biological molecules, not commonly detected using SERS have been detected using these new AuNR SERS sensors.

Similar to the SERS sensors, AuNRs have been immobilized on a glass surface [23] or other type of transparent substrate [82] for fabrication of LSPR sensors. For these applications the AuNRs are generally labeled with a specific binding receptor, such as antibody (Figure 14.8) for binding targets [23, 82, 83]. The sensors often display a redshift as the target ligand binds. This shift in wavelength can be clearly observed with the specific target binding in comparison to the nonspecific targets (Figure 14.9). Not only are these sensors selective but they also display a high sensitivity [23]. In one example, a single AuNR has been used to obtain a molecular detection limit of 18 streptavidin molecules [82].

14.4.1.2 Fabrication of Luminescence Sensors
A major difference in the fabrication of a luminescence sensor, when compared with SERS and LSPR sensors, is that a spacer is needed. The luminescence signal of dye molecules can be greatly enhanced by proximity to the AuNR surface. However, the luminescence is quenched if the distance between the dye molecules

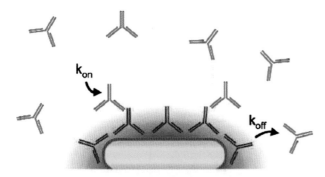

Figure 14.8 AuNR-based LSPR (localized surface plasmon resonance) sensor for an immunoassay. AuNRs are fixed onto a glass surface and coated with a capture antibody. As the sensor is exposed to the target antigens the LSPR wavelength displays a shift. Reprinted with permission from Reference [23]. Copyright 2008 The American Chemical Society.

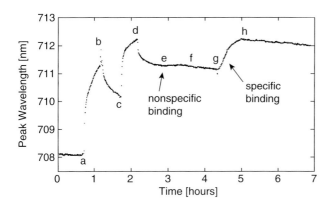

Figure 14.9 Wavelength shift in the LSPR sensor. Step a, a mixture of NHS and EDC were added to the NRs, at step c rabbit IgG was added. Steps b and d were washing steps. Step e was a nonspecific antimouse IgG. Steps f and h were a buffer rinse. Step g was the addition of antirabbit IgG where a strong specific binding effect was observed by the wavelength shift. Reprinted with permission from Reference [23]. Copyright 2008 The American Chemical Society.

and AuNRs is less than 10 nm. Thus, a spacer should be placed between the AuNR and the dye. This spacer can be made from proteins [69] (Figure 14.10), polymers, DNA [79], or other covalently immobilized molecules. By using nanorods, NIR dye signals were enhanced from approximately 7% to 74%, making weak signal NIR dyes suitable components for sensors [69].

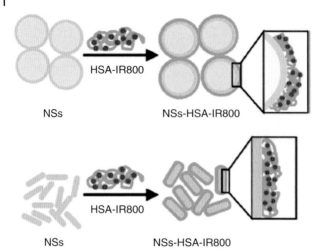

Figure 14.10 Schematic diagram of placing a spacer in the sensing component of a luminescence sensor. Nanoshells (NSs) and NRs coated with the HSA-IR800 complex. The blue line and red dots represent HSA and IR800, respectively. Reprinted with permission from Reference [69]. Copyright 2009 The American Chemical Society.

14.4.2
Bioimaging and Bioanalysis Based on Optical Measurements

Luminescence imaging is widely used in the analysis of biological samples. Usually, these methods are used for qualitative analysis and obtaining structural information of specimens. AuNRs can be used as direct signaling probes or as enhancement platforms in bioimaging. The detection principles are similar to those that employ other luminescence labeling reagents. Figure 14.11 shows an example of imaging blood vessels using AuNR probes by two-photon luminescence. Single AuNRs were observed passing through a blood vessel in a mouse [84].

For quantitative analysis, the absorption spectrometry of AuNRs is more appropriate. The absorption magnitude of AuNRs is proportional to the analyte concentration after AuNR probes bind to the targets. Owing to the high sensitivity provided by the AuNRs, trace amounts of analytes can be determined.

The absorption signal is usually very stable. In the rare case of signal variation, several surface treatments can improve the stability. PEG is a good candidate for coating AuNRs. Results have shown that PEG not only reduced the toxicity of AuNRs, it also increased the stability of the AuNRs. Figure 14.12 shows such an example. In this detection system, the absorption signal of the AuNRs was highly unstable. Surface treatments of AuNRs with phosphatidylcholine (PC) or CTAB had no effect and the signals reduced dramatically as the time passed by (Figure 14.12b and 14.12c). However, the absorption signal of PEG modified AuNRs remained constant for more than 10 min [22].

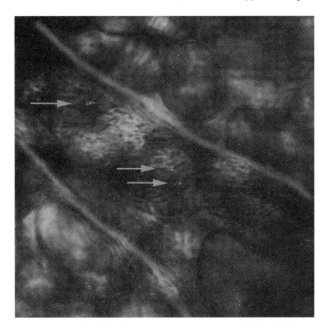

Figure 14.11 Two-photon luminescence image of individual AuNRs (marked with arrows) passing through a blood vessel in a mouse ear. The blood vessel walls have been highlighted for clarity. Reprinted with permission from Reference [84]. Copyright 2006 The Surface Science Society of Japan.

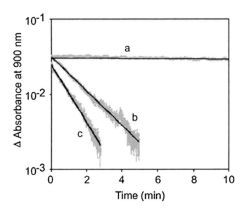

Figure 14.12 Absorption signals of surface modified AuNRs. (a) PEG-modified, (b) PC-modified, and (c) CTAB-modified AuNRs. Reprinted with permission from Reference [22]. Copyright 2008 John Wiley & Sons, Inc.

The signal enhancement provided by AuNRs is significant for improving sensitivity of optical signal-based analysis. A broad array of applications of AuNRs in bioanalysis is expected in the near future.

14.4.3
Bioanalysis Based on Electrochemical Measurements

Gold is a common material for making electrodes in electrochemical analysis. Both gold and glassy carbon electrodes have been modified with AuNRs to fabricate sensing platforms. These AuNR-modified electrodes have led to increased detection sensitivity, less fouling of the electrode surface, good reproducibility, and long-term stability [85–87].

The electrochemical methods used include electrochemical impedance spectroscopy (EIS) [88] and cyclic voltammetry (CV) [85]. Both high selectivity and low detection limit were obtained for protein detection when AuNR modified gold electrodes were used with EIS. This detection was label-free and presented a better analytical response than the unmodified electrodes [88]. By increasing the surface area of the electrode with a layer of AuNRs, the reduction peak in the CV curves became much more prominent compared to a smooth electrode surface [85]. However, the CV measurements still remained less sensitive than measurements made by EIS studies. These electrochemical reactions have shown good reproducibility [88].

Several biological samples have been detected using the AuNR-based electrochemical methods. In addition to proteins and DNA samples, some small molecules such as ascorbic acid and uric acid have also been detected with good selectivity [86]. Figure 14.13 shows a schematic diagram of gold nanowire-based sandwich assay for the detection of protein. A film of gold nanowires and ZnO

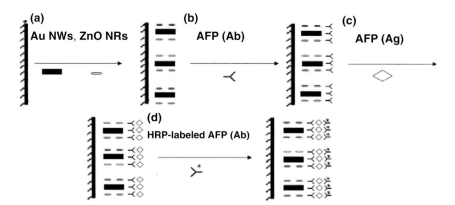

Figure 14.13 Schematic diagram of stepwise immunoassay using Au nanowires.
(a) Coating of AuNWs and ZnO NRs on an electrode; (b) AFP (antibody) loading; (c) AFP (antigen) loading; (d) HRP-labeled AFP (second antibody) loading.
Reprinted with permission from Reference [87]. Copyright 2008 Elsevier.

NRs were immobilized on a glassy carbon electrode surface where an antibody/target protein/labeled antibody sandwich structure was fabricated to determine α-1-fetoprotein (AFP). A detection limit of 0.1 ng ml^{-1} AFP was obtained [87].

14.5
Applications of Au Nanorods in Other Fields

In addition to trace analysis, AuNRs have also been applied to other research areas. Although the number of publications in these areas is less than bioanalysis discussed above, the applications represent viable directions in which AuNRs will have a future impact. In this section we present two examples of such applications: catalysis and photothermal therapy.

14.5.1
Au Nanorods as Supporting Material for Electrocatalyts

Determination of trace analytes requires not only high sensitivity and selectivity but also a fast reaction rate. Catalysts control the rates of chemical reactions and thereby control the yields of desired products over undesired by-products. Because catalysis is based on the reaction of molecules at specific surface sites, a high surface area will certainly provide more active sites, and thereby enhance catalytic activity. Thus, a high surface area is crucial for improving the efficiency of a catalyst.

AuNRs have been used as an effective supporting matrix for catalysts in electrochemical analysis. High surface-to-volume ratio is one of the major features of all nanomaterials. This feature provides AuNRs with a unique advantage in electrochemical reactions. Electrocatalysts can be dispersed onto the highly conductive AuNR matrix to speed up the electrochemical analysis. For instance, by depositing a sub-monolayer of Pt on AuNRs, the oxidation of formic acid is much faster than that using the same amount of Pt dispersed on carbon powder. Meanwhile, the formation of intermediate CO, which can poison the catalyst, was greatly suppressed due to the presence of AuNRs [89]. The AuNR array was used as a catalyst matrix as well [90]. An ultrathin layer-by-layer Pt/Ru/Pt or a Pt/Ru co-deposition layer was coated on the AuNR array to form an effective catalyst.

14.5.2
Au Nanorod-Based Photothermal Therapy

Gold nanomaterials have been studied for improving cancer diagnostics and therapy [91–93]. In fact, bioimaging and photothermal therapy can be incorporated into one application [25, 92, 93]. In one example [92], the AuNRs were first covered with a poly(styrenesulfonate) (PSS) polyelectrolyte that changed the surface charge of the AuNRs from positive to negative. Then, an antibody of anti-epidermal growth factor receptor (anti-EGFR) was adsorbed on the AuNR surface.

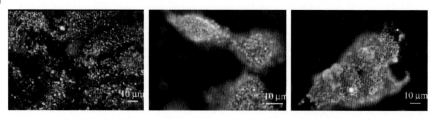

Figure 14.14 Light scattering images of anti-EGFR/AuNRs after incubation with cells for 30 min at room temperature. The orange to red is the most dominant color of AuNRs, corresponding to the surface plasmonic enhancement of the longitudinal oscillation in the NIR region. Reprinted with permission from Reference [92]. Copyright 2006 The American Chemical Society.

The antibody specifically targeted the malignant cells (Figure 14.14). Once bound to the malignant cells, the AuNRs were heated by laser irradiation for photothermal therapy.

Occasionally, selectivity is problematic for photothermal therapy of AuNRs. Slightly higher nonspecific binding occurred with AuNRs than Au nanospheres because the negative charge of the PSS on the surface had a higher affinity to both the malignant and nonmalignant cells [92]. Nevertheless, specific binding and selective destruction of the malignant cells has been achieved. One example showed that deltorphin-labeled AuNRs selectively bound to target cells quickly. Once the labeled AuNRs were introduced to a mixture of different cells, the AuNRs recognized malignant δOR proteins and selectively bound to the cells with δOR. Then, a laser for the photothermal therapy was applied, resulting in deactivation of only the δOR cells [93].

14.6
Conclusions

AuNRs are promising photoactive nanomaterials in trace analysis. Several methods for synthesis of AuNRs have been developed. Electrochemical synthesis including hard template and soft template methods produce AuNRs with uniform size and shape. The seed-mediate growth method is the most popular AuNR synthesis approach at present. Photochemical synthesis is used less at this point but has a great potential in the future considering its simplicity.

As direct optical probes, AuNRs can be used for both luminescence and visible absorption sensing applications although the latter method is more frequently used. When used as a luminescence probe, its luminescence intensity is related to the aspect ratio. A higher aspect ratio gives a much stronger luminescence signal. The most important use of AuNRs is for signal enhancement. Several pathways of signal enhancement are attractive, including SERS, LSPR and luminescence enhancement.

The ability to promote high signal enhancement leads to high sensitivity for trace analysis. Fabrication of sensors employing AuNRs is a typical way to take advantage of these properties for the analysis of targets. In addition, bioanalysis with AuNRs can be accomplished with optical- and electrochemical-based methods. Both methods have broad application in biomedical and biological fields.

In addition to trace analysis, AuNRs have shown advantages in other applications, including catalysis and photothermal therapy. Owing to unique properties of AuNRs, high signaling ability and low toxicity, a broad array of applications of AuNRs is expected in the future.

Acknowledgments

This work was supported by the National Science Foundation under grant CHE-0616878 and CHE0911472, NSF EPS-0814442, and NSF REU grant CHE-0552762, North Dakota Water Resource Research Institute Graduate Student Assistantship, and seed awards from the North Dakota DOE EPSCoR.

References

1 Xu, S., Hartvickson, S., and Zhao, J.X. (2008) *Langmuir*, **24** (14), 7492–7499.
2 Dal, H., Wong, E.W., Lu, Y.Z., Fan, S., and Lieber, C.M. (1995) *Nature*, **375**, 769–772.
3 Yu, S.-S.C., Lee, C.-L., and Wang, C.R.C. (1997) *J. Phys. Chem. B*, **101**, 6661–6664.
4 Leevy, W.M., Gammon, S.T., Jiang, H., Johnson, J.R., Maxwell, D.J., Jackson, E.N., Marquez, M., Piwnica-Worms, D., and Smith, B.D. (2006) *J. Am. Chem. Soc.*, **128**, 16476–16477.
5 Wang, L.Q., Peng, X.J., Zhang, W.B., Yin, F., Cui, J.N., and Gao, X.Q. (2005) *Chin. Chem. Lett.*, **16** (3), 341–344.
6 Johnson, H.E. and Aikens, C.M. (2009) *J. Phys. Chem. A*, **113**, 4445–4450.
7 Zhang, Y. and Lee, C.-T. (2009) *J. Phys. Chem. C*, **113**, 5920–5923.
8 Lee, S.-D., Kim, Y.-S., Yi, M.-S., Choi, J.-Y., and Kim, S.-W. (2009) *J. Phys. Chem. C*, **113**, 8954–8958.
9 Schierhorn, M., Boettcher, S.W., Ivanovskaya, A., Norvell, E., Sherman, J.B., Stucky, G.D., and Moskovits, M. (2008) *J. Phys. Chem. C*, **112**, 8516–8520.
10 Zhai, T., Fang, X., Bando, Y., Liao, Q., Xu, X., Zeng, H., Ma, Y., Yao, J., and Golberg, D. (2009) *ACS Nano*, **3**, 949–959.
11 Mohammad, A.M., Awad, M.I., El-Deab, M.S., Okajima, T., and Ohsaka, T. (2008) *Electrochim. Acta*, **53**, 4351–4358.
12 Jiang, D., Cao, L., Su, G., Liu, W., Qu, H., Sun, Y., and Dong, B. (2009) *J. Mater. Sci.*, **44**, 2792–2795.
13 Foss, C.A., Jr., Hornyak, G.L., Stockert, J.A., and Martin, C.R. (1992) *J. Phys. Chem.*, **96**, 7497–7499.
14 Jirage, K.B., Hulteen, J.C., and Martin, C.R. (1997) *Science*, **278**, 655–658.
15 Martin, C.R. (1994) *Science*, **266**, 1961–1966.
16 Li, Z., Kubel, C., Parvulescu, V.I., and Richards, R. (2008) *ACS Nano*, **2**, 1205–1212.
17 Moon, J. and Wei, A. (2005) *J. Phys. Chem. B*, **109**, 23336–23341.
18 Sander, M.S. and Tan, L. (2003) *Adv. Funct. Mater.*, **13**, 393–397.
19 Thurn-Albrecht, T., Schotter, J., Kastle, G.A., Emley, N., Shibauchi, T., Krusin-Elbaum, L., Guarini, K., Black, C.T., Tuominen, M.T., and Russell, T.P. (2000) *Science*, **290**, 2126–2129.
20 Park, S., Lim, J., Chung, S., and Mirkin, C.A. (2004) *Science*, **303**, 348–351.
21 Chang, S., Shih, C., Chen, C., Lai, W., and Wang, C.R.C. (1999) *Langmuir*, **15**, 701–709.

22 Niidome, T., Akiyama, Y., Shimoda, K., Kawano, T., Mori, T., Katayama, Y., and Niidome, Y. (2008) *Small*, **4**, 1001–1007.
23 Mayer, K.M., Lee, S., Liao, H., Rostro, B.C., Fuentes, A., Scully, P.T., Nehl, C.L., and Hafner, J.H. (2008) *ACS Nano*, **2**, 687–692.
24 Murphy, C.J., Gole, A.M., Stone, J.W., Sisco, P.N., Alkilany, A.M., Goldsmith, E.C., and Baxter, S.C. (2008) *Acc. Chem. Res.*, **41**, 1721–1730.
25 Huff, T.B., Hansen, M.N., Tong, L., Zhao, Y., Wang, H., Zweifel, D.A., Cheng, J.-X., and Wei, A. (2007) *Proc. SPIE*, **6448**, 64480D-1–64480D-9.
26 Kim, F., Song, J.H., and Yang, P. (2002) *J. Am. Chem. Soc.*, **124**, 14316–14317.
27 Murphy, C.J., Sau, T.K., Gole, A.M., Orendorff, C.J., Gao, J., Gou, L., Hunyadi, S.E., and Li, T. (2005) *J. Phys. Chem. B*, **109**, 13857–13870.
28 Jana, N.R., Gearheart, L., and Murphy, C. (2001) *J. Adv. Mater.*, **13**, 1389–1393.
29 Jana, N.R., Gearheart, L., and Murphy, C.J. (2001) *J. Phys. Chem. B*, **105**, 4065–4067.
30 Nikoobakht, B., Wang, Z.L., and El-Sayed, M.A. (2000) *J. Phys. Chem. B*, **104**, 8635–8640.
31 Smith, D.K. and Korgel, B.A. (2008) *Langmuir*, **24**, 644–649.
32 Pan, B., Cui, D., Oxkan, C., Xu, P., Huang, T., Li, Q., Chen, H., Liu, F., Gao, F., and He, R. (2007) *J. Phys. Chem. C*, **111**, 12572–12576.
33 Seo, S.S., Wang, X., and Murray, D. (2009) *Ionics*, **15**, 67–71.
34 Hubert, F., Testard, F., and Spalla, O. (2008) *Langmuir*, **24**, 9219–9222.
35 Perez-Juste, J., Liz-Marzan, L.M., Carnie, S., Chan, D.Y.C., and Mulvaney, P. (2004) *Adv. Funct. Mater.*, **14**, 571–579.
36 Melikyan, A. and Minassian, H. (2008) *Chem. Phys. Lett.*, **452**, 139–143.
37 Kou, X., Zhang, S., Yang, Z., Tsung, C.-K., Stucky, G.D., Sun, L., Wang, J., and Yan, C. (2007) *J. Am. Chem. Soc.*, **129**, 6402–6404.
38 Tsung, C.-K., Kou, X., Shi, Q., Zhang, J., Yeung, M.H., Wang, J., and Stucky, G.D. (2006) *J. Am. Chem. Soc.*, **128**, 5352–5353.
39 Ni, W., Kou, X., Yang, Z., and Wang, J. (2008) *ACS Nano*, **2**, 677–686.
40 Carbo-Argibay, E., Rodriques-Gonzalez, B., Pacifico, J., Pastoriza-Santos, I., Perez-Juste, J., and Liz-Marzan, L.M. (2007) *Angew. Chem. Int. Ed.*, **46**, 8983–8987.
41 Ojha, A.K., Chandra, G., and Roy, A. (2008) *Nanotechnology*, **19**, 1–8.
42 Wijaya, A. and Hamad-Schifferli, K. (2008) *Langmuir*, **24**, 9966–9969.
43 Jebb, M., Sudeep, P.K., Pramod, P., Thomas, K.G., and Kamat, P.V. (2007) *J. Phys. Chem. B*, **111**, 6839–6844.
44 Nandanan, E., Jana, N.R., and Ying, J.Y. (2008) *Adv. Mater.*, **20**, 2068–2073.
45 Hartland, G.V. (2004) *Phys. Chem. Chem. Phys.*, **6**, 5263–5274.
46 Kelly, K.L., Coronado, E., Zhao, L.L., and Schatz, G.C. (2003) *J. Phys. Chem. B*, **107**, 668–677.
47 Yang, J., Wu, J., Wu, Y., Wang, J., and Chen, C. (2005) *Chem. Phys. Lett.*, **416**, 215.
48 Xia, Y. and Halas, N.J. (2005) *MRS Bull.*, **30** (5), 338–348.
49 Stuart, D.A., Haes, A.J., Yonzon, C.R., Hicks, E.M., and Van Duyne, R.P. (2005) *IEE Proc. Nanobiotechnol.*, **152** (1), 13–32.
50 Haes, A.J., Haynes, C.L., McFarland, A.D., Schatz, G.C., Van Duyne, R.P., and Zou, S. (2005) *MRS Bull.*, **30** (5), 368–375.
51 Lakowicz, J.R. (2006) *Plasmonics*, **1** (1), 5–33.
52 Willets, K.A. and Van Duyne, R.P. (2007) *Annu. Rev. Phys. Chem.*, **58**, 267–297.
53 Mulvaney, S.P. and Keating, C.D. (2000) *Anal. Chem.*, **72**, 145R–157R.
54 Campion, A. and Kambhampati, P. (1998) *Chem. Soc. Rev.*, **27**, 241–250.
55 Krug, J.T., II., Wang, G.D., Emory, S.R., and Nie, S. (1999) *J. Am. Chem. Soc.*, **121**, 9208–9214.
56 Emory, S.R. and Nie, S. (1998) *J. Phys. Chem. B*, **102**, 493–497.
57 Emory, S.R., Haskins, W.E., and Nie, S. (1998) *J. Am. Chem. Soc.*, **120**, 8009–8010.
58 Wiley, B., Sun, Y., Chen, J., Cang, H., Li, Z.-Y., Li, X., and Xia, Y. (2005) *MRS Bull.*, **30** (5), 356–361.
59 Jana, N.R. and Pal, T. (2007) *Adv. Mater.*, **19** (13), 1761–1765.

60 Barnard, A.S. and Curtiss, L.A. (2007) *J. Mater. Chem.*, **17**, 3315–3323.
61 Nikoobakht, B., Wang, J., and El-Sayed, M.A. (2002) *Chem. Phys. Lett.*, **366**, 17.
62 Nikoobakht, B. and El-Sayed, M.A. (2003) *J. Phys. Chem. A*, **107**, 3372–3378.
63 Guo, H.Y., Ruan, F.X., Lu, L.H., Hu, J.W., Pan, J.G., Yang, Z.L., and Ren, B. (2009) *J. Phys. Chem. C*, **113** (24), 10459–10464.
64 Orendorff, C.J., Gearheart, L., Janaz, N.R., and Murphy, C.J. (2006) *Phys. Chem. Chem. Phys.*, **8**, 165–170.
65 Hu, X., Cheng, W., Wang, T., Wang, Y., Wang, E., and Dong, S. (2005) *J. Phys. Chem. B*, **109**, 19385–19389.
66 Liao, Q., Mu, C., Xu, D.-S., Ai, X.-C., Yao, J.-N., and Zhang, J.-P. (2009) *Langmuir*, **25**, 4708–4714.
67 Wang, C., Chen, Y., Wang, T., Ma, Z., and Su, Z. (2008) *Adv. Funct. Mater.*, **18**, 355–361.
68 Yang, Z., Ni, W., Kou, X., Zhang, S., Sun, Z., Sun, L.-D., Wang, J., and Yan, C.-H. (2008) *J. Phys. Chem. C*, **112**, 18895–18903.
69 Bardhan, R., Grady, N.K., Cole, J.R., Joshi, A., and Halas, N.J. (2009) *ACS Nano*, **3**, 744–752.
70 Anger, P., Bharadwaj, P., and Novotny, L. (2006) *Phys. Rev. Lett.*, **96**, 113002.
71 Horimoto, N.N., Imuraa, K., and Okamoto, H. (2008) *Chem. Phys. Lett.*, **467**, 105–109.
72 Lakowicz, J.R. (2005) *Anal. Biochem.*, **337**, 171–194.
73 Chance, R.R., Prock, A., and Silbey, R. (1974) *J. Chem. Phys.*, **60**, 2744–2748.
74 Aslan, K., Leonenko, Z., Lakowicz, J.R., and Geddes, C.D. (2005) *J. Phys. Chem. B.*, **109** (8), 3157–3162.
75 Aslan, K., Lakowicz, J.R., and Geddes, C.D. (2005) *J. Phys. Chem. B*, **109** (13), 6247–6251.
76 Lakowicz, J.R., Geddes, C.D., Gryczynski, I., Malicka, J., Gryczynski, Z., Aslan, K., Lukomska, J., Matveeva, E., Zhang, J., Badugu, R., and Huang, J. (2004) *J. Fluoresc.*, **14** (4), 425–441.
77 Mohamed, M.B., Volkov, V., Link, S., and El-Sayed, M.A. (2000) *Chem. Phys. Lett.*, **317**, 517–523.
78 Eustis, S. and El-Sayed, M. (2005) *J. Phys. Chem. B*, **109**, 16350–16356.
79 Li, C.-Z., Male, K.B., Hrapovic, S., and Luong, J.H.T. (2005) *Chem. Commun.*, **31**, 3924–3926.
80 Boyd, G.T., Yu, Z.H., and Shen, Y.R. (1986) *Phys. Rev. B*, **33**, 7923–7936.
81 Wang, H., Huff, T.B., Zweifel, D.A., He, W., Low, P.S., Wei, A., and Cheng, J.X. (2005) *Proc. Natl. Acad. Sci. U. S. A.*, **102**, 15752–15756.
82 Nusz, G.J., Curry, A.C., Marinakos, S.M., Wax, A., and Chilkoti, A. (2009) *ACS Nano*, **3**, 795–806.
83 Nusz, G.J., Marinakos, S.M., Curry, A.C., Dahlin, A., Hook, F., Wax, A., and Chilkoti, A. (2008) *Anal. Chem.*, **80**, 984–989.
84 Wei, A. (2006) *e-J. Surf. Sci. Nanotech.*, **4**, 9–18. http://www.sssj.org/ejssnt/
85 Lee, S.-J., Anandan, V., and Zhang, G. (2008) *Biosens. Bioelectron.*, **23**, 1117–1124.
86 Liu, H. and Tian, Y. (2008) *Electroanalysis*, **20**, 1227–1233.
87 Lu, X., Bai, H., He, P., Cha, Y., Yang, G., Tan, L., and Yang, Y. (2008) *Anal. Chim. Acta*, **615**, 158–164.
88 Wasowicz, M., Viswanathan, S., Dvornyk, A., Grzelak, K., Kludkiewicz, B., and Radecka, H. (2008) *Biosens. Bioelectron.*, **24**, 284–289.
89 Wang, S., Kristian, N., Jiang, S., and Wang, X. (2008) *Electrochem. Commun.*, **10**, 961–964.
90 Yoo, S.-H. and Park, S. (2008) *Electrochcim. Acta*, **53**, 3656–3662.
91 Huang, X., Jain, P.K., El-Sayed, I.H., and El-Sayed, M.A. (2007) *Nanomedicine*, **2**, 681–693.
92 Huang, X., El-Sayed, I.H., Qian, W., and El-Sayed, M.A. (2006) *J. Am. Chem. Soc.*, **128**, 2115–2120.
93 Black, K.C., Kirkpatrick, N.D., Troutman, T.S., Xu, L., Vagner, J., Gillies, R.J., Barton, J.K., Utzinger, U., and Romanowski, M. (2008) *Mol. Imaging*, **7**, 50–57.

Index

a

absorption spectra, gold nanorods   359, 365, 366
activated carbon   191, 206
– chemically-modified   207–209
"activated gold"   32
adsorption
– carbon monoxide   232
– lead   197, 198
– mercury in groundwater   196, 197
– organic molecules   239
– oxygen   233
– titania   223, 235–237
– water vapor   233, 234
adsorption energy   302, 303
adsorption isotherms   196, 213
affinity-based electrochemical biosensors   61, 62
affinity-based electrophoresis   120, 121
AFM *see* atomic force microscopy (AFM)
aggregation   270–272
– in porous media   272, 273
– quantum dots   48
alkaline hydrothermal treatment   226
alkylphosphonic acids   143
alloyed quantum dots   35
alloying
– "alloying point"   35
– and electrocatalysis   302, 303
alumina   82, 103
– aggregation   271, 272
– in gold nanorod synthesis   361
– nanoporous anodic membranes   342, 343, 352, 353
aluminosilicates   212–214
amidation   349, 350
analysis *see* bioanalysis; electrochemical analysis; fluorescence resonance energy transfer (FRET) analysis; trace analysis

anodic stripping voltammetry   66, 67, 69, 306, 307
anodization, titanium   225, 226
Anthozoa   13
antibacterials   143, 145, 269, 273–275
antibodies
– antigen–antibody immunoreactions   61, 62
– in displacement immunosensors   176–179
– α-fetoprotein antibody   80, 81
– gold nanorods   376–378
– quantum dot conjugates   173, 176
antigens   61, 62, 177, 376
apoferritin   75–77
aptamer biosensors   72
aqueous phase colloidal synthesis   41
aqueous samples, titania analysis   235–240
aqueous systems
– particle aggregation in   270
– trace-metal collection from   191–216
arsenic   305–307
"artificial atoms"   31
"artificial peroxidase" electrodes   166, 167
arylene-ethynylene tetracycle   171, 172
ascorbic acid   311, 312
aspect ratio   359, 363–365, 368, 370
atomic force microscopy (AFM)
– atmospheric ENPs   276
– electrochemical detection   331–333
– gold nanoparticles   289, 290
– PMMA nanopores   342, 353
atomic layer deposition   226
attachment efficiency   273
autofluorescence   13, 46
avalanche photodiode   324
avidin-functionalized silica   80, 81
azobenzene chromophores   22, 23

Trace Analysis with Nanomaterials. Edited by David T. Pierce and Julia Xiaojun Zhao
Copyright © 2010 WILEY-VCH Verlag GmbH & Co. KGaA, Weinheim
ISBN: 978-3-527-32350-0

b

band gap energy, quantum dots 33, 36, 37, 172
BCP nanopores *see* block copolymer (BCP) nanopores
benzenethiols 100
bifunctional nanoparticles 259
binding affinities (K_d) 141, 193–197, 199, 202
binding properties
– molecularly imprinted polymers 148–152
– rebinding target analytes 140, 141
bioanalysis
– gold nanorods 359, 360, 372
– optical measurements basis of 374–376
– sorbent performance comparisons 197–200
– using electrochemical detection 333–377
– using micro- and nanofluidic systems 111–127
bioimaging
– optical measurements basis of 374–376
– photoswitchable nanoprobes for 3–26
biological analytes, determination 310–313
biomacromolecules 48, 321, 351–355
biometallization 66
biomolecule–nanomaterial labels 63
biopharmaceuticals, molecularly imprinted polymers for 143–145
biorecognition module 173, 174
biosamples *see* bioanalysis
biosensors 103–106 *see also* electrochemical biosensors; immunosensors; sensors
– aptamer 72
– definition 61
– electron transport in nanomaterials 89–107
– enzyme 61, 62, 66, 67
– gold nanorod 372–374
– modular 173–175
– optical 230, 240, 245, 246, 248
– properties 90
– quantum dots 51, 52
– titania applications 240–246
biothiols 51, 52
biotinylated detection probe 66, 67, 77
bismuth 300–302
bisphenol A (BPA) 146, 153
block copolymer (BCP) nanopores 344–346
– gold nanoelectrode fabrication 348–352, 362, 363
– molecular permeation through 347–355
– surface functionalization 346, 347
bottom-up synthesis, quantum dots 40

bovine serum albumin (BSA) 45, 52, 53
"building-block" approach 263, 264
bulk continuum voltammetry 93–95
bulk detection 161, 163

c

cadmium detection 264
cadmium selenide 32, 33, 35, 37, 38
– fluorescence detection/quenching 265, 268
– near-infrared detection 46–48
– synthesis 40
– zinc sulfide quantum dots 5, 6
cadmium sulfide 32, 33, 37, 38
– electrochemical analysis 50, 51, 70, 71, 74, 75
– fluorescence enhancement and quenching 42
cadmium telluride 32, 33, 35, 37–39
– chemiluminescence analysis 48–50
– electrochemical immunoassays 73
– near-infrared detection 46–48
– synthesis 40, 41
cancer diagnostics and therapy 377, 378
capacitance (C) 90–93, 107, 320
capacity 193, 196
capillary electrochromatography (CEC) 119, 120
capillary electrophoresis (CE) 115, 165, 268
capillary gel electrophoresis (CGE) 115, 120
capillary zone electrophoresis (CZE) 119, 120
carbon fiber electrodes (CFEs) 327–329
carbon monoxide 231, 232
– annealing 298
– oxidation 296–298
carbon nanotubes (CNTs) 305, 306
– as electrochemical biosensors 73–75
– monolithic membranes 342, 343
– multi-walled 165, 233, 308
– single-walled 165, 166, 325, 333
carbon sorbents, nanoporous 206–211
carcinoembryonic antigen (CEA) 50, 51, 70, 73
catalysis *see* electrocatalysis
catalytic-based biosensors 61
catechols 49
cell-to-cell communication 126
cellomics 123, 124
cetyltrimethylammonium bromide (CTAB) 363–375
chamber stationary PCR devices 112, 113
charging energy 91

chelation therapy 198
Chelex-100 197, 198
chemical etching 329, 331, 342, 344, 361, 362
chemical vapor deposition (CVD) 228, 348–352
chemiluminescence (CL) 48–50, 146, 147, 162, 327
– electrogenerated 167–169, 327
chemiresistors 97–103
chemochromic optical sensor 230
chemosensors
– electron transport in nanomaterials 89–107
– metal nanoparticle films basis 98–101
– metal oxide nanoparticles basis 101–103
– in nanoparticle assemblies 97
– quantum dots 51, 52
chloramphenicol 145, 146
chloromethylation, activated carbon 208
cholesterol 143
chromium 309
chronoamperometry 307–310
circular wells 111
clean-up 144, 145, 151, 152
"click" chemistry 259, 260
clinoptilolite 212
CNTs see carbon nanotubes (CNTs)
cobalt oxide nanoparticles 305, 306
colloid suspensions, stability 270
colloidal nanoparticles see semiconductor nanocrystals
colloidal quantum dots see semiconductor nanocrystals
colorimetric detection and sensing 235, 236, 256–264
colorimetric tests 162, 166
competitive immunoassays 62
compressive strain 302, 303
computational modelling 141–143
condensation nucleus counter 276
conductance, single nanopipette-based 333–335
conducting polymer films 295, 296, 311
conductive immunosensors 69
confocal imaging 335, 355
confocal laser scanning microscopy (CLSM) 10, 14
convection-driven PCR devices 113, 114
copper salts 32
– colorimetric detection 263
– quantum dots 42, 48
core–shell nanoparticles, hydrophobic–hydrophilic 7, 8, 15, 16, 20

core–shell quantum dots 31, 32, 34–36, 39
costs, nanostructured materials 216
coulomb blockade effect 91, 92
Coulter counter 343
covalent molecular imprinting 138
cresyl violet 324
critical voltage bias 91, 92
crosslinkers, molecular imprinting 139, 140
crosslinking precipitation 95
crystal structures, quantum dots 33
CVD see chemical vapor deposition (CVD)
cyclic voltammetry 292, 293, 295, 309, 310, 312, 313, 346, 376
cylindrical nanopores 341–347

d
d-band theory 302, 303
Debye–Scherrer formula 37
diarylethenes 9–11, 24
diclofenac 150, 151
differential pulse voltammetry 93, 94, 312, 313
diffraction limit 3
diffusion layer 285
dimercaptosuccinic acid DMSA-Fe_3O_4 196–198, 202–204
dimethoate insecticide 141, 142
dipyridamole 146, 147
direct bandgap semiconductors 264, 265
direct fluorescence response, quantum dots 42, 43
direct synthesis modification 41
displacement immunosensors 176–179
DNA
– amplification 113, 114
– analysis micro- and nanofluidic tools 116–118
– denaturation, annealing and extension 113
– detection using titania 245, 246
– localized delivery 334, 335
– separation, hybridization and sample preparation 115, 116
DNA electrochemical detection
– apoferritin 76
– carbon nanotubes 74
– liposomes 77, 78
– metal nanopraticles 64–68
– nanoparticle assemblies 105, 106
– nanowires 82
– point-of-care diagnosis 83
– quantum dots 70–72
– silica 79, 80

DNA hybridization 61, 62, 64–68, 71, 74–76, 82, 115, 116
DNA microarrays 115, 116
DNAzymes 260–263, 265, 266
dopamine 311, 312, 333
double potential-pulse voltammetry 287, 301, 302
Dronpa 11–13
droplet-based PCR device 114, 115
dual-color fluorescent nanoprobes 13–19
dummy molecular imprinting polymer 145
dyes see also spiropyrans (SP)
– fluorescein 19, 124, 262
– luminescence enhancement 369
– luminescence sensors 372–374
– Rhodamine dye 11, 39, 267
– single-chromophore nanoparticles 17–19
dynamic light scattering (DLS) 276, 277

e
E2 see 17β-estradiol (E2)
electric field gradient focusing (EFGF) 121, 122
electrical/electronic conductivity 96, 97, 99, 105, 106, 320
– titania 224, 230, 231
"electroactive" silica 80
electroanalysis 285–313
electrocatalysis
– inorganic analyte determination 305–310
– nanoparticle composition dependence 302–304
– nanoparticle shape dependence 299–302
– nanoparticle size effects 296–299
– organic/biological analyte determination 310–313
electrocatalysts 68, 377
electrochemical analysis 50–52, 70, 71, 74, 75 see also DNA electrochemical detection; electrochemical immunoassays
– by atomic force microscopy 331–333
– bioanalysis 333–335, 376, 377
– biological systems, nanodomains of 333–335
– heavy metals 203, 204, 206
– nanoelectrodes for 328–332
– single-nanoparticle 326–328
electrochemical biosensors 61–63
– carbon nanotubes as 73–75
– nanomaterial labels used 63–82
– for point-of-care diagnosis 82–84
– silica nanoparticles in 79–81, 105
electrochemical fluorimetry 144
electrochemical imaging 330–332

electrochemical immunoassays 62
– apoferritin 76, 77
– liposomes 78
– metal nanoparticle labels 68–70
– point-of-care diagnosis 83, 84
– quantum dot labels 72, 73
– silica 80, 81
electrochemical immunosensors 68–70, 79
electrochemical impedance spectroscopy 376
electrochemical oxidation 244, 305, 306
electrochemical reactions 285
electrochemical sensors see also electrochemical biosensors; electrochemical immunosensors
– explosives 162, 164–167
electrochemiluminescence 49, 50, 52
electrochemistry 320
– single-molecule 321–326
electrodes 320 see also nanoelectrodes
– "artificial peroxidase" 166, 167
– magnetic and electromagnetic 203
– metal nanoparticle
– – analytical applications 304–313
– – conducting polymer films 295, 296
– – direct deposition on 289–293
– – geometry 296–304
– – modified 285–313
– – three-dimensional deposition 293–295
– nanomaterial modified 165, 166
– seed-mediated synthesis on 286–289, 308, 309, 364–366
electrogenerated chemiluminescence (ECL) 167–169, 327
electrolyte concentration 350, 351
electron hopping 95, 96, 98, 99
electron transfer kinetics 320, 321, 328
electron transport, in nanomaterials 89–107
electron tunneling 96, 98, 99
electronic energy levels/properties, quantum dots 36–38
electrophoresis
– affinity-based 120, 121
– capillary 115, 165, 268
– capillary gel 115, 120
– capillary zone 119, 120
– gel 115, 118, 120
electrophysiology 126
electroreduction 243
electrospray ionization 122
electrostatic interactions/attraction 348–350, 366
emission spectra, quantum dots 38, 39

endocrine-disrupting chemicals (EDCs) 135, 136, 151
endocytosis 18
energy dispersive X-ray analysis 276
engineered nanoparticles (ENPs) 255, 256
– characterization 276, 277
– environmental detection 275–278
– environmental fate 269–275
– environmental sensing 256–269
enoxacin 46, 47
entropic trapping 116, 117
environmental fate, engineered nanoparticles 269–275
environmental samples, sorbent performance comparisons 194–197
environmental sensing, using engineered nanoparticles 256–269
enzymes 52
– biosensors/bioassays 61, 62, 66, 67
– electrochemical detection 74
EosFP 13–15
17α-estradiol 135
17β-estradiol (E2) 135–137
– continuing development advances 155
– MIP submicron particles binding with 148–152
estrogens 135–137
estrone (E1) 135, 152
17α-ethynylestradiol (EE2) 135, 137, 149
excitation spectra, quantum dots 38, 39
exciton 265
exciton Bohr radius 34, 37
explosives
– categories 161–163
– structures and abbreviations 164
– trace detection 161–182

f
far-field fluorescence imaging 3, 4, 19–22
faradaic (redox-induced) current 320–323, 350, 351
ferritin 75–77, 351
ferrocene 348
α-fetoprotein antibody 80, 81
field detection/screening 166, 181, 191, 192, 255
field-emission scanning electron microscopy (FE-SEM) 287, 291
field-emission transmission electron microscopy (FETEM) 292
field flow fractionation (FFF) 277
flame ionization detector (FID) 100, 101
flow-injection analysis 48
flow-through PCR devices 113

fluorescein 19, 124, 262
fluorescence-based detection 264–267
fluorescence enhancement 42, 43
fluorescence (FL) spectroscopy 162, 166, 169–178
fluorescence imaging 3, 20, 21, 274, 323, 324–326 *see also* confocal laser scanning microscopy (CLSM); far-field fluorescence imaging; photoswitchable fluorescent nanoprobes
fluorescence quenching 42, 43, 169–176, 267–269
fluorescence resonance energy transfer (FRET) analysis 5, 7, 11, 173, 175, 266–268
– photoswitching 15–17
– single-molecule 323, 326, 334, 335
– using quantum dots 44, 45
fluorescent nanofibril films 171, 172
fluorescent polymer porous films 169–171
fluorescent semiconductor nanocrystals *see* semiconductor nanocrystals
fluorophore nanoparticles 4
formic acid oxidation 299, 300
fulvic acid 272
functionalized nanoparticles 24
– magnetic 200–206
– sensor applications 107
– silica 20, 80, 81, 194–200
– thiol 194–200

g
gallium salts 33, 34, 37
gas chromatography (GC) 100, 101, 162, 268
gas-phase samples, analysis 229–234
gas sensors 101–103, 180
gel electrophoresis 115, 118, 120
glass nanopore electrode 330
glassy carbon electrodes (GCEs) 52, 291–293, 305, 306, 308, 311–313
glucose detection 241, 242
gold nanoelectrodes 322, 376
– block copolymer nanopores 348–352, 362, 363
gold nanoparticles
– citrate-capped 257, 258, 271, 272, 290, 291
– DNAzyme detectors 260–263
– electrochemical detection of DNA 64, 65, 68, 105, 106
– electrochemical immunoassays 68–70, 83, 84
– electron transport in 89, 92–95, 98–100
– environmental detection 257–260
– fluorescence quenching 267–269
– history 32

– inorganic analyte determination   307–310
– layer-by-layer assemblies   294, 295
– miniaturized immunosensors   103, 104
– organic/biological analyte determination   311–313
– single-molecule   324
– two-dimensional arrays   287–292
gold nanorods   359
– applications in trace analysis   372–377
– electrocatalytic supports   377
– electrochemical synthesis   360–364
– luminescence enhancement   370, 371
– photochemical synthesis   364
– photothermal therapy   377, 378
– seed-mediated growth   364–366
– sensors   372–374
– signal enhancement   367–371
green fluorescent protein (GFP)   11, 13, 39
green to red fluorescence conversion   13–15
growth overpotential   287
GT-73   196–198

h

hard template   360–362
heavy metal collection   194–200
– functionalized magnetic nanoparticles   200–206
HEK-293 cells   8, 14, 16, 17, 19
HeLa cells   19, 20
heterogeneous binding models   149, 150
heterogeneous immunoassays   62
hexacyanoferrate(III)-loaded apoferritin   75, 76
hexamethylene triperoxide diamine (HMTD)   162, 163, 166, 167
high performance liquid chromatography (HPLC)   119, 137, 144, 162, 239, 240
– fluorescence detection (FD)   136, 141, 148
high-resolution imaging, nanoelectrodes for   328–332
highest occupied molecular orbital–lowest unoccupied molecular orbital (HOMO–LUMO)   89, 93–95
horseradish peroxidase (HRP)   66, 80, 167, 168
human rhino virus (HRV)   352, 353
humic acid   272, 273
humidity, and water vapor adsorption   233, 234
hybridization   233, 235, 242, 243 see also DNA hybridization
hydrodynamic radius   277
hydrogen, sensing   230–232
hydrogen peroxide   49, 50, 243–245
– electrocatalysis   307, 308
– environmental effects   258
– gold nanorod synthesis   365
hydrogen sulfide   307
hydrophobic–hydrophilic core–shell nanoparticles   7, 8, 15, 16, 20

i

ICP-MS see inductively-coupled plasma mass spectrometry (ICP-MS)
immobilization
– DNA strands   61, 68, 105, 106
– EosFP   14
– proteins   240, 241
immunoassays see electrochemical immunoassays
immunochromatographic strips (ISs)   83, 84
immunoglobulin G (IgG)   48, 49, 52, 53, 69, 103, 104
immunosensors
– conductive   69
– displacement immunosensors   176–179
– electrochemical   68–70, 79
– miniaturized   103, 104
imprint step   214
imprinting, molecular recognition   138–143
in vivo optical imaging   8, 9, 34
in vivo sensing   333
indirect bandgap semiconductors   264, 265
indirect sensing   89
indium salts   33, 34
indium tin oxide (ITO) electrode   49, 287, 288, 290, 291, 308, 309
– single-molecules   324, 325
inductively-coupled plasma mass spectrometry (ICP-MS)   199, 202, 205, 276, 277
inorganic analytes, determination   305–310
inorganic engineered nanoparticles   255
insulin   123
interleukin-6 (IL-6)   68
ion exchangers   191
ion-imprinted polymers   214, 215
ion-sensitive field-effect transistor (ISFET)   237–239
ionic strength   194, 195, 270, 271
iron oxide nanoparticles
– aggregation   272
– DMSA-modified   196–198, 202–204
– electrocatalysis   310
– heavy metal capture and detection   200, 201
– photoswitchable fluorescence   24, 25

isoelectric focusing (IEF) 118–120
isomerization
– photochromic diarylethenes 9–11
– photochromic spiropyrans 7–9
isotachophoresis (ITP) 121
isotopic labeling 277
isoxicam 136, 137
ITO electrode *see* indium tin oxide (ITO) electrode

k
Kaede 13, 14

l
lab-on-a-chip 122
"lab-on-QDs" 173
label-based/label-free electrochemical biosensors 62
laser-induced fluorescence (LIF) detection 123, 124
lattice parameters 32, 33
layer-by-layer (LbL) electrostatic assembly 293–295, 311
leaching step 214
lead salts 33, 34
– adsorption kinetics 197, 198
– colorimetric detection 262, 263
– magnetic nanoparticles collection 204
ligand effect 303
ligands 35, 36, 41, 90, 96, 98, 99
light scattering 46, 48, 49
linear scan voltammograms 301
liposomes 77–79, 124
liquid chromatography-mass spectrometry (LC-MS) 143, 145
localized delivery and imaging 333–335
localized surface plasmon resonance (LSPR) 256, 257, 367
– biosensors 372, 373
luminescence
– enhancement 369–371
– imaging 374, 375
– power 370
– sensors 372–374
luminol 167–169
lysis 124, 125
lysosomes 20

m
macroporous gels 155
magnetic beads 153, 167, 168
magnetic nanoparticles
– heavy metal capture and collection 200–206
– photoswitchable 22–25

magnetic resonance imaging (MRI) 22
maltose binding protein (MBP) 44
manganese-doped zinc sulfide 46, 47, 175
mass spectrometry (MS) 122, 162
– ICP-MS 199, 202, 205, 276, 277
– LC-MS 143, 145
mass transport, nanoelectrodes 329, 330
matrix-assisted pulsed laser evaporation (MAPLE) 228
MCSP nanoparticles 17–19
mercaptosuccinate 311
mercury 322
mercury salts 33, 51
– adsorption in groundwater 196, 197
– colorimetric detection 235, 236, 260–262, 264
– fluorescence detection 266
– fluorescence quenching 267, 268
merocyanine (MC) 5, 6, 18
– in bioimaging 20, 21
– FRET-based photoswitching 15
– isomerization 7–9
mesoporous carbons, templated 209–211
mesoporous silica, thiol-functionalized 194–200
mesoporous titania 225, 235–237
metal ion extraction 235–237
metal nanoparticles 63–70 *see also individual metal nanoparticles*
– deposition of three-dimensional films 293–295
– deposition on bare electrodes 289–293
– electrocatalysis, geometric factors 296–304
– electrode-modified, analytical applications 304–313
– electrodes modified with 285–313
– electron transport through 90–97
– in explosives 165
– films, sensors based on 98–101
– titania hybrids 227, 228
– two-dimensional arrays, fabrication 286–293
– voltammetry in solution 92–95
metal oxide nanoparticles
– chemosensors based on 101–103
– in explosives 165
– gas sensors 180
metallic phosphate-loaded apoferritin 75–77
methacrylic acid 139
methane 231
methanol oxidation 299, 300, 303, 304
methotrexate 144

methyl-carbamate pesticides 141
methylthiotriazine herbicides 147
metsulfuronmethyl 146
micellar electrokinetic chromatography (MEKC) 119–121
microcantilever sensors 162, 178–180
microchannels 111
microelectrodes, nanoparticle collision 326, 327
microemulsion copolymerization 7, 9, 15, 17
microfluidics 111–127
micro-orifice uniform deposition impactor (MOUDI) 276
microspheres (microbeads)
– gold nanoparticles 64, 65
– quantum dot conjugations 52–54
military explosives 161, 162
Miller index 301, 302
MIPs see molecularly imprinted polymers (MIPs)
MISPE see molecularly imprinted solid-phase extraction (MISPE)
modular arm 173–175
modular biosensor 173–175
molecular dynamics simulations 142
molecular permeation, block copolymer nanopores 347–355
molecular recognition, imprinting 138–143
molecularly imprinted polymeric microspheres (MIPMs) 146, 151
molecularly imprinted polymers (MIPs) 136, 137, 214
– biopharmaceuticals and toxins, analytical applications 143–145
– continuing development advances 153–156
– current progress 152, 153
– imprinting of 138–143
– submicron particles binding with E2 148–152
– submicron particles preparation 146–148
molecularly imprinted solid-phase extraction (MISPE) 137, 140, 142–144, 151
molecule-like voltammetry 93–96
monolithic cage structures, silica-based 263, 264
monolithic membranes, carbon nanotubes 342, 343
monolithic nanoporous membranes see nanoporous monolithic membranes
monolithic nanoporous sensors 263, 264
monomers, molecular imprinting 139, 140
MSP nanoparticles 17–19

multi-chamber stationary PCR devices 113
multi-walled carbon nanotubes (MWCNTs) 165, 233, 308
multiple surface functionalities 259
multiplex electrochemical detection 71, 72, 83, 84
multipore membranes 343
mycotoxin 145, 154
myoglobin 294, 295

n

n-type metal oxides 101, 102
NACs see nitroaromatics (NACs)
"nano-on-micro" assay 52–54
nanoarrays 285–293
nanocluster size 101–103
nanoelectrodes 319–321
– electrochemical detection and high-resolution imaging 328–332
– fabrication 328, 329
– mass transport near 329, 330
– single-molecule electrochemistry using 321–323
nanofibers 155, 156
nanofibril films, fluorescent 171, 172
nanofilms see also solid-state films
– conducting polymer 295, 296, 311
– polymer porous fluorescent 169–171
– thickness 103, 228
– three-dimensional 293–295
– titania 228, 229, 232, 241, 243, 244
nanofilter columns 116
nanofluidics 111–127
nanohybrids, titania-based 226–228
nanomaterial modified electrodes 165, 166
nanomaterials see also engineered nanoparticles (ENPs); nanoparticles
– in electrochemical biosensors/bioassays 61–84
– electron transport 89–107
– titania-based 223–247
– trace detection of high explosives with 161–182
nanoparticles see also engineered nanoparticles (ENPs)
– electrocatalysis
– – composition and 302–304
– – shape and 299–302
– – size and 296–299
– permeation through BCP nanopores 351–355
– photoswitchable magnetization properties 22–24
nanopipette-based conductance 333–335

nanopore diameter/size   342, 350, 351, 353–355
nanoporous alumina   361
nanoporous anodic alumina membranes (NAAMs)   342, 343, 352, 353
nanoporous carbon based sorbents   206–211
nanoporous monolithic membranes   341–343
– block copolymer-derived   344–346, 361–363
– molecular permeation   347–355
– surface functionalization   346, 347
nanoporous monolithic sensors   263, 264
nanorods   81, 82, 287 see also gold nanorods
nanotubes see also carbon nanotubes (CNTs)
– titania   225, 226, 230, 241
nanowires   81, 82, 84, 329, 362, 376
near-field optical microscopy   330–332
near-infrared (NIR) emitters   34, 46–48, 359, 371
nitramines   162–164
nitrate esters   162–164
nitrated explosives   161–166, 169–171
see also 2,4,6-trinitrotoluene
nitride quantum dots   34
nitroaromatics (NACs)   162–166, 169–172
nitrogen oxides   308, 309
nitrophenol adsorption   137
noble metal nanoparticles   256–260 see also gold nanoparticles; silver nanoparticles
non-competitive immunoassays   62
non-faradaic (charging) current   320
non-imprinted polymers (NIPs)   137, 139–141, 148, 149, 152
noncovalent molecular imprinting   138, 140
nucleic acids
– analysis   112–118
– electrochemical detection   73, 74

o

octanethiols   98, 100
n-octylamine   23
on-chip protein pre-concentration   121, 122
on-off fluorescent nanoprobes   5–13
optical biosensors   230, 240, 245, 246, 248
optical microscopy   20, 105, 323, 328, 330–332, 334
optical properties
– engineered nanoparticles   255
– quantum dots   38, 39
organic analytes, determination   310–313
organic compounds
– supernanostructures   175, 176
– titania analysis   237–240

organometallic synthesis, quantum dots   40, 41
output resistance change   101
overgrowth, gold nanorods   365, 366
oxidative stress   273–275
oxide annealing   298
oxygen, sensing   233, 234
oxygen reduction reaction   296–297, 303

p

p-type metal oxides   101
PAMAM see polyamidoamine (PAMAM)
PANI see polyaniline (PANI)
patch-clamp recording   126, 127
PCR devices see polymerase chain reaction (PCR) devices
PEDOT see poly(3,4-ethylenedioxythiophene) (PEDOT)
PEG see poly(ethylene glycol) (PEG)
pencil graphite electrode (PGE)   73, 74
pentaerythritol tetranitrate (PETN)   163, 164, 167
performance comparisons
– sorbents for biological samples   197–200
– sorbents for environmental samples   194–197
permeability, block copolymer nanopores   347–355
permselective transport   348
peroxide-based explosives   162–164, 166, 167
Perrin equation   266
perylene diimide (PDI)   15, 16
pesticides   141, 153, 154, 239, 266
pH
– and binding affinity   194–196, 202
– and permeability regulation   348, 349
– and zeta potential   271, 272
phenols   268
phenoxyacetic herbicides   141, 145
phenyl self-assembled monolayer on a mesoporous support (phenyl-SAMMS™)   197
phosphorescence   45–47, 245
photoactuated unimolecular logical switching attained reconstruction (PULSAR) microscopy   3, 4, 19–21
photochemical synthesis   364
photochromic diarylethenes   9–11
photochromic spiropyrans see spiropyrans (SP)
photoluminescence (PL)   47, 50, 173, 233, 234, 265, 321
photoswitchable fluorescent nanoprobes   4–22

– for bioimaging 19–22
– dual-color 13–19
– single-color (on-off) 5–13
photoswitchable (fluorescent) proteins
– green to red conversion 13–15
– sizes 39
– structural conversion 11–13
photoswitchable magnetic nanoparticles 22–25
photothermal microcantilever technique 179, 180
photothermal microscopy 276
photothermal therapy, gold nanorod-based 377, 378
phototoxicity 26
photovoltaic effect 223, 245
piezoelectric quartz crystal microbalance (QCM) 154
piezoelectric quartz crystal (PQC) sensor 153
pirimicarb pesticide 153, 154
PLA-PDMA-PS *see* polylactide-polydimethylacrylamide-polystyrene (PLA-PDMA-PS)
plastic explosives 162, 163
platinum nanoelectrodes 322, 326, 327
platinum nanoparticles 68, 296–304
– inorganic analyte determination 305, 306, 308
Plieth model 297
Pluronic copolymer F127 (PEO-PPO-PEO) 24
PMMA *see* polystyrene-poly(methylmethacrylate) (PS-PMMA)
point-of-care diagnosis 82–84
poly(3,4-ethylenedioxythiophene) (PEDOT) 295, 296, 311, 312
poly(4-aminothiophenol) (PATP) 292
polyacrylamide gel electrophoresis (PAGE) 118
poly(acrylic acid) (PAA) 139, 140
polyamidoamine (PAMAM) 288, 294, 295
polyaniline (PANI) 295, 296, 311
polycarbonate membranes 342
polyelectrolyte shells 66
polyester membranes 342
poly(ethylene glycol) (PEG) 352, 363, 374, 375
poly(guanine)-functionalized silica 80, 81
polylactide-polydimethylacrylamide-polystyrene (PLA-PDMA-PS) 344–346, 348

polymer porous films, fluorescent 169–171
polymerase chain reaction (PCR) devices
– miniaturization 112–115
– separation, DNA hybridization and sample preparation 115, 116
– in single-cell analysis 125
polymerization step 214
polymers *see also* block copolymer (BCP) nanopores
– conducting polymer films 295, 296, 311
– ion-imprinted 214, 215
– permeation through BCP nanopores 351–355
– template-specific 138
polyoxometalate PMo_{12} 291, 292, 296
polystyrene beads (PSBs) 54, 65, 167–169
poly(styrene)-*block*-poly(2-vinylpyridine) (PS-*b*-P2VP) 288–290
polystyrene-poly(ethylene oxide) (PS-PEO) 344, 345
polystyrene-polyethylene (PS-PE) 344, 345, 353, 354
polystyrene-polyisoprene-polylactide (PS-PI-PLA) 344–346, 355
polystyrene-polylactide (PS-PLA) 344–346
polystyrene-poly(methylmethacrylate) (PS-PMMA) 342, 344–347, 362
– nanoparticles/polymers/biomacromolecules permeation through 351–353
– permeability regulation 348–350
– small molecules permeation through 347, 348
– supporting electrolyte concentration 350, 351
porogen solvents 140, 147
porous media, engineered nanoparticles in 272, 273
porous silicon (PSi) sensors 170, 171
porphyrin 171, 172
post-synthesis modification 41
potential of zero total charge (PZTC) 297
potentiometric immunoassays 72, 73
potentiometric sensors 154
potentiostatic deposition 288, 309
preconcentration
– molecularly imprinted polymers 144, 145, 152
– on-chip protein 121, 122
– trace-metal collection 192, 193, 198–200, 203–205

proteins
- analysis 118–123
- arrays 122
- electrochemical detection 72–74, 376, 377
- immobilization 240, 241
- integrated microfluidics for analysis 122, 123
- on-chip pre-concentration 121, 122
- protein A 103, 104
- separations 118–121
Prussian Blue 166, 167
PULSAR see photoactuated unimolecular logical switching attained reconstruction (PULSAR) microscopy

q
quantized double-layer (QDL) charging 93–96
quantum confinement 36–38, 172, 224, 264, 265
quantum dots (QDs) see core–shell quantum dots; semiconductor nanocrystals
quartz-crystal microbalance (QCM) 239
quenching efficiency 171, 172, 175

r
radioactive isotopes 277
Raman spectroscopy see surface-enhanced Raman scattering spectroscopy (SERS)
Rayleigh–Bénard PCR device 113, 114
Rayleigh light scattering (RLS) analysis 48, 49
rebinding, target analytes 140, 141
receptor function 101
redox (faradaic) current 320, 321, 323, 350, 351
redox molecules 319, 321, 322, 336, 341, 348, 349
redox probe-loaded apoferritin 75
resazurin 325
restricted access media-molecularly imprinted polymer (RAM-MIP) 147, 153
reverse type-I core–shell quantum dots 35
reversibility 15, 90
- titania sensors 247
reversible saturable optically linear fluorescent transitions (RESOLFT) microscopy 3, 20
reversible switchable fluorescent proteins (RSFPs) 11–13
Rhodamine dye 11, 39, 267

room-temperature phosphorescence (RTP) 45–47
ruthenium 304

s
Sauerbrey equation 239
scanning electrochemical microscopy (SECM) 321, 322, 326, 327, 329–333
scanning electron microscopy (SEM) 139, 140, 142, 148, 241, 242, 276
- FE-SEM 287, 291
- nanoporous membranes 342, 363
scanning ion conductance microscopy (SICM) 334
scanning probe microscopy (SPM) 328
scanning tunneling microscopy (STM) 91
Scatchard plot analysis 141, 150
seal resistances 127
seed-mediated synthesis 286–289, 308, 309, 364–366
selectivity
- biosensors 90
- MIPs 144, 150
- titania nanomaterials 232, 239, 243, 244
- trace metal collection 191, 192, 211, 213
self-assembly 66, 67, 74, 95, 228, 344 see also thiol self-assembled monolayer on a mesoporous support (SH-SAMMS™)
semiconducting oxide nanoparticles 101–103
semiconductor nanocrystals 31, 32, 54, 55
- characteristics 35–39
- classifications 32–35
- electrochemical detection of DNA 70–72
- fluorescence detection 265
- fluorescence modulation 5–7
- fluorescence quenching 267–269
- fluoroimmunoassays using 176
- quenching sensors based on 172–175
- synthesis and surface chemistry 40, 41
- trace analysis using 42–54
semi-covalent molecular imprinting 138
sensitivity
- biological sampling 118
- gas sensors 101–103
- titania sensors 231, 233, 246
- trace metal collection 191, 192
sensors see also biosensors; chemosensors; electrochemical sensors; immunosensors
- colorimetric 260
- functionalized nanoparticles 107
- gas sensors 101–103, 180

– gold nanorod   372–374
– microcantilever   162, 178–180
– porous silicon   170, 171
– potentiometric   154
– quenching   172–175
– solid-state films   98–101, 169–171
– surface modification   103, 263, 332
SERS *see* surface-enhanced Raman scattering spectroscopy (SERS)
sewage treatment *see* wastewater treatment
signal amplification/enhancement   70, 73, 367–371, 368
silica-based monolithic cage structures   263, 264
silica nanoparticles
– colorimetric sensor   260
– dye luminescence   369
– in electrochemical biosensors   79–81, 105
– hybrids   228, 245, 275
– mesoporous carbons from   209
– molecularly imprinted   142
– photoswitchable   11, 20
– protein pre-concentration   121
– thiol-functionalized mesoporous   194–200
silicon micro-extraction chip (SMEC)   122
silver nanoparticles   65–67, 69, 105, 258
– in gold nanorod synthesis   363, 364
– inorganic analyte determination   307, 308
– synthesis   287
– toxicity   273, 275
single-cell analysis, microfluidics for   123–127
single chamber PCR devices   112
single-chromophore fluorescent nanoparticles   17–19
single-color (on-off) fluorescent nanoprobes   5–13
single electron transfer   91–93
single-molecule electrochemistry   321–326
single-molecule spectroelectrochemistry (SMS-EC)   321, 323–326, 328
single-molecule spectroscopy   328
single-nanoparticle electrochemical detection   326–328
single nanopipette-based conductance   333–335
single-nucleotide polymorphisms (SNPs)   65
single-pore membranes   343
single-walled carbon nanotubes (SWCNTs)   165, 166, 325, 333
size-exclusion chromatography   341
size-selective purification, quantum dots   40, 41

"smart" nanoparticles   26
soft template   362–364
soil medium   272, 273
sol–gel systems   155, 172, 225, 307
solid-phase extraction (SPE)   139, 140, 143, 144, 237, 239
solid-phase extraction (SPE) sorbents   191, 192, 200 *see also* thiol self-assembled monolayer on a mesoporous support (SH-SAMMS™)
– trace-metal collection and analysis   192, 193
solid-state films   95–97
– sensors based on   98–101, 169–171
sorbents *see also* solid-phase extraction (SPE) sorbents
– costs   216
– ion-imprinted polymers   214, 215
– nanoporous carbon based   206–211
– performance comparisons for biological samples   197–200
– performance comparisons for environmental samples   194–197
– zeolites   212–214
spacers   369, 372–374
spin coating   121, 228, 229, 344
spiropyrans (SP)   5, 6, 7, 329, 330
– in bioimaging   19–21
– FRET-based photoswitching   15–17
– isomerization   7–9
– single-chromophore nanoparticles from   17, 18
square wave voltammetry   69, 70, 307, 310
stable isotopes   277
stepwise self-assembly   67, 95
strain effect   302, 303
supernanostructures, organic   175, 176
superparamagnetism   24, 25, 200, 201, 203, 310
surface charge   270–272, 275, 277
surface –COOH groups   346, 349
surface dopants   102, 103
surface-enhanced Raman scattering spectroscopy (SERS)   162, 180, 181, 304, 368, 369
– sensors   372
surface functionality   201, 256, 268, 271, 273, 275
– BCP-derived cylindrical nanopores   346, 347
surface modification
– gold nanorods   366, 375
– quantum dots   41
– sensors   103, 263, 332

surface plasmon resonance (SPR)   162, 256, 257, 323, 367, 369, 370
– spectroscopy   258, 259
surface-to-volume ratio   89, 97, 377
surfactants   31, 227, 228, 300, 362–364, 366
SWCNTs *see* single-walled carbon nanotubes (SWCNTs)
synergistic catalysis   308
syringe filters   152

t

target analytes, rebinding   140, 141
TEM *see* transmission electron microscopy (TEM)
templating
– gold nanorods   360–364
– mesoporous carbons   209–211
– mesoporous titania   225
– specific polymers   138
– titania nanotubes   226
tensile strain   302, 303
TEPMS *see* track-etched polymer membranes (TEPMs)
terrorists   162
tethering, nanoparticles   291, 292
theophylline   79
thiol self-assembled monolayer on a mesoporous support (SH-SAMMS™)   194–199
thiols
– biothiols   51, 52
– gold reaction with   289, 290
– mesoporous silica   194–200
– nanoporous carbon   208
– octanethiols   98, 100
– surface chemistry   192, 215
thorium detection   263
three-dimensional nanofilms   293–295
thymidine-thymidine (T-T) mismatch   260–262, 266
tin   102, 103, 264, 307
titania nanofilms   228, 229, 232, 241, 243, 244
titania nanohybrids   226–228
titania nanomaterials   223–247
– aqueous samples analysis   235–240
– chemical analysis applications   229–246
– synthesis   224–229
– toxicity   273, 274
titania nanoneedles   242
titania nanoparticles   225, 230
titania nanotubes   225, 226, 230, 241
titania sensors   231, 233, 246, 247
TNT *see* 2,4,6-trinitrotoluene (TNT)

toluene   98
top-down synthesis, quantum dots   40
total internal reflection fluorescence (TIRF)   334
toxicity
– engineered nanoparticles   269, 273–275
– surfactants   363
toxins
– molecularly imprinted polymers for   143–145
– mycotoxin   145, 154
– phenols   268
trace analysis   90
– E2 in wastewater treatment   150–152
– gold nanorods in   372–377
– high explosives with nanomaterials   161–182
– micro- and nanofluidic systems using   111–127
– MIPs for biopharmaceuticals and toxins   143–145
– titania nanomaterials   229–246
– using quantum dots   42–54
trace-metal collection
– from aqueous systems   191–216
– functionalized magnetic nanoparticles   200–206
– solid-phase extraction sorbents   192, 193
– thiol functionalized silica   194–200
track-etched polymer membranes (TEPMs)   341, 342, 348, 352, 353
transducer function   101
transducers   61, 63
– carbon nanotube-based   73
– modification with metal nanoparticles   68
transmission electron microscopy (TEM)   274, 275, 290, 292, 300, 365, 366
triacetone triperoxide (TATP)   162, 163, 166, 167
triethylamine (TEA)   148, 151, 152
1,3,5-trinitrobenzene (TNB)   45, 174, 175
2,4,6-trinitrotoluene (TNT)   45, 162–166
– chemiluminescence   167–169
– fluorescence sensing   169, 170, 172–175, 178
tris(2,2′-bipyridine)cobalt(III)-doped silica   79, 80
two-dimensional nanoarrays   286–293
two-photon luminescence   371, 375

u

ultrasonication   150, 151
uranium detection   263

v

vapor response isotherms   98
vapor sorption   98, 99
voltammetric biosensors   240–245
voltammetry   320, 333 *see also* anodic stripping voltammetry; cyclic voltammetry
– differential pulse voltammetry   93, 94, 312, 313
– double potential-pulse voltammetry   287, 301, 302
– metal nanoparticles in solution   92–95
– square wave voltammetry   69, 70, 307, 310

w

wastewater treatment   136, 137, 144, 206
– continuing development advances   153–155
– current progress   152, 153
– trace analysis of E2 in   150–152

water contaminants   135–137
water vapor, sensing   233, 234

z

zearalenone (ZON)   154
zeolites   212–214
zero-dimensional materials *see* semiconductor nanocrystals
zeta potential   270–273, 277
zinc selenide   33, 35, 37
zinc sulfide   33, 35, 37
– cadmium selenide quantum dots   5, 6
– fluorescence quenching   268
– Mn-doped   46, 47, 175
zinc telluride   33, 37
zwitterionic polymer membranes   121